Geometric Orbifold Cohomology

Topological phases of quantum materials and brane charges in M-theory are measured by extraordinary cohomology theories defined on orbifold spacetimes. Geometric Orbifold Cohomology presents a modernized and enhanced formulation of these theories, establishing a rigorous framework for nonabelian and differential cohomology in the setting of higher geometry. Motivated by cutting-edge problems in mathematical physics—specifically the analysis of M-brane charges and topological insulators—this monograph bridges the gap between classical equivariant topology and modern cohesive homotopy theory. It fills a critical gap in the literature by offering a unified perspective where orbifold geometry is treated synthetically via modal operators in higher topos theory, allowing for a precise treatment of geometric singularities and differential structures.

- **Accessible Foundations**: Begins with a streamlined reconstruction of twisted nonabelian orbifold cohomology and orbifold K-theory using the language of topological groupoids and stacks, offering a pedagogical entry point for newcomers.
- **Higher Topos Theory**: Provides a detailed exposition of the transition from classical perspectives to the modern context of cohesive higher topos theory and global equivariant unstable homotopy theory.
- **Synthetic Orbifold Geometry**: Lays out a powerful synthetic differential theory of higher orbifold Cartan geometry, utilizing systems of modal operators to rigorously capture orbi-singularities and proper equivariant homotopy types.
- **Unification of Methods**: Unifies orbifold geometry with classical differential geometry and cohomology, compatibly enhancing the picture with modalities that reflect proper equivariant structures.
- **Concrete Applications**: Showcases the theory's utility through the analysis of tangentially twisted nonabelian orbifold cohomology, specifically J-twisted orbifold Cohomotopy, applied to "Hypothesis H" and the classification of topological phases.

This volume is designed to serve a dual purpose: the first part acts as an invitation for advanced graduate students and beyond, in mathematics and theoretical physics, leading them from basic topological charges to modern research applications. The subsequent parts provide a comprehensive resource for academic researchers in algebraic topology, differential geometry, and string theory interested in a cutting edge formulation of geometric cohomology and its applications to quantum systems.

Geometric Orbifold Cohomology

Hisham Sati and Urs Schreiber

CRC Press
Taylor & Francis Group
Boca Raton London New York

CRC Press is an imprint of the
Taylor & Francis Group, an **informa** business
A CHAPMAN & HALL BOOK

First edition published 2027
by CRC Press
2385 NW Executive Center Drive, Suite 320, Boca Raton FL 33431

and by CRC Press
4 Park Square, Milton Park, Abingdon, Oxon, OX14 4RN

CRC Press is an imprint of Taylor & Francis Group, LLC

ISBN: 978-1-041-14751-0 (hbk)
ISBN: 978-1-041-14752-7 (pbk)
ISBN: 978-1-003-67596-9 (ebk)

DOI: 10.1201/9781003675969

Typeset in TeXGyreTermesX font
by KnowledgeWorks Global Ltd.

Publisher's note: This book has been prepared from camera-ready copy provided by the authors.

Contents

Preface

The mathematical theory presented here – geometric nonabelian cohomology formulated modally in cohesive ∞-toposes – goes back to [Sc09][SSS12][Sc13], motivated by applications to topological phases/charges in physics (cf. [SS25d][SS25i] and part I of this book). It has seen a number of developments since (cf. [FSS23][Ch24][SS26c]; a recent expository digest is in [Sc25]), but a dedicated monograph had been missing. This book is a first step in this direction.

The seed of this book lay in developments in 2019, when we started finding evidence [FSS20][FSS21a] for the hypothesis ("Hypothesis H") that something called *flux-quantization of 11D super-gravity* (reviewed in [SS25d][SS25j]) on spacetime manifolds should take place in the generalized cohomology theory named *tangentially twisted unstable 4-Cohomotopy*. But we also found [SS20][BSS19] that in the neighborhood of orbifold-singularities the relevant cohomology theory should really be *proper equivariant unstable 4-Cohomotopy in the RO-degree given by the singularity's isotropy representation*. The proof that the latter is indeed a special case of the former, when the notion of "tangential twist" is generalized from manifolds to orbifolds, became the seed text (now Thm. 11.2.6, see Fig. 1.7 on p. 21) that eventually grew into this book.

However, while a first draft of the book circulated – which (as referees rightly remarked) did not dwell on traditional orbifold cohomology theories such as orbifold K-theory – we became absorbed with first discovering and then developing the relation of Hypothesis H to quantum materials relevant for topological quantum computing this via orbifold K-theory in the manner of [SS23b] (now the topic of chapter 5 of this book) whose natural discussion via cohesion in ∞-toposes meanwhile turned into a monograph of its own [SS26c].

Concretely, Hypothesis H has led (applied to *M5-brane probes of orbifolds* [SS25e][SS25f]) to an understanding [SS25h][SS26a][SS25i] of symmetry-protected anyonic topological order in quantum materials known as *crystalline fractional Chern insulators*. Connecting this back to the now traditional K-theory classification of topological phases of matter requires analyzing the Boardman homomorphism from unstable Cohomotopy to K-theory, but generalized to an operation between orbifold cohomology theories twisted by the corresponding crystallographic point group. Establishing this is now the content of § 5.2.

The eventual ambition of these applications – to resolve beyond the equivariant topology also the more fine-grained geometric (Riemannian, conformal, supersymmetric, ...) aspects of orbifolds and their differential cohomology – motivates the detailed development of modal orbifold *geometry* in part IV. The eponymous paradigm of this book is that when (differential, Cartan) geometry is formulated "synthetically" via modalities in cohesive ∞-toposes, then it *automatically* generalizes from manifolds to orbifolds (and further to étale ∞-stacks) and thereby provides a powerful and conceptually neat unification of orbifold geometry and of orbifold cohomology with classical differential geometry and with classical differential cohomology.

Hisham Sati
Urs Schreiber
February 2026

Authors

Hisham Sati is Professor of Mathematics at NYU Abu Dhabi and the founding director of the Center for Quantum and Topological Systems. His interdisciplinary research spans mathematical physics, algebraic topology, and differential geometry, and their interactions through fundamental physical theories. He has delivered the Adams Memorial Lecture in Topology.

Urs Schreiber is Research Scientist at NYU Abu Dhabi, specializing in the mathematical foundations of quantum field theory. His work applies algebraic topology and geometric homotopy theory to fundamental physics, including topological quantum technology. He is the creator of the nLab, a research wiki for math and physics.

Thanks

For useful discussion on an early precursor of this book, the authors thank Vincent Braunack-Mayer, David Carchedi, David Jaz Myers, Dmitri Pavlov, and Charles Rezk. For suggesting the addition of further examples and applications we thank an anonymous referee.

1

Introduction

Topological phases of quantum materials, and generally brane charges in quantum gravity, are measured by extraordinary cohomology of orbifolded spaces, and fragilely/microscopically by nonabelian cohomology twisted by tangential and other geometric orbifold structure. These applications, with more or less direct significance for cutting-edge experimental physics, motivate the further development of the ancient algebro-topological concept of *cohomology* to a robust theory of *extraordinary nonabelian twisted geometric orbifold cohomology*.

After surveying the general outlook and key example applications in part I, this book begins gently, in part II, with a pedagogical development of a streamlined construction of topological twisted nonabelian orbifold, cohomology, realized concretely as connected components of topological slice mapping groupoids. We showcase in this manner a neat novel construction of twisted orbi-orientifold K-theory.

To refine this notion of twisted nonabelian orbifold cohomology and capture geometric (tangential) orbifold structure, the second half of the book transitions to more abstract modern geometric homotopy theory (higher topos theory). We provide motivation for and an introduction to this transition in part III.

Using this more abstract perspective, we introduce in part IV *singular-cohesive ∞-toposes* which faithfully embed manifolds, orbifolds, and higher etale stacks and orbispaces into a modal synthetic geometry. This allows for the elegant axiomatization of higher orbi-differential Cartan geometry

Again by forming connected components of suitable slice mapping stacks, but now in the singular-cohesive ∞-topos, this yields, in part V relevant notions of generalized twisted nonabelian geometric orbifold cohomology. A key result established in this generality is that tangentially twisted cohomology of orbifolds reduces in the vicinity of G-orbifold singularities to RO(G)-graded equivariant cohomology Thm. 11.2.6, (see Fig. 1.7 on p. 21). We close by detailing the construction of twisted nonabelian (unstable) orbifold Cohomotopy in this manner, which in the motivating applications is the fragile/microscopic classifier for phases of Chern insulator materials and M-branes.

DOI: 10.1201/9781003675969-1

1.1 Motivation

1.1.1 Orbifolds

Where a *manifold* is a space that looks locally like a Cartesian space $\mathbb{R}^n$ (cf. [Lee12]), so an *orbifold* ([Sa56][Sa57][Th80][Hae84], review in [MM03][Ka08, §6][BG08, §4][IKZ10]) is, more generally, a space that looks locally like the quotient (suitably understood) of an $\mathbb{R}^n$ by the action of a finite group G of diffeomorphisms. Here the G-action may have fixed-points which in the quotient become *singular* points, such as a crease in a piece of paper or the tip of a cone, cf. Fig. 1.1:

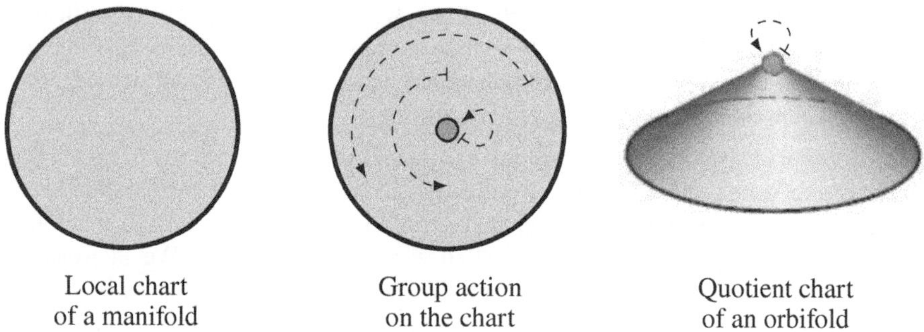

Figure 1.1 Orbifold charts.

The purpose of orbifold structure is to generalize the differentiable structure of manifolds to allow for singular loci of such form: *orbi-singularities.*

A basic class of examples is provided by the orbifold quotients $\mathbb{T}^n /\!\!/ \mathbb{Z}_2$ of the n-torus $\mathbb{T}^n := \mathbb{R}^n/\mathbb{Z}^n$ by the $\mathbb{Z}_2$-action that swaps the signs of the canonical coordinates:

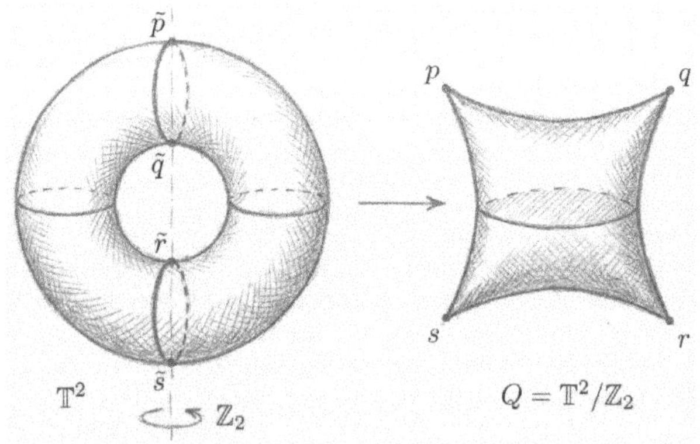

Figure 1.2 The pillowcase orbifold $\mathbb{T}^2 /\!\!/ \mathbb{Z}_2$ (graphics following [Dr11, Fig. 2.6]).

$\mathbb{T}^0 /\!\!/ \mathbb{Z}_2$ is – in a sense which we will discuss in detail – the *classifying* or *delooping* groupoid $\mathbf{B}\mathbb{Z}_2$ (1.14): A single point but equipped with a non-trivial involution.

$\mathbb{T}^1 /\!\!/ \mathbb{Z}_2$ plays a central role in Hořava-Witten theory [HW96].

$\mathbb{T}^2 /\!\!/ \mathbb{Z}_2$ is known as the *pillowcase orbifold* (cf. Fig. 1.2).

$\mathbb{T}^3 /\!\!/ \mathbb{Z}_2$ does not have a special name but appears as the space of effective crystal momenta for quantum materials with time-reversal symmetry (cf. §xy).

$\mathbb{T}^4 /\!\!/ \mathbb{Z}_2$ is the time-honored *Kummer surface* (cf. [Do20] and Fig. 1.3).

$\mathbb{T}^5 /\!\!/ \mathbb{Z}_2$ is the background of *MO5-planes M-theory* (cf. [Wi96][SS20]).

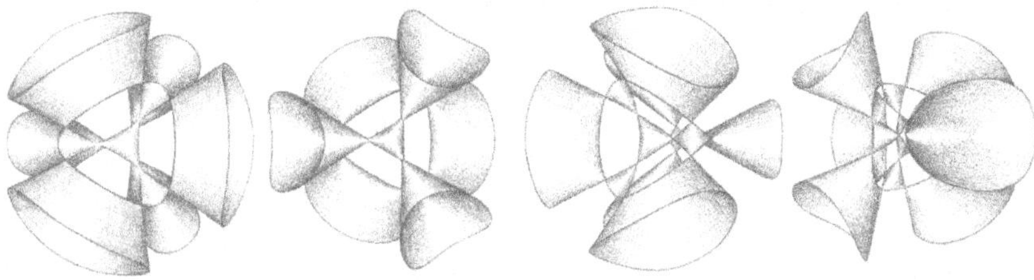

Figure 1.3 The Kummer orbifold $\mathbb{T}^4 /\!\!/ \mathbb{Z}_2$. Shown is a sequence of projections to 2d of the 4-dimensional structure (following [Do20]).

Thus, orbifolds have become commonplace in mathematics (e.g. [BLP05][Rat06, §13][JY11]) and play a decisive role in theoretical physics (see [AMR02]), notably so in string/M theory ([DHVW85][DHVW86][BL99] [SS20]), and in solid-state physics ([JBD96][Jo99][GT19][SS23b][SS22b]) – in fact, it is via "geometric engineering" of quantum field theories [KKV97] (cf. [Ka98, §1.4.1][DZ23]), on the branes of string/M-theory, probing orbi-singularities, that modern descriptions [SS25h] of strongly correlated quantum materials arise [SS25e][SS25f]. This book is to provide rigorous but practical mathematical framework and tools, notably for such applications.

This is not, a priori, immediate: Definition of the *geometric homotopy theory* (cf. §IV) and thus of the *geometric cohomology* (§V) of orbifolds may appear subtle and elusive, as witnessed by the convoluted history of the concept (cf. [Le08, Intro.][IKZ10, §1]). In fact, the issue had remained somewhat open, as we proceed to recall:

1.1.2 Orbifolds as étale stacks?

A proposal popular among Lie theorists [MP97] (see [Mo02][Le08][Am12]) is to regard an orbifold with local charts $G_i \curvearrowright U_i$ (8.37) as

- the étale groupoid, in particular, a Lie groupoid (see [MM03][TX06]) or topological groupoid (see [CPRST14]);
- equivalently, the étale geometric stack, in particular, the differentiable or topological stack ([Car11][Car19][Gi13])

obtained by gluing the corresponding *homotopy quotient stacks* $U_i /\!\!/ G_i$ (1.15).

This proposal is directly modeled (explicitly so in [Jo12, §8]) on the concept of Deligne-Mumford stacks in algebraic geometry ([DM69], review in [Kr09]) and extends to a concept of general étale ∞-stacks [Car20][Car16]. It relies on the fact that étale stacks, in their role as homotopy-theoretic generalizations of sheaves, fully capture geometric aspects (via generalized sheaf cohomology [Bro73], see [NSS12a]), while in their role as geometric refinements of classifying spaces, they support *Borel equivariant cohomology* (see [Tu11]). However, Borel cohomology is coarser than the *proper equivariant cohomology* that is generally relevant in theory and in applications:

1.1.3 Proper equivariant cohomology

Proper equivariant cohomology[1] formulated in equivariant homotopy theory (review in [Blu17][May96]) is obtained by refining the purely homotopy-theoretic nature of Borel cohomology by the geometric ("cohesive", see §1.2) nature of fixed loci (see Ex. 9.2.27) of topological group actions – hence by the characteristic nature of orbifold geometry – as encoded in the category of orbits of the equivariance group (recalled in §8.1.8).

The proper equivariant version of ordinary cohomology is known as *Bredon cohomology* [Bre67a][Bre67b] (review in [Blu17, §1.4][tD79, §7]); beyond that, there is a wealth of proper equivariant generalized cohomology theories (Def. 8.1.23) such as *equivariant K-theory* [Se68][AS69] (which is proper equivariant by [ASe04, §6 & A3.2][FHT11, A.5][DL98]) and *equivariant Cohomotopy theory* [Se71][tD79, §8][SS20][BSS19].

However, if orbifolds are modeled just by étale stacks, then their proper equivariant cohomology remains, by and large, invisible. This is true even for Chen-Ruan orbifold cohomology:

1.1.4 Traditional orbifold cohomology and its shortcomings

Given an orbifold $\mathcal{X}$, we write (see §1.2) $\cup\mathcal{X}$ for the étale stack underlying it, and[2] $\int\cup\mathcal{X}$ for its geometric realization or classifying space (often denoted $B\mathcal{X}$). In the case that $\mathcal{X}$ is the global quotient orbifold of a G-space X, this is the homotopy type of the *Borel construction* so that we may generally call $\int\cup\mathcal{X}$ the *Borel space* of the orbifold. Now, traditional orbifold cohomology is [ALR07, p. 38] just the ordinary cohomology (e.g. singular cohomology) of this Borel space; hence it is *Borel cohomology*:

$$\overset{\substack{\text{traditional}\\\text{orbifold cohomology}}}{H^\bullet_{\mathrm{trad}}(\mathcal{X}, A)} \;:=\; \overset{\text{Borel cohomology}}{\underset{\substack{\text{singular}\\\text{cohomology}}}{H^\bullet_{\mathrm{sing}}}\big(\underset{\substack{\text{Borel}\\\text{space}}}{\int\cup\mathcal{X}}, A\big)}\,. \tag{1.1}$$

This can be considered with any kind of coefficients A, notably in the generality of local coefficient systems [MP99], but it always remains an invariant of just the

[1]We follow [DHLPS19] with the terminology "proper equivariant cohomology", see Remark 10.2.48, using it to distinguish from *naïve* or *Borel* equivariance.

[2]The "esh"-symbol "ʃ" stands for *shape* [Sc13, 3.4.5][Sh15, 9.7], following [Bo75], which for well-behaved topological spaces is another term for their *homotopy type* [Lu09a, 7.1.6][Wa17, 4.6]; see Ex. 9.1.19.

Borel space. Moreover, for a coefficient ring that inverts the order of the isotropy groups of $\mathcal{X}$, hence in particular for rational, real and complex number coefficients $A \in \{\mathbb{Q}, \mathbb{R}, \mathbb{C}\}$, the purely torsion cohomology of the orbifold's finite isotropy groups becomes invisible, and traditional orbifold cohomology reduces further (e.g. [ALR07, Prop. 2.12]) to an invariant of just the shape $\int \vee\mathcal{X}$ of the singular quotient space $\vee\mathcal{X}$ (the "coarse moduli space") underlying the orbifold (often denoted $|\mathcal{X}|$):

$$\underset{\text{orbifold cohomology}}{\overset{\text{traditional rational}}{}} H^\bullet_{\text{trad}}(\mathcal{X}, \mathbb{Q}) \;\simeq\; \underset{\text{singular cohomology}}{H^\bullet_{\text{sing}}}\overset{\text{ordinary cohomology}}{}\big(\underset{\text{naïve/coarse quotient space}}{\textstyle\int \vee \mathcal{X}}, \mathbb{Q}\big)\,. \tag{1.2}$$

It is in this form that orbifold cohomology was originally introduced (in [Sa56, Thm. 1], following [Ba54], reviewed in [ALR07, 2.1]).

Of course, it did not go unnoticed that this coarse notion of orbifold cohomology is insensitive to the actual nature of orbifolds. In reaction (and motivated by algebraic constructions [DHVW85][DHVW86] on 2d conformal field theories interpreted as describing strings propagating on orbifold spacetimes), Chen and Ruan famously proposed a new orbifold cohomology theory in [CR04]. But in fact Chen-Ruan cohomology of an orbifold is (see [Cl14, p. 4,7] for review) just Satake's coarse cohomology (1.2), but applied to the corresponding "inertia orbifold" (cf. [LU04b][SS24]) $\mathbf{Map}(\int S^1, \mathcal{X})$ of maps from the shape of the circle:

$$\overset{\text{Chen-Ruan orbifold cohomology}}{H^\bullet_{\text{CR}}(\mathcal{X})} \;\simeq\; \overset{\text{traditional orbifold cohomology}}{H^\bullet_{\text{trad}}\big(\underset{\text{inertia orbifold}}{\mathbf{Map}(\textstyle\int S^1, \mathcal{X})}, \mathbb{C}\big)}\,. \tag{1.3}$$

Still, it turns out that, for global G-quotient orbifolds $\mathcal{X} = \curlyvee(X /\!\!/ G)$, Chen-Ruan cohomology is equivalent to a proper equivariant cohomology theory, namely to Bredon cohomology with coefficient system given specifically by:

$$A_{\text{CR}} \;:\; G/H \;\longmapsto\; \text{ClassFunctions}(H, \mathbb{C})\,. \tag{1.4}$$

This was observed in [Mo02, p. 18], using [Hu90, Thm. 5.5] with [IIo88, Prop. 6.5 b)]:

$$\overset{\text{Chen-Ruan cohomology}}{\mathbf{H}^\bullet_{\text{CR}}\big(\underset{\text{global quotient orbifold}}{\curlyvee(X/\!\!/G)}, \mathbb{C}\big)} \;\simeq\; \overset{\text{Bredon cohomology}}{H^\bullet_G\big(X, \underset{\text{specific system of coefficients (1.4)}}{A_{\text{CR}}}\big)}. \tag{1.5}$$

Thus the success of Chen-Ruan cohomology (surveyed in [ALR07, §4,5]) highlights the relevance of proper equivariance in orbifold cohomology. At the same time, this means that to detect the full proper equivariant homotopy type of orbifolds, one needs an orbifold cohomology theory that induces Bredon coefficient systems more general than (1.4); and, in fact, one that subsumes also generalized equivariant cohomology theories such as equivariant K-theory. In [AR03] the authors *define* orbifold K-theory to be the equivariant K-theory of any global quotient presentation (see also [ARZ06][BU09][HW11]):

$$\overset{\text{traditional orbifold K-theory}}{K^\bullet\big(\underset{\text{global quotient orbifold}}{\curlyvee(X/\!\!/G)}\big)} \;:=\; \overset{\text{equivariant K-theory}}{K^\bullet_G(X)}\,. \tag{1.6}$$

This has been justified for this specific case of K-theory by checking explicitly [PS10, Prop. 4.1] that the right-hand side of (1.6) is independent of the choice of global quotient presentation on the left. However, in general, this approach of circumventing an intrinsic definition of orbifold cohomology by just defining it to be equivariant cohomology of global quotient presentations is, besides being somewhat unsatisfactory, in need of justification:

1.1.5 Orbifolds in global equivariant homotopy theory?

That orbifold cohomology should also capture proper equivariant cohomology was suggested explicitly in [PS10]. However, the fundamental issue remained that a quotient presentation $\mathcal{X} \simeq \prec(X /\!\!/ G)$ of an orbifold is not intrinsic to the orbifold, similarly to a choice of coordinate atlas, while in equivariant cohomology theory, the equivariance group G is traditionally taken to be fixed. But this suggests [Schw17, Intro.][Schw18, p. ix-x] (details in [Ju20]) that the right context for orbifold cohomology is "global" equivariant homotopy theory [Schw18] (following [HG07] and originally motivated from patterns seen in genuine equivariant stable homotopy theory [Se71][LMS86]) where the equivariance group G is allowed to vary in a prescribed class of groups. On the other hand, plain global homotopy theory retains no geometric information!

1.1.6 The open problem

The open problem is thus to set up a mathematical theory of *proper orbifold cohomology* which unifies:

- **(i)** the higher *geometric* and *differential* aspects of orbifolds captured by geometric/differential homotopy theory; and
- **(ii)** the *singular* (equivariant) aspects of orbifolds captured by proper and global equivariant homotopy theory.

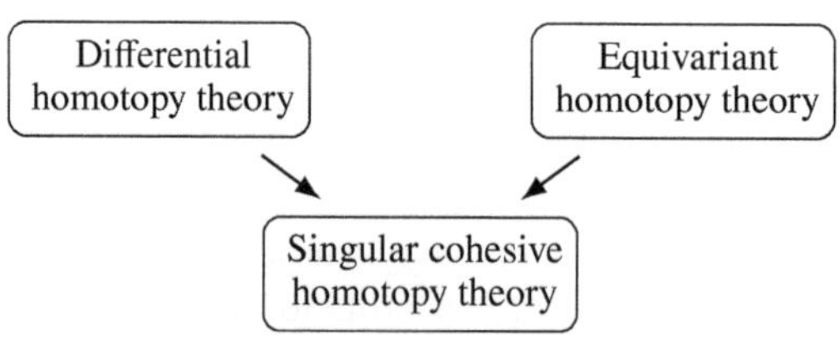

To achieve this, we turn to higher topos theory ([TV05][Lu09a][Re10], see §8.1) as the ambient foundational homotopy theory in which we formulate higher orbigeometry by means of systems of cohesive modalities ([Sc13][Sc19][Sc25][SS26c], see §9, §10):

1.1.7 Higher toposes, where higher geometry takes place

For our purposes, a higher topos is a universe of generalized & higher geometric spaces (exposition in [Sc25]):

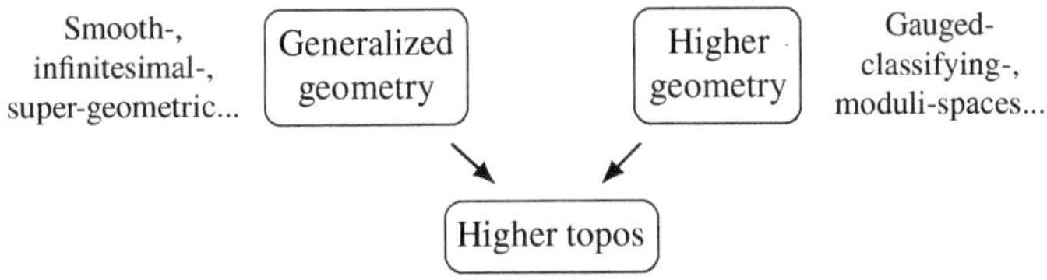

(a) Here "**generalized geometry**" refers to what Grothedieck called *functorial geometry* [Gr65] (review in [DG80]), which he urged in [Gr73] should supersede any point-set (locally ringed) definition of geometric spaces (further amplified by Lawvere, cf. [La86][La91]): The idea is to define spaces operationally – much as envisioned in physics – by how they may be *probed* by a small category of *probe spaces* (*affine spaces* or *charts*, see [FSS14][JSSW19][Sc25][GS25] and Def. 9.1.10), such as

$$\mathrm{Chrt} \;=\; \left\{ \begin{array}{ll} \mathrm{CrtSpc} & \text{(Def. 8.1.5)} \\ \mathrm{JetCrtSp} & \text{(Def. 9.1.24)} \\ \mathrm{SuperCrtSp} & \text{(Def. 9.1.43)} \\ \mathrm{Snglrt} & \text{(Def. 9.2.1)} \\ \mathrm{SnglrChrt} & \text{(Lem. 9.2.16)} \\ \ldots, & \end{array} \right. \tag{1.7}$$

and then to encode a would-be generalized ("target"-)space $\mathcal{X}$ by assigning to each $\Sigma \in \mathrm{Chrt}$ the collection

$$\underset{\text{probe space}}{\Sigma} \;\longmapsto\; \underset{\substack{\text{collection of probinges of} \\ \text{generalized space } \mathcal{X} \text{ by } \Sigma}}{\mathcal{X}(\Sigma) \;:=\; \big\{\text{“}\,\Sigma \to \mathcal{X}\,\text{”}\big\}} \tag{1.8}$$

of geometric (e.g. smooth, super-geometric, etc.) maps into $\mathcal{X}$, where the quotation marks indicate that, at this point of bootstrapping $\mathcal{X}$ into existence, the category in which these *probings* are actual maps is yet to be specified. To that end, one observes that a minimal set of consistency conditions on such an abstract assignment (1.8) to be anything like collections of maps into a space $\mathcal{X}$ are:

(i) Functoriality of probes. For every map ϕ of Chrt there is an operation of "pre-composition of probe maps by ϕ":

$$\begin{array}{ccc} \substack{\text{map of} \\ \text{probe spaces}} & \substack{\text{pre-composition operation} \\ \text{on collections of probes}} & \\ \Sigma_1 & \mathcal{X}(\Sigma_1) & \\ \Big\downarrow{\scriptstyle \phi} & \Big\uparrow{\scriptstyle \mathcal{X}(\phi) \,=\, \text{“}(-)\circ\phi\text{”}} & \\ \Sigma_2 & \mathcal{X}(\Sigma_2) & \end{array} \qquad \text{such that} \qquad \begin{array}{c} \mathcal{X}(\phi_2) \circ \mathcal{X}(\phi_1) \\ \simeq \\ \mathcal{X}(\phi_2 \circ \phi_1)\,. \end{array} \tag{1.9}$$

(ii) Gluing of probes. If $\{ U_i \twoheadrightarrow \Sigma \}_{i \in I}$ is a cover of $\Sigma \in \mathrm{Chrt}$ by several $U_i \in \mathrm{Chrt}$, then probings of $\mathcal{X}$ by Σ should be equivalent to those tuples of probinges by the U_i, which are coherently identified on intersections:

$$\mathcal{X}(\Sigma) \;\simeq\; \left\{ \begin{array}{c} \text{Tuples of probes } U_i \longrightarrow \mathcal{X} \\ \text{identified on intersections } U_i \cap U_j \\ \text{compatibly on } U_i \cap U_j \cap U_k \\ \text{etc.} \end{array} \right\}. \tag{1.10}$$

In the jargon of topos theory (see [MLM92][Joh02]), condition (1.9) says that the collection $\mathcal{X}(-)$ of probes of $\mathcal{X}$ is a *pre-sheaf* on Chrt, while condition (1.10) says that this is in fact a *sheaf*. Hence the category of generalized geometric spaces probeable by Chrt is the category of sheaves (the *Grothendieck topos*) on Chrt:

$$\underset{\text{geometric spaces}}{\overset{\text{topos of generalized}}{}} \;\; \mathrm{GnrlzdSpc} \;:=\; \mathrm{Shv}\big(\mathrm{Chrt}\big) \;\; \underset{\text{on site of charts}}{\overset{\text{category of sheaves}}{}} \tag{1.11}$$

Now, every $\Sigma \in \mathrm{Chrt}$ is itself canonically regarded as a generalized space $y(\Sigma) \in \mathrm{GnrlzdSpc}$, by taking its probes to be those given by morphisms of Chrt (this is the *Yoneda embedding*[3], recalled as Prop. 8.1.53):

$$\underset{y(\Sigma)}{\overset{\substack{\text{chart regarded as} \\ \text{generalized space}}}{}} \;:\; \Sigma' \longmapsto \overset{\substack{\text{collection of its} \\ \Sigma'\text{-shaped probes}}}{\big\{\Sigma' \to \Sigma\big\}} \;=:\; \mathrm{Chrt}(\Sigma', \Sigma) \tag{1.12}$$

Hence we have completed the bootstrap construction of generalized spaces $\mathcal{X}$ in (1.8) if we may remove the quotation marks there, hence if for $\mathcal{X} \in \mathrm{GnrlzdSpc}$ there is a natural equivalence

$$\underset{\mathcal{X}(\Sigma)}{\overset{\substack{\Sigma\text{-shaped} \\ \text{probes of } \mathcal{X}}}{}} \;\simeq\; \overset{\text{actual maps from } y(\Sigma) \text{ to } \mathcal{X}}{\big\{y(\Sigma) \to \mathcal{X}\big\}} \;:=\; \mathrm{GnrlzdSpc}\big(y(\Sigma), \mathcal{X}\big)\,. \tag{1.13}$$

That this is indeed the case is the statement of the *Yoneda lemma* (recalled as Prop. 8.1.54), which thus implies consistency and existence of generalized geometry!

(b) On the other hand, "**higher geometry**" (see [FSS14][FSS19][JSSW19] for exposition and applications) refers to the refinement of the above theory of generalized geometric spaces, where the collection of probes (1.8) of a generalized space is not necessarily just a set, but may be a set equipped with equivalences between its elements (a *gauged set*), and with higher order equivalences (higher gauge transformations) between these, etc. – called an ∞-*groupoid* (typically modeled as a Kan simplicial set, see [GJ99, I.3]). For example, for $X \in \mathrm{Set}$ and G a discrete group acting on X, the corresponding *action groupoid* (Ex. 8.1.31) consists of the elements $x \in X$, but equipped with an equivalence between x_1 and x_2 for every group element

[3] Shown here for sub-canonical Grothendieck topologies on Chrt, which is the case in all examples of interest here.

whose action takes x_1 to x_2:

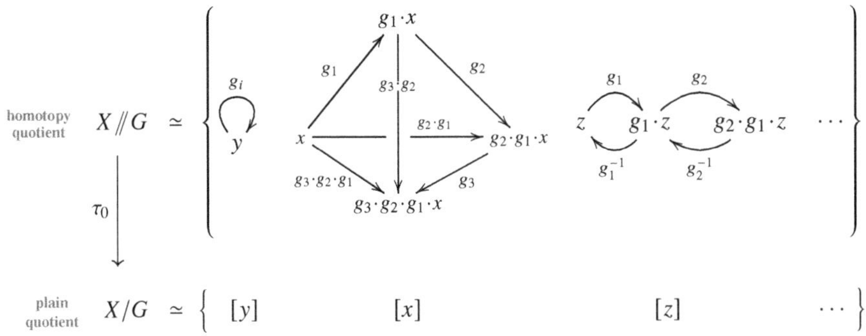

This is a model for the *homotopy quotient* of X by G, which resolves the plain quotient X/G (the set of equivalence classes) by remembering not only *that* but *how* pairs of elements are equivalent. More precisely, the action groupoid remembers the *graph* and *syzygies* of the G-action, encoded in its Kan simplicial *nerve* (Ex. 8.1.87):

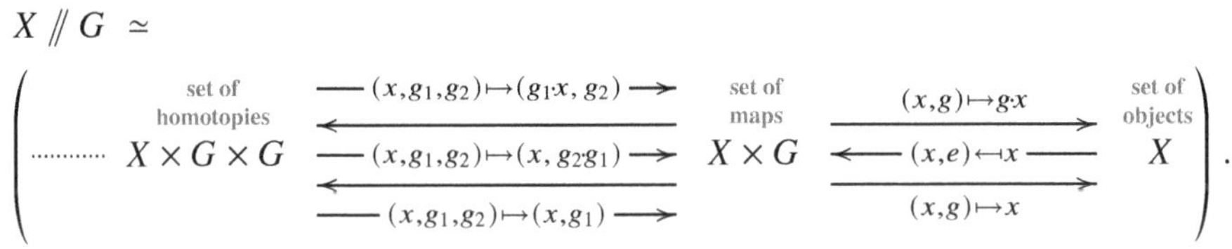

In particular, if an element $y \in X$ is fixed by the group action, then in the homotopy quotient it appears as the one-object *delooping groupoid* of G:

$$\mathbf{B}G \;\equiv\; * \,/\!\!/\, G \;\simeq\; \left\{ \overset{g}{\circlearrowright}_{*} \;\middle|\; g \in G \right\}. \tag{1.14}$$

More generally, if $X \in \mathrm{Chrt}$ in the list (1.7) is equipped with the action of a discrete group G, then we obtain a higher generalized space $\mathcal{X} := X /\!\!/ G$ whose ∞-groupoid of Σ-shaped probes (1.8) is the action groupoid of the induced action on the set of Σ-shaped probes of X (the following formula is for contractible charts, Lemma 9.1.13):

$$\underset{\text{global quotient orbifold}}{X /\!\!/ G} \;:\; \Sigma \;\longmapsto\; \underset{\text{groupoid of its }\Sigma\text{-shaped probes}}{(X /\!\!/ G)(\Sigma) \;:=\; X(\Sigma) /\!\!/ G \;=\; \mathrm{Chrt}(\Sigma, X) /\!\!/ G}\,. \tag{1.15}$$

Such a higher generalized space with collections of probes (1.8) being groupoids, and satisfying the appropriate gluing condition (1.10), may be called a 2-*sheaf* or *sheaf of groupoids* [Br93] on Chrt, in generalization of (1.11), but is commonly known as a *stack* [DM69][Gi72][Jar01][Ho08], following *champ* [Gi66]. Generally, a higher generalized space with ∞-groupoids of probes is thus an ∞-*sheaf* or ∞-*stack* on Chrt, in generalization of (1.11):

$$\underset{\infty\text{-topos}}{\mathbf{H}} \;:=\; \underset{\substack{\text{higher category of}\\ \text{higher generalized spaces}}}{\mathrm{HigherGnrlzdSpc}} \;:=\; \underset{\infty\text{-category of }\infty\text{-stacks}}{\mathrm{Shv}_\infty\big(\underset{\substack{\infty\text{-site of}\\ \text{probe spaces}}}{\mathrm{Chrt}}\big)}\,. \tag{1.16}$$

1.1.8 Higher topos theory

The theory of ∞-stacks originates with [Bro73], developed in [Jar87][Jar96] (survey in [Jar15]) and brought into the more abstract form in [TV05][Lu09a][Re10] (introduction in [Re19]). While the theory has a reputation of being intricate, this is really a reflection of its simplicial models and hence of the richness of its implications, while – on the contrary – finitary constructions internal to ∞-toposes behave so very well that they may naturally be formulated [Sh19] in a kind of programming language now known as *homotopy type theory* [UFP13]. While we will not dwell on this here, we do focus on elegant internal constructions. For some of these, a homotopy type-theoretic formulation has already been explored in the literature, cf. Table 1.1.

Table 1.1 Existing formalizations.

Theory internal to an ∞-topos	**Formulation in ordinary math**		**Formulation in ho-type theory**
Galois theory	§8.2	[NSS12a]	[BvDR18]
Modalities & cohesion	§9.1	[SSS12][Sc13]	[RSS17][Sh15]
Étale ∞-stacks	§10.2	[KS17]	[Ch24][CRi20]
Cohomology	§11	[SSS12][NSS12a]	[Cav15][BH18]

In particular, a key aspect of our treatment here is that we capture (orbi-)geometry in terms of systems of adjoint *modal operators* on the ambient higher topos:

1.1.9 Dual modalities in an ∞-Topos

In view of the above every ∞-topos $\mathbf{H}$ may be thought of as a homotopy theory of generalized geometric spaces of a certain nature. In order to narrow back in, among these very generalized spaces, onto those which are relatively tame, we may, in the spirit of [La91][La94][La07], axiomatize qualities of geometric spaces (such as being *discrete*, *smooth*, *étale*, *reduced*, *bosonic*, *singular*, etc.) via the systems of (co-)reflective sub-∞-categories $\mathbf{H}_{\bigcirc}, \mathbf{H}_{\square}, \cdots \subset \mathbf{H}$, that the objects with these properties (should) form inside $\mathbf{H}$ [SSS12][Sc13]:

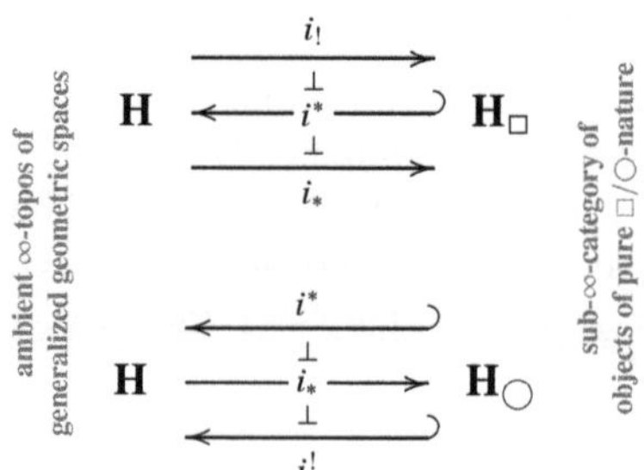

(1.17)

These *reflections* induce systems of adjoint (co-)projection operators $\bigcirc \dashv \square : \mathbf{H} \to \mathbf{H}$, the associated *idempotent (co-)monads*:

$$\bigcirc := i^* \circ i_! \,, \quad \square := i^* \circ i_* \qquad \text{or} \qquad \bigcirc := i^* \circ i_* \,, \quad \square := i^! \circ i_* \,, \tag{1.18}$$

to which we refer as *modal operators* or just *modalities* [Sc13][RSS17][Co20]. These are idempotent (Prop. 8.1.45),

$$\bigcirc\bigcirc X \simeq \bigcirc X \,, \qquad \square\square X \simeq \square X \,, \tag{1.19}$$

which means that they act like *projecting out* certain qualitative aspects of generalized spaces, while them being adjoint means that they project out an *opposite pair* of such qualities. Therefore, their (co-)unit transformations $\eta^{\square}$ (8.72) and $\epsilon^{\bigcirc}$ (8.73) exhibit every $X \in \mathbf{H}$ as carrying a quality intermediate to these two opposite extreme aspects [LR03, p. 245]:

$$\underset{\substack{\text{pure}\\ \bigcirc\text{-aspect}}}{\bigcirc X} \xrightarrow{\;\epsilon^{\bigcirc}_X\;} \underset{\substack{\text{generalized}\\ \text{geometric space}}}{X} \xrightarrow{\;\eta^{\square}_X\;} \underset{\substack{\text{pure}\\ \square\text{-aspect}}}{\square X} \,. \tag{1.20}$$

It turns out that by axiomatizing, this way, that every space X has a pair of opposite extreme aspects $\bigcirc X$ and $\square X$ to it, the spaces X themselves are forced to behave like carrying the kind of extra geometric structure which may be in between these opposites.

1.1.10 Differential topology in an ∞-topos

For example, any adjoint modality $\flat \dashv \sharp$ (see Def. 9.1.1) that contains the initial modality $\varnothing \dashv *$ (which globally projects to the initial and the terminal object, respectively) acts like projecting out *discrete* and *purely continuous* (co-discrete, chaotic) aspects of a space. Consequently, the existence of such a modality on $\mathbf{H}$ exhibits each space $X \in \mathbf{H}$ as carrying quality intermediate to these extremes; hence, in this example, as equipped with a kind of *topology* (see [Sh15, §3], following [La94]).

We observe here that extending this basic example to a larger system of adjoint modalities allows to abstractly encode the presence of differential geometric structure (Def. 9.1.22) and of super-geometric structure (Def. 9.1.42) in a powerful abstract way.

1.1.11 Generalized cohomology in an ∞-topos

At the same time, ∞-toposes may be understood as naturally embodying the ultimate notion of *generalized cohomology theories* (following [SSS12][NSS12a][Sc13]) subsuming and combining all of the examples listed in Table 1.2 on p. 12.

Namely, all these cohomology theories become "representable" in ∞-topos theory, meaning that their cohomology classes are simply the (homotopy) equivalence classes of maps in the ∞-topos to a given *classifying object* $A \in \mathbf{H}$:

Concretely, for $X, A \in \mathbf{H}$ a pair of objects, with X regarded as a domain "space" and A as the "coefficients" of cohomology, then *A-cohomology of X* is embodied by the morphisms from X to A:

(i) a morphism $X \xrightarrow{c} A$ is a *cocycle*;

(ii) a homotopy $X \underset{c_2}{\overset{c_1}{\Downarrow}} A$ is a *coboundary*;

(iii) the homotopy groups of the cocycle space

$$H^{-n}(X, A) := \pi_n \mathbf{H}(X, A) \simeq \pi_0 \mathbf{H}(X, \Omega^n A) \qquad (1.21)$$

are the *cohomology sets* of X with coefficients in A. (Here $\Omega^n(-)$ is the n-fold based looping operation.)

This is the *intrinsic cohomology theory* of the ∞-topos $\mathbf{H}$ – we discuss various examples in §11.

Table 1.2 Generalized cohomology and ∞-toposes

Flavor of cohomology	**realized in ∞-toposes**	
Sheaf hyper-cohomology	non-discrete ∞-toposes	[Bro73]
Non-abelian cohomology	general ∞-toposes	[SSS12][NSS12a, 3]
Twisted non-abelian cohomology	slice ∞-toposes	Prp. 8.1.62, Rem. 8.2.21
Twisted extraordinary cohomology	tangent ∞-toposes	Ex. 8.1.67, Rem. 8.2.23
Differential cohomology	cohesive ∞-toposes	Def. 9.1.1, Rem. 9.1.21
Étale cohomology	elastic ∞-toposes	Def. 9.1.22, Def. 11.2.1
Superspace cohomology	solid ∞-toposes	Def. 9.1.42, Rem. 9.1.46
Proper equivariant cohomology	singular ∞-toposes	Def. 9.2.3, Rem. 11.1.4, Thm. 11.1.9

1.1.12 Geometric orbifold cohomology in cohesive ∞-toposes

In summary, this suggests that the otherwise thorny question of how to conceive of

(i) orbifold geometry with its delicate interplay of differential geometric with equivariant and homotopy theoretic aspects

(ii) orbifold cohomology sensitive to this peculiar orbi-geometry while unconstrained by it in generality finds a natural and powerful solution when orbifolds are understood as objects of ∞-toposes equipped with systems of adjoint geometric modalities. This approach we lay out in the present book.

1.2 Results

With this machinery in hand, we develop the following results.

1.2.1 Axiomatic orbifold geometry in modal homotopy theory

By the above, to formulate proper orbifold cohomology, we ask for ∞-toposes (1.16) equipped with a system of adjoint modalities (1.18) that capture both aspects of proper orbifold cohomology:

(i) the *geometric* (differential, étale) aspect

(ii) the *singular* (proper equivariant) aspect.

Table 1.3 Modalities for Singular Super-Geometry (§9). The table shows the symbols used for the modalities in the main text, and indicates the modal geometric quality which they express. For example a space X is *discrete* or *bosonic* or *singular* iff it is equivalent to its purely discrete aspect $X \simeq \flat X$, or its purely bosonic aspect, $X \simeq \overset{\rightsquigarrow}{X}$, or its purely singular aspect $X \simeq$ ᐯX, respectively.
The modality τ_n expresses that an object contains higher gauge transformation only of degree $\leq n$.

τ_n **n-groupoidal**		
ʃ **shaped**	♭ **discrete**	♯ **continuous**
$\Re$ **reduced**	$\Im$ **étale**	$\mathcal{L}$ **locally constant**
⇉ **even**	⇝ **bosonic**	Rh **rheonomic**
ᐯ **singular**	∪ **smooth**	⋎ **orbi-singular**

1. The geometric aspect of orbifold theory. In order to formulate, in suitable ∞-toposes, the **(a)** differential topology, **(b)** differential geometry, and **(c)** super-geometry of orbifolds (hence of manifolds, super-manifolds, super-orbifolds, etc.) in their smooth guise as étale ∞-stacks (1.16), we consider a corresponding progression of adjoint modalities (1.18), which starts out in the form of the "axiomatic cohesion" of [La07], on to a second layer that contains a "de Rham shape" operation $\Im$ as considered in [Si96][ST97], and then to a third layer which captures super-geometry in a new axiomatic way [Sc13]:

Table 1.4 Progression of the geometric modalities.

id $\dashv$	id				
$\vee$	$\vee$				
$\rightrightarrows$ $\dashv$	$\rightsquigarrow$ $\dashv$	Rh			for super-geometry in solid ∞-toposes (Def. 9.1.42)
	$\vee$	$\vee$			
	$\Re$ $\dashv$	$\Im$ $\dashv$	$\&$		for differential geometry in elastic ∞-toposes (Def. 9.1.22)
		$\vee$	$\vee$		
		$\int$ $\dashv$	$\flat$ $\dashv$	$\sharp$	for differential topology in cohesive ∞-toposes (Def. 9.1.1)
			$\vee$	$\vee$	
			$\varnothing$ $\dashv$	$*$	

The key observation then is how this system of geometric cohesion goes along with a parallel system of cohesive modalities axiomatizing orbi-singular structure [Re14]:

2. The singular aspect of orbifold theory. Envision the picture of an orbifold singularity $\curlyvee$ and a mathematical magnifying glass held over the singular point. Under this magnification, one sees the resolved singular point as a *fuzzy fattened point*, to be denoted $\overset{G}{\curlyvee}$. Removing the magnifying glass, what one sees with the bare eye depends on how one squints:

(i) The physicists (like [BL99, §1.3]) and the classical geometers (like [IKZ10][Wat15]) say that they see an actual singular point, such as the tip of a cone $\llcorner$. This is the *plain quotient* $\overset{G}{\curlyvee} := */G = *$, a point.

(ii) The higher geometers (like [MP97] [CPRST14]) say that still they see the smooth G-action on that point, hence a smooth (stacky) geometry $\cup$. This is the *homotopy quotient* $\overset{G}{\curlyvee} := * /\!\!/ G = \mathbf{B}G$ (1.14), a groupoid.

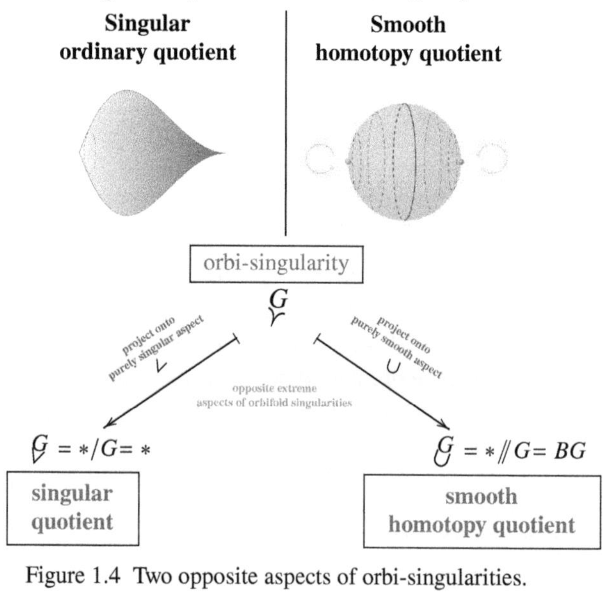

Figure 1.4 Two opposite aspects of orbi-singularities.

We observe in §9.2 that just this pair of dual perspectives is captured by the cohesive structure on global equivariant homotopy theory that had been found in [Re14] but whose conceptual interpretation had remained open [Re14, Footnote 8]. We find that the resulting system of modalities $\vee \dashv \cup \dashv \curlyvee$ serves to neatly axiomatize the nature of orbi-singularities and, hence, of orbispaces.

In combination with the previous geometric modalities, we thereby obtain the axiomatic infrastructure for our theory of geometric orbifold cohomology. This general abstract backdrop we lay out in Part IV.

Using this general cohesive axiomatics, in part V we begin (§10) by (re-)developing the differential geometry of orbifolds in this language, starting in the broader generality of "∞-orbifolds", namely of étale ∞-stacks.

1.2.2 Differential geometry of étale ∞-stacks

We present, in §10.2, a general theory of higher differential geometry formulated internally to these elastic ∞-toposes (§9.1). This deals with étale ∞-stacks locally modeled on any group ∞-stack V ("V-folds", Def. 10.2.2). For the special case $V = (\mathbb{R}^n, +)$, this subsumes ordinary manifolds (Ex. 10.2.5) and ordinary étale Lie groupoids (Ex. 10.2.6). For V a super-symmetry group (9.126), this produces a theory of super-orbifolds (Ex. 10.2.8), capturing, for instance, those that appear as target spaces in superstring theory (e.g. [PR04][GIR08]) and M theory [HSS18], or those that appear as moduli spaces of super-Riemann surfaces [Ra87][LBR88][Wi12][CV17].

The different incarnations of geometric orbifolds in the resulting modal language are summarized by the following diagram:

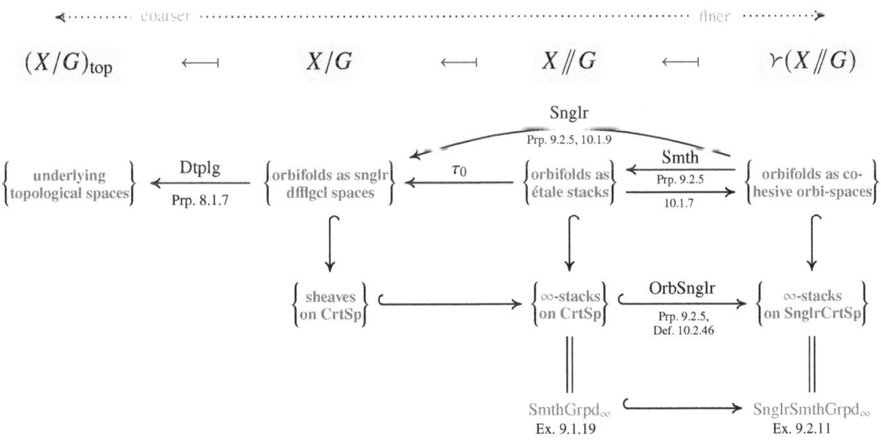

1.2.3 Bundles and Gerbes on étale ∞-stacks

With orbifolds, in their incarnation as étale stacks, thus embedded into a fully-fledged ∞-topos, the general theory of (equivariant) ∞-bundles [SS26c][NSS12a] [NSS12b] (see §8.2) applies to provide the theory of *fiber bundles* on orbifolds (cf. [LGTX04][Se06][BG08]) and of *gerbes* on orbifolds (cf. [LU04a][Car10] [BX][TT14]) naturally generalized to higher, to non-abelian, and to twisted gerbes on orbifolds.

1.2.4 Differential cohomology of étale ∞-stacks

Moreover, since the intrinsic cohomology theory of cohesive ∞-toposes is differential cohomology (Rem. 9.1.21), this realization of étale ∞-stacks within cohesive ∞-toposes immediately provides their (generalized, nonabelian) *differential cohomology theory* (via [FSS23, §9], reviewed in [SS26d], following [SSS12][FSS14][Sc13][BNV13]). This includes, in particular (as made explicit in [PR19]), the Borel-equivariant/orbifold Deligne cohomology considered in [KT18] (which, for finite groups, coincides with [LU03][Gom05]), given, in low degrees, by:

(i) gerbes with connection, hence including what in string theory is known as the *discrete torsion* classification of the B-field on orbifolds [Va86][VW95] [Sha00a][Sha00c][Sha02][Ru03]; and

(ii) 2-gerbes with connection, hence including what in M-theory is known as the *discrete torsion* classification of the C-field on orbifolds [Sha00b][Se01][dB+02, §4.6.2].

1.2.5 Proper equivariant enhancement of geometric homotopy theory

We enhance all of the above to a theory of properly orbi-singular spaces, formulated internally to "singular-elastic" ∞-toposes (§9.2), where a natural notion of orbi-singularization $\prec$ (Prop. 9.2.5) promotes (Def. 10.2.46) any such ∞-category of étale ∞-groupoids faithfully to its *proper* orbifold version (Rem. 10.2.48). This detects geometric fixed point spaces (Def. 9.2.25) in the sense of proper equivariant homotopy theory. We show (Prop. 10.1.2, Lem. 10.1.7) that the cohesive shape (Def. 9.1.1) of the orbi-singularization of an étale groupoid is its incarnation as an orbispace in global equivariant homotopy theory, in the sense of [HG07][Re14][Kö16][Schw17] (Rem. 10.1.1).

1.2.6 The proper 2-category of orbifolds

One model for the axioms of singular-cohesive homotopy theory is the ∞-topos of *singular-smooth ∞-groupoids* (established as Ex. 9.1.19, 9.2.11). In this model, the proper 2-category (Rem. 10.2.48) of orbifolds $\mathcal{X}$, either Lie theoretically (Ex. 10.1.11) or topologically (Ex. 10.1.12), is equivalent, via passage to

(i) their purely smooth aspect $\cup\mathcal{X}$, to the traditional 2-category of étale stacks (Prop. 10.1.9),

(ii) while their purely singular aspect $\vee\mathcal{X}$ gives the underlying singular coarse quotient space (Prop. 10.1.6).

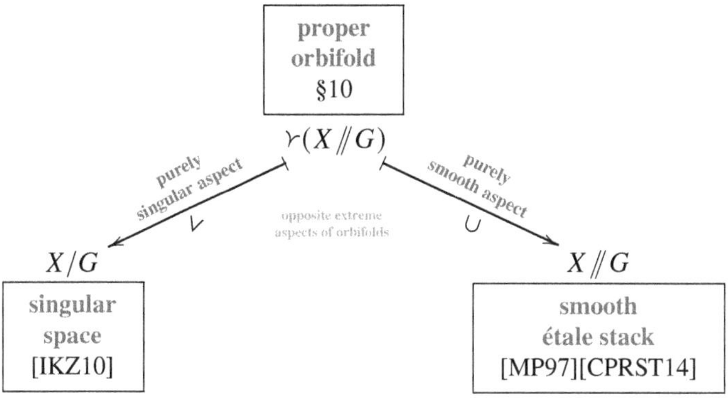

Figure 1.5 Two opposite aspects of orbifolds.

1.2.7 Cartan geometry of étale ∞-stacks

In this higher cohesive topos-theoretic formulation we find all fundamental phenomena of differential geometry naturally generalized to étale ∞-stacks, hence, in particular, to orbifolds, as indicated in Table 1.5 (p. 19).

1.2.8 Orbifold étale cohomology

Based on this, we give a natural general definition of *étale cohomology of V-étale ∞-stacks* (Def. 11.2.1), hence in particular of *orbifold étale cohomology*, which is sensitive to the above (integrable) G-structures and hence to geometry/special holonomy on orbifolds. For example, in the case of complex structure, this orbifold étale cohomology subsumes traditional notions of complex-geometric orbifold cohomology such as orbifold Dolbeault cohomology [Ba54][Ba56][CR04][Fe03] and orbifold Bott Chern cohomology [An12][Ma05]. Abstractly, orbifold étale cohomology is the intrinsic cohomology (1.21) of integrably G structured étale ∞-stacks when regarded in the slice ∞-topos (Prop. 8.1.62) over the G-Haefliger stack (Def. 10.2.33) via the classifying map of their G-structure (Prop. 10.2.35). As such, orbifold étale cohomology is the progenitor of *tangentially twisted* (proper) orbifold cohomology (Def. 11.2.3, Def. 11.2.5), to which we turn next.

1.2.9 Proper equivariant cohomology

While the proper 2-category of orbifolds is equivalent to the traditional one of orbifolds as étale stacks, its full embedding into an ambient singular-cohesive ∞-topos (§9.2) provides for more general coefficient objects, each of which is guaranteed to produce a proper orbifold Morita-class invariant (Rem. 10.2.48). Our first main Theorem 11.1.9 shows that the intrinsic cohomology (1.21) of orbifolds, regarded in singular-cohesive homotopy theory (Def. 9.2.3), subsumes all proper G-equivariant cohomology theories: Bredon cohomology with any coefficient system, as well as proper equivariant generalized cohomology theories.

1.2.10 Traditional orbifold cohomology

In particular, Prop. 10.1.2 and Theorem 11.1.9 imply, via [Ju20] (Remark 10.1.1), that proper orbifold cohomology in singular-cohesive homotopy theory subsumes Chen-Ruan orbifold cohomology, via (1.5), and Adem-Ruan orbifold K-theory, via (1.6). Hence, it also subsumes Freed-Hopkins-Teleman orbifold K-theory [FHT11] (reviewed in [Nu13, §3.2.2]) and Jarvis-Kaufmann-Kimura's "full orbifold K-theory" [JKK05][GHHK08] for orbifolds with global quotient presentations (by [FHT11, Prop. 3.5] and [JKK05, 3], respectively). Moreover, singular-cohesion provides a natural transformation $\cup\mathcal{X} \xrightarrow{\epsilon^{\cup}_{\mathcal{X}}} \curlyvee\mathcal{X}$ which restricts this proper orbifold cohomology to the underlying étale stack, where it reduces to traditional Borel orbifold cohomology (1.1) and, in particular, to Satake cohomology (1.2) (see also, e.g., [ADG11][BNSS18]).

1.2.11 Twisted orbifold cohomology

All these cohomology theories generalize to their twisted versions (e.g., local coefficients for ordinary cohomology, as in [MP99], or twisted K-theory, as in [AR03]), by passage to slices of the ambient singular-cohesive ∞-topos (Remark 8.2.21). In particular, slicing of orbifolds over $\curlyvee^{\mathbb{Z}_2}$ via their orientation bundle promotes them (Ex. 11.1.10) to *orientifolds* [DFM11][FSS15, 4.4][SS20].

1.2.12 Revisiting twisted orbifold K-theory

For illustration, we redevelop twisted orbifold K-theory in this language, using the result of its stacky representability from [SS26c, Ex. 4.5.4].

Table 1.5 Differential orbi-geometry in differential cohesive ∞-toposes

	§10.2	**Cartan geometry for étale ∞-stacks**	Literature for ordinary orbifolds
(i)	Def. 10.2.14	***Frame bundles***	[MM03, p. 42]
(ii)	Def. 10.2.24	***G-structures***	[Wo16][Zh06][BZ03]
	Def. 10.2.30, Def. 10.2.31	*-locally / -globally integrable*	
(ii.a)	Ex. 10.2.32	***Geometric structures***	[Ap00, §1.8][Wo16]
		- Riemannian structure	[Bo92][HM04][Rat06] [BZ07][He09a] [He09b][Ak12][Kan13] [BDP17][Lan18]
		- Flat structure	[BDP17][Ref06] [IU12, §8][SS20]
		- Complex structure	[SW99][FS07]
		- Symplectic structure	[Ve00][Go01][DE05] [HM12][CP14][Ch17] [RC19]
		- Lorentzian structure	[HS91][Ne02][LMS02a] [LMS02b][BR07] [ZR12]
		- Pseudo-Riemannian structure	[Mel09][Zh18][BZ19]
		- Conformal structure	[Ap98][Ap00]
		- CR-structure	[DM02]
		- Hypercomplex structure	[BGM98]
(ii.b)	Ex. 10.2.32	***Special holonomy***	[Jo00][CT05]
		- Kähler structure	[Fu83][Je97][Ab01] [BBFMT16]
		- Calabi-Yau structure	[Ro91][Jo98][Jo99a] [Jo99b][Jo00, §6.5.1] [St10][RZ11][CDR16]
		- Quaternionic Kähler	[GL88][Jo00, §7.5.2]
		- Hyper-Kähler struc.	[BD00]
		- G_2*-structure*	[Jo00, §11][Rei15]
		- Spin(7)*-structure*	[Jo00, §13][Ba07]
(iii)	Def. 10.2.29	***Local isometries***	[BZ07]
(iv)	Def. 10.2.33	***Haefliger stacks***	[Hae71][Hae84] [Car19, §2.2, §3][Car16]).
(v)	Def. 10.2.36	***Tangential structures***	[Wee18][Pa20]
(v.a)	Ex. 10.2.40	***Higher Spin-structures***	
		- Orientation	[Dr94]
		- Spin structure	[Ve96][Ac01] [BGR07][DLM02]
		- Spin^c *structure*	[Du96, §14]
		- String structure	[PW88][LU04a][LU06]
		- Fivebrane structure	[BL12] (cf. [SSS09][SSS12])

1.2.13 Proper orbifold étale cohomology

Finally, we promote (Def. 11.2.5) orbifold étale cohomology, in its guise as tangentially twisted cohomology, to a *proper* orbifold cohomology theory in the above sense (Rem. 10.2.48). Our **second main theorem** 11.2.6 shows that this *proper orbifold étale cohomology* unifies:

(i) ($\cup$) étale cohomology (Def. 11.2.1) of smooth V-folds (Def. 10.2.2),

(ii) ($\flat$) proper equivariant cohomology (Def. 11.1.2) of flat orbifolds (Def. 10.2.41), i.e., of their flat frame bundles (Prop. 10.2.42).

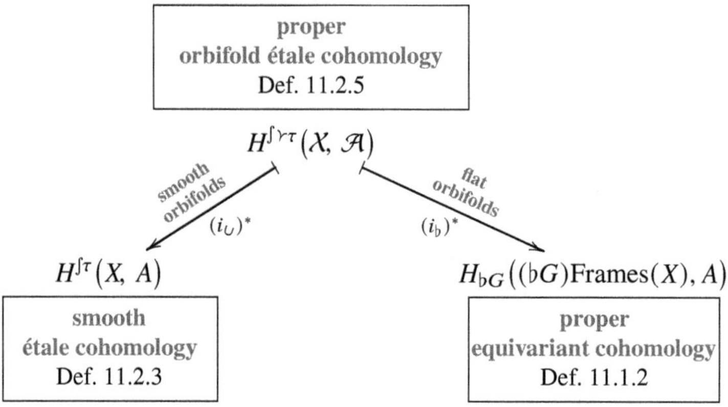

Figure 1.6 Two opposite aspects of étale cohomology.

1.2.14 Tangentially twisted orbifold Cohomotopy

We construct a fundamental class of examples of such proper orbifold étale cohomology theories, which we call *tangentially twisted* (Def. 11.2.18). Their coefficients are *Tate spheres* (Def. 11.2.9), in the sense of (unstable) motivic homotopy theory (Ex. 11.2.10), with twisting via an intrinsic *Tate J-homomorphism* (Def. 11.2.14). Specified to ordinary orbifolds (Ex. 11.2.19), Theorem 11.2.6 shows that tangentially twisted orbifold Cohomotopy subsumes:

(i) ($\cup$) tangentially twisted Cohomotopy theory of smooth but curved spaces, as introduced in [FSS20][FSS21a].

(ii) ($\flat$) RO-graded equivariant Cohomotopy theory of flat orbifolds, as discussed in [SS20][BSS19].

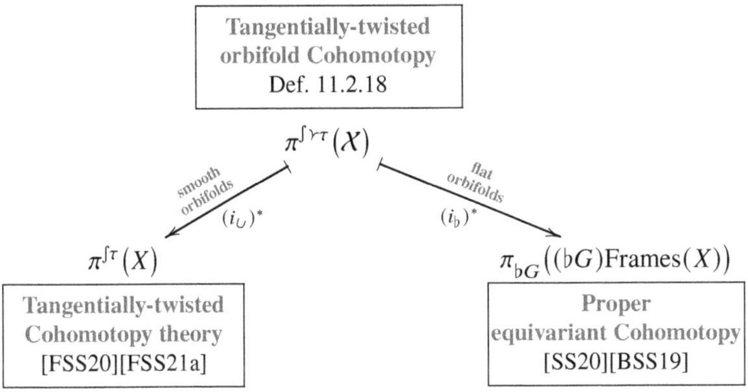

Figure 1.7 Two aspects of tangentially twisted orbifold Cohomotopy. Theorem 11.2.6 shows that the natural notion of tangential twist of Cohomotopy, when generalized to orbifolds, reduces near orbi-singularities to proper equivariant Cohomotopy in the RO-degree of the singularity's isotropy representation.

We conclude in Remark 11.2.20 on the impact of this unification.

1.2.15 Outlook on differential orbifold cohomology

While

(i) generalized differential cohomology on smooth manifolds [HS05] is well-understood (see [Bu12]) and

(ii) plain global equivariant cohomology has been established [Schw18] and understood to provide proper orbifold cohomology ([Ju20], see Remark 10.1.1),

their combination to (generalized, global) *proper equivariant differential cohomology* has remained elusive. Explicit constructions have been explored for the case of equivariant/orbifold differential K-theory [SV07][BS09][Or09], but even these theories do not seem to be well-understood yet [BS12, p. 47]. What has been missing is a coherent theoretical framework for proper equivariant differential cohomology: Since

(a) differential cohomology is the intrinsic cohomology (1.21) of cohesive ∞-toposes (by Remark 9.1.21) and

(b) proper equivariant cohomology is the intrinsic cohomology of ∞-toposes over a (global) orbit category (by Remark 11.1.4),

proper equivariant differential cohomology should be the intrinsic cohomology of ∞-toposes that combine these two properties. This is exactly what our notion of singular-cohesive ∞-toposes expresses (Def. 9.2.3), as confirmed by Theorem 11.1.9. For example, in singular-cohesive ∞-toposes there exists the (global) proper equivariant version of twisted differential non-abelian cohomology [FSS23], now given by homotopy fiber products parametrized over Snglrt (Def. 9.2.1). Hence, singular-cohesive ∞-toposes constitute a coherent framework in which to discuss proper equivariant/orbifold differential cohomology in general. We will develop this elsewhere.

We briefly comment on another related approach in the literature:

1.2.16 Proper ∞-categories of general étale ∞-stacks

Another general theory of étale ∞-stacks has been presented in [Car20], generalizing an elegant characterization of étale 1-stacks due to [Car19] by following the discussion of derived Deligne-Mumford stacks conceived as structured ∞-toposes in [Lu09b]. This approach proceeds externally via characterizing the *sites* (recalled as Prop. 8.1.57) which present ∞-toposes of étale ∞-stacks; and is thus complementary to the internal perspective proceeding from inside an ambient ∞-topos which we are presenting here. We briefly indicate the relation between the two:

- The approach in [Car20] is to pick an ∞-site of PrbSpc (denoted "$\mathcal{L}$" there) which is equipped with a suitable notion of which of its 1-morphisms qualify as being étale maps (the external version of our notion Def. 9.1.28). The inclusion i of the wide subcategory on these étale morphisms induces, by left Kan extension, a pair of adjoint ∞-functors ($i_! \dashv i^*$) between the corresponding ∞-stack ∞-toposes, and the étale ∞-stacks are then characterized as those in the essential image of the left adjoint $i_!$. This is shown on the right of the following diagram:

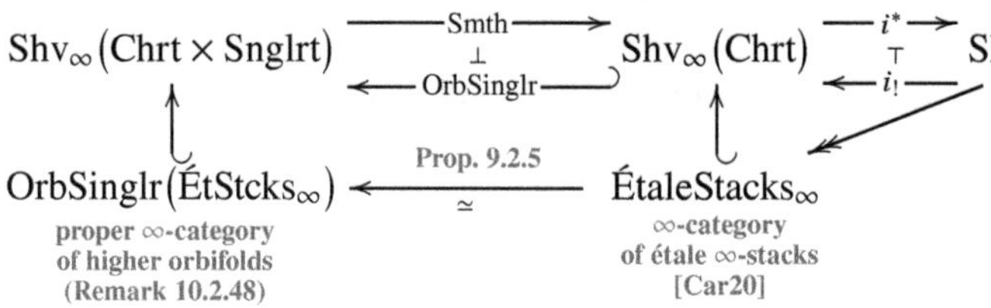

(1.22)

- Following Remark 10.2.48, we may and should enhance this construction to the *proper ∞-category of higher orbifolds* Def. 9.2.3, Def. 10.2.46, as shown on the left in (1.22).
- In fact, the archetypical example of PrbSpc considered in [Car20] is SmoothManifolds (Def. 8.1.9), in which case the left-hand side of (1.22) is the singular-cohesive ∞-topos of our Ex.s 9.1.19, 9.2.11, containing the proper (Def. 10.2.46) ∞-category of orbi-$\mathbb{R}^n$-folds in our Ex. 10.2.6.
- On the other hand, a general ∞-topos Shv_∞(PrbSpc) is not going to be cohesive (Def. 9.1.1) or even elastic (Def. 9.1.22). This means that various nice geometric properties, which we derive here, of objects in the proper ∞-category of higher orbifolds, are not guaranteed to exist in the general setup of [Car20]. Notably, the theory of frame bundles on orbifolds, according to Prop. 10.2.14, and the main theorem on the induced étale cohomology of orbifolds (Theorem 11.2.6) crucially use the internal modal logic of singular-cohesive and singular-elastic ∞-toposes as in §9, which may not exist, or not exist completely, for any given site of PrbSpc as in [Car20].

Part I

Invitation: Topological Charges

2

Generalized Cohomology

A key motivation for the development of generalized cohomology of manifolds and orbifolds is its application to the description of *topological phases* and *topological charges* in quantum physics.

2.1 Topological Quantum

In algebraic topology and homotopy theory, *cohomology* is, quite generally, about *deformation classes* of structures fibered over spaces [nL-AT][FSS23, §I][SS26c]. Our central perspective is that this is ultimately encoded via *classifying maps* (cf. Fig. 2.1) in the notion of *homotopy* (recalled in § 4.1.4), namely of continuous deformations between continuous maps $f, g : X \to \mathcal{A}$, making them have the same *homotopy class* $\pi_0(-)$:

$$X \underset{g}{\overset{f}{\rightrightarrows}} Y \;\;(\Downarrow) \qquad \text{exists iff} \qquad [f] = [g] \;\in\; \pi_0\,\mathrm{Map}(X, Y)\,. \tag{2.1}$$

In physics, cohomology describes global dynamical invariants of quantum systems: Strongly interacting quantum materials in their ground states may occupy globally non-trivial configurations classified by cohomology classes of the space occupied by the sample (or dually of its Brillouin space of crystal momenta). Moreover, in the "geometric engineering" of such quantum systems on higher-dimensional gravitating "branes" [nL-GeEn][SS25f][GSS25b], older arguments suggest that some form of generalized cohomology measures the charges of higher gauge field fluxes sourced by such branes (cf. § 2.1.1 for more).

These suggestions follow up on the classical observation, going back to a famous insight by Dirac from a century ago, that the totality of ordinary magnetic flux through surfaces is classified (in modern parlance) by ordinary integral cohomology [Al85][Fra97, §16.4e][SS25d, §2.1].

It was a major development (first in brane physics [nL-DK], then in topological quantum materials [nL-QK]) to realize that more general topological charges may plausibly be classified more accurately in "extraordinary" or "generalized" cohomology theories (in a sense going back to Whitehead, cf. [nL-WC]), more fine-grained

DOI: 10.1201/9781003675969-2

than ordinary cohomology – such as notably in the famous example of topological "K-cohomology theory" (traditionally just called topological *K-theory*), cf. [nL-K].

Ordinary cohomology	Whitehead-generalized stable cohomology	Fully generalized nonabelian cohomology
$H^n(X; A) \simeq \pi_0 \operatorname{Map}(X, \underbrace{B^n A}_{\text{Eilenberg-MacLane space}})$	$E^n(X) \simeq \pi_0 \operatorname{Map}(X, \underbrace{E^n}_{\text{stage in a spectrum of spaces}})$	$H(X; \mathcal{A}) \simeq \pi_0 \operatorname{Map}(X, \underbrace{\mathcal{A}}_{\text{any space}})$

$$\operatorname{Map}(X, \mathcal{A}) = \{X \dashrightarrow \mathcal{A}\}$$

Figure 2.1 The unifying perspective on general *cohomology* is as homotopy classes $\pi_0(-)$ of maps Map(−, −) into a *classifying space* $\mathcal{A}$.[1]

But there are yet more fine-grained *unstable* generalized cohomology theories called *non-abelian cohomology* ([Toë02, Def. 6.0.6][Sc09, Def. 2.3][Lu14, Def. 6], cf. [FSS23, §1][SS25g, §1]), which properly capture also non-linear Gauss laws of flux densities [SS25d, SS25b]. In full generality, all these notions of cohomology are neatly understood as being about homotopy classes (2.1) of classifying maps to some *classifying space*, see Fig. 2.1.

2.1.1 Choices of Charge Cohomology

The mathematical situation in § 2.1 highlights a general question (largely open) in physics theory building:

Which generalized (nonabelian) cohomology theories reflect the topological phases/charges of a given quantum system, both microscopically as well as at some level of coarse graining?

This is the question for the choice of *flux quantization* [SS25d]. Here we discuss this question in parallel for three situations, shown in Fig. 2.3, which are quite distinct at face value but turn out to be intimately related.

---

[1]More generally (in Fig. 2.1), we are to consider *pointed* spaces – the base point of the domain X regarded as its "point at infinity" and the base point of the coefficient space $\mathcal{A}$ regarded as zero – and maps $c \in \operatorname{Map}^*(X, \mathcal{A})$ that preserve these base-points – thus literally implementing the condition that solitonic charges *vanish at infinity*, $c(\infty) = 0$ (cf. [SS25d, §2.2][SS25b, §A.2][SS23-Mf, Ntn. 3.3]). We disregard this extra structure here just for brevity of the exposition; the pointed generalization of all statements is straightforward.

Similarly, another important generalization: For *geometric* cohomology (such as étale cohomology or differential cohomology), the classifying space $\mathcal{A}$ is generalized to a *moduli stack* **A**. This is particularly relevant for refining the discussion here from topological sectors to actual gauge field configurations (cf. [SS25d, §3.3][SS25a]), and this generalization is what we turn to in parts IV and V.

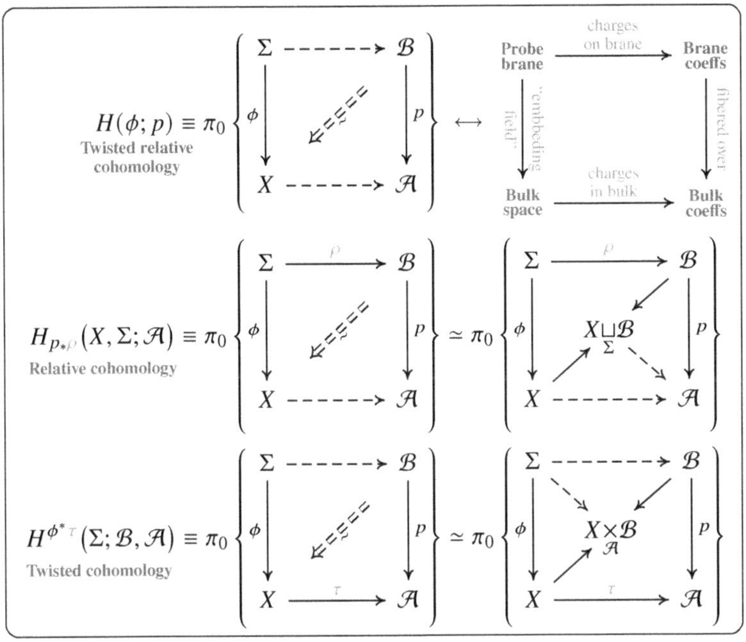

Figure 2.2 The general notion of *twisted relative non-abelian cohomology*, in evident generalization of Fig. 2.1, has cocycles given by squares of maps commuting up to specified homotopy.[2]
On the right, the expression "$(-) \sqcup_{(-)} (-)$" denotes a homotopy pushout and "$(-) \times_{(-)} (-)$" a homotopy pullback. The ordinary notion of *relative cohomology* is hence recovered for ϕ a cofibration and $\mathcal{B} \equiv *$, whence $X \sqcup_\Sigma \mathcal{B} \simeq X/\Sigma$. But in general, the homotopy pushout is richer (cf. Fig. 2.6).

Traditionally, topological phases and brane charges have been conjectured to be classified by stable generalized cohomology theories (with classifying spaces given by stages of spectra E, cf. again Fig. 2.1) such as:

- *ordinary cohomology* (cf. [SS25d, Ex. 3.10]),
- *complex K-cohomology* (cf. [nL-DK][nL-QK]),
- *elliptic cohomology* (cf. [Se88, §6][SK04][SK05][ST11][Hua20][Hua25]),
- *Morava K-cohomology* (cf. [SK04, SW05]).

Curiously, all of these proposed cohomology theories are *complex oriented* (cf. Fig. 2.4) – which is remarkable in view, the phenomenon presented in § 2.2.

On the other hand, closer analysis reveals [SS26a, SS26b]/[FSS20, GSS25a, GSS25a] (following [FSS17][Sa18, §2.5], reviewed in [FSS19][SS25d, §4.4]) that the most fine-grained microscopic description of topological insulator phases and of brane charges in D=5/D=11 supergravity is in a non-abelian cohomology theory,

[2]All 2-dimensional diagrams we show, here and in the following, are filled by homotopies (2.1), but we display only some of these homotopies explicitly, for emphasis.

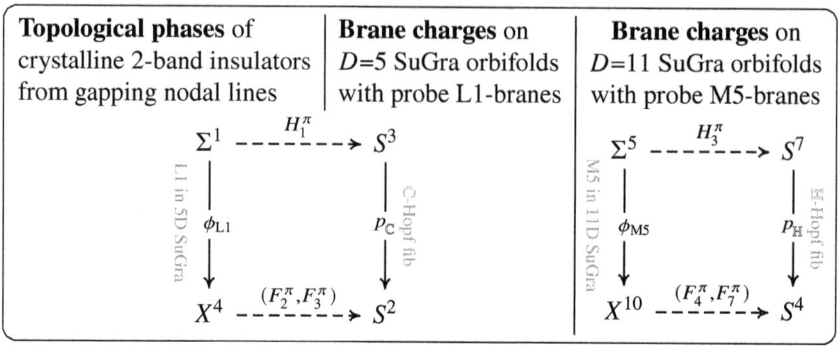

Figure 2.3 We discuss here the refined cohomological description [SS26b]/[FSS20, FSS21a] of microscopic charges in 5D/11D Supergravity with probe L1/M5-branes, minimally flux-quantized in Cohomotopy relatively twisted by the Hopf fibration (5.123). We find (cf. Fig. 2.6) that measurement of these unstable charges in a stable cohomology theory E is equivalent to complex/quaternionic four-ten-dimensional orientation in E – such as exists in particular on complex K-theory, E = KU (discussed in §5).

namely in *Cohomotopy* $\pi^n(-)$ [STHu59, §VII][FSS23, Ex. 2.7],

$$\pi^n(-) := \pi_0 \, \mathrm{Map}(-; S^n)\,, \tag{2.2}$$

specifically in 2-Cohomotopy/4-Cohomotopy, whose classifying space is the 2-sphere/ 4-sphere S^2/S^4 – and that in the presence of gapped nodal lines and of probe L1/M5-branes this becomes *twisted Cohomotopy relative to the brane* (recalling Fig. 2.2) classified by the complex/quaternionic Hopf fibration $p_{\mathbb{C}}/p_{\mathbb{H}}$ (5.123) – this is shown in Fig. 2.5.

This "proper" flux quantization in Cohomotopy – properly reflecting the nonlinear Gauss laws of the flux densities – has the remarkable consequence that it reflects the actual phase space structure of the (globally completed) gauge fields [SS25a]. Moreover, when spacetime has an "M-fiber" $\mathbb{R}^1_{\mathrm{M}}$ and the probe p-brane "wraps" it, in that

$$\phi : \overbrace{\mathbb{R}^1_{\mathrm{M}} \times \Sigma^{p-1}}^{\Sigma^p} \xrightarrow{\;\mathrm{id}\times\phi'\;} \overbrace{\mathbb{R}^1 \times X^{d-1}}^{X^d} \tag{2.3}$$

then:

The topological quantum observables on solitonic field configurations are completely determined from the flux quantization law. [SS25b, SS26b]

(Concretely, with the flux quantization given by its classifying space $\mathcal{A}$, the algebra of topological quantum observables is the Pontrjagin algebra of the mapping space of a Cauchy surface of spacetime into $\mathcal{A}$.)

In summary then, the microscopic flux quantization of 5D/11D supergravity with probe L1/M5 branes that we consider corresponds to the following set of charges in

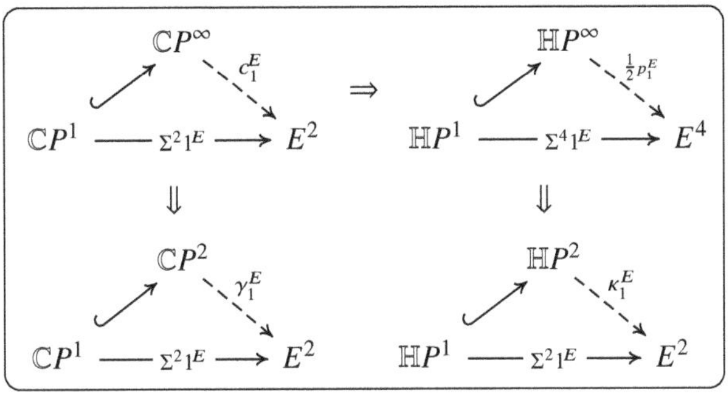

Figure 2.4 Notions of universal orientations of fibers of vector bundles in a multiplicative stable cohomology theory E (cf. [SS23-Mf, §3.8]):
Top left: A *complex orientation* is a choice of 2-class c_1^E on $\mathbb{C}P^\infty$ which restricts to the unit on $\mathbb{C}P^1 \simeq S^2$ (cf. [TK06, §3.2][Ko96, §4.3]).
Top right: A *quaternionic orientation* is a choice of 4-class $\frac{1}{2}p_1^E$ on $\mathbb{H}P^\infty$ restricting to the unit on $\mathbb{H}P^1 \simeq S^4$ (cf. [TK06, §3.9]).
Bottom row: A complex orientation *over 4-dimensions* is a choice of 2-class γ_1^E on $\mathbb{C}P^2$ restricting to the unit on $\mathbb{C}P^1$; and a quaternionic orientation *over 10-dimensions* is a choice of 4-class κ_1^E on $\mathbb{H}P^2$ restricting to the unit on $\mathbb{H}P^1$ [Hop84, §1.2][SS23-Mf, §3.8].
Note that complex orientations in $4k + 2$ dimensions induce quaternionic orientations in the same dimension [SS23-Mf, Thm. 3.99].

twisted relative Cohomotopy:

$$
\begin{aligned}
H(\phi'_{\mathrm{L1}}; p_{\mathbb{C}}) &= \pi_0 \left\{ \begin{array}{ccc} \Sigma^0 & \overset{H_1^\pi}{\dashrightarrow} & S^3 \\ \big\downarrow{\scriptstyle \phi'_{\mathrm{L1}}} & & \big\downarrow{\scriptstyle p_{\mathbb{C}}} \\ X^3 & \underset{F_3^\pi}{\overset{F_2^\pi}{\dashrightarrow}} & S^2 \end{array} \right\}, \\
H(\phi'_{\mathrm{M5}}; p_{\mathbb{H}}) &= \pi_0 \left\{ \begin{array}{ccc} \Sigma^4 & \overset{H_3^\pi}{\dashrightarrow} & S^7 \\ \big\downarrow{\scriptstyle \phi'_{\mathrm{M5}}} & & \big\downarrow{\scriptstyle p_{\mathbb{H}}} \\ X^9 & \underset{F_7^\pi}{\overset{F_4^\pi}{\dashrightarrow}} & S^4 \end{array} \right\}.
\end{aligned}
\tag{2.4}
$$

2.1.2 Cohomology Operations on Charges

Cohomotopy is rich and may contain more information than necessary in a given situation. A nonabelian *cohomology operation* [FSS23, Def. 2.3] from n-Cohomotopy to a stable cohomology theory E, hence a natural transform $\pi^n(-) \to E^n(-)$, may be understood as a *coarse graining* or extraordinary *character map*, which retains less but potentially more pertinent information. By the Yoneda lemma, such cohomology operations are given simply by postcomposition with maps $o : S^n \to E^n$ between the corresponding classifying spaces:

$$\begin{array}{ccc} \pi^n(X) & \xrightarrow{\qquad o_* \qquad} & E^n(X) \\ \big[X \dashrightarrow S^n\big] & \longmapsto & \big[X \dashrightarrow S^n \xrightarrow{\;o\;} E^n\big]. \end{array} \tag{2.5}$$

The fundamental (but most coarse) example is the (real, for our purpose) *Chern-Dold character* [nL-CD][FSS23, §7] on Cohomotopy (seen through its stabilization), which extracts its degree=n class in $\mathbb{R}$-rational cohomology

$$\begin{array}{ccc} \pi^n(X) & \xrightarrow{\qquad\qquad} & H^n(X;\mathbb{R}) \\ \big[X \dashrightarrow S^n\big] & \longmapsto & \big[X \dashrightarrow S^n \xrightarrow{\Sigma^n 1} B^n\mathbb{R}\big] \end{array} \tag{2.6}$$

by composing with the $\mathbb{R}$-rational unit class

$$\Big[S^n \xrightarrow{\Sigma^n 1} B^n\mathbb{R}\Big] = 1 \in \mathbb{R} \simeq H^n(S^n;\mathbb{R}). \tag{2.7}$$

In view of flux quantization of the F_2/F_4 flux in 5D/11D supergravity, this character map (2.6) witnesses how Cohomotopy indeed quantizes the total flux in that it forces its de Rham class to be the rational image of a Cohomotopy class (cf. [SS25d, §3]).

This notion of nonabelian cohomology operations has an evident generalization [FSS23, Def. 3.6] to twisted relative nonabelian cohomology (Fig. 2.2), where a cohomology operation is thus given by "pasting" (4.27) of homotopy-commuting squares, in our case as follows:

$$\begin{array}{ccc} H(\phi;p) & \xrightarrow{\qquad\qquad} & E^n\big(X^d, \Sigma^p\big) \\ \left[\begin{array}{ccc} \Sigma^p & \dashrightarrow & S^{2n-1} \\ {\scriptstyle \phi}\big\downarrow & & \big\downarrow{\scriptstyle p} \\ X^d & \dashrightarrow & S^n \end{array}\right] & \longmapsto & \left[\begin{array}{ccccc} \Sigma^p & \dashrightarrow & S^{2n-1} & \longrightarrow & * \\ {\scriptstyle \phi}\big\downarrow & & {\scriptstyle p}\big\downarrow & \swarrow & \big\downarrow{\scriptstyle 0} \\ X^d & \dashrightarrow & S^n & \xrightarrow{\;o\;} & E^n \end{array}\right]. \end{array} \tag{2.8}$$

With the left-hand side of (2.8) understood as the set of microscopic brane charges (Fig. 2.5), we may think of the character map (2.8) as "measuring" or coarse-graining these to the extent reflected in the given stable E-cohomology.

2.2 Charge Measurement

The first main observation we highlight now (following [SS23-Mf, §2.8, §3.8]) is the following, whose proof is shown in Fig. 2.6:

Measuring (2.8) *in a stable cohomology theory E, the fragile charges of 2-band insulators in the presence of nodal lines, or the microscopic brane charges in 5D/11D supergravity in the presence of probe L1/M5-branes* (see Fig. 2.5), *is equivalent to having four/ten-dimensional complex/quaternionic E-orientation* (Fig. 2.4).

Example 2.2.1. Every elliptic curve C (over any base ring) entails an elliptic cohomology theory E_C (cf. [Se88, (5.2)]), which is complex oriented (cf. [Se88, Ex. p. 197]) and hence also quaternionic-oriented (cf. [SS23-Mf, Prop. 3.98]). Therefore, the result of Fig. 2.6, in the situation (2.3) of an M5-brane wrapped on the M-fiber, says that a choice of an elliptic curve and of a sub-4-manifold determines an extraordinary character map from microscopic M-brane charges to the elliptic cohomology of the bulk spacetime relative to the brane locus:

$$\left.\begin{array}{rl} \text{elliptic curve} & C \\ \text{4-manifold} & \Sigma^4 \xrightarrow{\phi} X^9 \end{array}\right\} \Rightarrow H(\phi'_{\mathrm{M5}}; p_{\mathbb{H}}) \longrightarrow E^0_C(X^9, \Sigma^4)\,. \qquad (2.9)$$

This result is similar to the situation expected by informal arguments in [GPPV21].

Example 2.2.2. Consider specifically the situation of an M5-brane worldvolume of the form $\Sigma^{1,5} = \mathbb{R}^{1,2} \times S^3$ inside 11D Minkowski spacetime $\mathbb{R}^{1,10}$. Since the latter is equivalent to the point, in this situation the microscopically quantized charges of the 3-form field H_3 on $\Sigma^{1,5}$ are in plain 3-Cohomotopy, classified by the fiber of the $\mathbb{H}$-Hopf fibration, hence are given by some integer $c \in \mathbb{Z} \simeq \pi^3(S^3)$. Measuring this situation in a stable cohomology theory E which is complex-oriented in 10d by some h_3^E (according to Fig. 2.6), sees a charge H_3^E of the 3-form field H_3 classified by $\Omega E^4 \simeq E^3$ and being the c fold multiple of the generator there:

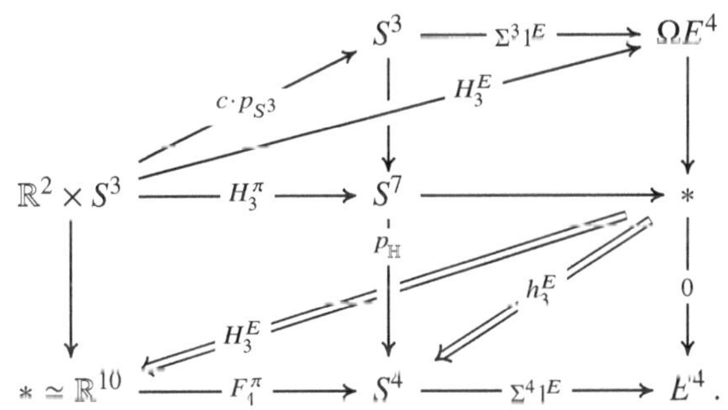

(2.10)

This means that, in this simple situation, the coarsened charges seen in E cohomology still reflect the full microscopic charges on the M5 in Cohomotopy iff the unit class $[1^E]$ is non-torsion.

2.2.1 Tangential Twisting and Orbifolding

Finally, all these considerations are to be generalized to include *tangential twisting* of the charge cohomology theory by the tangential G-structure of the spacetime domain.

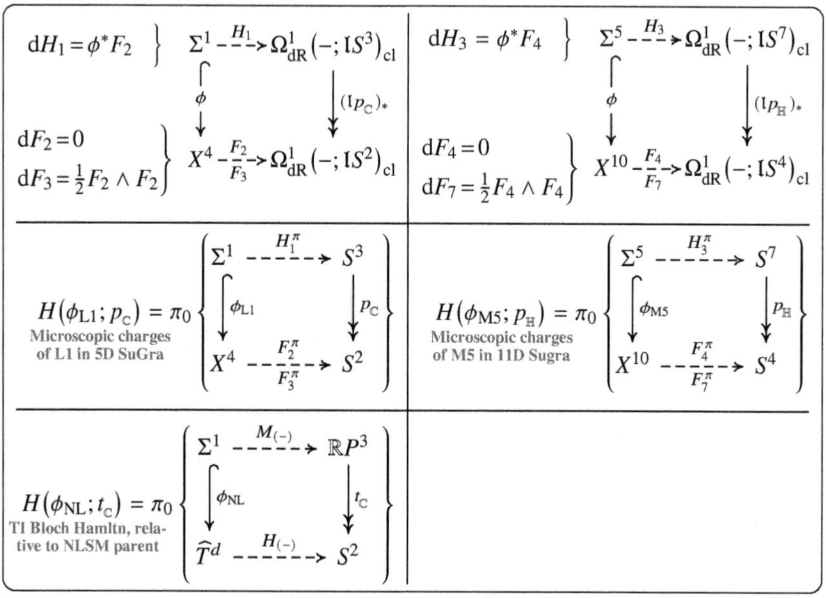

Figure 2.5 **1st row**: On a Cauchy surface X^4/X^{10} of spacetime and compatibly on the probe brane, $\Sigma^1/\Sigma^5 \hookrightarrow X^4/X^{10}$, the relative Gauss laws of the flux densities on probe L1/M5-branes in 5D/11D supergravity are equivalent [SS25a] to the closure condition (cl) on differential forms Ω_{dR} with coefficients in the real Whitehead L_∞-algebra $\mathfrak{l}p_{\mathbb{C}}/\mathfrak{l}p_{\mathbb{H}}$ of the $\mathbb{C}/\mathbb{H}$-Hopf fibration (cf. [FSS23, §5, §12]).
2nd row: The proper quantization of these fluxes is therefore [SS25d] in those twisted relative nonabelian cohomology theories whose classifying fibration p has the same $\mathfrak{l}p$. The minimal and hence most fine-grained choice among all these is the $\mathbb{C}/\mathbb{H}$ Hopf fibration $p_{\mathbb{C}}/p_{\mathbb{H}}$ itself. The dashed maps shown classify the charges under this twisted relative cohomotopical flux quantization [SS26b, §4.1.2]/[FSS20, §3.7]FSS19c[FSS21b].
3rd row: But other admissible proper flux quantization laws exist. For instance, the factorization $t_{\mathbb{C}}$ (5.124) of the $\mathbb{C}$-Hopf fibration through $\mathbb{R}P^3$ has the same relative real Whitehead L_∞-algebra, $\mathfrak{l}t_{\mathbb{C}} \simeq \mathfrak{l}p_{\mathbb{C}}$. Choosing this for flux quantization turns out to classify topological classes of Bloch Hamiltonians $H_{(-)}$ of 2-band topological insulators (TI) sensitive to the topology of the mass term $M_{(-)}$ which creates the TI phase from a parent nodal line topological semimetal phase (NLSM) – we discuss this in § 3.1.3.

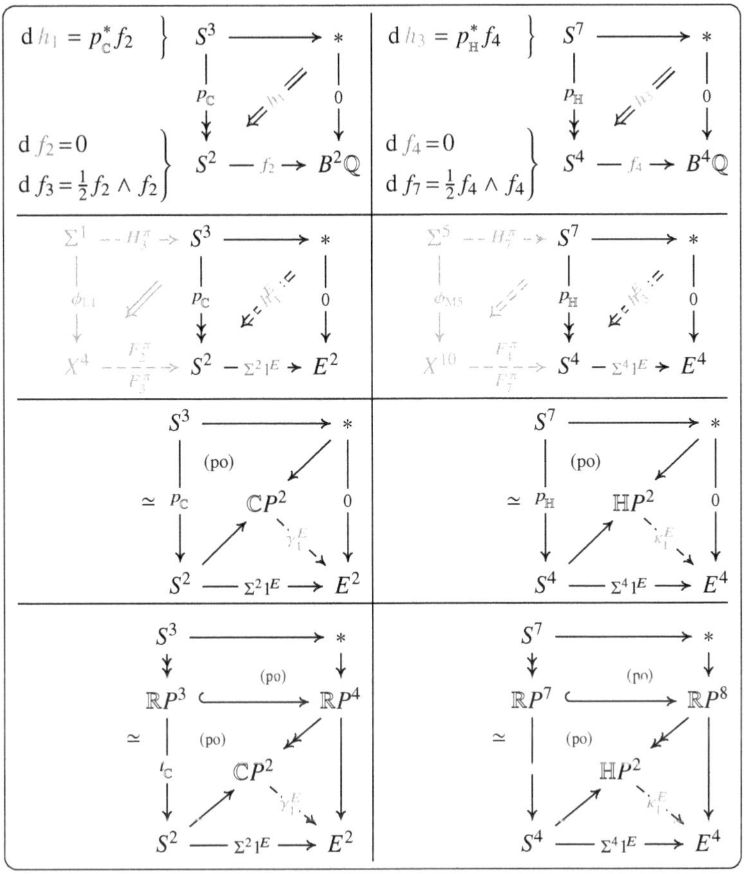

Figure 2.6 How measuring (2.8) in stable cohomology E the chiral flux on L1/M5-branes, microscopically in relative twisted Cohomotopy (Fig. 2.5), is equivalently a four/ten-dimensional $\mathbb{C}/\mathbb{H}$-orientation in E-cohomology (following [SS23-Mf, §2.8, §3.8]):
Top row: The generator h_1/h_3 of the Sullivan model of the $\mathbb{C}/\mathbb{H}$ Hopf fibration $p_{\mathbb{C}}/p_{\mathbb{H}}$, relative to that of the base, exhibits a null homotopy of the pullback of the generator f_2/f_4, the latter giving the unit map to the rational classifying space.
2nd row: Lifting this situation from rational cohomology to any multiplicative stable cohomology theory E means to ask for a null homotopy h_1^E/h_3^E of the pullback of the E-unit $\Sigma^2 1^E/\Sigma^4 1^E$. Indicated in gray is how this defines a character cohomology operation (2.8) from $p_{\mathbb{C}}$-twisted Cohomotopy, to E-cohomology relative to the probe L1/M5-brane, by forming pasting composites of homotopy squares (4.27).
3rd row: Factoring through the homotopy pushout (po) exhibits (Lem. 5.2.18) these null homotopies as equivalent to maps γ_1^E/κ_1^E from $\mathbb{C}P^2/\mathbb{H}P^2$ to E^2/E^4, whose restriction to $\mathbb{C}P^1/\mathbb{H}P^1$ is (homotopic to) the E-unit: These are four/ten-dimensional $\mathbb{C}/\mathbb{H}$-orientations in E-cohomology (cf. Fig. 2.4). We spell this out for $E = \mathrm{KU}$ in § 5.2.
4th row: By the pasting law (Prop. 4.1.4) and by Lem. 5.2.18, this statement remains true when the microscopic brane flux is quantized instead in $\mathbb{R}P^3/\mathbb{R}P^7$ (cf. 3rd row of Fig. 2.5), if now the stable coefficients are taken to be the pullback of the E-orientation to $\mathbb{R}^4/\mathbb{R}^8$:

Concretely, the $\mathbb{C}/\mathbb{H}$-Hopf fibration $p_{\mathbb{C}}/p_{\mathbb{H}}$ is equivariant with respect to a canonical $\mathrm{Spin}(3)/\mathrm{Spin}(5) \simeq \mathrm{U}(2)/\mathrm{Sp}(2)$ action (Def. Lem. 5.2.17), so that the microscopic brane charges may and should be [FSS20, FSS21a, SS21] twisted by tangential $\mathrm{U}(2)/\mathrm{Sp}(2)$-structure τ (Fig. 2.7). By our Thm. 11.2.6, this implies that the charges are in the correspondingly RO-graded equivariant cohomology in the vicinity of orbifold singularities. It is this tangentially twisted/equivariantized version of the construction of orientations as extraordinary characters that we establish in chapter 5.

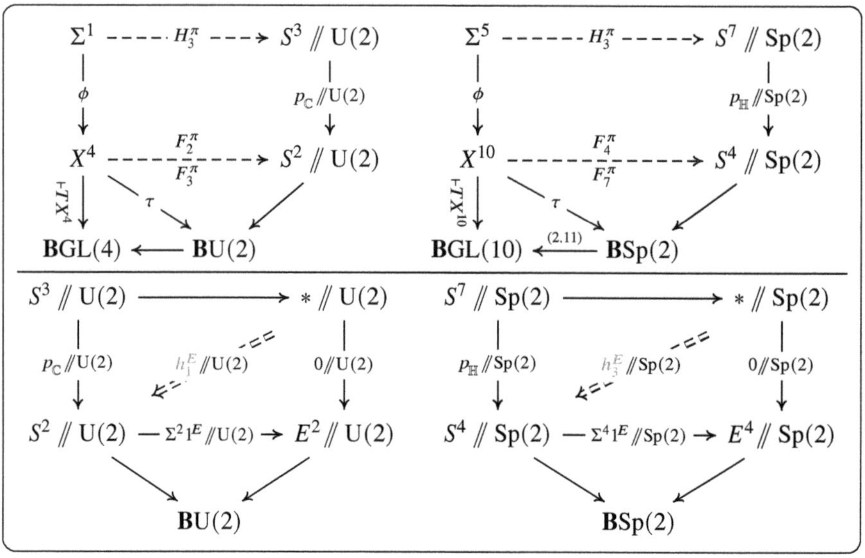

Figure 2.7 **Top row:** With coupling to background gravity taken into account – whose topological charges are encoded in the class of the tangent bundle TX – the relative brane charges are to be further twisted by tangential $\mathrm{SU}(2)/\mathrm{Sp}(2)$-structure τ, hence equivalently by $\mathrm{Spin}(3)/\mathrm{Spin}(5)$-structure on spacetime. [3]If we understand the homotopy quotients $(-) /\!\!/ (-)$ and deloopings $\mathbf{B}(-) \simeq * /\!\!/ (-)$ in topological groupoids (stacks), then this tangentially twisted cohomology automatically reduces to RO-graded equivariant cohomology in the vicinity of orbi-singularities (Thm. 11.2.6).
Bottom row: In this situation, the E-valued orientation character maps (Fig. 2.6) are to be equivariantized accordingly. This is what we construct, for $E = \mathrm{KU}$, in chapter 5.

[3]Beware the crucial subtlety (cf. Fig. 2.7) that $\mathrm{Spin}(5) \simeq \mathrm{Sp}(2)$ as abstract Lie groups, but that as subgroups of $\mathrm{Spin}(8)$ (and hence of the full $\mathrm{Spin}(1,10)$) they are *not* equivalent – but related by the less widely appreciated form of *triality* [FSS20, §2.3]:

$$\begin{array}{ccc} \mathrm{Spin}(5) & \xrightarrow{\sim} & \mathrm{Sp}(2) \\ \downarrow & & \downarrow \iota \\ \mathrm{Spin}(8) & \xrightarrow[\text{triality}]{\sim} & \mathrm{Spin}(8)\,. \end{array} \tag{2.11}$$

3

Phases & Branes

For motivation, we preview application of the construction in chapter 5 to the understanding of fragile topological phases and of microscopic charges in physical systems.

Here first some general words on the mathematical reflection of *band nodes* in crystalline quantum matter and of *monopole branes* in (higher) gauge quantum systems, in their charged singular version as well as in their *gapped* or *probe* incarnation that we are concerned with here (cf. [SS25d, §2.2]).

For the following, consider:

(i) X^d, a manifold representing either the momentum space of a crystalline quantum material or the physical space hosting (higher) gauge fields.

(ii) $\phi : \Sigma^p \hookrightarrow X^d$, a submanifold representing the spatial part of the *worldvolume* of a higher dimensional gauge monopole ("p-brane").

(iii) $\mathcal{A}$, the classifying space of gapped Bloch Hamiltonians or of gauge charges, respectively.

(iv) $G \curvearrowright (-)$, a smooth action of a Lie group on this data, being the crystalline point group symmetry or the orbifolding group, respectively.

Then the topological charge of ϕ being:

- an essential *band node*, where the Berry curvature would diverge

respectively:

- a singular *monopole p-brane*, where the bulk flux density would diverge,

is detected/measured by the G-equivariant $\mathcal{A}$-cohomology of the *complement* space $X^d \setminus \Sigma^p \subset X^d$ (disregarding here any further twists, just not to notationally overburden the discussion at this point):

$$H_G\big(X^d \setminus \Sigma^p; \mathcal{A}\big) = \pi_0\left\{ \begin{array}{ccc} \circlearrowleft^{G} & & \circlearrowleft^{G} \\ X^d \setminus \Sigma^p & \dashrightarrow & \mathcal{A} \end{array} \right\}. \tag{3.1}$$

The historical and archetypical example is that of a Dirac monopole $\phi : \{0\} \hookrightarrow \mathbb{R}^3$ whose magnetic charge is classified by $\mathcal{A} \simeq B\mathrm{U}(1)$ as

$$H\big(\mathbb{R}^3 \setminus \{0\}; B\mathrm{U}(1)\big) \simeq H^2(S^2; \mathbb{Z}) \simeq \mathbb{Z}, \tag{3.2}$$

where

$$S^2 \underset{\text{hmtpy}}{\simeq} S^2 \times \mathbb{R}_{>0} \simeq \mathbb{R}^3 \setminus \{0\} \tag{3.3}$$

DOI: 10.1201/9781003675969-3

arises as the 2-sphere around the monopole at the origin of space. The physical picture is hence that the cohomology of the complement $X^d \setminus \Sigma^p$ measures the charge reflected in the total field flux that emanates from the monopole brane (where its density diverges) and penetrates through the boundary of a tubular neighborhood.

Of course, if X^d itself has nontrivial cohomology, then this may contribute to the cohomology of the complement $X^d \setminus \Sigma^p$.

Now, when the band node gets *gapped* (by deforming the nature of the underlying quantum material), or respectively when the monopole brane is regarded in the *probe limit* where the backreaction of its (small) charge onto the ambient space is negligible and hence neglected, then this must mean that the charge is well-defined (non-divergent) on all of X^d, classified by a map $X^d \xrightarrow{\tau} \mathcal{A}$, *and* that as such it receives no contribution from the previous node/brane locus, hence the composite map $\phi^* c$

$$\begin{array}{ccc} \Sigma^p & & \\ {\scriptstyle \phi}\big\downarrow & \searrow^{\phi^* c} & \\ X^d & \xrightarrow[\tau]{} & \mathcal{A} \end{array} \tag{3.4}$$

has trivial class, in a suitable sense, namely that it can be equipped with whatever structure it is that reflects the undoing of the previous charge. For instance, for a band node, its undoing is its *gapping*, typically exhibited by a choice of *mass term*.

Whatever the trivialization process is, it will itself have a topological class which should have some classifying space $\mathcal{B}$, equipped with a fibration $\mathcal{B} \overset{p}{\twoheadrightarrow} \mathcal{A}$ encoding which trivializations $b \in \mathcal{B}$ concern which charges $p(b) \in \mathcal{A}$.

In conclusion, the topological data of ϕ, a *gapped* band node or *probe* brane, relative to a background charge c, should be classified by a map $\Sigma^p \dashrightarrow \mathcal{B}$ making this diagram commute:

$$\begin{array}{ccc} \Sigma^p & \dashrightarrow & \mathcal{B} \\ {\scriptstyle \phi}\big\downarrow & & \big\downarrow {\scriptstyle p} \\ X^d & \xrightarrow{\tau} & \mathcal{A} \end{array} \tag{3.5}$$

The deformation classes (relative homotopy classes) of such maps form the *twisted relative cohomology* of Fig. 2.2.

Thus, a pair of topological phases on X^d may superficially have the same topological class in $H_G(X^d; \mathcal{A})$, but in reality arise from a pair of topologically distinct gapping procedures $m_1 \neq m_2 \in H_G^{\phi^* c}(\Sigma^p; \mathcal{B}, \mathcal{A})$ (cf. Fig. 2.2) of a given mother phase. These twisted cohomology classes m on Σ^p hence witness that the two phases are not actually deformable into each other, after all.

We discuss the example of 2-band Chern insulators in § 3.1.3.

3.1 Chern Phases

Applying the orbifold cohomology construction of § 5.2.5 to the fragile crystalline topological phase [nL-UT] of 2-band Chern insulators [nL-CI] recovers exactly the

Bloch Hamiltonian (cf. [SS26a, SS25i]) and then its equivariant K-theory class (cf. [FM13][SS23b]). At gapped nodal lines, the construction in § 5.2.6 gives the corresponding *relative* K-classification, a new prediction.

We proceed to say this in more detail.

3.1.1 Bloch Hamiltonian Maps

In solid-state physics, *Bloch's theorem* (cf. [RS78, §XIII.16][Ser23, §5.1.3]) entails that the Hamiltonian operator H for single electrons propagating in a d-dimensional crystalline material is a direct integral

$$\begin{aligned} \mathcal{H} &\simeq \int_{\widehat{T}^d} \mathcal{H}_{\mathrm{Blch}}\, \mathrm{d}\mu \\ H &= \int_{\widehat{T}^d} H_{[\vec{k}]}\, \mathrm{d}\mu \;:\; \mathcal{H} \longrightarrow \mathcal{H} \end{aligned} \tag{3.6}$$

over crystal momenta $[\vec{k}]$ varying in the *Brillouin torus* (cf. [FM13, p. 52][Thi25, §2.1])

$$\widehat{T}^d \simeq \mathbb{R}^d / \mathbb{Z}^d , \tag{3.7}$$

of a continuous family of *Bloch Hamiltonians*

$$H_{(-)} : \widehat{T}^d \dashrightarrow \mathrm{End}(\mathcal{H}_{\mathrm{Blch}}) . \tag{3.8}$$

acting on a fiber Hilbert space $\mathcal{H}_{\mathrm{Blch}}$. These Bloch Hamiltonians have discrete real spectrum, and the graphs of eigenvalues of $H_{(-)}$ are called the *energy bands* (cf. [RS78, Fig. XIII.13][See04, §2]).

In an *insulator* ground state, electron states occupy the lowest $v \in \mathbb{N}$ of these bands below a given *Fermi energy*: the *valence bands*. Depending on external excitations, some number $c \in \mathbb{N}$ of further bands may be accessible to excited electrons, the *conduction bands*. Hence, in dependence on external parameters, the system's ground state and its accessible excitations are approximately described by finite-rank Bloch Hamiltonians, given by maps of this form (cf. [FM13, Prop. D.13]):

$$H_{(-)} : \widehat{T}^d \dashrightarrow \mathcal{B}(\mathbb{C}^{v+c}) \hookrightarrow \mathrm{End}(\mathcal{H}_{\mathrm{Blch}}) , \tag{3.9}$$

where we use the notation $\mathcal{B}(-)$ ("bounded operators") for notational brevity:

$$\mathcal{B}(\mathbb{C}^n) \simeq \mathrm{Mat}_{n\times n}(\mathbb{C}) . \tag{3.10}$$

Now for *gapped* ground states, hence with a positive *energy gap* between the valence and the conduction bands, the Bloch Hamiltonian map factors further through the subspace shown on the right here:

$$H_{(-)} : \widehat{T}^d \dashrightarrow \mathcal{B}(\mathbb{C}^{v+c})_{\mathrm{gap}} := \left\{ H \in \mathcal{B}(\mathbb{C}^{v+c}) \,\middle|\, \begin{array}{l} H^\dagger = H \\ \mathrm{Eig}_{<0}(H) \simeq \mathbb{C}^v \\ \mathrm{Eig}_{>0}(H) \simeq \mathbb{C}^c \end{array} \right\} , \tag{3.11}$$

where we now made explicit that the Bloch Hamiltonians are hermitian and we have chosen the origin of the energy scale to be the Fermi energy so that the v valence bands are those of negative energy. Quantum materials whose ground state fills the valence bundle of Bloch Hamiltonians of the form (3.11) with a *nontrivial homotopy class* $\left[H_{(-)}\right] \in \pi_0\mathrm{Map}\big(\widehat{T}^d, \mathcal{B}(\mathbb{C}^{\nu+c})_{\mathrm{gap}}\big)$ are called *topological insulators* [nL-TI]: Insulators because of the gap to the conduction band and "topological" because of the twist $\left[H_{(-)}\right]$ in the electron couplings which is locally trivial but globally non-trivial.

More generally, that a subgroup G of the point group of the crystal's space group (cf. [Arm88, §26][Ham15, §2]) is respected by (common jargon: "protects") the Bloch Hamiltonians means that there is a unitary representation of the point group on the Bloch quantum states

$$U : G \longrightarrow \mathrm{U}\big(\mathbb{C}^{\nu+c}\big), \tag{3.12}$$

such that (cf. [NS18][Sta20, §5.2])

$$\underset{g \in G}{\forall_{[\vec{k}] \in \widehat{T}^d}} \;:\; H_{g\cdot[\vec{k}]} = U(g) \circ H_{[\vec{k}]} \circ U(g)^{-1}. \tag{3.13}$$

We highlight that this says equivalently that the Bloch Hamiltonian map (3.11) is *G-equivariant* (4.13) with respect to the given point group action on the Brillouin torus (3.7) and the conjugation of the action of (3.12) on the Bloch Hamiltonians:

$$\widehat{T}^d \overset{H_{(-)}}{\dashrightarrow} \mathcal{B}\big(\mathbb{C}^{\nu+c}\big)_{\mathrm{gap}}\,. \qquad (\text{both sides carrying a } G\text{-action } \circlearrowleft^{G}) \tag{3.14}$$

Yet more generally, the crystallographic symmetries may be accompanied by *time reversal symmetries* in $\mathbb{Z}_2$ (cf. [SS23b, (32)]). This is exhibited by equipping the symmetry group with a homomorphism

$$\sigma : G \longrightarrow \mathbb{Z}_2 \tag{3.15}$$

and generalizing (3.13) to

$$\underset{g \in G}{\forall_{[\vec{k}] \in \widehat{T}^d}} \;:\; H_{g\cdot[\vec{k}]} = \begin{cases} U(g) \circ H_{[\vec{k}]} \circ U(g)^{-1} & \text{if } \sigma(g) = \mathrm{e} \\ U(g) \circ \overline{H_{[\vec{k}]}} \circ U(g)^{-1} & \text{otherwise}, \end{cases} \tag{3.16}$$

where $\overline{(-)}$ denotes component-wise complex conjugation.

In summary:

The parameters/couplings of d-dimensional G-crystalline (ν, c)*-band quantum materials vary in the equivariant mapping space* $\mathrm{Map}\big(\widehat{T}^d, \mathcal{B}(\mathbb{C}^{\nu+c})_{\mathrm{gap}}\big)^G$ (4.14).

3.1.2 Unstable topological phases of matter

A continuous deformation of the crystalline material – say by external tuning, heat, or other noise – changes the Bloch Hamiltonians (3.6) continuously, hence is an *equivariant homotopy* of the corresponding maps (3.8) and (3.9). That these deformations

(i) preserve the presence of an energy gap means that this homotopy, too, factors through the space (3.11) of gapped finite-rank Bloch Hamiltonians,

(ii) preserve the G-symmetry (3.14) (hence respect the symmetry protection) means that this is an equivariant homotopy (4.21):

$$
\begin{array}{ccc}
\overset{G}{\circlearrowright} & & \overset{G}{\circlearrowright} \\
\widehat{T}^d & \underset{H'_{(-)}}{\overset{H_{(-)}}{\rightrightarrows}} \Downarrow & \mathcal{B}(\mathbb{C}^{v+c})_{\mathrm{gap}}
\end{array}
\;:\;
\begin{array}{ccc}
\{0\} & \xrightarrow{\widehat{H_{(-)}}} & \\
\downarrow{\scriptstyle 0} & & \\
[0,1] & \dashrightarrow & \mathrm{Map}\big(\widehat{T}^d,\, \mathcal{B}(\mathbb{C}^{v+c})_{\mathrm{gap}}\big)^G, \\
\uparrow{\scriptstyle 1} & & \\
\{1\} & \xrightarrow{\widehat{H'_{(-)}}} &
\end{array}
\tag{3.17}
$$

and hence that the *deformation classes* of such gapped ground states are classified by the corresponding equivariant homotopy classes (4.21):

$$
(v, c)\mathrm{Phases}(d)^G := \pi_0 \,\mathrm{Map}\Big(\widehat{T}^d,\, \mathcal{B}\big(\mathbb{C}^{v+c}\big)_{\mathrm{gap}}\Big)^G . \tag{3.18}
$$

To get a better handle on this classification, and since these equivariant homotopy classes of maps depend only on the equivariant homotopy type of the classifying space, it is useful to pass to a tighter model of the latter. The following is immediate, but worth making explicit:

Lemma 3.1.1. *The classifying space* (3.11) *of* (v, c)*-gapped Bloch Hamiltonians is* $\mathrm{U}(\mathbb{C}^{v+c}) \rtimes \mathbb{Z}_2$*-equivariantly homotopy equivalent to the Grassmannian space* Gr_v^{v+c} (cf. [BZA24]) *of* v*-dimensional complex subspaces of* $\mathbb{C}^{v+c}$*:*

$$
\begin{array}{ccc}
\overset{\mathrm{U}(\mathbb{C}^{v+c})\rtimes\mathbb{Z}_2}{\circlearrowright} & & \overset{\mathrm{U}(\mathbb{C}^{v+c})\rtimes\mathbb{Z}_2}{\circlearrowright} \\
\mathcal{B}\big(\mathbb{C}^{v+c}\big)_{\mathrm{gap}} & \xrightarrow{\;\sim\;} & \mathrm{Gr}_v^{v+c} ,
\end{array}
\tag{3.19}
$$

where $\mathrm{U}(\mathbb{C}^{v,c})$ *acts by conjugation on the left and by left multiplication on the right, and* $\mathbb{Z}_2$ *acts by complex conjugation on both sides.*

Proof. Consider the following homeomorphic subspaces of $\mathcal{B}_v(\mathbb{C}^{v+c})$:

$$
\begin{array}{ccc}
\left\{ N \in \mathcal{B}\big(\mathbb{C}^{v+c}\big) \,\middle|\, \begin{smallmatrix} N^\dagger = N \\ \mathrm{Eig}_{-1} \simeq \mathbb{C}^v \\ \mathrm{Eig}_{+1} \simeq \mathbb{C}^c \end{smallmatrix} \right\} & \overset{\sim}{\longleftrightarrow} & \left\{ P \in \mathcal{B}\big(\mathbb{C}^{v+c}\big) \,\middle|\, \begin{smallmatrix} P^\dagger = P \\ P \circ P = P \\ \ker(P) \simeq \mathbb{C}^v \end{smallmatrix} \right\} \\
N & \longmapsto & \tfrac{1}{2}\big(1 + N\big) \\
2P - 1 & \longmapsfrom & P ,
\end{array}
\tag{3.20}
$$

where the homeomorphism evidently respects the given group action on both sides. The space of projectors on the right is, moreover, homeomorphic to the Grassmannian

$$
\begin{array}{ccc}
\left\{ P \in \mathcal{B}\big(\mathbb{C}^{v+c}\big) \,\middle|\, \begin{smallmatrix} P^\dagger = P \\ P \circ P = P \\ \ker(P) \simeq \mathbb{C}^v \end{smallmatrix} \right\} & \overset{\sim}{\longleftrightarrow} & \overbrace{\left\{ V \subset \mathbb{C}^{v+c} \,\middle|\, \dim(V) = v \right\}}^{\mathrm{Gr}_v^{v+c}} \\
P & \longmapsto & \ker(P) ,
\end{array}
\tag{3.21}
$$

and, again, this homeomorphism is evidently equivariant for the given group actions. Therefore, we are reduced to showing that the space of gapped Bloch Hamiltonians is equivariantly homotopy equivalent to the space of normalized Bloch Hamiltonians on the left of (3.20). To this end, consider the maps

$$
\begin{array}{ccccc}
 & N & \longleftarrow\!\shortmid & N & \\
\mathcal{B}(\mathbb{C}^{v+c})_{\mathrm{gap}} \equiv \left\{ H \in \mathcal{B}(\mathbb{C}^{v+c}) \,\middle|\, \begin{smallmatrix} H^\dagger = H \\ \mathrm{Eig}_{<0} \simeq \mathbb{C}^v \\ \mathrm{Eig}_{>0} \simeq \mathbb{C}^c \end{smallmatrix} \right\} & \begin{smallmatrix}\longleftarrow \\ \longrightarrow\end{smallmatrix} & \left\{ N \in \mathcal{B}(\mathbb{C}^{v+c}) \,\middle|\, \begin{smallmatrix} N^\dagger = N \\ \mathrm{Eig}_{-1} \simeq \mathbb{C}^v \\ \mathrm{Eig}_{+1} \simeq \mathbb{C}^c \end{smallmatrix} \right\} \\
H & \longmapsto & H \circ \sqrt{H^2}^{-1}, &
\end{array}
\tag{3.22}
$$

where, on the right, $\sqrt{-}$ denotes the unique positive definite operator square root, and $(-)^{-1}$ its inverse operator.

Now, the map going right-left-right is the identity, so we are reduced to showing that the map going left-right-left is equivariantly homotopic to the identity:

$$
\eta : \left(H \mapsto H \circ \sqrt{H^2}^{-1} \right) \Rightarrow \left(H \mapsto H \right). \tag{3.23}
$$

But such a homotopy is, for instance, given by

$$
\begin{array}{ccc}
\mathcal{B}(\mathbb{C}^{v+c})_{\mathrm{gap}} \times [0,1] & \xrightarrow{\quad \eta \quad} & \mathcal{B}(\mathbb{C}^{v+c})_{\mathrm{gap}} \\
(H, t) & \longmapsto & H \circ \left((1-t)\sqrt{H^2}^{-1} + t \right),
\end{array}
\tag{3.24}
$$

whose equivariance is again evident. □

In summary, we have seen that

The G-symmetry protected v-band topological phases of d-dimensional crystalline gapped quantum materials with access to c conduction bands are classified by G-equivariant nonabelian cohomology of the Brillouin torus with coefficients in Gr_v^{v+c}.

in that, with (3.18):

$$
(v, c)\mathrm{Phases}(d)^G \simeq H_G\big(\widehat{T}^d, \mathrm{Gr}_v^{v+c}\big). \tag{3.25}
$$

This classification is "unstable" (called "fragile" or "delicate"; cf. [nL-UT]) in that in its assumption of fixed finite numbers v and c of valence and conduction bands accessible by the system, its classification may break down when more bands become accessible to the system, in particular in the stable K-theoretic limit where $v, c \to \infty$.

Example 3.1.2 (Fragile crystalline 2-band insulator phases). The prominent case of crystalline *2-band Chern phases* (cf. [AF15][Ser23, §8-9] such as the *Haldane model*, cf. [Ser23, §8.3]) corresponds to setting $v \equiv c \equiv 1$ in (3.25) and $\sigma \equiv \mathrm{e}$ in (3.15). In this case the fragile classifying space (3.19) happens to be given by the 2-sphere

$$
\mathrm{Gr}_1^2 \simeq \mathbb{C}P^1 \simeq S^2, \tag{3.26}
$$

whence (3.25) says [SS26a, (17)][SS25i] that the most fine-grained *fragile crystalline Chern-phases* are classified by the equivariant form [Cru03, SS20] of *Cohomotopy* cohomology theory (cf. [STHu59, §VII][FSS23, Ex. 2.7])

$$
\pi^n(-) := \pi_0 \mathrm{Map}(-, S^2) \tag{3.27}
$$

in "degree" $n = 2$:

$$\begin{aligned}(1,1)\mathrm{Phases}(d)^G &\simeq \pi^2_G(\widehat{T}^d)\\ &\equiv \pi_0\,\mathrm{Map}(\widehat{T}^d,\ \mathbb{C}P^1)^G\,,\end{aligned} \tag{3.28}$$

hence by the connected components of the equivariant mapping space (4.14) to the 2-sphere.

This is the example on which we will focus now. While Chern phases have received considerable attention, in particular in their 2-band form, actual analysis of their fragile crystalline phases (3.28) seems to have found little to no attention before we brought up the issue in [SS26a, (17)][SS25i].

3.1.3 Mass Terms gapping Nodal Lines

Following Ex. 3.1.2, we go one step further and take account of the topological class of the process by which the 2-band Chern insulator phases in Ex. 3.1.2 are obtained by "gapping out" *nodal lines*. A nodal line in a *topological semimetal* (cf. [nL-TS]) is a curve in the Brillouin torus (3.7) over which the bulk energy gap between the valence and the conduction bands closes. A deformation of the material, reflected in a *mass term* being added to its Bloch Hamiltonian, may lift the band energy degeneracy over the previous nodal curve to turn the topological semimetal into a topological insulator (3.11).

Typically, such band nodes are "protected" by a symmetry (3.12), such as by $\mathbb{Z}_2$ "mirror symmetry" (cf. [Fa+16, §2.A][Ma+18]), which acts on the Brillouin torus by reflection of the x-coordinate (say):

$$\begin{array}{ccc}\mathbb{Z}_2\times\widehat{T}^3 & \longrightarrow & \widehat{T}^3\\ ([1],\ [k_x,k_y,k_z]) & \longmapsto & [-k_x,k_y,k_z]\end{array} \tag{3.29}$$

and acts on the bands as the X Pauli matrix (5.95):

$$-\mathrm{i}\gamma(\mathbf{i}) \equiv \begin{pmatrix}1 & 0\\ 0 & -1\end{pmatrix}. \tag{3.30}$$

The nodal line then lies in one of the mirror planes $[k_x] = 0$, say at $k_x = 0$, where the band symmetry (3.13) implies that the Bloch Hamiltonian, which may generally be expanded in the Pauli matrices (5.95) as

$$\begin{aligned}&H_{[\vec{k}]} = h_{[\vec{k}]} - \mathrm{i}\Big(h^x_{[\vec{k}]}\gamma(\mathbf{i}) + h^y_{[\vec{k}]}\gamma(\mathbf{j}) + h^z_{[\vec{k}]}\gamma(\mathbf{k})\Big) \in \mathcal{B}(\mathbb{C}^2)\,,\\ &h^{(-)}_{[\vec{k}]} \in \mathbb{R}\,,\end{aligned} \tag{3.31}$$

is of the form

$$H_{[0,k_y,k_z]} = h_{[0,k_y,k_z]} - \mathrm{i}h^x_{[0,k_y,k_z]}\gamma(\mathbf{i})\,, \tag{3.32}$$

with vanishing energy gap on the nodal line $\phi : S^1 \hookrightarrow \widehat{T}^3$,

$$h^x_{[0,k_y,k_z]} = 0 \quad\Leftrightarrow\quad [0,k_y,k_z] \in \phi(S^1) \subset \widehat{T}^3. \tag{3.33}$$

Conversely, the *mass term* perturbation which gaps this nodal curve (cf. [SS23b, p. 23]) typically needs to anti-commute (cf. [MF13, (4)][CS14, p. 8][FH21, Lem 9.55]) with the Bloch Hamiltonian there, breaking the mirror symmetry. With the mass term relevant only on a tubular neighborhood of the nodal curve, we may therefore consider it as localized on the nodal curve, where it hence is of the form

$$\begin{aligned} M_s &= -\mathrm{i}\big(m_s^y\,\gamma(\mathbf{j}) + m_s^z\,\gamma(\mathbf{k})\big) \\ (m_s^x, m_s^y) &\in \mathbb{R}^2 \setminus \{0\} \\ s &\in S^1\,, \end{aligned} \tag{3.34}$$

hence equivalently

$$\begin{aligned} M_s &= -m_s\, e^{\frac{\alpha_s}{2}\gamma(\mathbf{i})} \cdot \mathrm{i}\gamma(\mathbf{j}) \cdot e^{-\frac{\alpha_s}{2}\gamma(\mathbf{j})}\,, \\ m_s &\in \mathbb{R}_{>0}\,, \\ \alpha_s &\in \mathbb{R}\,. \end{aligned} \tag{3.35}$$

3.1.4 Classifying Fibration for Mass Terms

The above analysis (3.34) shows (which may not previously have been appreciated) that there is topology in the choice of mass term, classified by a winding number. (Mathematically, this is of just the form familiar from the *Su-Schrieffer-Heeger model*, cf. [AOP16, §1], but the physical phenomena described in both cases are different.)

Concretely, the above formula (3.35) makes manifest that, after picking any Bloch basis state for the valence bundle over the basepoint $[0] \in S^1$, the variation of the mass term along the nodal curve is equivalently reflected by the correspondingly varying family of its Bloch eigenstates, say

$$s \longmapsto e^{\frac{\alpha_s}{2}\gamma(\mathbf{i})}\begin{pmatrix} +1 \\ -1 \end{pmatrix} = \begin{pmatrix} +e^{+\mathrm{i}\frac{\alpha_s}{2}} \\ -e^{-\mathrm{i}\frac{\alpha_s}{2}} \end{pmatrix}. \tag{3.36}$$

We observe now that this family of Bloch states over the (previous, now gapped) nodal curve, reflecting the choice of gapping process (mass term) giving rise to a fully gapped 2-band insulator phase $H_{(-)} : \widehat{T}^d \to \mathbb{C}P^1$, is exactly the choice of a dashed arrow making the following diagram commute

$$\begin{array}{ccc} S^1 & \xrightarrow{\; s \,\mapsto\, \begin{pmatrix} +e^{+\mathrm{i}\frac{\alpha_s}{2}} \\ -e^{-\mathrm{i}\frac{\alpha_s}{2}} \end{pmatrix}\cdot\mathbb{Z}_2 \;} & \mathbb{R}P^3 /\!\!/ \mathbb{R} \\ {\scriptstyle\phi}\big\downarrow & & \big\downarrow{\scriptstyle t_{\mathbb{C}}} \;\; \downarrow \\ \widehat{T}^d & \xrightarrow{\quad H_{(-)} \quad} & \mathbb{C}P^1 /\!\!/ \mathbb{C}\,, \end{array} \tag{3.37}$$

where the map on the right is the factor of the complex Hopf fibration from (5.124).

To note here how the commutativity of this diagram accurately reflects the aspects of the gapping process relevant for topological classification: The composite left-bottom map identifies the fibers of the valence bundle (after the gapping) over the locus where the nodal gap closure had been, and the composite top-right map identifies these with exactly the fibers that the mass term has produced by shifting away the degenerate conduction band.

This indicates that the classification of topological insulator phases which takes into account their origin by gapping a given topological semimetal phase with a nodal curve ϕ, is given not just by plain Cohomotopy as in (3.28), but by its relative twisted version $H(\phi, t_{\mathbb{C}})^G$ from Fig. 2.6.

Note that the above dashed map (3.37) has codomain $\mathbb{R}P^3 \simeq S(\mathbb{C}^2)/\mathbb{Z}_2$ instead of $S(\mathbb{C}^2)$ itself because with the mass term (3.35) being a well-defined function on S^1, the Bloch states (3.36) are in general only periodic up to sign inversion. Explicitly, iff the winding number of the mass term (3.35) is *even* then the dashed map (3.37) factors via $S(\mathbb{C}^2) \twoheadrightarrow \mathbb{R}P^3$ (5.124) through the actual $\mathbb{C}$-Hopf fibration

$$\text{even mass term winding} \quad \Rightarrow \quad \begin{array}{ccc} S^1 & \xrightarrow{\; s \,\mapsto\, \left(\begin{smallmatrix} +e^{+\mathrm{i}\frac{\alpha_s}{2}} \\ -e^{-\mathrm{i}\frac{\alpha_s}{2}} \end{smallmatrix}\right) \;} & S(\mathbb{C}^2) \\ {\scriptstyle \phi}\big\downarrow & & \big\downarrow{\scriptstyle p_{\mathbb{C}}} \\ \widehat{T}^d & \xrightarrow{\quad H_{(-)} \quad} & \mathbb{C}P^1 . \end{array} \tag{3.38}$$

In conclusion, we find that 2-band gapped topological phases together with the gapping process that made room for it, of nodal curves ϕ in a parent semi-metal phase, are jointly classified by the relative twisted generalization of the (equivariant) Cohomotopy classification from Ex. 3.1.2 which is classified by the fibration $t_{\mathbb{C}}$ (5.124)

$$H_G(\phi;\, t_{\mathbb{C}}) = \pi_0 \left\{ \begin{array}{ccc} \Sigma^1 & \dashrightarrow & \mathbb{R}P^3 \\ {\scriptstyle \phi}\big\downarrow & & \big\downarrow{\scriptstyle t_{\mathbb{C}}} \\ \widehat{T}^d & \overset{H_{(-)}}{\dashrightarrow} & \mathbb{C}P^1 \end{array} \right\}^G . \tag{3.39}$$

This describes the (potentially) practically relevant physical situation where the topological insulator phase remains close enough to the parent nodal line semimetal phase that its deformations cannot reach insulator phases that arise from topologically distinct gapping processes.

3.1.5 Stabilization to K-Theory

Still more popular in the current literature than unstable/fragile band topology (3.25) is (cf. [nL-QK]) the coarser but *stable* classification of crystalline topological phases in equivariant K-theory (5.83).

Our construction in § 5.2 now gives, first of all, the coarsening cohomology operation from the fine-grained but fragile crystalline phases classified by equivariant 2-Cohomotopy (3.28) to coarse but stable classification by equivariant K-theory (5.83):

Given a rank=2 unitary representation U (3.12) of the crystal point group G, and hence the induced stable G-action on $\mathrm{Fred}_{\mathrm{gr}}$ as in Ntn. 5.2.27

$$G \xrightarrow{U} \mathrm{U}(2) \xrightarrow{\bigoplus_{\mathrm{N}} \gamma} \mathrm{U}(\mathcal{H}) \longrightarrow \mathrm{Aut}(\mathrm{Fred}_{\mathrm{gr}}) , \tag{3.40}$$

encoding the given band symmetry as per (3.13), composition with the map which is the tautological $\mathbb{C}P^1$-parameterized Fredholm operator (as in Lem. 5.2.29) gives the cohomology operation from fragile equivariant 2-Cohomotopy to stable equivariant K-theory, shown at the bottom here:

$$\begin{array}{ccc} \overset{\curvearrowleft G}{\mathbb{C}P^1} & \xrightarrow{\quad F^{\mathrm{taut}}_{\mathbb{C}P^1} \ominus F^{\mathbb{C}^1}_{\mathrm{triv}} \quad} & \overset{\curvearrowleft G}{\mathrm{Fred}_{\mathrm{gr}}} \\ \pi_0 \, \mathrm{Map}\big(\widehat{T}^d, \mathbb{C}P^1\big)^G & \xrightarrow{\quad \left(F^{\mathrm{taut}}_{\mathbb{C}P^1} \ominus F^{\mathbb{C}^1}_{\mathrm{triv}}\right)_* \quad} & \pi_0 \, \mathrm{Map}\big(\widehat{T}^d, \mathrm{Fred}_{\mathrm{gr}}\big)^G \\ \| & & \| \\ \pi^2_G\big(\widehat{T}^d\big) & \xrightarrow{\quad \mathrm{ch}^{\pi/K} \quad} & \mathrm{KU}_G\big(\widehat{T}^d\big)\,. \end{array} \tag{3.41}$$

Recalling the construction of this map from § 5.2, at $P \in \mathbb{C}P^1$ the tautological Fredholm operator is $F^{\mathrm{taut}}_P = F_P$ (5.144), which is the Fredholm operator incarnation (5.55) of the projector P. Unwinding the definitions, this recovers the normalized Bloch Hamiltonian in its Pauli matrix expansion (cf. [SS26c, §6.2.3]).

Secondly, we obtain the refinement of this stabilization construction to the relative twisted classification (3.39) of 2-band insulator phases sensitive to the gapping procedure from their nodal line semimetal parent phase: The corresponding cohomology operation is now given by forming the pasting composite according to Fig. 2.6 with the $\mathbb{R}P^3$-relative form of the homotopy h^{KU}_1 from Prop. 5.2.30:

$$\begin{array}{ccc} H_G\big(\phi;\, t_{\mathbb{C}}\big) & \xrightarrow{\qquad\qquad} & H_G\big(\phi;\, \gamma^{\mathrm{KU}}_1\big) \\ \left(\begin{array}{ccc} \Sigma^1 & \xrightarrow{M_{(-)}} & \mathbb{R}P^3 \\ {\scriptstyle \phi}\downarrow & & \downarrow{\scriptstyle t_{\mathbb{C}}} \\ \widehat{T}^d & \xrightarrow{H_{(-)}} & \mathbb{C}P^1 \end{array}\right) & \longmapsto & \left(\begin{array}{ccccc} \Sigma^1 & \xrightarrow{M_{(-)}} & \mathbb{R}P^3 & \hookrightarrow & \mathbb{R}P^4 \\ {\scriptstyle \phi}\downarrow & & \downarrow{\scriptstyle t_{\mathbb{C}}} & \overset{h^{\mathrm{KU}}_1}{\Swarrow} & \downarrow{\scriptstyle (f^{\mathbb{R}}_{\mathbb{C}})^* \gamma^{\mathrm{KU}}_1} \\ \widehat{T}^d & \xrightarrow[H_{(-)}]{} & \mathbb{C}P^1 & \xrightarrow[F^{\mathrm{tau}}_{\mathbb{C}P^1} \ominus F^{\mathbb{C}^1}_{\mathrm{triv}}]{} & \mathrm{Fred}_{\mathrm{gr}} \end{array}\right). \end{array} \tag{3.42}$$

3.2 M-Brane Charges

We here discuss (along the lines of Figs. 2.6 and 2.7) the measurement in topological K-theory of the charges on M5-brane worldvolumes (such as sourced by the singular self-dual string), which microscopically are in twisted relative Cohomotopy. This will also give us occasion to make precise and complete an old suggestion of [Ho98, (3.7, 3.17)][Wi01, p. 6-8] (see Rem. 3.2.11) for how to exhibit D6-brane charge in K-theory, namely for how to construct an explicit map from $S^3 \simeq \mathbb{R}^3_{\cup\{\infty\}}$ to the space of self-adjoint Fredholm operators, which represents, under Prop. 5.1.18, a generator of $\mathrm{KU}^1(S^3) \simeq \mathbb{Z}$.

3.2.1 Recap of abelian (D-)brane charges

In order to make contact with traditional discussion, we briefly recall the gist of the traditional idea of measuring (D-)brane charges in abelian cohomology (cf. [nL-DK]). While "well known", in its totality, the following is not always easy to glean from existing literature.

So let E be an abelian cohomology theory (cf. Fig. 2.1) such as:

- $E^n(-) \simeq H^n(-;A)$ – ordinary cohomology,
- $E^n(-) \simeq \mathrm{KU}^n(-)$ – complex topological K-theory.

Then:

- A **singular magnetic brane** is a substantial source of flux. In the absence of twistings, the total flux through any sphere enclosing the brane is the same (Gauß law); hence, the flux *density* on the enclosing spheres scales with a negative power of their radius and thus diverges at the would-be locus of the brane, which hence is a *singularity*. Therefore, the charge/total flux of singular branes is to be measured on the non-singular *complement* of their worldvolume. For flat branes, this complement is homotopy equivalent to the enclosing sphere:
$$\mathbb{R}^d \setminus \mathbb{R}^p \;\simeq\; \mathbb{R}^p \times \mathbb{R}_{>0} \times S^{d-p-1} \underset{\text{hmtpy}}{\sim} S^{d-p-1}\,. \tag{3.43}$$

  The original (theoretical) example is the *Dirac magnetic monopole* with $d = 3$ and $p = 0$ and with the magnetic flux that is sourced by the monopole measured in ordinary integral 2-cohomology:
$$H^2(\mathbb{R}^3 \setminus \mathbb{R}^0; \mathbb{Z}) \simeq H^2(S^2;\mathbb{Z}) \simeq \mathbb{Z}\,. \tag{3.44}$$

Table 3.1 Topologies on which to measure the total charge associated with singular (middle row) and probe D-branes (bottom row).

	Type IIA					Type IIB					
Dp-brane species	D0	D2	D4	D6	D8	D(-1)	D1	D3	D5	D7	D9
Enclosing sphere $S^{9-p-1} \sim \mathbb{R}^9 \setminus \mathbb{R}^p$	S^8	S^6	S^4	S^2	S^0	S^9	S^7	S^5	S^3	S^1	S^{-1}
Transverse space $S^{9-p} \sim \mathbb{R}^{9-p}_{\cup\{\infty\}} \times \mathbb{R}^p$	S^9	S^7	S^5	S^3	S^1	S^{10}	S^8	S^6	S^4	S^2	S^0

  Often overlooked is that the analogue remains true for singular type IIA/B D-branes (cf. [SS23a, Rem. 4.5]), hence for $d = 9$ and $p = 2k$ or $p = 2k + 1$, respectively, with the total RR-flux that they source measured in KU^0 or KU^1, respectively (cf. the second row of Tab. 3.1):
$$\begin{aligned} \mathrm{KU}^0(\mathbb{R}^9 \setminus \mathbb{R}^{p=2k}) &\simeq \mathrm{KU}^0(S^{9-2k-1}) \simeq \mathbb{Z}\,, \\ \mathrm{KU}^1(\mathbb{R}^9 \setminus \mathbb{R}^{p=2k+1}) &\simeq \mathrm{KU}^1(S^{9-2k}) \simeq \mathbb{Z}\,. \end{aligned} \tag{3.45}$$

Instead, the traditional literature insists that D-brane charge in type IIA/B is measured in $\mathrm{KU}^1/\mathrm{KU}^0$, respectively (degrees reversed) – but this statement refers to the charge not of substantial singular branes but of "probe branes", which is of different conceptual nature:

- An **electric probe brane** is like a fundamental particle in perturbation theory, hence not "back-reacted". The spacetime topology is hence unaffected by their presence, and their charge/number is the integral, in E^{d-p}, of a density on the compactified transverse space. For flat branes, this is homotopy equivalent to *another* sphere:

$$\mathbb{R}^p \times \big(\mathbb{R}^{d-p}_{\cup\{\infty\}}\big) \simeq \mathbb{R}^p \times \times S^{d-p} \underset{\text{hmtpy}}{\sim} S^{d-p} . \tag{3.46}$$

The archetypical example is electric fundamental particles (like electrons) whose total charge/number is measured in ordinary integral cohomology:

$$H^3\big(\mathbb{R}^0 \times \mathbb{R}^3_{\cup\{\infty\}};\mathbb{Z}\big) \simeq H^3\big(S^3;\mathbb{Z}\big) \simeq \mathbb{Z} . \tag{3.47}$$

Applied to D-branes (cf. the bottom row in Tab. 3.1), this gives the formulas found in the traditional literature [nL-DK] on "D-brane charge":

$$\begin{array}{ccccc} \mathrm{KU}^{9-2k}\big(\mathbb{R}^{2k} \times \mathbb{R}^{9-2k}_{\cup\{\infty\}}\big) & \simeq & \mathrm{KU}^1\big(S^{9-2k}\big) & \simeq & \mathbb{Z} \\ \mathrm{KU}^{9-2k-1}\big(\mathbb{R}^{2k+1} \times \mathbb{R}^{9-2k-1}_{\cup\{\infty\}}\big) & \simeq & \mathrm{KU}^0\big(S^{9-2k-1}\big) & \simeq & \mathbb{Z} \end{array} \tag{3.48}$$

It is, tacitly, this second set of formulas, for *probe* D-branes, which led to the proposal [Ho98, (3.7, 3.17)][Wi01, pp. 6-8] for formulas for D6-brane charge in $\mathrm{KU}^1(S^3)$ in terms of Fredholm operators parameterized over their transverse space $\mathbb{R}^3_{\cup\{\infty\}}$.

In § 3.2.3 we complete these old arguments. But what actually motivates us here is that (§ 3.2.2) the same mathematics also describes stable measurement of charges *on M5-branes* (such as of the self-dual 1-brane there), along the lines of § 2.2.

3.2.2 Charge on M5 probes of flat space

Consider the simple but important case of Fig. 2.7 where an M5 brane worldvolume $\Sigma^{1,5}$ probes flat Minkowski spacetime.[1]

Since this ambient bulk spacetime is contractible, its microscopic bulk charge is trivial and equivalently represented by a map constant on a point in S^4, whence the relative cohomology on the brane worldvolume is microscopically in the plain 3-Cohomotopy of Σ^5. Measuring this charge in relative KU^0 is equivalent to measuring

[1] But the following depends only on the pullback of the bulk Cohomotopy charge to the M5-worldvolume being trivial, $\phi^*(F_4^\pi, F_7^\pi) \simeq 0$, which is the case for instance also for holographic embeddings of M5-branes into $\mathrm{AdS}_7 \times S^4$, cf. [GSS25b].

it in $\Omega\mathrm{KU}^0 \simeq \mathrm{KU}^1$ by pushforward along the unit $\Sigma^3 1^{\mathrm{KU}} : S^3 \to \mathrm{KU}^1$

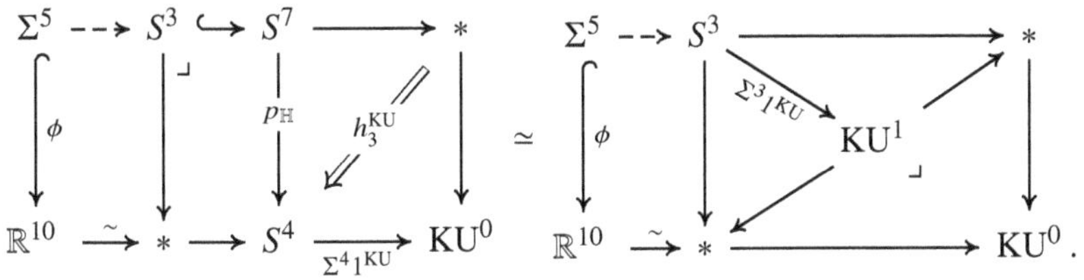

(3.49)

Example 3.2.1. The primary *singular* brane (cf. § 3.2.1) *on* the M5-worldvolume is the 1-brane known as the *self-dual string* or *M-string* [nL-MSt], which is the source of the self-dual H_3-flux (cf. [GSS25a, §3.3]): Its enclosing sphere (3.43) is the 3-sphere

$$\Sigma^5 \equiv \mathbb{R}^5 \setminus \mathbb{R}^1 \simeq \mathbb{R}^1 \times \mathbb{R}_0 \times S^3 \underset{\text{hmtpy}}{\sim} S^3 , \tag{3.50}$$

so that its microscopic brane charge in Cohomotopy is

$$\pi^3(\Sigma^5) \equiv \pi_0\{\Sigma^5 \dashrightarrow S^3\} \simeq \mathbb{Z}, \tag{3.51}$$

as it should be. And, in fact, this is already equal to the stable brane charge as seen in K-theory, in that the coarsening cohomology operation is an isomorphism, in this case:

$$\pi^3(\Sigma^5) \simeq \pi_0\{\Sigma^5 \dashrightarrow S^3\} \xrightarrow[\sim]{(\Sigma^3 1^{\mathrm{KU}})_*} \pi_0\{\Sigma^5 \dashrightarrow \mathrm{KU}^1\} \simeq \mathrm{KU}^1(S^3) . \tag{3.52}$$

In order to understand this M5-worldvolume charge measurement in K-theory more generally and more deeply, we proceed to construct an explicit model for $\Sigma^3 1^{\mathrm{KU}}$ as a map to self-adjoint Fredjolm operators:

3.2.3 The 3-sphere of self-adjoint Fredholm operators

By Prop. 5.1.18 we have

$$\{S^3 \dashrightarrow \mathrm{Fred}^{\pm}_{\mathbb{C}}\} \simeq \pi_0\, \mathrm{KU}^1(S^3) \simeq \mathbb{Z} . \tag{3.53}$$

We ask now for an essentially explicit formula for the map on the left that corresponds to $1 \in \mathbb{Z}$ on the right. Basic as this question is, it does not seem to have been discussed in the mathematical literature. But a partial suggestion for how to go about this construction may be understood to have been proposed by [Ho98, (3.7, 3.17)][Wi01, pp. 6-8], there with the aim of describing the K-theory charge of D6-branes (whose transversal space is $\simeq S^3$). We will now complete this suggestion with a rigorous solution. The key to that is the following Prop. 3.2.9 (of which [Ho98, (3.7)] may have been an echo).

Before discussing this proposition and its consequences, first recall some basic operator algebraic background:

Definition 3.2.2 (Spectrum of self-adjoint operators). The *spectrum* $\mathrm{spec}(A) \subset \mathbb{C}$ of a linear operator $A \in \mathcal{B}(\mathcal{H})$ is the subset of $\lambda \in \mathbb{C}$ for which $A - \lambda \cdot \mathrm{id}$ does not have a bounded linear inverse. For self-adjoint operators, where the spectrum is real

$$A^\dagger = A \quad \Rightarrow \quad \mathrm{spec}(A) \subset \mathbb{R}, \tag{3.54}$$

this is (cf. [Mor17, Rem. 9.15]) the union of

(i) the *discrete spectrum* $\mathrm{spec}_{\mathrm{dsc}}(A) \subset \mathrm{spec}(A)$, consisting of those eigenvalues λ of A which are

(a) isolated: there exists $\epsilon \in \mathbb{R}_{>0}$ such that $(\lambda - \epsilon, \lambda + \epsilon) \subset \mathbb{R}$ contains no other point of $\mathrm{spec}(A)$,

(b) of finite geometric multiplicity: the eigenspace $\ker(A - \lambda \cdot \mathrm{id})$ is finite-dimensional,

(ii) the *essential spectrum* which is remainder:

$$\mathrm{spec}_{\mathrm{ess}}(A) = \mathrm{spec}(A) \setminus \mathrm{spec}_{\mathrm{dsc}}(A)\,. \tag{3.55}$$

Definition 3.2.3. The *unitary Fredholm group* (cf. [nL-FG]) is the subgroup of unitary operators on $\mathcal{H}$ (5.1) which differ from the identity by a compact operator:

$$\mathrm{U}^c(\mathcal{H}) := \big\{ U \in \mathrm{U}(\mathcal{H}) \,\big|\, U - \mathrm{id} \in \mathcal{K}(\mathcal{H}) \big\}\,. \tag{3.56}$$

We will also be interested in the space

$$-\mathrm{U}^c(\mathcal{H}) := \big\{ U \in \mathrm{U}(\mathcal{H}) \,\big|\, U + \mathrm{id} \in \mathcal{K}(\mathcal{H}) \big\}\,, \tag{3.57}$$

which is not a group but as a space is homeomorphic to the unitary Fredholm group.

Lemma 3.2.4. *The group homomorphism from the stable unitary group to the unitary Fredholm group* (3.56)*, given by stabilizing unitary matrices*

$$\begin{array}{ccc} \bigcup_{n \in \mathbb{N}} \mathrm{U}(\mathbb{C}^n) & \longrightarrow & \mathrm{U}^c(\mathcal{H}) \\ (U, n) & \longmapsto & U \oplus \bigoplus_{\mathbb{N}_{>0}} \mathrm{id}_{\mathbb{C}^n} \end{array} \tag{3.58}$$

is a weak homotopy equivalence, in that it induces isomorphisms on all homotopy groups.

Notation 3.2.5. Recall from Rem. 5.1.19 that the non-contractible component of the space self-adjoint complex Fredholm operators, $\mathrm{Fred}^{+}_{\mathbb{C}}$, is that where these have both positive and negative essential spectrum (Def. 3.2.2). Denote the further subspaces with essential spectrum concentrated on ± 1 ([SSi69, (2.5)]) and with actual spectrum concentrated on ± 1 as follows, respectively:

$$\begin{array}{ccc} \mathcal{F}_0 & := & \Big\{ f \in \mathrm{Fred}^{+}_{\mathbb{C}} \,\Big|\, {\scriptstyle \mathrm{spec}(f) \atop = \{+1,-1\}} \Big\} \\ \big\downarrow & & \\ \mathcal{F} & := & \Big\{ f \in \mathrm{Fred}^{+}_{\mathbb{C}} \,\Big|\, {\scriptstyle \mathrm{spec}_{\mathrm{ess}}(f) = \{+1,-1\} \atop |f| = 1} \Big\} \\ \big\downarrow & & \\ \mathrm{Fred}^{+}_{\mathbb{C},*} & := & \Big\{ f \in \mathrm{Fred}^{+}_{\mathbb{C}} \,\Big|\, {\scriptstyle \mathrm{spec}_{\mathrm{ess}}(f) \cap \mathbb{R}_{>0} \neq \varnothing \atop \mathrm{spec}_{\mathrm{ess}}(f) \cap \mathbb{R}_{<0} \neq \varnothing} \Big\}. \end{array} \tag{3.59}$$

Remark 3.2.6. By functional calculus, the elements $f \in \mathcal{F}_0$ (3.59) satisfy:

$$f \in \mathcal{F}_0 \quad \Rightarrow \quad f^2 = \mathrm{id}\,. \tag{3.60}$$

Example 3.2.7. Let $D(\mathbb{H}_{\mathrm{im}}) \simeq D^3$ denote the closed unit ball in the space of imaginary quaternions (Def. 5.2.1). Under the representation of such quaternions $x \in \mathbb{H}_{\mathrm{im}}$

by Pauli matrices γ_x (Ex. 5.2.2), these give self-adjoint elements $-\mathrm{i}\gamma_x \in \mathcal{B}(\mathbb{C}^2)$ of operator norm $|\gamma_x| = |x| \in [0,1]$. Therefore, *stabilizing* these matrices by forming their direct sum with infinitely many copies of $\mathrm{diag}(+1,-1) \in \mathcal{B}(\mathbb{C}^2)$, under (5.2), they constitute elements of the space $\mathcal{F}$ (3.59). Precisely when $x \in D(\mathbb{H}_{\mathrm{im}})$ is on the boundary, hence, if $|x| = 1$, then this stabilized element is actually in $\mathcal{F}_0 \subset \mathcal{F}$. Hence, we have a commuting diagram of maps of this form:

$$\begin{array}{ccc} x & \longmapsto & -\mathrm{i}\gamma_x \oplus \bigoplus_{\mathbb{N}_{>0}} \begin{pmatrix} +1 & 0 \\ 0 & -1 \end{pmatrix} \\ D^3 \simeq D(\mathbb{H}_{\mathrm{im}}) & \xrightarrow{\ \mathrm{stab}\ } & \mathcal{F} \\ \uparrow & & \uparrow \\ S^2 \simeq \partial D(\mathbb{H}_{\mathrm{im}}) & \longrightarrow & \mathcal{F}_0\,, \end{array} \tag{3.61}$$

and therefore a map of quotient spaces

$$S^3 \simeq D(\mathbb{H}_{\mathrm{im}})/\partial \xrightarrow{\ \mathrm{stab}\ } \mathcal{F}/\mathcal{F}_0\,. \tag{3.62}$$

Our next goal is to show that this map represents the generator of $\mathrm{KU}^1(S^3) \simeq \mathbb{Z}$.

Lemma 3.2.8. *The second inclusion in* (3.59) *is a homotopy equivalence, as is the coprojection of the quotient by the first inclusion:*

$$\mathcal{F}/\mathcal{F}_0 \underset{\mathrm{hmtpy}}{\overset{\sim}{\twoheadleftarrow}} \mathcal{F} \underset{\mathrm{hmtpy}}{\overset{\sim}{\hookrightarrow}} \mathrm{Fred}^{+}_{\mathbb{C},*}\,. \tag{3.63}$$

Proof. The first statement (concerning the right map) is [SSi69, (2.5)]. We proceed to prove the second statement (concerning the left map). For that we invoke some basic homotopy theory which we have not reviewed here, but which may be found in standard textbooks, such as [AGP02].

Namely, we claim that the first inclusion in (3.59) is a *Hurewicz cofibration* ([AGP02, Def. 4.1.5]) including a *contractible space*:

$$* \underset{\mathrm{hmtpy}}{\overset{\sim}{\longleftarrow}} \mathcal{F}_0 \underset{\in\mathrm{Cof}}{\hookrightarrow} \mathcal{F}\,. \tag{3.64}$$

To see that $\mathcal{F}_0$ is contractible, note with (3.60) that, under passage to eigenspaces, the operators in this space correspond to choices of direct sum decompositions $\mathcal{H} \simeq \mathcal{H}_+ \oplus \mathcal{H}_-$. This shows that $\mathcal{F}_0$ is a Grassmannian homeomorphic to

$$\mathcal{F}_0 \simeq \mathrm{U}(\mathcal{H})/\big(\mathrm{U}(\mathcal{H}) \times \mathrm{U}(\mathcal{H})\big)\,, \tag{3.65}$$

and hence its contractibility is a consequence of Kuiper's theorem (5.16).

To see that we have a cofibration: Since all spaces of operators in question are metric spaces (as subspaces of bounded operators with the operator norm topology), they are *perfectly normal Hausdorff* spaces (cf. [AGP02, Def. 4.1.13]). Moreover, $\mathcal{F}_0 \subset \mathcal{F}$ is a closed subspace (being the preimage of $\{0\}$ under the map $\mathcal{F} \to \mathcal{B}(\mathcal{H}) : f \mapsto f^2 - \mathrm{id}$). Therefore, it is sufficient (by [AGP02, Thm. 4.1.14]) to see that the inclusion is a *strong deformation retract of a neighborhood* [AGP02, Def. 4.1.11]. That neighborhood may be taken to be the invertible operators among $\mathcal{F}$, and the retraction may then be given by functional calculus, shifting all points in the spectrum to their sign in $\{+1,-1\}$.

This implies the claim by the general fact (by [AGP02, Thm. 4.2.1]) that the quotient coprojection of a cofibrantly included contractible space is a homotopy equivalence. □

Proposition 3.2.9 (The Atiyah-Singer exponential map). *The following exponential map from the self-adjoint Fredholm operators in $\mathcal{F}$ (3.59) to minus the unitary Fredholm group $-\mathrm{U}^c(\mathcal{H})$ (3.57) is a homotopy equivalence:*

$$\exp\big(\mathrm{i}\pi(-)\big) : \mathcal{F} \xrightarrow[\text{hmtpy}]{\sim} -\mathrm{U}^c(\mathcal{H})\,, \tag{3.66}$$

which as such descends to the quotient by $\mathcal{F}_0$:

$$\exp\big(\mathrm{i}\pi(-)\big) : \mathcal{F}/\mathcal{F}_0 \xrightarrow[\text{hmtpy}]{\sim} -\mathrm{U}^c(\mathcal{H})\,. \tag{3.67}$$

Proof. The first statement is [SSi69, Prop. 3.3]. To see that the map as such passes to the quotient, as claimed in the second statement, recall that $f \in \mathcal{F}_0$ implies that $f^2 = \mathrm{id}$ (3.60), whence Euler's formula gives that:

$$f^2 = \mathrm{id} \quad \Rightarrow \quad \begin{cases} \exp\big(\mathrm{i}\pi f\big) = \cos(\pi)\cdot \mathrm{id} + \sin(\pi)\cdot f \\ \qquad\qquad\;\; = -\mathrm{id}\,. \end{cases} \tag{3.68}$$

Therefore, we have a commuting diagram

$$\begin{array}{ccc} \mathcal{F} & \xrightarrow[\text{hmtpy}]{\sim} & -\mathrm{U}^c(\mathcal{H}) \\ {\scriptstyle \wr}\big\downarrow{\scriptstyle \text{hmtpy}} & \nearrow & \\ \mathcal{F}/\mathcal{F}_0\,, & & \end{array} \tag{3.69}$$

where the top map is a homotopy equivalence by (3.66) while the left map is a homotopy equivalence by (3.63). By the "2 out of 3"-property of homotopy equivalence, this implies the claim that also the diagonal map is a homotopy equivalence. □

Now we are ready to prove:

Proposition 3.2.10. *Under the equivalence of Prop. 5.1.18, the generator of* $\mathrm{KU}^1(S^3)$ *is represented by forming stabilized Pauli matrices* (3.62)*:*

$$\begin{array}{ccccccc} \pi_0\left\{S^3 \dashrightarrow \mathcal{F}/\mathcal{F}_0\right\} & \underset{(3.63)}{\simeq} & \pi_0\left\{S^3 \dashrightarrow \mathrm{Fred}^{+}_{\mathbb{C}}\right\} & \underset{(5.71)}{\simeq} & \mathrm{KU}^1(S^3) & \simeq & \mathbb{Z} \\ \left[D(\mathbb{H}_{\mathrm{im}}) \overset{\mathrm{stab}}{\dashrightarrow} \mathcal{F}\right] & & & \longmapsto & & & 1\,. \end{array} \tag{3.70}$$

Proof. Since the exponential map (3.67) is a homotopy equivalence, it is sufficient to see that the composite

$$\begin{array}{ccccc} S^3 \simeq D(\mathbb{H}_{\mathrm{im}})/\partial & \xrightarrow{\quad \mathrm{stab} \quad} & \mathcal{F}/\mathcal{F}_0 & \xrightarrow{\quad \exp(\mathrm{i}\pi(-)) \quad} & -\mathrm{U}^c(\mathcal{H}) \\ x & \longmapsto & -\mathrm{i}\gamma_x \oplus \bigoplus_{\mathrm{N}} \begin{pmatrix} +1 & 0 \\ 0 & -1 \end{pmatrix} & \longmapsto & e^{\pi\gamma_x} \oplus \big(-\mathrm{id}_{\bigoplus_{\mathrm{N}} \mathbb{C}^2}\big) \end{array} \tag{3.71}$$

represents the generator of $\pi_3\big(\mathrm{U}^c(\mathcal{H})\big) \simeq \mathbb{Z}$. But, as seen from its components shown in the second row, this map is, up to stabilization, just the exponential map from the unit ball in the Lie algebra $\mathfrak{su}(2)$ onto $\mathrm{SU}(2) \simeq S^3$ and as such represents the generator of $\pi_0\{S^3 \dashrightarrow S^3\} \simeq \mathbb{Z}$, by Lem. 3.2.4. □

Remark 3.2.11. The terms $-\mathrm{i}\gamma_x$ and $e^{\pi\gamma_x}$ in (3.71) clearly want to correspond with the expressions in [Ho98, (3.7, 3.17)][(3.2)][Wi01], under the syntactic translation

$$-\mathrm{i}\gamma_x \;\leftrightarrow\; \frac{\vec{\sigma}\cdot\vec{x}}{|x|}\,.$$

In [Ho98], it had remained open in which spaces these expressions are meant to take values and how they actually represent the claimed K-theory classes. In particular, [Wi01] pointed out that some argument for trivialization of these expressions at $|x| = 1$ was missing, and suggested that the resolution has to do with Kuiper's theorem. But the spaces in which this would happen were still not declared, nor was a reason given for why the result represents a class in $\mathrm{KU}^1(S^3)$, and specifically the generator. We suggest that Prop. 3.2.10 fills these gaps and thereby completes this old argument.

Part II

Topological Orbifold Cohomology

4

Nonabelian Cohomology

We give a pedagogical and practical new account of a streamlined theory of *topological stacks* (in § 4.1, as a faithful fragment of the smooth ∞-groupoids briefly recalled in Thm. 9.1.19), neatly supporting a general notion of *twisted orbifold cohomology* (in Def. 4.2.5).

Here "topological stacks" (cf. [Car12][SS26c, §4.2]) refers to the *geometric homotopy theory* (cf. Thm. 9.1.19) of *groupoids* (cf. [IR21] and § 4.1.5) with topological structure (*topological groupoids*, cf. [Mac87, §II][SS26c, §2.2.1]), subject to *Morita equivalences* (discussed in § 4.1.6). This may be viewed as the first-stage enhancement of classical general topology to include *gauge transformations* between points of topological spaces, such as the isotropy group actions in orbifolds and the quantum symmetry actions in spaces of Fredholm operators, which in fibered combination make for the *twisted orbifold K-theory* discussed in § 5.1.

For perspective, afterwards we briefly indicate (in Thm. 9.1.19) how this theory of topological stacks is a full fragment of the more encompassing *cohesive homotopy theory* of *smooth ∞-groupoids* ([SS26c, §4.3], going back to [SSS12, §3.1][Sc13]) which we discuss in more detail in § 9.1.1. This is in the general context of "geometric homotopy theory" (*∞-topos theory* [TV05, Lu09a][FSS23, §1]) – see exposition for mathematical physicists in [Sc25].

4.1 Topological Orbifolds

Discussion of topological stacks in traditional literature may tend to look mysterious to the newcomer and cumbersome to the expert. We spell out an approach (in specialization of Thm. 9.1.19) which is transparent and practically useful.

4.1.1 Topological Spaces

To set up notation, first some quick paragraphs on topological spaces in general. (Beware that from § 4.1.2 on we will be entirely concerned only with the special case of *D-topological spaces*.)

For X a topological space, we write

- $\pi_0 X$ for its set of path-connected components,
- $\flat X$ for its underlying set of points,

DOI: 10.1201/9781003675969-4

both regarded as discrete topological spaces, if necessary. We write:

- "$*$" for the singleton space (the *point*),
- "$\varnothing$" for the empty set regarded as a topological space.

The archetypical topological spaces for our purposes (cf. the next § 4.1.2) are the *Cartesian spaces* $\mathbb{R}^n$ with their usual Euclidean topology. For a pair of topological spaces (X, Y), an arrow $X \to Y$ denotes a *continuous function* between them, called a *map*, for short. On every space X there is the identity map $X \xrightarrow{\mathrm{id}} X$. A *homeomorphism* is a map f which has an inverse map f^{-1}, denoted $f : X \xrightarrow{\sim} Y$, such that $f^{-1} \circ f = \mathrm{id}$ and $f \circ f^{-1} = \mathrm{id}$. A most basic but important example for our purpose is the homeomorphy of Cartesian spaces with their own open balls of any radius $\epsilon \in \mathbb{R}_{>0}$:

$$\begin{array}{ccc} \mathbb{R}^n & \xrightarrow{\ \sim\ } & \mathbb{D}^n_\epsilon := \left\{ x \in \mathbb{R}^n \,\middle|\, |x| < \epsilon \right\} \\ x & \longmapsto & \epsilon \frac{x}{1+|x|} \,. \end{array} \tag{4.1}$$

A *diagram* of maps is always meant/understood to *commute* in that all composite maps between any pair of spaces in the diagram are equal. Notably, a *commuting square* is

$$\begin{array}{ccc} Y & \xrightarrow{\ c\ } & B \\ {\scriptstyle \phi}\downarrow & & \downarrow{\scriptstyle p} \\ X & \xrightarrow{\ b\ } & A \end{array} \quad \Leftrightarrow \quad p \circ c = b \circ \phi \,. \tag{4.2}$$

The following are some *universal constructions* on topological spaces (cf. [Sc17a, §I.6]) that we need:

Definition 4.1.1. For $X \xrightarrow{f_1} B \xleftarrow{f_2} Y$ a pair of coincident maps, their *fiber product*, or *pullback* (pb) of one along the other, is $X \underset{B}{\times} Y \xrightarrow{(p_1, p_2)} X \times Y$, unique up to compatible homeomorphism, which makes the following bottom right square of maps commute (4.2), and universally so in that it uniquely factors (shown by the dashed map) every other completion to a commuting square:

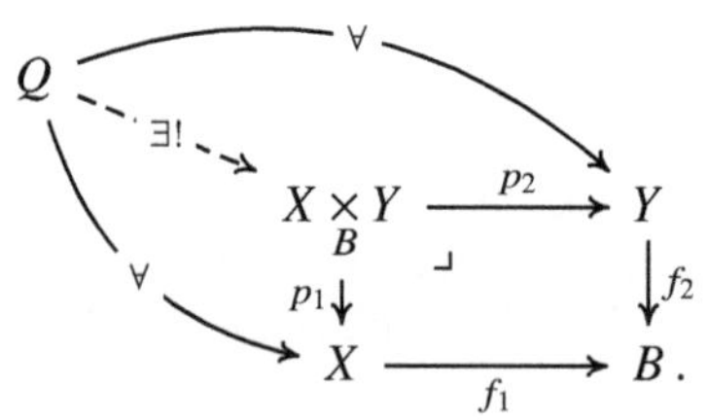

(4.3)

Dually, for $X \xleftarrow{f_1} T \xrightarrow{f_2} Y$ a pair of co-emanent maps, their *cofiber product*, or *pushout* (po) of one along the other, is $X \overset{B}{\sqcup} Y \xleftarrow{(q_1, q_2)} X \times Y$, unique up to compatible homeomorphism, which makes the following top-left square of maps commute (4.2) and universally so in that it uniquely factors (shown by the dashed map) every other completion to a commuting square:

$$\begin{array}{ccccc} T & \xrightarrow{f_2} & Y & & \\ {\scriptstyle f_1}\downarrow & \ulcorner & \downarrow{\scriptstyle q_2} & \searrow^{\forall} & \\ X & \xrightarrow[q_1]{} & X \overset{Y}{\sqcup} Y & \overset{\exists !}{\dashrightarrow} & Q\,. \\ & \searrow & \underset{\forall}{\longrightarrow} & \nearrow & \end{array} \tag{4.4}$$

Definition 4.1.2. For a pair of parallel maps $T \underset{f_2}{\overset{f_1}{\rightrightarrows}} X$, their *coequalizer* is $X \xrightarrow{q} X/T$, unique up to compatible homeomorphism, which makes the following horizontal composites agree, and universally so in that it uniquely factors (shown by the dashed map) every other such coequalizing completion:

$$\begin{array}{ccccc} T & \underset{g}{\overset{f}{\rightrightarrows}} & X & \xrightarrow{q} & X/T\,. \\ & \searrow^{\forall} & {\scriptstyle\forall}\downarrow & \swarrow_{\exists !} & \\ & & Q & & \end{array} \tag{4.5}$$

Concretely, X/T is the quotient space by the smallest equivalence relation $\sim$ on X for which $f_1(x) \sim f_2(x)$ for all $x \in X$.

Example 4.1.3 (cf. [BT82, p. 217][AGP02, §3.1]). For $n \in \mathbb{N}$ a pushout (4.4) along the boundary inclusion of the closed n-dimensional unit ball, $S^{n-1} = S(\mathbb{R}^n) \hookrightarrow D(\mathbb{R}^n) = D^n$, is called an *n-cell attachment*

$$\begin{array}{ccc} S(\mathbb{R}^n) & \hookrightarrow & D(\mathbb{R}^n) \\ {\scriptstyle f}\downarrow & \ulcorner & \downarrow \\ X & \hookrightarrow & X \cup_f D^n \end{array} \tag{4.6}$$

with *attaching map* f. A topological space that arises from $\varnothing$ via (possibly transfinite) sequences of such n-cell attachments is called a *cell complex* and a *CW-complex* if the cell dimension n is increasing monotonically in the process.

Proposition 4.1.4 (Pasting law, cf. [nL-PL]). *Given a "pasting" diagram of commuting squares,* (4.2) $\begin{array}{ccccc} \cdot & \to & \cdot & \to & \cdot \\ \downarrow & & \downarrow & & \downarrow \\ \cdot & \to & \cdot & \to & \cdot \end{array}$ *then:*

- **(i)** *If the right square is a pullback* (4.3)*, then the left square is so iff the total rectangle is.*
- **(ii)** *If the left square is a pushout* (4.4)*, then the right square is so iff the total rectangle is.*

4.1.2 D-Topological Spaces

The topological spaces that one actually cares about in geometry are *probeable* (cf. [Sc25]) by *Cartesian spaces* $\mathbb{R}^n$ $(n \in \mathbb{N})$ with their standard Euclidean topology, in that they are *D-topological* (cf. [nL-DT][SS26c, Ntn. 4.3.19]):

Definition 4.1.5. A topological space X is called *D-topological* (traditionally: *Delta-generated* or *numerically generated*) if their subsets $S \subset X$ are open iff their preimages under all maps of the form $\mathbb{R}^n \to X$ are open in Cartesian space.

Basic classes of examples D-topological spaces include, in increasing generality:

- topological manifolds,
- cell complexes (Ex. 4.1.3),
- retracts of cell complexes,

and thereby all the cofibrant spaces of algebraic topology (cf. [nL-CM]).

More generally, all pushouts (4.4) (and generally: colimits) of D-topological spaces are themselves D-topological.

Moreover, the topology of topological spaces A may be refined to their induced D-topology

$$X_D \xrightarrow{x \mapsto x} X \tag{4.7}$$

without changing the maps into it out of D-topological spaces:

$$X \text{ is D-topological} \quad \Rightarrow \quad \{X \dashrightarrow A\} \simeq \{X \dashrightarrow A_D\}. \tag{4.8}$$

In particular, this means that for the discussion of cohomology in terms of maps from manifolds/CW-complexes into classifying spaces (as surveyed in § 2.1), the topology on the classifying spaces may without restriction be taken to be refined to their D-topology.

For some pairs (X, Y) of exotic D-topological spaces, their fiber products $X \times_B Y$ (4.3) may fail to be D-topological – but the D-topologization $(X \times_B Y)_D$ (4.7) still satisfies the universal property (4.3) among D-topological spaces, and hence is the correct fiber product (generally: limit) in the category of D-topological spaces.

Finally, the central property of the category of D-topological spaces is its *Cartesian closure*, meaning that for (X, Y), a pair of D-topological spaces, the set

$$\mathrm{Map}(X, Y) \equiv \{X \dashrightarrow Y\} \tag{4.9}$$

of all maps $X \to Y$ becomes a D-topological space (with the D-topologization of the *compact-open topology*) such that there are natural homeomorphisms (cf. [SS26c, Rem. 2.0.17])

$$\begin{array}{ccc} \mathrm{Map}(Z \times X, Y) & \xrightarrow[\sim]{\widetilde{(-)}} & \mathrm{Map}(Z, \mathrm{Map}(X, Y)) \\ f(-,-) & \longmapsto & f(-)(-)\,. \end{array} \tag{4.10}$$

In summary, this says that, in this technical sense, D-topological spaces form a *convenient category of topological spaces* (cf. [nL-CS][§1.1.1]SS25-EBund). Therefore, we declare that:

Notation 4.1.6. From now on, *by "topological spaces" we mean "D-topological spaces" (Def. 4.1.5)*. In consequence, we say "*topological group*" (4.11) for "*D-topological group*" and "*topological groupoid*" (in § 4.1.5) for "*D-topological groupoid*", etc.

4.1.3 Spaces with group action

Definition 4.1.7 (cf. [SS26c, §2.1]). For G a *topological group*, hence a topological space equipped with maps of the form

$$G \times G \xrightarrow{(-)\cdot(-)} G\,, \quad * \xrightarrow{\mathrm{e}} G\,, \tag{4.11}$$

and for X a topological space, a (left) *topological G-action on X* is a map

$$\begin{array}{ccc} G \times X & \xrightarrow{G \curvearrowright X} & X \\ (g,x) & \mapsto & g \cdot x\,, \end{array} \quad \text{s.t. } \forall_{x \in X} \begin{cases} \mathrm{e} \cdot x = x \\ \forall_{g_1, g_2 \in G} : (g_2 \cdot g_1) \cdot x = g_2 \cdot (g_1 \cdot x)\,. \end{cases} \tag{4.12}$$

One also says that $G \curvearrowright X$ is a *G-space*, for short. For a pair $G \curvearrowright X$, $G \curvearrowright Y$ of these, an *equivariant map* f between them is

$$f \in \mathrm{Map}(X,Y) \;\; \text{s.t.} \;\; \forall_{\substack{x \in X \\ g \in G}} \; f(g \cdot x) = g \cdot f(x)\,. \tag{4.13}$$

We denote the subspace of the mapping space (4.9) on the G-equivariant maps as follows:

$$\left\{ \begin{array}{ccc} \overset{G}{\curvearrowright} & & \overset{G}{\curvearrowright} \\ X & \overset{f}{\dashrightarrow} & Y \end{array} \right\} \equiv \mathrm{Map}(X,Y)^G \subset \mathrm{Map}(X,Y)\,. \tag{4.14}$$

Here for $G \curvearrowright X$ a G-space, the notation

$$X^G := \big\{ x \in X \,\big|\, \forall_{g \in G}\, g \cdot x = x \big\} \subset X \tag{4.15}$$

indicates the *G-fixed subspace*.

For example, the plain mapping space between G-spaces becomes itself a G-space by the *conjugation action*

$$\begin{array}{ccc} G \times \mathrm{Map}(X,Y) & \longrightarrow & \mathrm{Map}(X,Y) \\ (g, f(-)) & \longmapsto & g^{-1} \cdot f(g \cdot -) \end{array} \tag{4.16}$$

and its G-fixed points (4.15) are precisely the G-equivariant maps (4.14).

The most basic examples of G-spaces are the *coset spaces* for subgroups $H \subset G$,

$$G/H := \big\{ g \cdot H \,\big|\, g \in G \big\} \tag{4.17}$$

equipped with their inherited G-action:

$$\begin{array}{ccc} G \times G/H & \longrightarrow & G/H \\ (g', g \cdot H) & \mapsto & g' \cdot g \cdot H\,. \end{array} \tag{4.18}$$

We come back to this in (4.67).

4.1.4 Homotopy

A *homotopy* (cf. [FF2016, §3]) between a pair of parallel maps $f, g : X \to Y$ is a continuous deformation between them,

$$X \underset{g}{\overset{f}{\Downarrow \eta}} Y \; : \; [0,1] \longrightarrow \mathrm{Map}(X,Y)\,, \tag{4.19}$$

namely, a continuous path between the corresponding points $(\widetilde{f}, \widetilde{g})$ in the mapping space (4.9), hence a map η fitting into this commuting diagram of maps:

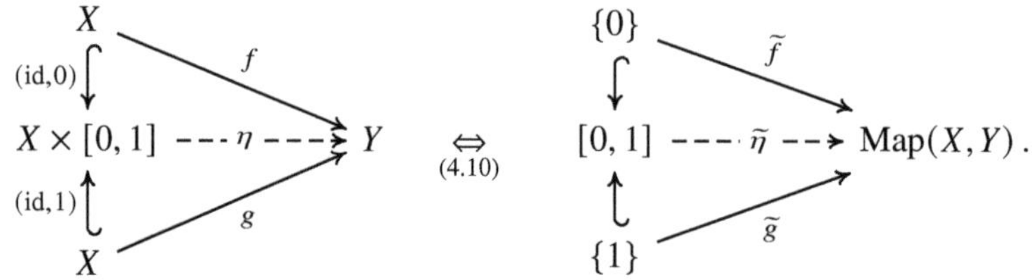

(4.20)

When the spaces are equipped with G-action (4.12), then an *equivariant homotopy* between equivariant maps (4.13) is a homotopy (4.20) running inside the equivariant mapping space (4.14):

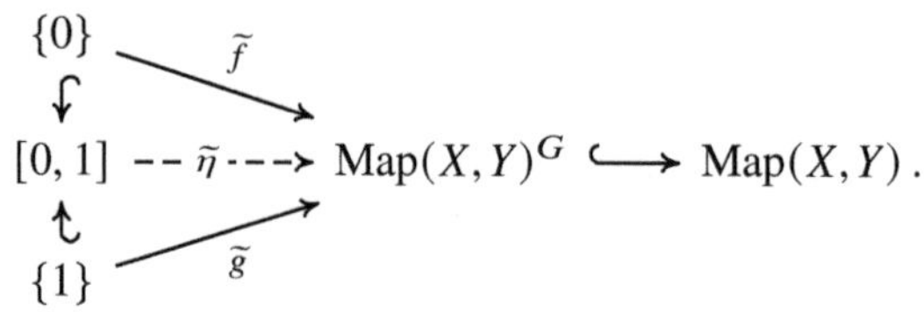

(4.21)

A map is a *homotopy equivalence* to be denoted

$$f : X \xrightarrow[\text{hmtpy}]{\sim} Y \tag{4.22}$$

if there exists a reverse map $\overline{f} : Y \to X$ and homotopies (4.19) of this form:

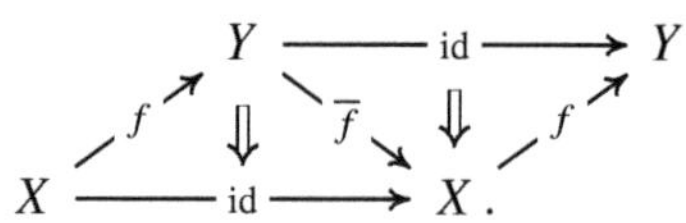

(4.23)

The *vertical composition* of a composable pair of homotopies (4.19) is the evident concatenation of these paths of maps

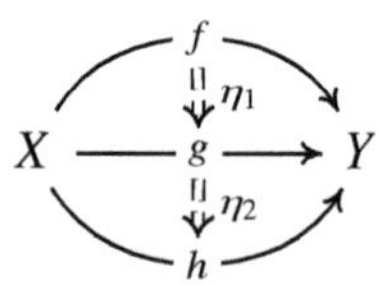

$$: \quad (x, s) \mapsto \begin{cases} \eta_1(x, s) & \text{if } s \in [0, \frac{1}{2}] \\ \eta_2(x, s) & \text{if } s \in [\frac{1}{2}, 1], \end{cases} \tag{4.24}$$

while *horizontal composition* of homotopies by maps is the evident actual composition of component maps:

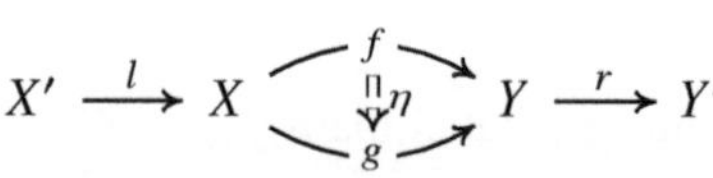

$$: \quad (x', s) \mapsto r \circ \eta(-, s) \circ l(x') . \tag{4.25}$$

Combining this, one obtains horizontal composition of homotopies themselves, as

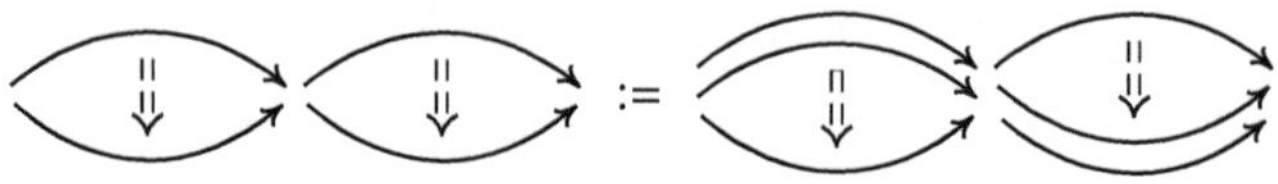

(4.26)

(or the other way around, which is different but higher-order homotopic), such as in the important special case of *pasting composites* of "square" homotopies:

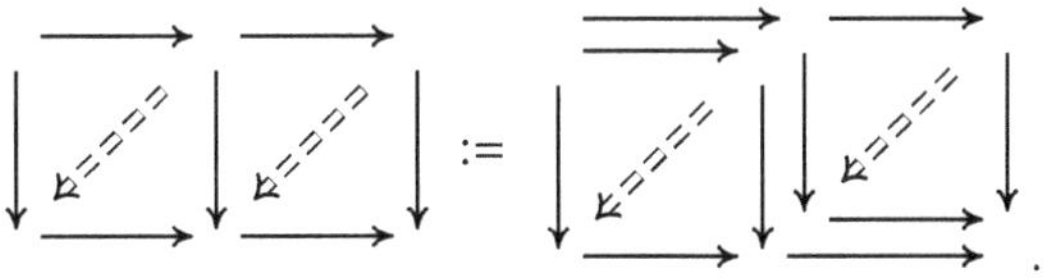

(4.27)

The higher homotopy theory of such 2- and higher-dimensional diagrams of maps and homotopies is usefully captured by *model category structure on topological spaces* (cf. [nL-CM], for review in our context see [FSS23, §1]). Here we proceed with making explicit only the most minimum amount of technology necessary at this point.

4.1.5 Topological Groupoids

A *groupoid* (cf. [We96][Sc17a, §2.1][IR21][Sc25, p. 6]) is a "set with gauge transformations" between its elements. For example, the phase space of a gauge theory is a groupoid, whose "objects" are the gauge field configurations and whose "morphisms" are the actual gauge transformations between them. This example is actually a *Lie groupoid*, hence with smooth structure on its sets of objects and morphisms (in physics this is best known for infinitesimal gauge transformations only, which gives the underlying *Lie algebroid* whose Chevalley-Eilenberg algebra is known as the *BRST complex*; cf. [Sc17b, §10].)

For the time being, we disregard smooth structure (we turn to this instead in Thm. 9.1.19) and consider groupoids in the broad generality where they are equipped with any topological structure (which here means: any *D-topological structure*, by Ntn. 4.1.6, whence the following is about *D-topological groupoids*):

Definition 4.1.8. A *topological groupoid* $\mathcal{X}$ (cf. [Mac87, §II.1][SS26c, Ntn. 2.2.1]) is a topological space $\mathrm{Mor}(\mathcal{X})$ of "morphisms" and a subspace $\mathrm{Obj}(\mathcal{X})$ of "objects" (identity morphisms) – equipped with continuous maps of this form:

$$\mathrm{Mor}(\mathcal{X}) \;{}_s{\times}_t\; \mathrm{Mor}(\mathcal{X}) \xrightarrow{(-)\circ(-)} \mathrm{Mor}(\mathcal{X}) \circlearrowleft^{i} \;\underset{\xrightarrow{\;t\;}}{\overset{\xrightarrow{\;s\;}}{\xleftarrow{\;e\;}}}\; \mathrm{Obj}(\mathcal{X})\,, \qquad (4.28)$$

such that

(i) $s \circ e = t \circ e = \mathrm{id}$, which means that the morphisms form a reflexive graph over the subspace of objects,

(ii) the *composition* operation $(-) \circ (-)$ – of a morphism f whose *target* object $t(f)$ coincides with the *source* object $s(g)$ of another morphism g – is associative, and unital with respect to the *identity morphisms* e_x on objects x and has *inverse morphisms* given by i.

For notational transparency, when identities and inverses are understood, it is often useful to denote topological groupoids by the set of generic pairs of composable morphisms and their composites, like this:

$$\mathcal{X} \equiv \left\{ \begin{array}{ccccc} & & y & & \\ & {}^{f}\nearrow & & \searrow^{g} & \\ x & & \xrightarrow{\quad g\circ f \quad} & & z \end{array} \;\middle|\; \begin{array}{l} x, y, z \in \mathrm{Obj} \\ f, g \in \mathrm{Mor} \end{array} \right\} . \tag{4.29}$$

Given a topological groupoid $\mathcal{X}$, we say:

- The coequalizer (4.5) of the source and taget map is its space of *isomorphism classes*:
$$[\mathcal{X}]_0 := \mathrm{Obj}(\mathcal{X}) / ({}_s\!\sim_t) \, . \tag{4.30}$$
- The fiber of the combined source/target map
$$\mathrm{Mor}(\mathcal{X}) \xrightarrow{(s,t)} \mathrm{Obj}(\mathcal{X})^2 \tag{4.31}$$
over a single object $x := (x, x) \in \mathrm{Obj}(\mathcal{X})^2$ is the *isotropy group* (or *automorphism group* or *stabilizer group*) of x
$$\mathcal{X}_x := \left\{ \underset{x}{\overset{g}{\circlearrowleft}} \;\middle|\; g \in \mathrm{Mor}(\mathcal{X}) \right\} , \tag{4.32}$$
with topological group structure (4.11) inherited from the restriction of the topological groupoid structure.

Remark 4.1.9. There are several natural variants and generalizations of the *topological groupoids* of Def. 4.1.8:

(i) A *Lie groupoid* (cf. [Mac87][MM03]) is a topological groupoid whose spaces of objects and morphisms are equipped with the structure of smooth manifolds, whose structure maps are smooth maps, and whose source and target maps are submersions, so that their fiber product (of composable morphisms) also inherits the structure of a smooth manifold.

Most of the following examples and discussion apply to Lie groupoids just as well. Exceptions are mapping objects, starting with Ex. 4.1.17, which may be "too large" to be smooth manifolds (nor even Fréchet manifolds, for that matter).

(ii) More generally, one may consider *diffeological groupoids*, whose morphism space is equipped with the structure of a *diffeological space* (cf. [IZ13][SS26c, Ntn. 4.3.15] and our Def. 8.1.6) and whose structure maps are smooth maps with respect to that diffeological structure. This class faithfully subsumes both D-topological groupoids and Lie groupoids and is closed under all operations discussed here.

(iii) Fully generally, as far as groupoids in differential topology are concerned, one may consider *smooth groupoids* [Sc13, Def. 1.2.252][Egg14], whose space of morphisms is equipped with the structure of a *smooth set* [Sc13, Def. 1.2.16, 1.3.58][GS25][Sc25][SS26c, Ntn. 4.3.15][IM25] and whose structure maps are smooth with respect to that. This class faithfully subsumes all of the above but also contains "non-concrete" groupoids, like the moduli stacks of $\mathbf{B}\Gamma_{\mathrm{conn}}$ of Γ-principal bundles with connection ([Sc13, Prop. 1.2.107][FSS14][BSS18, Ex. 2.11], in variation of the plain moduli stack $\mathbf{B}\Gamma$ discussed in Exs. 4.1.12 and 4.1.33 and Thm. 4.1.13).

(iv) From this point on it is natural to generalize, finally, to higher smooth groupoids, namely to *smooth ∞-groupoids* [Sc13, §4.4][SS26c, §4.3], which is what we indicate in Thm. 9.1.19 and discuss more comprehensively in § 9.1.1.

For the remainder of this section on topological groupoids, we disregard all this further generality just for the pedagogy of the exposition. The inclined reader is invited to make the evident substitutions.

Example 4.1.10. Given a topological space X, it may be regarded as a topological groupoid whose only morphisms are identities:

$$\left(X \xrightarrow{\ \mathrm{id}\ } X \underset{\mathrm{id}}{\overset{\mathrm{id}}{\rightrightarrows}} X \right) = \{ x \,|\, x \in X \}, \tag{4.33}$$

(with $\mathrm{id}: X \leftarrow X$ the identity-assigning map)

and we denote this groupoid still by "X".

Example 4.1.11. The *interval groupoid* I has two objects, Obj = $\{0, 1\}$, and a single morphism and its inverse between these (hence, with the identity morphisms, a total of four morphisms):

$$I = \{ 0 \rightleftarrows 1 \}. \tag{4.34}$$

Example 4.1.12 (cf. [SS26c, Ex. 2.2.6])**.** For Γ a topological group, its *delooping groupoid* is the topological groupoid with a single object, Γ worth of morphisms, composition given by the group operation $(-)\cdot(-) : \Gamma \times \Gamma \to \Gamma$, and inversion given by group inverses $(-)^{-1} : \Gamma \to \Gamma$, hence:

$$\mathbf{B}G := \left(\Gamma \times \Gamma \xrightarrow{(-)\cdot(-)} \Gamma \underset{\longrightarrow}{\overset{\longrightarrow}{\xleftarrow{\ e\ }}} * \right) = \left\{ \begin{array}{c} \bullet \xrightarrow{\gamma_1} \bullet \xrightarrow{\gamma_2} \bullet \\ \bullet \xrightarrow{\gamma_2 \cdot \gamma_1} \bullet \end{array} \,\middle|\, \gamma_i \in \Gamma \right\}. \tag{4.35}$$

(with $(-)^{-1}$ acting on Γ)

When regarded as a *topological stack*, in § 4.1.6, this simple delooping groupoid is (a representation of) the *moduli stack of principal* Γ*-bundles* (cf. Thm. 4.1.13 and Ex. 4.1.33), in fact of *equivariant principal* Γ*-bundles* (cf. Ex. 4.1.37).

Example 4.1.13. For X a topological space Γ a topological group (4.11), the groupoid

$$\Gamma\mathrm{PrnBdl}(X) \equiv \left\{ \begin{array}{c} P_1 \xrightarrow{\gamma} P_2 \xrightarrow{\gamma'} P_3 \\ P_1 \xrightarrow{\gamma' \circ \gamma} P_3 \end{array} \right\} \tag{4.36}$$

has as objects the (discrete set of) Γ-principal bundles P over X (cf. [Hus94, §4.3][RS17, §1.1][Nak18, §9][SS26c, Ntn. 2.0.25]) and as morphisms their Γ-equivariant bundle homomorphisms,

$$\begin{array}{ccc} P_1 & \xrightarrow{\ \gamma\ } & P_2 \\ & \searrow \quad \swarrow & \\ & X & \end{array} \tag{4.37}$$

(with Γ acting on P_1 and P_2)

hence the *gauge transformations* if we think of these bundles as charge sectors of gauge fields.

Over the point, this reduces to the delooping groupoid from Ex. 4.1.12:

$$\Gamma\mathrm{PrnBdl}(*) = \mathbf{B}\Gamma. \tag{4.38}$$

Example 4.1.14 (cf. [SS26c, Ex. 2.2.6])**.** Given an action $G \circlearrowright X$ (Def. 4.1.7), its *action groupoid* (or *homotopy quotient*) is

$$
\begin{aligned}
G \backslash\!\backslash X \simeq & \left(G \times G \times X \xrightarrow{(g_2, g_1, x) \mapsto (g_2 \cdot g_1, x)} G \times X \underset{(g,x) \mapsto g \cdot x}{\overset{(g,x) \mapsto x}{\underset{\longleftarrow\, (\mathrm{e},x) \leftarrow\!\shortmid x \,-}{\rightrightarrows}}} X \right) \\
= & \left\{ \begin{array}{ccc} & g_1 \cdot x & \\ {\scriptstyle g_1} \nearrow & & \searrow {\scriptstyle g_2} \\ x & \xrightarrow{g_2 \cdot g_1} & g_2 \cdot g_1 \cdot x \end{array} \;\middle|\; \begin{array}{l} x \in X \\ g_1, g_2 \in G \end{array} \right\}.
\end{aligned}
\tag{4.39}
$$

For an action $G \circlearrowright *$ on the point, the corresponding action groupoid (4.39) is the delooping groupoid (4.35):

$$
G \backslash\!\backslash * = \mathbf{B}G = \left\{ \bullet \overset{g}{\circlearrowright} \;\middle|\; g \in G \right\}.
\tag{4.40}
$$

Example 4.1.15. Given a topological groupoid $\mathcal{X}$, its *underlying topologically discrete* groupoid $\flat \mathcal{X}$ has the same objects, morphisms, and structure maps as $\mathcal{X}$, but for the discrete topology:

$$
\flat \mathcal{X} := \left(\flat \mathrm{Mor}(\mathcal{X}) \; {}_s\!\times_t \; \flat \mathrm{Mor}(\mathcal{X}) \xrightarrow{\;\circ\;} \flat \mathrm{Mor}(\mathcal{X}) \overset{i}{\circlearrowright} \underset{t}{\overset{s}{\underset{\longleftarrow \mathrm{e} -}{\rightrightarrows}}} \flat \mathrm{Obj}(\mathcal{X}) \right).
\tag{4.41}
$$

Example 4.1.16. Given a topological space X, its *fundamental groupoid*, $\textstyle\int_1 X$, is the topological groupoid with objects the discrete set of points of X and morphisms the homotopy classes of continuous paths between fixed endpoints, with composition by concatenation $(-) \star (-)$ of paths:

$$
\coprod_{\substack{x,y,z \\ \in X}} \begin{array}{c} \pi_0\Big(\mathrm{Map}\big([0,1], X\big)_x^y\Big) \\ \times \\ \pi_0\Big(\mathrm{Map}\big([0,1], X\big)_y^z\Big) \end{array} \xrightarrow[{[\gamma_1 \star \gamma_2]}]{([\gamma_1], [\gamma_2]) \,\mapsto} \coprod_{x,y \in X} \pi_0\Big(\mathrm{Map}\big([0,1], X\big)_x^y\Big) \underset{\left(x \overset{[\gamma]}{\rightsquigarrow} y\right) \mapsto y}{\overset{\left(x \overset{[\gamma]}{\rightsquigarrow} y\right) \mapsto x}{\underset{\longleftarrow\, [\mathrm{cnst}_x] \leftarrow\!\shortmid x \,-}{\rightrightarrows}}} \flat X \,,
\tag{4.42}
$$

where

$$
\begin{array}{ccc}
\mathrm{Map}\big([0,1], X\big)_x^y & \longrightarrow & \mathrm{Map}\big([0,1], X\big) \\
\big\downarrow & \lrcorner & \big\downarrow {\scriptstyle (\mathrm{ev}_0, \mathrm{ev}_1)} \\
* & \xrightarrow{\;(x,y)\;} & X \times X
\end{array}
\tag{4.43}
$$

is the space of paths with endpoints (x, y). Hence, in the notation (4.29), a fundamental groupoid looks like this:

$$
\textstyle\int_1 X = \left\{ \begin{array}{ccc} & y & \\ {\scriptstyle x \overset{[\gamma_1]}{\rightsquigarrow} y} \nearrow & & \searrow {\scriptstyle y \overset{[\gamma_2]}{\rightsquigarrow} z} \\ x & \xrightarrow{x \overset{[\gamma_1 \star \gamma_2]}{\rightsquigarrow} y} & z \end{array} \right\}.
\tag{4.44}
$$

The isomorphism classes (4.30) of a fundamental groupoid are the connected components of the topological space

$$
\Big[\textstyle\int_1 X\Big]_0 = \pi_0 X \,.
\tag{4.45}
$$

Example 4.1.17. For $\mathcal{X}$ a topological groupoid (4.28) and U a space, then forming mapping spaces (4.9) from U into the component spaces of $\mathcal{X}$,

$$\mathrm{Map}(U,\mathcal{X}) := \left(\begin{array}{c} \mathrm{Map}\big(U, \mathrm{Mor}(\mathcal{X})\big) \;{}_{s_*}\!\times_{t_*} \mathrm{Map}\big(U, \mathrm{Mor}(\mathcal{X})\big) \\ \Big\downarrow{\scriptstyle \circ_*} \\ \mathrm{Map}\big(U, \mathrm{Mor}(\mathcal{X})\big) \underset{t_*}{\overset{s_*}{\underset{\xleftarrow{\;e_*\;}}{\rightrightarrows}}} \mathrm{Map}\big(U, \mathrm{Obj}(\mathcal{X})\big) \end{array} \right), \tag{4.46}$$

gives a topological groupoid which may be thought of as the groupoid of *U-parameterized* objects of $\mathcal{X}$.

If here $U \equiv \mathbb{R}^n$ is a Cartesian space (4.1), we also call $\mathrm{Map}(\mathbb{R}^n, \mathcal{X})$ the space of *n-dimensional plots* of $\mathcal{X}$. Consider the quotient spaces

$$\mathrm{Map}\big(\mathbb{G}^n, -\big) := \mathrm{Map}\big(\mathbb{R}^n, -\big) / \sim_n \tag{4.47}$$

(where "$\mathbb{G}$" is for "germ", which is not an actual topological space itself but defined via the above formula) by the equivalence relation $\sim_n$ which identifies a pair of maps $\phi, \phi' : \mathbb{R}^n \to (-)$ if they agree on any open ball $\mathbb{D}^n_\epsilon$ (4.1) around the origin:

$$\phi \sim_n \phi' \quad \Leftrightarrow \quad \exists\, \epsilon \in \mathbb{R}_{>0}, \;\; \phi|_{\mathbb{D}^n_\epsilon} = \phi'|_{\mathbb{D}^n_\epsilon} . \tag{4.48}$$

This construction extends to topological groupoids $\mathcal{X}$ as in (4.46) to yield what we may call the *groupoid of stalks of n-dimensional plots* of $\mathcal{X}$:

$$\mathrm{Map}(\mathbb{G}^n,\mathcal{X}) := \left(\begin{array}{c} \mathrm{Map}\big(\mathbb{G}^n, \mathrm{Mor}(\mathcal{X})\big) \;{}_{s_*}\!\times_{t_*} \mathrm{Map}\big(\mathbb{G}^n, \mathrm{Mor}(\mathcal{X})\big) \\ \Big\downarrow{\scriptstyle \circ_*} \\ \mathrm{Map}\big(\mathbb{G}^n, \mathrm{Mor}(\mathcal{X})\big) \underset{t_*}{\overset{s_*}{\underset{\xleftarrow{\;e_*\;}}{\rightrightarrows}}} \mathrm{Map}\big(\mathbb{G}^n, \mathrm{Obj}(\mathcal{X})\big) \end{array} \right). \tag{4.49}$$

Definition 4.1.18. Given a pair $(\mathcal{X}, \mathcal{Y})$ of topological groupoids (Def. 4.1.8), a *continuous functor* (or *topological functor*) $\mathcal{X} \xrightarrow{F} \mathcal{Y}$ between them is a map $\mathrm{Mor}(\mathcal{X}) \xrightarrow{F_1} \mathrm{Mor}(Y)$ which homomorphically respects all the structure (4.28):

$$\begin{array}{ccc} \mathcal{X} & \xrightarrow{\quad F \quad} & \mathcal{Y} \\ & & \\ x & & F_0(x) \\ \downarrow{\scriptstyle f} & \mapsto & {\scriptstyle F(f)}\downarrow \\ y & & F_0(y) \\ \downarrow{\scriptstyle g} & \mapsto & {\scriptstyle F(g)}\downarrow \\ z & & F_0(z) \end{array} \qquad \begin{array}{c} x \xrightarrow{g \circ f} z \\ F_0(x) \xrightarrow{F(g) \circ F(f)} F_0(z) \end{array} \tag{4.50}$$

(Here $F_0 : \mathrm{Obj}(\mathcal{X}) \to \mathrm{Obj}(\mathcal{Y})$ is the restriction of F to identity morphisms identified with objects.)

Definition 4.1.19. A continuous ("natural") *transformation* between parallel continuous functors (Def. 4.1.18) is a map

$$\mathcal{X} \underset{F_2}{\overset{F_1}{\underset{\Downarrow \eta}{\rightrightarrows}}} \mathcal{Y} \;:\; \mathrm{Obj}(\mathcal{X}) \longrightarrow \mathrm{Mor}(\mathcal{Y}) \tag{4.51}$$

whose values make these diagrams commute:

$$\begin{array}{ccc} x_1 & & F_1(x_1) \xrightarrow{\eta(x_1)} F_2(x_1) \\ \forall f \big\downarrow & : & F_1(f)\big\downarrow \qquad\qquad \big\downarrow F_2(f) \\ x_2 & & F_1(X_2) \xrightarrow{\eta(x_2)} F_2(x_2)\,. \end{array} \tag{4.52}$$

The *vertical composition* of composable such transformations is by composition of their component morphisms:

$$\mathcal{X} \underset{F_3}{\overset{F_1}{\underset{\Downarrow \eta_2}{\overset{\Downarrow \eta_1}{\xrightarrow{F_2}}}}} \mathcal{Y} \;:\; x \longmapsto F_1(x) \xrightarrow{\eta_1(x)} F_2(x) \xrightarrow{\eta_2(x)} F_3(3)\,, \tag{4.53}$$

while the *horizontal composition* with functors is

$$\mathcal{X}' \xrightarrow{L} \mathcal{X} \underset{F_2}{\overset{F_1}{\Downarrow \eta}} \mathcal{Y} \xrightarrow{R} \mathcal{Y}' \;:\; x' \longmapsto R\Big(\eta\big(L_0(x')\big)\Big)\,. \tag{4.54}$$

Example 4.1.20. There is a close relation between homotopies between topological spaces (4.19) and transformations between topological groupoids (4.51). Concretely, any homotopy between topological spaces gives a transformation between their fundamental groupoids (Ex. 4.1.16).

Example 4.1.21. Between delooping groupoids (Ex. 4.1.12), continuous functors (Def. 4.1.18) are continuous group homomorphisms, hence continuous linear representations if the second group is linear:

$$\Big\{\mathbf{B}G \xrightarrow{\mathbf{B}\rho} \mathbf{B}\mathrm{U}(\mathcal{H})\Big\} \simeq \Big\{\rho \in \mathrm{Rep}(G,\mathcal{H})\Big\}\,. \tag{4.55}$$

Transformations (Def. 4.1.19) between these are *intertwiners* of representations.

Definition 4.1.22. For a pair of topological groupoids $\mathcal{X}, \mathcal{Y}$ (Def. 4.1.8), their *functor groupoid*, $\mathrm{Func}(\mathcal{X},\mathcal{Y})$, is the topological groupoid whose objects are the continuous functors $\mathcal{X} \to \mathcal{Y}$ (Def. 4.1.18), topologized as a subspace of $\mathrm{Map}\big(\mathrm{Mor}(\mathcal{X}), \mathrm{Mor}(\mathcal{Y})\big)$ (4.9), and whose morphisms are the continuous transformations (4.51), topologized as the product space of that with $\mathrm{Map}\big(\mathrm{Obj}(\mathcal{X}), \mathrm{Mor}(\mathcal{Y})\big)$. Composition and inversion of transformations are given by the composition of inversion of their component functions (4.52). Hence in the notation (4.29):

$$\mathrm{Func}(\mathcal{X},\mathcal{Y}) = \left\{ \quad \right\}. \tag{4.56}$$

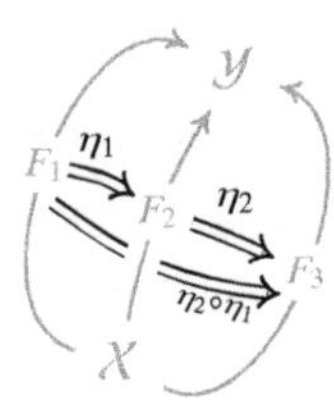

4.1.6 Topological Stacks

If we think – as we may and should – of topological groupoids as topologized "sets with gauge transformations between their elements", then some of them ought to be "the same up to gauge fixing," and yet no invertible continuous functors exist between them (Fig. 4.1 illustrates a simple example, a special case of Ex. 4.1.28).

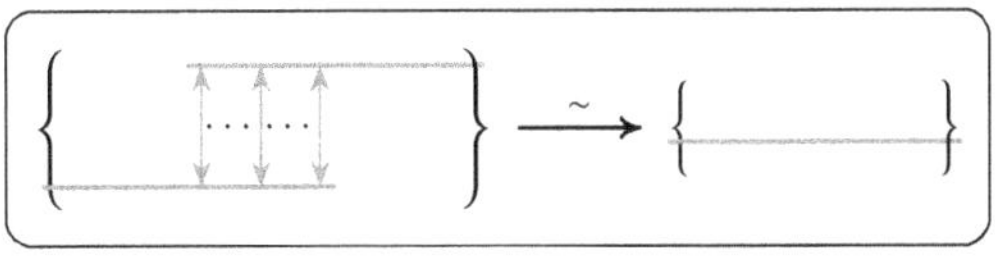

Figure 4.1 Indicated on the left is a topological groupoid whose space of objects is the disjoint union of two intervals but whose morphisms uniquely connect – and thereby uniquely identify – a subinterval of points in either component (a *Čech groupoid*, cf. Ex. 4.1.28). Indicated on the right is the topological groupoid whose space of objects is the result of gluing these two intervals along this subinterval and which has no non-identity morphisms. The evident topological functor from the left to the right is an *equivalence* of topological groupoids (Def. 4.1.23), but no continuous functor can serve as its inverse.

This means that topological functors by themselves are too rigid as a notion of "maps" between topological groupoids and that the actual *maps* (Def. 4.1.24) must subsume inverses to those topological functors that ought to be equivalences of topological groupoids:

Definition 4.1.23. A topological functor (Def. 4.1.18) is an *equivalence*, denoted

$$F : \mathcal{X} \xrightarrow{\sim} \mathcal{Y}\,, \tag{4.57}$$

if for all $n \in \mathbb{N}$ the induced functor on stalks of n-dimensional plots (4.49),

$$F_* : \flat\mathrm{Map}(\mathbb{G}^n, \mathcal{X}) \longrightarrow \flat\mathrm{Map}(\mathbb{G}^n, \mathcal{X})\,, \tag{4.58}$$

is:

(i) surjective on isomorphism classes (4.30) of objects ("essentially surjective"):

$$\underset{n\in\mathbb{N}}{\forall} \;:\; \big[\flat\mathrm{Map}(\mathbb{G}^n, \mathcal{X})\big]_0 \xrightarrow{\;\;F_*\;\;} \big[\flat\mathrm{Map}(\mathbb{G}^n, \mathcal{Y})\big]_0\,, \tag{4.59}$$

(ii) bijective on morphisms between pairs of objects ("fully faithful"):

$$\underset{n\in\mathbb{N}}{\forall}\;\underset{x,y\in\mathrm{Obj}(\mathcal{X})}{\forall} : \mathrm{Mor}\Big(\flat\mathrm{Map}(\mathbb{G}^n, \mathcal{X})\Big)^{\mathrm{y}}_{\mathrm{x}} \xrightarrow[\sim]{F_*} \mathrm{Mor}\Big(\flat\mathrm{Map}(\mathbb{G}^n, \mathcal{Y})\Big)^{F_*(\mathrm{y})}_{F_*(\mathrm{x})}\,, \tag{4.60}$$

where

$$\mathrm{Mor}(-)^y_x := * \;{}_x\!\times_s \mathrm{Mor}(-)\;{}_t\!\times_y \;*\,. \tag{4.61}$$

denotes the subspace of morphisms between a given pair of objects.

Definition 4.1.24. A (Morita) *map* between topological groupoids (as opposed to a plain continuous functor, Def. 4.1.18) is a span of continuous functors (Def. 4.1.18), with the left one an equivalence (Def. 4.1.23):

$$\mathcal{X} \xleftarrow{\;\sim\;} \widehat{\mathcal{X}} \xrightarrow{\;F\;} \mathcal{Y}\,. \tag{4.62}$$

In particular, a (Morita) *equivalence* between topological groupoids is a span of continuous functors which are equivalences

$$\mathcal{X} \xleftarrow{\sim} \widehat{\mathcal{X}} \xrightarrow{\sim} \mathcal{Y} \,. \tag{4.63}$$

When considered up to (Morita) equivalence, topological groupoids are also referred to as *topological stacks* (which is terrible terminology but completely standard); cf. §9.1.19.

Remark 4.1.25. Equivalent topological groupoids (Def. 4.1.24) have homeomorphic spaces $[-]_0$ of isomorphism classes (4.30):

$$\mathcal{X} \xrightarrow{\sim} \mathcal{Y} \quad \Rightarrow \quad [\mathcal{X}]_0 \xrightarrow{\sim} [\mathcal{Y}]_0 \,. \tag{4.64}$$

Example 4.1.26. If a continuous functor $F : \mathcal{X} \longrightarrow \mathcal{Y}$ is a *homotopy equivalence* (better terminology would be "transformation equivalence", cf. Thm. 4.1.20 and (4.23)) in that there exists a reverse continuous functor $\overline{F} : \mathcal{Y} \longrightarrow \mathcal{X}$ and continuous transformations (Def. 4.1.19) of this form:

$$\begin{array}{ccccc} & \mathcal{Y} & \xrightarrow{\mathrm{id}} & \mathcal{Y} & \\ {}^{F}\nearrow & \Downarrow & \searrow^{\overline{F}} \quad \Downarrow & \nearrow^{F} & \\ \mathcal{X} & \xrightarrow{\mathrm{id}} & \mathcal{X} & & \end{array} \tag{4.65}$$

then it is an equivalence in the sense of Def. 4.1.23.

Proof. After passage to n-dimensional stalks of plots (4.49), this reduces to the basic statement of category theory that essentially surjective and fully faithful functors are equivalences (cf. [Mac98, p. 93]). □

Example 4.1.27. For $H \overset{\iota}{\hookrightarrow} G$ a topological subgroup (4.11), consider the coset space (4.17)

$$G/H = \{ g \cdot H \,|\, g \in G \} \,, \tag{4.66}$$

with its canonical left G-action (4.18). Then the corresponding homotopy quotient (4.39) is equivalent (Def. 4.1.23) to the delooping of H (Ex. 4.1.12):

$$\begin{array}{ccc} \mathbf{B}H & \xrightarrow{\sim} & G \backslash\!\backslash G/H \\ \left(\begin{array}{c} \bullet \\ {\scriptstyle h}\big\downarrow \\ \bullet \end{array} \right) & \longmapsto & \left(\begin{array}{c} \mathrm{e} \cdot H \\ {\scriptstyle h}\big\downarrow \\ \mathrm{e} \cdot H \end{array} \right) . \end{array} \tag{4.67}$$

A key class of examples of equivalences of topological groupoids (Def. 4.1.23) which are not homotopy equivalences (Ex. 4.1.26) are projections out of Čech groupoids (cf. Fig. 4.1 and Lem. 4.1.29),

Example 4.1.28 (Čech groupoids)**.** For a topological manifold X equipped with an *open cover*

$$\mathcal{U} := \Big\{ U_i \underset{\text{open}}{\overset{\iota_i}{\hookrightarrow}} X \Big\}_{i \in I}, \quad \coprod_i U_i \xrightarrow{(\iota_i)_{i \in I}} X \,, \tag{4.68}$$

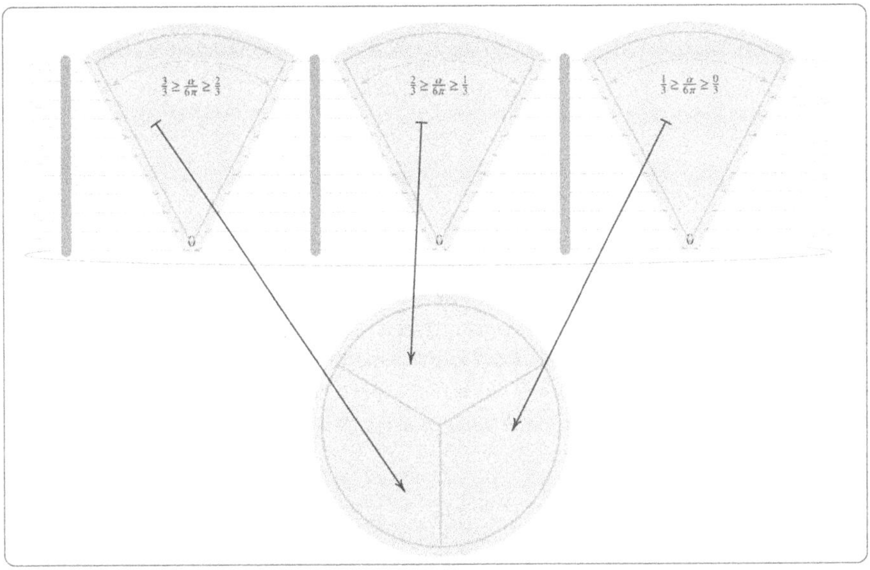

Figure 4.2 **Top row:** The Čech groupoid (Ex. 4.1.28) of a good (Def. 4.1.30) open cover of the open disk $\mathbb{D}^2_{1+\epsilon}$. The space of objects is shown in light gray, and the space of non-identity morphisms is shown in dark gray (α denotes an angular coordinate function).
Bottom row: The original disk, as a groupoid with only identity morphisms (Ex. 4.1.10), receiving the canonical projection functor (4.71) from the Čech groupoid.

then the corresponding *Čech groupoid* (cf. Fig. 4.2) is the topological groupoid (4.28) given by

$$X_{\mathcal{U}} := \left(\coprod_{i,j,k} U_{ijk} \xrightarrow{(x,i,j,k) \mapsto (x,i,k)} \underset{i,j}{\coprod} U_{ij} \overset{(x,i,j) \mapsto (x,j,i)}{\circlearrowleft} \begin{array}{c} \xrightarrow{(x,i,j) \mapsto (x,i)} \\ \xleftarrow{(x,i,i) \leftarrow\!\shortmid (x,i)} \\ \xrightarrow[(x,i,j) \mapsto (x,j)]{} \end{array} \coprod_i U_i \right), \tag{4.69}$$

where we abbreviate $U_{ij} := U_i \cap U_j$ and $U_{ijk} := U_i \cap U_j \cap U_k$, and where (x,i) denotes a point $x \in X$ but regarded as the corresponding point of U_i, etc. Hence in the notation (4.29), a Čech groupoid looks like this:

$$X_{\mathcal{U}} = \left\{ \begin{array}{ccccc} & & (x,j) & & \\ & \overset{(x,i,j)}{\nearrow} & & \overset{(x,j,k)}{\searrow} & \\ (x,i) & & \xrightarrow{\quad (x,i,k) \quad} & & (x,k) \end{array} \right\}. \tag{4.70}$$

This comes with a continuous functor (Def. 4.1.18) to the original space (regarded as a groupoid per Ex. 4.1.10), which is (see Lem. 4.1.29) an equivalence (Def. 4.1.23):

$$\begin{array}{ccc} X_{\mathcal{U}} & \xrightarrow{\;\;\sim\;\;} & X \\ (x,i) & & x \\ \big\downarrow{\scriptstyle (x,i,j)} & \longmapsto & \Vert \\ (x,j) & & x\,. \end{array} \tag{4.71}$$

To appreciate the relevance of this fact, consider a topological group Γ and observe that there is a unique continuous functor $X \to \mathbf{B}\Gamma$ to its delooping groupoid (Ex. 4.1.12), and that unique functor is trivial (constant). But, due to the equivalence (4.71), *maps* of this form, in the sense of Def. 4.1.24, subsume the continuous functors out of the Čech groupoid (4.69) of any open cover, and these are identified with the cocycles, relative to $\mathcal{U}$, of nonabelian *Čech cohomology* $H^1(X;\Gamma)$ (for which cf. [Wed16, §7][Al85, §4]):

$$\begin{array}{ccccc} X & \xleftarrow{\;\;\sim\;\;} & X_{\mathcal{U}} & \xrightarrow{\;\;\gamma\;\;} & \mathbf{B}\Gamma \\ x & & (x,i) & & \bullet \\ \Vert & \longleftarrow\!\shortmid & \big\downarrow{\scriptstyle (x,i,j)} & \longmapsto & \big\downarrow{\scriptstyle \gamma_{ij}(x)} \\ x & & (x,j) & & \bullet \\ \Vert & \longleftarrow\!\shortmid & \big\downarrow{\scriptstyle (x,j,k)} & \longmapsto & \big\downarrow{\scriptstyle \gamma_{jk}(x)} \\ x & & (x,k) & & \bullet \end{array} \tag{4.72}$$

(with the composite $(x,i,k) : (x,i) \to (x,k)$ mapping to $\gamma_{ik}(x)$)

while their transformations (Def. 4.1.19) are identified with Čech coboundaries:

$$\begin{array}{ccc} X_{\mathcal{U}} \underset{g'}{\overset{g}{\rightrightarrows}} \mathbf{B}G & & \text{(with } h : g \Rightarrow g' \text{)} \\ (x,i) & \longmapsto & \bullet \xrightarrow{\;h_i(x)\;} \bullet \\ \big\downarrow{\scriptstyle (x,i,j)} & & \big\downarrow{\scriptstyle \gamma_{ij}(x)} \qquad \big\downarrow{\scriptstyle \gamma'_{ij}(x)} \\ (x,j) & \longmapsto & \bullet \xrightarrow[\;h_j(x)\;]{} \bullet\,. \end{array} \tag{4.73}$$

We generalize this example further as Ex. 4.1.34. But first we make explicit the argument we just used, since it is an instructive illustration of the general machinery at play here:

Lemma 4.1.29. *The above continuous functor* (4.71) *is indeed an equivalence of topological groupoids* (Def. 4.1.23).

Proof. First to see that the functor is essentially surjective (4.59) on n-dimensional stalks: For $\mathbf{x} : \mathbb{R}^n \to X$ a continuous map (a *plot*), the open cover property of $\mathcal{U}$ (4.68) implies that there exists $i \in I$ and an open neighborhood $U_x \subset X$ of $\mathbf{x}(0)$ such that $U_x \subset U_i$. Therefore, the preimage $\mathbf{x}^{-1}(U_x)$ is an open neighborhood of 0 in $\mathbb{R}^n$, and by the Euclidean topology on $\mathbb{R}^n$ this contains an open ball $D^n_\epsilon \subset \mathbf{x}^{-1}(U_x)$ around 0, such that $\mathbf{x}|_{D^n_\epsilon}$ factors through U_i. Precomposed with any map $\mathbb{R}^n \xrightarrow{\sim} \mathbb{D}^n_\epsilon$

which is the identity on a smaller open ball $D^n_{\epsilon'} \subset D^n_{\epsilon}$, this is a plot $(\mathbf{x}, i)$ of objects of the Čech groupoid whose germ maps to the germ of the given $\mathbf{x}$.

Then to see that the functor is fully faithful (4.60) on n-dimensional stalks: Given a pair of plots of the Čech groupoid, $\mathbf{x}_i : \mathbb{R}^n \to U_i$ and $\mathbf{x}_j : \mathbb{R}^n \to U_j$, there are two cases: regarded as maps to X their germs either coincide – in which case there is a unique morphism between them in X regarded as a groupoid (Ex. 4.1.10), namely the identity – or their germs do not coincide, in which case the set of morphisms between them in X is empty. We need to see that the same two cases hold for morphisms between the germs of these plots regarded in the Čech groupoid. In the second case this is immediate from the definition, while in the first case it follows from the open cover property that the germs of $\mathbf{x}_i$ and $\mathbf{x}_j$ both factor through $U_i \cap U_j$, which constitutes the required unique morphism between them in the Čech groupoid. □

4.1.7 Mapping Stacks and Nonabelian Cohomology

Definition 4.1.30 (Good open cover)**.** An open cover $\{U_i \overset{\iota_i}{\underset{\text{open}}{\hookrightarrow}} X\}_{i \in I}$ (4.68) of an n-dimensional manifold is *good* if all finite intersections of its patches are either empty or homeomorphic to $\mathbb{R}^n$:

$$\forall_{k \in \mathbb{N}_{\geq 1}} \forall_{i_i, \cdots, i_k \in I} \;:\; U_{i_1} \cap \cdots \cap U_{i_k} \simeq \left\{ \begin{array}{l} \varnothing \text{ or} \\ \mathbb{R}^n . \end{array} \right. \tag{4.74}$$

We say that this is *differentially good* if these homeomorphisms (4.74) exist even as *diffeomorphisms*.

For general topological manifolds, the existence of good open covers is not known, but we have:

Lemma 4.1.31 (cf. [BT82, Thm. 5.1][FStS12, Prop. A.1])**.** *Every smooth manifold admits a differentiably good open cover* (Def. 4.1.30).

It turns out (in Thm. 9.1.19) that the Čech groupoids (Ex. 4.1.28) of *good* open covers (Def. 4.1.30) are "fine enough" (technical term: *cofibrant*) to represent all maps out of topological manifolds into delooping groupoids (Ex. 4.1.12). Therefore:

Definition 4.1.32. Let X be a topological manifold which admits a good open cover $\mathcal{U}$ (Def. 4.1.30, such as any smooth manifold does, by Lem. 4.1.31), and let Γ be a topological group, then the *mapping stack*[1] from X to the delooping groupoid $\mathbf{B}\Gamma$ (4.35) is, up to equivalence, the functor groupoid Func$(-,-)$ (Def. 4.1.22) into $\mathbf{B}\Gamma$ out of the Čech groupoid $X_{\mathcal{U}}$ (Ex. 4.1.28) of the good open cover $\mathcal{U}$:

$$\begin{aligned} \mathrm{Map}(X, \mathbf{B}\Gamma) &:= \mathrm{Func}(X_{\mathcal{U}}, \mathbf{B}\Gamma) \\ &= \{X \dashrightarrow \mathbf{B}\Gamma\}, \end{aligned} \tag{4.75}$$

[1]In the literature, the mapping stack (4.75) may also be called the *derived internal hom*, or similar. We are tacitly using here that $X_{\mathcal{U}}$ is cofibrant ([SS26c, Ex. 4.3.42]) and that $\mathbf{B}\Gamma$ is fibrant ([SS26c, Lem 4.3.30]) in the local projective model structure of simplicial presheaves over the site of Cartesian spaces. This is discussed in Thm. 9.1.19.

where in the second line we are showing a more suggestive notation which highlights again that this is to be thought of as maps out of X itself, in the sense of Def. 4.1.24.

The underlying topologically discrete groupoid (4.41) we denote by a boldface $\mathbf{H}(-,-)$:

$$\mathbf{H}(X, \mathbf{B}\Gamma) := \flat\,\mathrm{Map}(X, \mathbf{B}\Gamma)\,, \tag{4.76}$$

since this is (cf. [Jar09]) the *cocycle groupoid* (objects are Čech cocycles, morphisms are coboundaries) whose isomorphism classes (4.30) are the *nonabelian cohomology sets* $H(-,-)$, as discussed in Ex. 4.1.33.

Example 4.1.33 (Ordinary nonabelian cohomology)**.** We have seen in Ex. 4.1.28 (4.72) and (4.73) that (4.75) is the groupoid whose objects are 1-cocycles and whose morphisms are coboundaries in nonabelian Čech cohomology $H^1(X;\Gamma)$, relative to the cover $\mathcal{U}$ (for which cf. [Wed16, §7][Al85, §4]). In general, for these sets of cocycles modulo coboundaries to give the full cohomology set $H^1(X;\Gamma)$, one has to take their colimit over cover refinements. But in the presence of a good open cover, it is sufficient to evaluate there! (For the case of abelian groups Γ, this is [Bre97, §4.18], in general, it follows immediately from the cofibrancy of the Čech groupoid.) Therefore we have:

$$H^1(X;\Gamma) \simeq \big[\mathbf{H}(X, \mathbf{B}\Gamma)\big]_0 \equiv \big[\flat\mathrm{Map}(X, \mathbf{B}\Gamma)\big]_0\,. \tag{4.77}$$

In words: The isomorphism classes of the mapping stack from a space X to $\mathbf{B}\Gamma$ are identified with the nonabelian 1-cohomology of X with coefficients in Γ, and in components this identification incarnates as Čech cohomology.

These classes are equivalently the *isomorphism classes* of Γ*-principal bundles* over X. In fact, before passing to isomorphism classes,

$$\Gamma\mathrm{PrnBdl}(X) \simeq \mathbf{H}(X, \mathbf{B}\Gamma) \tag{4.78}$$

is equivalent (cf. Ex. 4.1.45 and [SS26c, §5.1.2]) to the groupoid of Γ-principal bundles (Thm. 4.1.13).

This means that the delooping groupoid $\mathbf{B}\Gamma$ serves not just as a coefficient object for nonabelian cohomology, in (4.77), but as the *moduli stack* of Γ-principal bundles (cf. [NSS12a][FSS14][SS26c]).

On the other hand, there is also the *classifying space* of Γ ([Mil67], cf. [SS26c, Prop. 3.3.4]), the "bar construction" or "topological realization" $|-|$ of $\mathbf{B}\Gamma$ (cf. [SS26c, Ntn. 2.2.28]), traditionally denoted as:

$$B\Gamma := |\mathbf{B}\Gamma|\,. \tag{4.79}$$

Over paracompact topological spaces X (such as manifolds and CW complexes, Ex. 4.1.3), we have that homotopy classes of maps into BG coincide with transformation classes of maps into $\mathbf{B}G$ and hence give the same nonabelian cohomology (cf. [RS17, Thm. 3.5.1][SS26c, Thm. 5.1.13]):

$$\pi_0\mathrm{Map}(X,\, BG) \underset{(4.45)}{=} \big[\smallint\mathrm{Map}(X, BG)\big]_0 \simeq \big[\flat\mathrm{Map}(X, \mathbf{B}G)\big]_0 \underset{(4.77)}{=} H^1(X, G)\,. \tag{4.80}$$

To note here the two kinds of classifying objects that appear:

- $\mathbf{B}\Gamma$ is a topological groupoid with trivial space of objects,

- $B\Gamma$ is, as a topological space, a topological groupoid with only identity morphisms.

Hence, general topological groupoids, regarded as classifying objects for cohomology, unify these two extremes. We will see this in action from § 4.1.8 on.

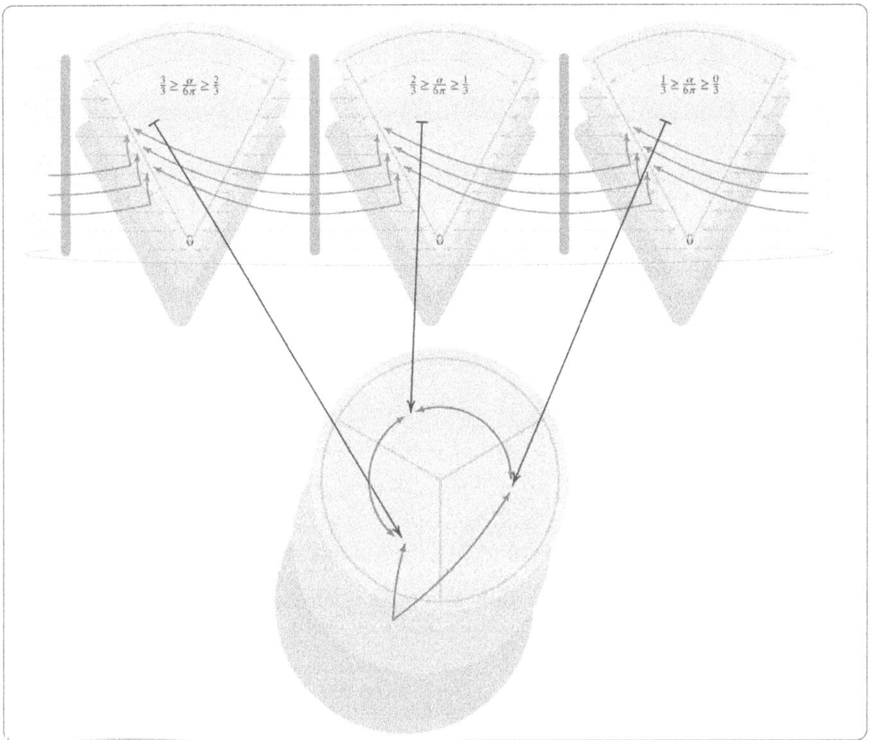

Figure 4.3 In equivariant generalization of Fig. 4.2:
Top row: The equivariant Čech groupoid (Ex. 4.1.34) of a (good, Def. 4.1.30) equivariant open cover of the open disk $\mathbb{D}^2_{1+\epsilon}$ equipped with rigid $\mathbb{Z}_3$-rotation action. The space of objects is shown in light gray, the space of non-identity morphisms is shown in darker shades of gray.
Bottom row: The $\mathbb{Z}_3$-action groupoid (4.39) of the disk, receiving the canonical projection functor (4.84) from the equivariant Čech groupoid.

In equivariant generalization of Ex. 4.1.28, we have:

Example 4.1.34 (Equivariant Čech groupoids). For $G \curvearrowright X$ a topological group action (Def. 4.1.7), consider an open cover $\mathcal{U} := \big\{U_i \overset{\iota_i}{\hookrightarrow} X\big\}_{i \in I}$ (4.68) which is *equivariant* in that the disjoint union of its patches is equipped with an action $G \circlearrowright \coprod_i U_i$, and compatibly so in that the covering map is equivariant (4.13):

$$\begin{array}{ccc} {}^{G}\circlearrowright & & {}^{G}\circlearrowright \\ \coprod_i U_i & \xrightarrow{(\iota_i)_{i\in i}} & X \end{array} \tag{4.81}$$

then the corresponding *equivariant Čech groupoid* is

$$(G\backslash\!\backslash X)_{\mathcal{U}} = \left\{ \begin{array}{ccccc} & & (g_1 \cdot x,\, g_1 \cdot j) & & \\ & \nearrow^{((x,i,j),\, g_1)} & & \searrow^{((g_1\cdot x,\, g_1\cdot j,\, g_1\cdot k),\, g_2)} & \\ (x,i) & & \xrightarrow{((x,i,j),\, g_2\cdot g_1)} & & (g_2 \cdot g_1 \cdot x,\, g_2 \cdot g_1 \cdot k) \end{array} \right\}, \tag{4.82}$$

where the general morphisms may usefully be thought of as composites of pure Čech morphisms and pure gauge morphisms, in either order:

$$\begin{array}{ccccc} & \nearrow^{((x,i,j),e)} & (x,j) & \searrow^{((x,j,j),g)} & \\ (x,i) & \longrightarrow & \big((x,i,j),g\big) & \longrightarrow & (g\cdot x,\, g\cdot j)\,. \\ & \searrow_{((x,i,i),g)} & (g\cdot x,\, g\cdot i) & \nearrow_{((g\cdot x,\, g\cdot i,\, g\cdot j),e)} & \end{array} \tag{4.83}$$

In the equivariant generalization of (4.71), the canonical functor from the equivariant Čech groupoid to the original action groupoid is an equivalence (Def. 4.1.23, cf. [SS26c, Ex. 4.3.45])

$$\begin{array}{ccc} (G\backslash\!\backslash X)_{\mathcal{U}} & \xrightarrow{\sim} & G\backslash\!\backslash X \\ (x,i) & & x \\ \big\downarrow{\scriptstyle ((x,i,j),g)} & \longmapsto & {\scriptstyle g}\big\downarrow \\ \big((g\cdot x,\, g\cdot j)\big) & & g\cdot x\,, \end{array} \tag{4.84}$$

so that, in generalization of (4.72), a Morita map now is a cocycle in nonabelian *equivariant Čech cohomology* (cf. [SS26c, Rem. 4.3.46]):

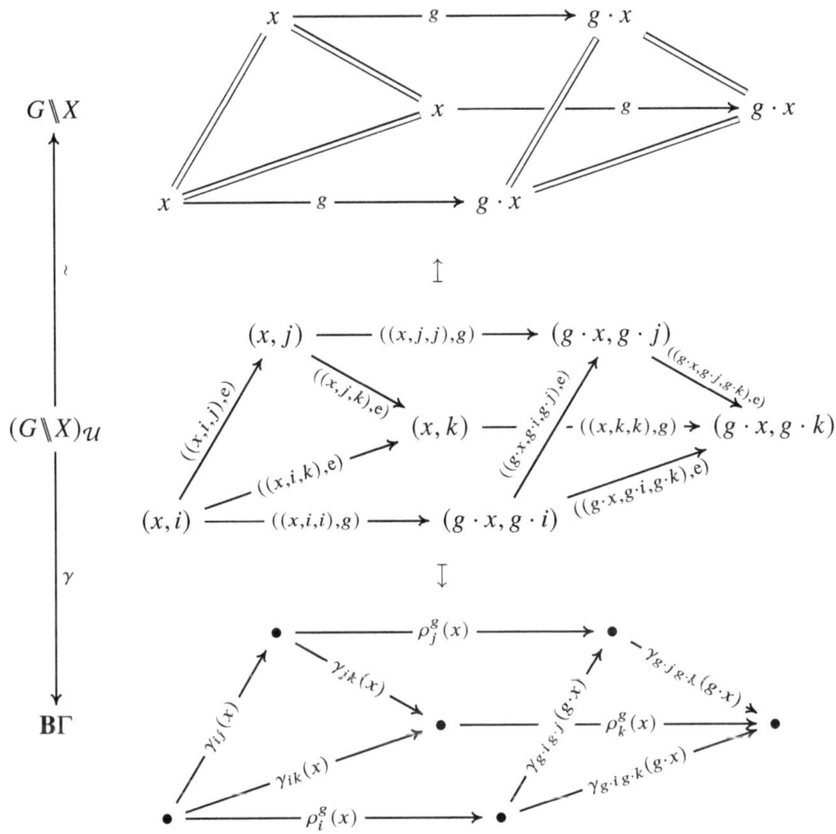

(4.85)

In equivariant generalization of Lem. 4.1.31, we have:

Lemma 4.1.35 ([Ya14][2]). *For G a finite group, X a smooth manifold, and $G \circlearrowright X$ smooth action* (Def. 4.1.7), *there exists a good open cover of X* (Def. 4.1.30) *which is equivariant* (4.81).

In equivariant generalization of Def. 4.1.32, we have:

Definition 4.1.36. For $G \circlearrowright X$ a topological G-space (Def. 4.1.7) that admits a good open cover $\mathcal{U}$ (Def. 4.1.30) which is G-equivariant (4.81) (such as is the case for G, a finite group acting smoothly on a smooth manifold G, by Lem. 4.1.35), and for Γ a

[2] In fact, [Ya14] proves something much stronger than Lem. 4.1.35, namely that good equivariant *regular* covers exist (cf. also [SS26c, Def. 2.1.24]). This is needed when expressing equivariance in terms of systems of H-fixed loci for subgroups $H \subset G$, cf. [SS26c, §6.2.1].

topological group, we say that

$$\begin{aligned} \mathrm{Map}\big(G\backslash\!\!\backslash X,\, \mathbf{B}\Gamma\big) &:= \mathrm{Func}\big((G /\!\!/ X)_{\mathcal{U}},\, \mathbf{B}\Gamma\big) \\ &\equiv \big\{G\backslash\!\!\backslash X \dashrightarrow \mathbf{B}\Gamma\big\} \end{aligned} \tag{4.86}$$

is the (up to equivalence) *mapping stack* out of the homotopy quotient $G\backslash\!\!\backslash X$ (4.39) into the delooping $\mathbf{B}\Gamma$ (Ex. 4.1.12), where $(G\backslash\!\!\backslash X)_{\mathcal{U}}$ is the equivariant Čech groupoid from Ex. 4.1.34.

The underlying topologically discrete groupoid (4.41) we denote again by boldface $\mathbf{H}(-,-)$ (4.76):

$$\mathbf{H}\big(G\backslash\!\!\backslash X,\, \mathbf{B}\Gamma\big) := \flat\, \mathrm{Map}\big(G\backslash\!\!\backslash X,\, \mathbf{B}\Gamma\big)\,, \tag{4.87}$$

since this is now the *cocycle groupoid* of ordinary nonabelian *equivariant Čech cohomology*, see Ex. 4.1.37.

In generalization of Ex. 4.1.33 we have:

Example 4.1.37 (Equivariant nonabelian cohomology and equivariant bundles [SS26c]). In the situation of Def. 4.1.36, we have that

$$G\mathrm{Equ}\Gamma\mathrm{PrnBdl}(X) \simeq \mathbf{H}\big(G\backslash\!\!\backslash X,\, \mathbf{B}\Gamma\big) \tag{4.88}$$

is equivalently the groupoid of *G-equivariant Γ-principal bundles* on X (see Ex. 4.1.45 for the construction), and its connected components (hence equivalently the isomorphism classes of these bundles) are the *equivariant nonabelian Čech cohomology* of X:

$$H^1_G(X;\, \Gamma) \simeq \big[\mathbf{H}(G\backslash\!\!\backslash X,\, \mathbf{B}\Gamma)\big]_0 \tag{4.89}$$

These constructions of mapping stacks (4.75) and (4.86) do not actually depend, up to equivalence, on the use of good Čech groupoids; these are just a particularly nice choice (when they exist) of general *cofibrant resolutions*, of which a larger class is the following:

Definition 4.1.38 (cf. [SS26c, Prop. 4.2.37]). A Dugger-*cofibrant resolution* of a topological groupoid $\mathcal{X}$ is an equivalence (Def. 4.1.23)

$$\widehat{\mathcal{X}} \xrightarrow{\ \sim\ } \mathcal{X} \tag{4.90}$$

with a topological groupoid $\widehat{\mathcal{X}}$ for which:

- all three component spaces $\mathrm{Obj}(\widehat{\mathcal{X}})$, $\mathrm{Mor}(\widehat{\mathcal{X}})$, $\mathrm{Mor}(\widehat{\mathcal{X}})\ {}_s\!\times_t \mathrm{Mor}(\widehat{\mathcal{X}})$ are homeomorphic to disjoint unions of Cartesian spaces (4.1),
- the identity-morphisms including maps like $\mathrm{Obj}(\widehat{\mathcal{X}}) \overset{e}{\hookrightarrow} \mathrm{Mor}(\widehat{\mathcal{X}})$ and $\mathrm{Mor}(\widehat{\mathcal{X}})\ {}_s\!\times_t \mathrm{Obj}(\widehat{\mathcal{X}}) \xrightarrow{\mathrm{id}\, {}_s\times_t\, e} \mathrm{Mor}(\widehat{\mathcal{X}})\ {}_s\!\times_t \mathrm{Mor}(\widehat{\mathcal{X}})$, etc., are the inclusions of disjoint summands.

Therefore, in further generalization of Def. 4.1.36, we set:

Definition 4.1.39. For $\widehat{\mathcal{X}} \xrightarrow{\sim} \mathcal{X}$ a Dugger-cofibrant resolution (Def. 4.1.38) of a topological groupoid and Γ a topological group, then the *mapping stack* from $\mathcal{X}$ to the delooping $\mathbf{B}\Gamma$ is

$$\begin{aligned} \mathrm{Map}(\mathcal{X}, \mathbf{B}\Gamma) &:= \mathrm{Func}(\widehat{\mathcal{X}}, \mathbf{B}\Gamma) \\ &\equiv \{\mathcal{X} \dashrightarrow \mathbf{B}\Gamma\}. \end{aligned} \tag{4.91}$$

Lemma 4.1.40. *The mapping stack* $\mathrm{Map}(\mathcal{X}, \mathbf{B}\Gamma)$ *in Def. 4.1.39 is well-defined in that it depends, up to Morita equivalence* (Def. 4.1.24), *only on the equivalence class of* $\mathcal{X}$*:*

$$\mathcal{X}' \xrightarrow[\sim]{F} \mathcal{X} \quad \Rightarrow \quad \mathrm{Map}(\mathcal{X}, \mathbf{B}\Gamma) \xrightarrow[\sim]{F^*} \mathrm{Map}(\mathcal{X}', \mathbf{B}\Gamma). \tag{4.92}$$

Proof. Discussed in Fact 8.1.17 of Thm. 9.1.19. □

4.1.8 Slice mapping stacks and Twisted cohomology

Definition 4.1.41. Say that a continuous functor F (Def. 4.1.18) between topological groupoids is a *global fibration*, to be denoted as on the left here:

$$\begin{array}{c} \mathcal{X} \\ {\scriptstyle F}\big\downarrow{\scriptstyle \in \mathrm{Fib}} \\ \mathcal{Y} \end{array} \quad \Leftrightarrow \quad \begin{array}{ccc} \forall x & \overset{\exists \widehat{f}}{\dashrightarrow} & t(\widehat{f}) \\ \big\downarrow & & \big\downarrow \\ s(f) & \xrightarrow{\forall f} & t(f) \end{array} \tag{4.93}$$

if, as indicated on the right, for all

- $n \in \mathbb{N}$, $f : \mathbb{R}^n \to \mathrm{Mor}(\mathcal{Y})$, a plot of morphisms in $\mathbb{Y}$,
- $x : \mathbb{R}^n \to \mathrm{Obj}(\mathcal{Y})$, a plot of objects of $\mathcal{Y}$ such that $F(x) = s(f)$,

there exists $\widehat{f} : \mathbb{R}^n \to \mathcal{X}$ such that $F(\widehat{f}) = f$ and $s(\widehat{f}) = x$ (a "lift").

Example 4.1.42. For an action $\Gamma \circlearrowright X$ (Def. 4.1.7), its homotopy quotient (Def. 4.1.14) comes with a canonical functor to $\mathbf{B}\Gamma$ (4.40), which is a global fibration (4.93):

$$\begin{array}{c} \Gamma \backslash\!\backslash X \\ {\scriptstyle p^{\Gamma \circlearrowright X}_{\mathrm{univ}}}\big\downarrow{\scriptstyle \in \mathrm{Fib}} \\ \mathbf{B}\Gamma \end{array} \qquad \begin{array}{c} x \xrightarrow{\;\gamma\;} \gamma \cdot x \\ \big\downarrow \\ \bullet \xrightarrow{\;\gamma\;} \bullet\,. \end{array} \tag{4.94}$$

Here the subscript notation indicates that this simple fibration is in fact the (stacky) *universal* Γ*-associated* X*-fiber bundle* and even the universal *globally equivariant* such: see Ex. 4.1.45.

Definition 4.1.43 (Homotopy fiber product of topological groupoids). Given a pair of coincident continuous functors, $\mathcal{X} \xrightarrow{F_1} \mathcal{B} \xleftarrow{F_2} \mathcal{Y}$, the corresponding fiber product (Def. 4.1.1) of their morphism spaces canonically inherits the structure of a topological groupoid, the *fiber product of topological groupoids*:

$$\begin{array}{ccc} \mathcal{X} \underset{\mathcal{B}}{\times} \mathcal{Y} & \longrightarrow & \mathcal{Y} \\ \big\downarrow & \lrcorner & \big\downarrow{\scriptstyle F_2} \\ \mathcal{X} & \underset{F_1}{\longrightarrow} & \mathcal{B}, \end{array} \tag{4.95}$$

hence

$$\begin{aligned} \mathrm{Mor}(\mathcal{X} \times_{\mathcal{B}} \mathcal{Y}) &= \mathrm{Mor}(\widehat{\mathcal{X}}) \underset{\mathrm{Mor}(\widehat{\mathcal{B}})}{\times} \mathrm{Mor}(\widehat{\mathcal{Y}}), \\ \mathrm{Obj}(\mathcal{X} \times_{\mathcal{B}} \mathcal{Y}) &= \mathrm{Obj}(\widehat{\mathcal{X}}) \underset{\mathrm{Obj}(\widehat{\mathcal{B}})}{\times} \mathrm{Obj}(\widehat{\mathcal{Y}}), \end{aligned} \tag{4.96}$$

etc., and the structure maps on $\mathcal{X} \times_{\mathcal{B}} \mathcal{Y}$ are induced entry-wise from the given ones on $\mathcal{X}$, $\mathcal{Y}$, and $\mathcal{B}$.

If here F_1 or F_2 are global fibrations (Def. 4.1.41) then this is a representation of the (up to equivalence) *homotopy fiber product* of the corresponding topological stacks, or the *homotopy pullback* of one either along the other.[3]

Example 4.1.44. The homotopy fiber (Def. 4.1.43) of $G\backslash\!\!\backslash X \to \mathbf{B}G$ (4.94) is X (via Ex. 4.1.10):

$$\begin{array}{ccc} X & \longrightarrow & G\backslash\!\!\backslash X \\ \big\downarrow & \lrcorner & \big\downarrow {\scriptstyle \in \mathrm{Fib}} \\ * & \xrightarrow{\exists!} & \mathbf{B}G \end{array} \tag{4.97}$$

Conversely ([SS26c, Prop. 4.2.77]), every fibration over $\mathbf{B}G$ with fiber X exhibits a G-action on X this way! More generally (cf. [SS26c, Prop. 4.2.79]), for $\phi : G' \to G$ a homomorphism of topological groups, and noting that this induces from the G action on X also G'-action on X, via

$$\begin{array}{ccc} G' \times X & \longrightarrow & X \\ (g', x) & \longmapsto & \phi(g') \cdot x, \end{array} \tag{4.98}$$

then the homotopy quotient by G' is the following homotopy pullback of that by G:

$$\begin{array}{ccc} G'\backslash\!\!\backslash X & \longrightarrow & G\backslash\!\!\backslash X \\ \big\downarrow & \lrcorner & {\scriptstyle p}\big\downarrow {\scriptstyle \in \mathrm{Fib}} \\ \mathbf{B}G' & \xrightarrow{\mathbf{B}\phi} & \mathbf{B}G. \end{array} \tag{4.99}$$

Example 4.1.45 (Stacky universal fiber bundles)**.** For $G \circlearrowright X$ a finite group action on a smooth manifold, and for $(G\backslash\!\!\backslash X)_{\mathcal{U}} \xrightarrow{\sim} G\backslash\!\!\backslash G$ the Čech groupoid resolution (Ex. 4.1.34) of an equivariant good open cover (Lem. 4.1.35), and given a map $G\backslash\!\!\backslash X \xrightarrow{\gamma} \mathbf{B}\Gamma$ to the delooping of a topological group Γ, hence an equivariant Čech cocycle (4.85), then the homotopy pullback (Def. 4.1.43) along γ of the homotopy quotient $\Gamma\backslash\!\!\backslash\Gamma$ (4.94) is, up to equivalence (homotopy G-quotient of), the G-equivariant Γ-principal bundle[4] $\mathcal{P} \xrightarrow{p} X$ corresponding to γ under the Čech theory of Exs. 4.1.33 and 4.1.37:

$$\begin{array}{ccccc} G\backslash\!\!\backslash \mathcal{P} & \xleftarrow{\sim} & G\backslash\!\!\backslash \widehat{\mathcal{P}} & \longrightarrow & \Gamma\backslash\!\!\backslash\Gamma \\ \big\downarrow {\scriptstyle G\backslash\!\!\backslash p} & & \big\downarrow & \lrcorner & {\scriptstyle p^{\Gamma \circlearrowright \Gamma}_{\mathrm{univ}}}\big\downarrow {\scriptstyle \in \mathrm{Fib}} \\ G\backslash\!\!\backslash X & \xleftarrow{\sim} & (G\backslash\!\!\backslash X)_{\mathcal{U}} & \xrightarrow{\gamma} & \mathbf{B}\Gamma. \end{array} \tag{4.100}$$

---

[3] We are tacitly using here that stackification preserves finite homotopy limits so that a representative for the homotopy fiber product in topological stacks is given already by a homotopy fiber product of topological groupoids, hence already by a fiber product with a global fibration, not necessarily a local fibration.

[4] With the conventions used here, the principal bundle $\mathcal{P}$ in (4.100) carries a *right* Γ-action.

It is straightforward but instructive to unwind the definitions to see that this diagram reduces in components exactly to the traditional constructions relating principal bundles to their Čech data. A comprehensive discussion is in [SS26c, §5.1].

In fact this situation is universal in the following way: Given a Γ-principal bundle $\mathcal{P}$ and any action $\Gamma \circlearrowright F$ then there is classically the *$\mathcal{P}$-associated F-fiber bundle*

$$\mathcal{P} \otimes_\Gamma F := \big\{ (\pi, \phi) \in \mathcal{P} \times F \big\} \big/ \big(\forall_{\gamma \in \Gamma}\; (\pi \cdot \gamma, \phi) \sim (\pi, \gamma \cdot \phi) \big) , \tag{4.101}$$

and this is equivalently just the pullback of $\Gamma \backslash\!\backslash F$ (4.94) along the cocycle γ for $\mathcal{P}$ (cf. [SS26c, §5.2.2]):

$$\begin{array}{ccccc} G \backslash\!\backslash (\mathcal{P} \otimes_\Gamma F) & \xleftarrow{\;\sim\;} & G \backslash\!\backslash \widehat{(\mathcal{P} \otimes_\Gamma F)} & \longrightarrow & \Gamma \backslash\!\backslash F \\ {\scriptstyle G \backslash\!\backslash p} \big\downarrow {\scriptstyle \in \mathrm{Fib}} & & \big\downarrow & \lrcorner & {\scriptstyle p^{\Gamma \circlearrowright F}_{\mathrm{univ}}} \big\downarrow {\scriptstyle \in \mathrm{Fib}} \\ G \backslash\!\backslash X & \xleftarrow{\;\sim\;} & (G \backslash\!\backslash X)_{\mathcal{U}} & \xrightarrow{\;\gamma\;} & \mathbf{B}\Gamma . \end{array} \tag{4.102}$$

In variation of Def. 4.1.32, we set:

Definition 4.1.46 (Slice mapping stack, cf [SS26c, 4.2.66])**.** Given

- $\Gamma \circlearrowright Y$ a topological group action (Def. 4.1.7),
- $\widehat{\mathcal{X}} \xrightarrow{\sim} \mathcal{X}$ a cofibrant resolution (Def. 4.1.38),
- $\widehat{\mathcal{X}} \xrightarrow{\tau} \mathbf{B}\Gamma$ a continuous functor (the *twist*),

then the *slice mapping stack* from $\mathcal{X}$ to the homotopy quotient $\Gamma \backslash\!\backslash Y$ over $\mathbf{B}\Gamma$ (4.94) is this fiber product (4.95):

$$\begin{array}{ccc} \mathrm{Map}(\mathcal{X}, \Gamma \backslash\!\backslash Y)_{\mathbf{B}\Gamma} := \mathrm{Func}(\widehat{\mathcal{X}}, \Gamma \backslash\!\backslash Y) \; {}_{p_*}\!\!\times_{\widetilde{\tau}} * & \longrightarrow & \mathrm{Func}(\widehat{\mathcal{X}}, \Gamma \backslash\!\backslash Y) \\ \big\downarrow \quad \lrcorner & & \big\downarrow {\scriptstyle p_*} \\ * & \xrightarrow{\;\tau\;} & \mathrm{Func}(\widehat{\mathcal{X}}, \mathbf{B}\Gamma) . \end{array} \tag{4.103}$$

Following (4.75), we will more suggestively denote this also as follows:

$$\mathrm{Map}(\mathcal{X}, \Gamma \backslash\!\backslash Y)_{\mathbf{B}\Gamma} \equiv \left\{ \begin{array}{ccc} & & \Gamma \backslash\!\backslash Y \\ & \nearrow & \big\downarrow {\scriptstyle p} \\ \mathcal{X} & \xrightarrow{\;\tau\;} & \mathbf{B}\Gamma \end{array} \right\} . \tag{4.104}$$

Remark 4.1.47. A slice mapping stack of the form Def. 4.1.46 is just a topological space in that it is a topological groupoid with only identity morphisms (Ex. 4.1.10). This is because the fiber product (4.103) enforces the horizontal composition (4.54) of the transformations between slice maps with the projection map to be the identity on τ:

$$\begin{array}{ccc} & \overset{f}{\dashrightarrow} & \Gamma \backslash\!\backslash Y \\ & \Downarrow{\scriptstyle \eta} \;\; \overset{f'}{\nearrow} & \big\downarrow {\scriptstyle p} \\ \widehat{\mathcal{X}} & & \mathbf{B}\Gamma \end{array} \;=\; \widehat{\mathcal{X}} \underset{\tau}{\overset{\tau}{\rightrightarrows}} \mathbf{B}\Gamma , \quad (\text{with } \Vert \text{ between the two } \tau) \tag{4.105}$$

This means that the component morphisms of $p \cdot \eta$ all have to be the identity $\bullet \xrightarrow{\mathrm{e}} \bullet$, which here means (4.94) that already the component maps of η have to be identities, $x \xrightarrow{\mathrm{e}} x$.

Example 4.1.48. Given a G-equivariant Γ-associated F-fiber bundle according to (4.102) in Ex. 4.1.45, then the slice mapping stack into it is equivalently its space $\Gamma_X(-)^G$ of equivariant *sections* of this bundle:

$$
\begin{aligned}
\Gamma_X\big(\mathcal{P}\otimes_\Gamma F\big)^G &\simeq \left\{ \begin{array}{ccc} & & G\backslash\!\!\backslash(\mathcal{P}\otimes_\Gamma F) \\ & \nearrow & \big\downarrow {\scriptstyle G\backslash\!\!\backslash p} \\ G\backslash\!\!\backslash X & = & G\backslash\!\!\backslash X \end{array} \right\} \\
&\simeq \left\{ \begin{array}{ccc} & & \Gamma\backslash\!\!\backslash F \\ & \nearrow & \big\downarrow {\scriptstyle p^{\Gamma \circlearrowright F}_{\mathrm{univ}}} \\ G\backslash\!\!\backslash X & \xrightarrow{\gamma} & \mathbf{B}\Gamma \end{array} \right\}.
\end{aligned}
\tag{4.106}
$$

Here the first equivalence follows because, by the slicing, dashed functors out of any (Čech) resolution actually have to factor through $G\backslash\!\!\backslash X$, while the second follows either by direct inspection or abstractly by the universal property of the homotopy pullback in (4.102).

Example 4.1.49 (cf. [SS26c, Prop. 4.2.77]). For a pair of G-spaces $G \circlearrowright X$ and $G \circlearrowright Y$, the slice mapping stack (Def. 4.1.46) between their homotopy quotients over $\mathbf{B}G$ (via Ex. 4.1.44), which is a topological space by Rem. 4.1.47, is naturally homeomorphic to the equivariant mapping space (4.14) between X and Y:

$$
\begin{array}{ccc}
\mathrm{Map}(X,Y)^G & \xrightarrow{\;\sim\;} & \mathrm{Map}\big(G\backslash\!\!\backslash X, G\backslash\!\!\backslash Y\big)_{\mathbf{B}G} \\
\left(\begin{array}{c} \overset{G}{\circlearrowright} \quad\; \overset{G}{\circlearrowright} \\ X \xrightarrow{\;f\;} Y \end{array} \right) & \longmapsto & \left(\begin{array}{ccc} G\backslash\!\!\backslash X & \xrightarrow{G\backslash\!\!\backslash f} & G\backslash\!\!\backslash Y \\ & \searrow \; \mathbf{B}G \; \swarrow & \end{array} \right).
\end{array}
\tag{4.107}
$$

This follows because the slicing over $\mathbf{B}G$ entails that all Čech morphisms in any resolution $\widehat{G\backslash\!\!\backslash X}$ of $G\backslash\!\!\backslash X$ have to map to identity morphisms in $G\backslash\!\!\backslash Y$, whence all Morita maps come from plain functors $G\backslash\!\!\backslash X \to G\backslash\!\!\backslash Y$.

In particular:

- An *equivariant homotopy* (4.21) is equivalently a homotopy taking values in the slice mapping stack on the right of (4.107).

$$
\begin{array}{ccc}
\{0\} & \xrightarrow{\tilde f} & \\
\downarrow & & \\
[0,1] & \xrightarrow{\tilde\eta} & \mathrm{Map}(X,Y)^G \\
\uparrow & & \\
\{1\} & \xrightarrow{\tilde g} &
\end{array}
\quad\Leftrightarrow\quad
\begin{array}{ccc}
\{0\} & \xrightarrow{\widetilde{G\backslash\!\!\backslash f}} & \\
\downarrow & & \\
[0,1] & \xrightarrow{\tilde\eta} & \mathrm{Map}\big(G\backslash\!\!\backslash X, G\backslash\!\!\backslash Y\big)_{\mathbf{B}G} \\
\uparrow & & \\
\{1\} & \xrightarrow{\widetilde{G\backslash\!\!\backslash g}} &
\end{array}.
\tag{4.108}
$$

- For

$$
G \circlearrowright X \text{ a trivial action} \quad\Leftrightarrow\quad G\backslash\!\!\backslash X = X \times \mathbf{B}G\,,
\tag{4.109}
$$

so that G-equivariant maps from X to Y are equivalently (cf. [SS26c, Ex. 1.8]) maps from X into the fixed locus $Y^G \subset Y$ (4.15), it follows that

$$\left\{ \begin{array}{ccc} & & G\backslash\!\!\backslash Y \\ & \nearrow & \downarrow \\ X \times \mathbf{B}G & \xrightarrow{\mathrm{pr}_2} & \mathbf{B}G \end{array} \right\} \simeq \left\{ X \dashrightarrow Y^G \right\}. \tag{4.110}$$

In mild but crucial generalization of Ex. 4.1.49, we also have:

Example 4.1.50. Given a homomorphism of topological groups $\phi : G' \longrightarrow G$, and actions $G' \circlearrowright X$ and $G \circlearrowright Y$, then the slice mapping stack (Def. 4.1.46) between their homotopy quotients sliced over $\mathbf{B}G$ (as in Ex. 4.1.44, for $G'\backslash\!\!\backslash X$ via $\mathbf{B}\phi$) is a topological space naturally homeomorphic to the G'-equivariant maps from X to Y, the latter with its G'-action induced via ϕ (4.99):

$$\left\{ \begin{array}{ccc} G'\backslash\!\!\backslash X & \dashrightarrow & G\backslash\!\!\backslash X \\ \downarrow & & \downarrow \\ \mathbf{B}G' & \xrightarrow{\mathbf{B}\phi} & \mathbf{B}G \end{array} \right\} \simeq \left\{ \begin{array}{cccc} G'\backslash\!\!\backslash X & \dashrightarrow & G'\backslash\!\!\backslash Y \longrightarrow & G\backslash\!\!\backslash X \\ & \searrow \quad \swarrow & & \downarrow \\ & \mathbf{B}G' & \xrightarrow{\mathbf{B}\phi} & \mathbf{B}G \end{array} \right\}. \tag{4.111}$$

As indicated on the right, this follows from Ex. 4.1.49 by the universal property (4.3) of the pullback.

In view of (4.80), we thereby obtain a natural reformulation of *equivariant cohomology* as a form of twisted cohomology of topological stacks (and it is this reformulation which naturally generalizes to a notion of orbifold cohomology, see Thm. 4.2.6 in Def. 4.2.5):

Proposition 4.1.51. *For a pair of topological G-actions $G \circlearrowright X$ and $G \circlearrowright A$, the G-equivariant cohomology of X with coefficients in A – in the style as on the left hand side of* (4.80) —, *is:*

$$H_G(X, A) := \pi_0 \left\{ \begin{array}{ccc} & & G\backslash\!\!\backslash A \\ & \nearrow & \downarrow \\ G\backslash\!\!\backslash X & \longrightarrow & \mathbf{B}G \end{array} \right\} \simeq \pi_0 \left\{ \begin{array}{ccc} {}^{G}\!\!\circlearrowright & & {}^{G}\!\!\circlearrowright \\ X & \dashrightarrow & A \end{array} \right\}. \tag{4.112}$$

Proof. By Ex. 4.1.49. □

More generally, in twisted generalization of (4.80) in Ex. 4.1.33, we have:

Definition 4.1.52. In the situation of Def. 4.1.46, we say that the connected components of the space which is, by Rem. 4.1.47, the slice mapping stack of Def. 4.1.46, are the *τ-twisted cohomology* of $\mathcal{X}$ with coefficients in Y:

$$H^{\tau}(\mathcal{X}; Y) := \Big[\textstyle\int \mathrm{Map}(\mathcal{X}, \Gamma\backslash\!\!\backslash Y)_{\mathbf{B}\Gamma} \Big]_0 \underset{(4.45)}{=} \pi_0 \left\{ \begin{array}{ccc} & & \Gamma\backslash\!\!\backslash Y \\ & \nearrow & \downarrow p \\ \mathcal{X} & \xrightarrow{\tau} & \mathbf{B}\Gamma \end{array} \right\}. \tag{4.113}$$

Remark 4.1.53. When the twist τ in (4.113) is trivial, in that it factors through the point, then this twisted cohomology (4.113) reduces, via (4.97) and (4.111), to the ordinary nonabelian cohomology as on the left of (4.80):

$$\left\{ \begin{array}{ccccc} & & & & \Gamma\backslash\!\!\backslash Y \\ & & \nearrow & & \downarrow{\scriptstyle p} \\ \mathcal{X} & \longrightarrow & * & \longrightarrow & \mathbf{B}\Gamma \\ & \llcorner & \underset{\tau}{\longrightarrow} & \lrcorner & \end{array} \right\} \simeq \left\{ \begin{array}{ccccc} & & Y & \longrightarrow & \Gamma\backslash\!\!\backslash Y \\ & \nearrow & \downarrow & \lrcorner & \downarrow{\scriptstyle p} \\ \mathcal{X} & \longrightarrow & * & \longrightarrow & \mathbf{B}\Gamma \\ & \llcorner & \underset{\tau}{\longrightarrow} & \lrcorner & \end{array} \right\} \simeq \big\{ \mathcal{X} \dashrightarrow Y \big\} . \tag{4.114}$$

In contrast, to obtain the twisted generalization of the right-hand side of (4.80) requires considerably more structure, namely passage, via a *twisted Elmendorf theorem* ([SS26c, Thm. 6.2.3]), to a perspective (cf. [SS26c, §6.2.1]) where topological groupoids are probed not just by Cartesian plots (as indicated in Ex. 4.1.17) but also by deloopings $\mathbf{B}K$ of all finite groups K. We discuss this in § 9.2.

Lemma 4.1.54. *A (Morita) equivalence* (4.63) *of topological groupoids induces*

- *a bijection between their equivalence classes of twists* (Def. 4.1.46),
- *an isomorphism in the correspondingly twisted cohomology* (Def. 4.1.52)*:*

$$\begin{array}{ccc} \mathcal{X}' & \xrightarrow[\sim]{F} & \mathcal{X} \\ & {}_{\tau'}\searrow \quad \swarrow_{\tau} & \\ & \mathbf{B}\Gamma & \end{array} \quad \Rightarrow \quad H^{\tau'}(\mathcal{X}'; Y) \xrightarrow{\sim} H^{\tau}(\mathcal{X}; Y) . \tag{4.115}$$

Proof. The first statement follows by Lem. 4.1.40. With this, the second follows on general abstract grounds discussed in Thm. 9.1.19 along the following lines: In this commuting diagram,

$$\begin{array}{ccccc} \mathrm{Map}(\mathcal{X}, \Gamma\backslash\!\!\backslash Y)_{\mathbf{B}\Gamma} & \longrightarrow & \mathrm{Map}(\mathcal{X}, \Gamma\backslash\!\!\backslash Y) & \xrightarrow[F^*]{\sim} & \mathrm{Map}(\mathcal{X}', \Gamma\backslash\!\!\backslash Y) \\ \downarrow & \lrcorner & \downarrow{\scriptstyle p_*} & & \downarrow{\scriptstyle p_*} \\ * & \xrightarrow{\;\;\widetilde{\tau}\;\;} & \mathrm{Map}(\mathcal{X}, \mathbf{B}\Gamma) & \xrightarrow[F^*]{\sim} & \mathrm{Map}(\mathcal{X}', \mathbf{B}\Gamma) , \\ \llcorner & & \underset{\widetilde{\tau'}}{\longrightarrow} & & \lrcorner \end{array} \tag{4.116}$$

the left square is a homotopy pullback by definition (4.103), and the right square is because both its horizontal maps are equivalences, by Lem. 4.1.40. Therefore, the total rectangle is a homotopy pullback by the homotopy pasting law. This implies an equivalence between $\mathrm{Map}(\mathcal{X}, \Gamma\backslash\!\!\backslash Y)_{\mathbf{B}\Gamma}$ and $\mathrm{Map}(\mathcal{X}', \Gamma\backslash\!\!\backslash Y)_{\mathbf{B}\Gamma}$, and that implies the claim by (4.64). □

4.2 Orbifold Cohomology

We have seen in § 4.1.8 that – when seen from the perspective of the geometric homotopy theory of topological groupoids – *G-equivariant cohomology* (Prop. 4.1.51) is a form of *twisted cohomology* (Def. 4.1.52), namely with coefficient fibration a homotopy quotient projection $G\backslash\!\!\backslash Y \to \mathbf{B}G$ and twist *also* such a projection, $G\backslash\!\!\backslash X \to \mathbf{B}G$. If we here just allow the twist to be a more general map of topological groupoids $\widehat{\mathcal{X}} \xrightarrow{\tau} \mathbf{B}G$, then we immediately have a notion of twisted cohomology on topological groupoids, which restricts (by Ex. 4.1.50) to G'-equivariant cohomology on all subgroupoids of $\mathcal{X}$ that look like global homotopy quotients!

In particular, this immediately gives a good notion of twisted cohomology of *orbifolds* (see Def. 4.2.1), namely of topological groupoids that admit an open cover by Cartesian homotopy quotients $G' /\!\!/ \mathbb{R}^n$, for varying finite groups G'.

4.2.1 Orbifolds as Groupoids

With the language of topological and Lie groupoids (Def. 4.1.8 and Rem. 4.1.9) in hand (§ 4.1), there is a beautifully transparent and powerful definition of *orbifolds* (recall Fig. 1.1). These are the *proper étale* groupoids (cf. Def. 4.2.1), meaning essentially that orbifolds are those groupoids that *locally* look like homotopy quotients $G \backslash\!\!\backslash \mathbb{D}^n$ (Def. 4.1.14) of finite groups G acting continuously/smoothly on open disks $\mathbb{D}^n$, in generalization of how a manifold locally looks like a plain open disk $\mathbb{D}^n$. This observation is due to [MP97, pp. 15][MP99, §4][Mo02] with further details spelled out in [Le08][Am12][CPRST14].

Definition 4.2.1. A (topological/smooth) *orbifold* is a topological/Lie groupoid $\mathcal{X}$ (Def. 4.1.8 and Rem. 4.1.9) such that:

- *Étale*:
  - the spaces of objects and morphisms are equipped with the structure of topological/smooth manifolds,
  - the source or target map, and with it, necessarily all the structure maps (4.28) are local homeomorphisms/diffeomorphisms.
- *Proper*: the combined source/target map (4.31) is *proper*, which with the previous item means that this map preserves closed subsets and has compact fibers.

The space $[\mathcal{X}]_0$ of isomorphism classes (4.30) of the groupoid $\mathcal{X}$ is called the underlying *coarse space* of the orbifold. Conversely, given a (paracompact Hausdorff) topological space X then an *orbifold structure* on X is a choice of proper étale groupoid $\mathcal{X}$ and of a homeomorphism $[\mathcal{X}]_0 \xrightarrow{\sim} X$. Proper étale groupoids which are equivalent (Def. 4.1.23) have homeomorphic coarse spaces (4.64) and are regarded as presenting the same orbifold structure.

Remark 4.2.2. Given an orbifold $\mathcal{X}$ (Def. 4.2.1) then for all $x \in \mathrm{Obj}(\mathcal{X})$, the isotropy groups $\mathcal{X}_x$ (4.32) are discrete by étaleness and compact by properness of (s,t), hence are *finite groups*. Furthermore, properness of (s,t) together with the manifoldness of $\mathrm{Obj}(\mathcal{X})$ implies that the space of isomorphism classes $[\mathcal{X}]_0$ (4.30) is paracompact and Hausdorff.

Example 4.2.3 (Global homotopy quotient orbifolds). Consider X a topological/smooth manifold, G a Lie group, and $G \circlearrowright X$ a continuous/smooth action (Def. 4.1.7). Then the homotopy quotient $G \backslash\!\!\backslash X$ (4.39) is an orbifold (Def. 4.2.1) under the following conditions of increasing generality:

- G is finite:
  in this case $G \backslash\!\!\backslash X$ is called a *very good orbifold*,

- G is discrete acting properly discontinuously with finite stabilizers: in this case $G \backslash\!\backslash X$ is called a *good orbifold*,
- G is compact Lie, acting properly with finite stabilizers: $G \backslash\!\backslash X$ of this form but not of the previous forms are called *bad orbifolds*.

Example 4.2.4 (Spindle orbifold). For $n_+, n_1 \in \mathbb{N}_{\geq 1}$ a pair of positive integers which are coprime, $\gcd(n_+, n_-) = 1$, then the (n_+, n_-)*-spindle orbifold* (cf. Ex. 4.2.4) is the result of gluing the conical global quotients $\mathbb{Z}_{n_\pm} \backslash\!\backslash \mathbb{D}^2_1$ (Ex. 4.2.3), with respect to $n_\pm$-fold rotation action on the disc, along a joint open annulus at their smooth ends. When $n_\mp = 1$ but $n_\pm > 1$ then the spindle reduces to what is called the *teardrop orbifold*. When $n_+ = 1 = n_i$ then the spindle orbifold reduces to the 2-sphere manifold.

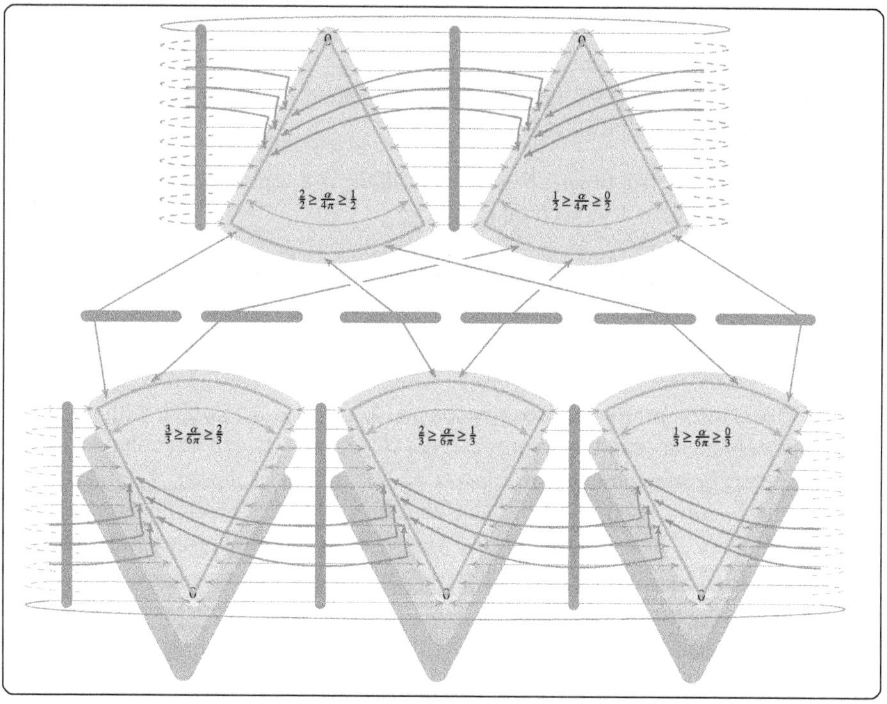

Figure 4.4 Indicated is a proper étale Lie groupoid which represents the $(2, 3)$-spindle orbifold (Ex. 4.2.4) and is Dugger-cofibrant (Def. 4.1.38). Shown in light gray is the space of objects, while the darker shades of gray are summands of the space of non-invertible morphisms. On the bottom we have the equivariant Čech groupoid resolution (Ex. 4.1.34) of the action groupoid $\mathbb{Z}_3 \backslash\!\backslash \mathbb{D}^2_{1+\epsilon}$ (from Fig. 4.3), and on the top the analogous equivariant Čech groupoid for $\mathbb{Z}_2 \backslash\!\backslash \mathbb{D}^2_{1+\epsilon}$. Indicated in the middle are (the component spaces of) further gluing morphisms which identify all the outer strips with each other and thereby glue the bottom conical orbifold of order 3 to the top conical orbifold of order 2 to form the $(2, 3)$-spindle. To note that the composition of any two composable such gluing morphisms gives a translation morphism by either group action.

We indicate two explicit proper étale groupoids representing the spindle orbifold:

- A minimal model for the (n_+, n_-)-spindle has

$$\begin{aligned} \mathrm{Obj} :=& \mathbb{C}_+ \sqcup \mathbb{C}_- \\ \mathrm{Mor} :=& \big(\mathbb{C}_+ \times \mathbb{Z}_{n_+} \;\sqcup\; \mathbb{C}_+ \times \mathbb{Z}_{n_+}\big) \\ &\sqcup \Big(\big(\mathbb{C}^\times_{+-} \times \mathbb{Z}_{n_+} \times \mathbb{Z}_{n_-}\big) \sqcup \big(\mathbb{C}^\times_{-+} \times \mathbb{Z}_{n_-} \times \mathbb{Z}_{n_+}\big)\Big) \end{aligned} \tag{4.117}$$

(where $\mathbb{C}^\times := \mathbb{C} \setminus \{0\}$, and the subscripts on $\mathbb{C}$ are just to index the disjoint summands) and, with the abbreviation

$$q_\pm := e^{2\pi \mathrm{i}/n_\pm} \tag{4.118}$$

the source/target maps internal and weightless gluing morphisms are

$$\begin{array}{ccc} \mathrm{Mor} & \xrightarrow{\quad (s,t) \quad} & \mathrm{Obj}^2 \\ \big((z, \pm), [k_\pm]\big) & \longmapsto & \big((z, \pm), (z q_\pm^{k_\pm}, \pm)\big) \\ \big((v, \pm\mp), [0], [0]\big) & \longmapsto & \big((v^{n_\pm}, \pm), (v^{-n_\mp}, \mp)\big) . \end{array} \tag{4.119}$$

Finally, the general gluing morphism is the unique composite

$$\begin{aligned} &\big((v, \pm\mp), [k_\pm], [k_\mp]\big) \\ \equiv\;& \big((v^{-n_\mp}, \mp), [k_\mp]\big) \circ \big((v, \pm, \mp), [0], [0]\big) \circ \big((v^{n_\pm} \cdot q^{-k_\pm}, \pm), [k_\pm]\big) , \end{aligned} \tag{4.120}$$

which defines either side by the other (meaning that the gluing morphisms are a groupoid torsor from either side over these action groupoids), and its source/target map is

$$(s, t) \;:\; \big((v, \pm\mp), [k_\pm], [k_\mp]\big) \mapsto \big((v^{-n_\mp} \cdot q^{+k_\mp}, \pm), (v^{n_\pm} \cdot q^{-k_\pm}, \mp)\big) . \tag{4.121}$$

In the special case of $n_\pm = 1$ this definition reduces to the Čech groupoid (Ex. 4.1.28) corresponding to the standard open cover of the Riemann sphere by two copies $\mathbb{C}_\pm$ with double overlap $\mathbb{C}^\times \subset \mathbb{C}_+$ identified inside the second copy via $z \mapsto z^{-1}$.

- For the purpose of computing mapping stacks (§ 4.1.7) out of the spindle orbifold, we want a presentation by a Dugger-cofibrant groupoid (Def. 4.1.38). But the above minimal model fails to be Dugger-cofibrant since the factor $\mathbb{C}^\times$ of the space of gluing morphisms is (not contractible and hence) not diffeomorphic to $\mathbb{R}^2$.

  One obtains a Dugger-cofibrant spindle by considering the disjoint union of the Dugger-cofibrant Čech groupoids for the two cones $\mathbb{Z}_{n_\pm} \backslash \mathbb{D}^2_{1+\epsilon}$ from Fig. 4.3 and gluing the collars of all their segments to each other by a summand-wise contractible space of gluing morphisms.

  This construction is indicated in Fig. 4.4 for the case $(n_+, n_-) = (3, 2)$.

4.2.2 Twisted Orbifold Cohomology

We saw in § 4.1.8 a fairly general definition of (nonabelian) twisted cohomology of topological groupoids, and in § 4.2.1 that orbifolds are a special case of topological groupoids. Therefore, we immediately obtain a notion of *twisted orbifold cohomology*, inheriting all the good properties that we saw hold in general for twisted nonabelian cohomology formulated via mapping stacks.

Therefore, in discussing twisted orbifold cohomology here we are essentially reduced to summarizing previous statements on the twisted nonabelian cohomology of general topological groupoids – which may serve as a concise review of the above discussion and to highlight how phenomena discussed in the literature by other means are elegantly reproduced by our general approach to cohomology via mapping stacks.

This also means that, at this level of plain (twisted nonabelian) cohomology, the actual *geometry* of orbifolds – their smooth structure and possibly further geometric structure such as Riemannian, holomorphic, conformal structure, ... – plays no role. The refinement to *geometric orbifold cohomology* which *is* sensitive to further geometric structure is discussed in chapter 10.

First, in the specialization of Def. 4.1.52 we have:

Definition 4.2.5 (Twisted orbifold cohomology)**.** For $\Gamma \circlearrowright Y$ a topological group action (Def. 4.1.7) on a topological space Y (to play the role of the *classifying space* for the orbifold cohomology theory), then the *twisted orbifold cohomology* with coefficients in Y, of an orbifold $\mathcal{X}$ equipped with a map (the twist) $\mathcal{X} \xrightarrow{\tau} \mathbf{B}\Gamma$, is the τ-twisted cohomology of the orbifold groupoid $\mathcal{X}$ according to Def. 4.1.52, hence the connected components of the slice mapping stack (Def. 4.1.46, which is an ordinary space, by Rem. 4.1.47):

$$H^{\tau}(\mathcal{X};Y) := \Big[\textstyle\int \mathrm{Map}\big(\widehat{\mathcal{X}},\, \Gamma\backslash\!\!\backslash Y\big)_{\mathbf{B}\Gamma}\Big]_0 = \pi_0 \left\{ \begin{array}{ccc} & & \Gamma\backslash\!\!\backslash Y \\ & \nearrow & \downarrow p \\ \mathcal{X} & \xrightarrow{\tau} & \mathbf{B}\Gamma \end{array} \right\}. \tag{4.122}$$

Theorem 4.2.6. *Twisted orbifold cohomology* (Def. 4.2.5) *has the following properties:*

(i) *It is an invariant of the (Morita) equivalence class of the orbifold.*

In particular, if $G\backslash\!\!\backslash X$ and $G'\backslash\!\!\backslash X'$ are two global quotient presentations (Ex. 4.2.3) *of the same orbifold class, then their orbifold cohomology agrees, and their twisted cohomology agrees for corresponding twists.*

(ii) *If on a global quotient presentation* (Ex. 4.2.3) *the twist is just the canonical projection $G\backslash\!\!\backslash X \xrightarrow{p} \mathbf{B}\Gamma$, then twisted orbifold cohomology reduces to the G-equivariant cohomology of X.*

Proof. The first statement is Lem. 4.1.54, the second is Prop. 4.1.51. □

Remark 4.2.7. The basic properties in Thm. 4.2.6 have a somewhat more convoluted history in the literature on orbifold cohomology by other means (reviewed in § 1.1.4) than the mapping stack approach used here, for more discussion of which see also [SS26c, Rem. 6.2.4].

5

Orbifold K-Theory

We describe, in § 5.1, an elegant model of *twisted orbifold K-theory* based on the general geometric homotopy formulation of twisted orbifold cohomology (from Def. 4.2.5) via mapping stacks between topological groupoids.

Then we explicitly construct (§ 5.2) in this model the low-dimensional equivariant orientation character map, a specialization of the general discussion in § 2.2. Here for orbifold K-theory, this turns out to be a glorified incarnation of the equivariant trivialization of the pullback of the tautological line bundle along the Hopf fibration, which we review in § 5.1.5.

5.1 Orbi K-Cohomology

We recall and expand on the elegant model of *twisted orbifold K-theory* from [SS23b, §2.2][SS26c, Ex. 6.2.5], complementing traditionally more component-based definitions in the literature (cf. [AR03][LU04a][TXLG04][JKK05][ALR07, Ch 3][FHT11][Gom17]). Our model is based on the picture (surveyed in chapter 2 and Fig. 2.1, reviewed in §§ 4.1.7 and 4.1.8) of generalized cohomology as being about *homotopy classes of classifying maps*: in the present case, between topological groupoids/stacks (reviewed in § 4.1) from the domain orbifold $\mathcal{X}$ to the universal $\mathrm{PU}(\mathcal{H})$-structured Fredholm space bundle that classifies twisted equivariant K-theory according to [ASe04], but:

- **(i)** with the domain $\mathcal{X}$ regarded as the étale groupoid incarnation of orbifolds, cf. Def. 4.2.1,
- **(ii)** with the classifying fibration promoted (5.77) from a topological bundle over the ordinary *classifying space* $B\mathrm{PU}(\mathcal{H})$ to a stacky universal fiber bundle (Ex. 4.1.45) over the topological *moduli stack* $\mathbf{B}\mathrm{PU}(\mathcal{H}) \simeq * /\!\!/ \mathrm{PU}(\mathcal{H})$ (cf. Exs. 4.1.12 and 4.1.34), in fact over its further extension by PCT symmetries (§5.1.2) to the moduli stack $\mathbf{B}\mathrm{QS}$ of *quantum symmetries* (§ 5.1.2).

This construction utilizes that, as homotopy types, the space of Fredholm operators is a classifying space of complex K-theory (Prop. 5.1.16) but as an actual topological space, it moreover carries a strict topological group action of quantum symmetries $\mathrm{QS} \twoheadrightarrow \mathrm{PU}(\mathcal{H})$ $\mathrm{QS} \circlearrowright \mathrm{Fred}_{\mathrm{gr}}$, which implements (Prop. 5.1.20, following [SS23b,

DOI: 10.1201/9781003675969-5

§2.2] inspired by [FM13]) the grading and the flavors of topological K-theory (KU, KO, KR).

We see in § 5.2 how the model lends itself neatly to the construction of the four/ten-dimensional equivariant $\mathbb{C}/\mathbb{H}$-orientations of KU (Fig. 2.6), where we obtain the relevant Fredholm operators and unitary operators essentially "tautologically" from the $\mathbb{C}/\mathbb{H}$-operator algebraic construction of the $\mathbb{C}/\mathbb{H}$-Hopf fibration (Def. 5.2.15).

5.1.1 Hilbert Space and PCT

Notation 5.1.1 (The Hilbert space)**.** In all of the following, we consider:

(i) A countably infinite-dimensional complex Hilbert space

$$\mathcal{H} \in \mathrm{Hilb}_{\mathbb{C}}, \tag{5.1}$$

as such unique up to isomorphism; in particular, isomorphic to the direct sum, as Hilbert spaces, of countably many copies of the standard k-dimensional complex vector space:

$$\forall k \in \mathbb{N}_{\geq 0} \;:\; \mathcal{H} \simeq \bigoplus_{n \in \mathbb{N}} \mathbb{C}^k . \tag{5.2}$$

(ii) A *real structure* on $\mathcal{H}$ (5.1), hence a *complex anti-linear* involution and as such self-adjoint (cf. [Uhl16, Def. 3.1])

$$\begin{array}{ccc} \mathcal{H} & \xrightarrow{T} & \mathcal{H}, \\ \psi & \mapsto & T\psi \\ \mathrm{i}\psi & \mapsto & -\mathrm{i}T\psi \end{array} \qquad \begin{array}{l} T^2 = \mathrm{id} \\ \langle -, T- \rangle = \overline{\langle T-, - \rangle}, \end{array} \tag{5.3}$$

which under any of the above identifications (5.2) we may take to be given by ordinary complex conjugation on the $\mathbb{C}$ summands.

(iii) A *quaternionic structure* on $\mathcal{H}$ (5.1), hence a *complex anti-linear* endomorphism, and as such self-adjoint (cf. [Uhl16, Def. 3.1]) and squaring to minus the identity:

$$\begin{array}{ccc} \mathcal{H} & \xrightarrow{J} & \mathcal{H}, \\ \psi & \mapsto & J\psi \\ \mathrm{i}\psi & \mapsto & -\mathrm{i}J\psi \end{array} \qquad \begin{array}{l} J^2 = -\mathrm{id} \\ \langle -, J- \rangle = -\overline{\langle J-, - \rangle}, \end{array} \tag{5.4}$$

which under any of the above identifications (5.2) for even k we may take to be given on the $\mathbb{C}^2$ summands by

$$\begin{array}{ccc} \mathbb{C}^2 & \longrightarrow & \mathbb{C}^2 \\ \begin{pmatrix} z_1 \\ z_2 \end{pmatrix} & \mapsto & \begin{pmatrix} -z_2^* \\ z_1^* \end{pmatrix}. \end{array} \tag{5.5}$$

(iv) The group generated by T (5.3), to be denoted

$$\mathbb{Z}_2^{(t)} := \{\mathrm{id}, T\}. \tag{5.6}$$

(v) The fixed locus of this involution (5.3) and (5.6), which is the countably infinite-dimensional *real* Hilbert space, to be denoted

$$\mathcal{H}^{\mathbb{Z}_2^{(t)}} \in \mathrm{Hilb}_{\mathbb{R}} . \tag{5.7}$$

(vi) The $\mathbb{Z}_2$-*graded Hilbert space* whose homogeneous summands are both (5.1), to be denoted

$$\mathcal{H}_{\mathrm{gr}} := \mathcal{H} \ominus \mathcal{H} , \tag{5.8}$$

which inherits a real structure via (5.3).

(vii) The *grading involution* on $\mathcal{H}_{\mathrm{gr}}$ (5.8), to be denoted:

$$\begin{array}{ccc} \mathcal{H}_{\mathrm{gr}} & \xrightarrow{P} & \mathcal{H}_{\mathrm{gr}} \\ \begin{pmatrix} \psi_+ \\ \psi_- \end{pmatrix} & \mapsto & \begin{pmatrix} \psi_- \\ \psi_+ \end{pmatrix}, \end{array} \tag{5.9}$$

which satisfies

$$P^2 = \mathrm{id}, \quad P^\dagger = P, \quad P \circ T = T \circ P . \tag{5.10}$$

(viii) The group generated by

$$C := TP , \tag{5.11}$$

to be denoted

$$\mathbb{Z}_2^{(c)} := \{\mathrm{id}, C\} . \tag{5.12}$$

(ix) The diagonal subgroup

$$\begin{array}{ccccc} \mathbb{Z}_2^{(p)} & := & \{\mathrm{id}, P\} & \hookrightarrow & \mathbb{Z}_2^T \times \mathbb{Z}_2^C \\ & & P & \mapsto & TC , \end{array} \tag{5.13}$$

(x) so that we have these two isomorphic incarnations of the *PCT-group*:

$$\begin{array}{ccc} \mathbb{Z}_2^{(t)} \times \mathbb{Z}_2^{(p)} & \xrightarrow{\sim} & \mathbb{Z}_2^{(t)} \times \mathbb{Z}_2^{(p)} \\ T & \mapsto & T \\ P & \mapsto & CT . \end{array} \tag{5.14}$$

Definition 5.1.2 (Projective unitary groups).

(i) We write

$$\mathrm{U}(\mathcal{H}) \in \mathrm{Grp}(\mathrm{Top}) \tag{5.15}$$

for the topological group of unitary linear operators on $\mathcal{H}$ (5.1), equipped with the norm topology.[1] This is famously contractible, by Kuiper's theorem (cf. [nL-Ku]):

$$\mathrm{U}(\mathcal{H}) \underset{\mathrm{hmtpy}}{\simeq} * . \tag{5.16}$$

(ii) We write

$$\mathrm{PU}(\mathcal{H}) := \mathrm{U}(\mathcal{H})/\mathrm{U}(1) \quad \in \mathrm{Grp}(\mathrm{Top}) \tag{5.17}$$

for the topological quotient group of (5.15) by its subgroup of operators acting by multiplication with a complex number of unit norm. The quotient coprojection

[1] More ambitiously (cf. [ASe04]), one equips $\mathrm{U}(\mathcal{H})$ (5.15) with the strong/weak operator topology or compact-open topology, which are all equal here but strictly coarser than the norm topology, see [SS26c, Ex. 2.3.19] for further pointers.

is a locally trivial U(1)-principal bundle [Sim70, Thm. 1]:

$$\begin{array}{ccc} \mathrm{U}(1) & \hookrightarrow & \mathrm{U}(\mathcal{H}) \\ & & \downarrow \\ & & \mathrm{PU}(\mathcal{H})\,. \end{array} \tag{5.18}$$

(iii) The group $\mathbb{Z}_2^{(t)}$ (5.6) acts compatibly on these groups:

$$\overset{\mathbb{Z}_2^{(t)}}{\circlearrowright}\mathrm{U}(1) \hookrightarrow \overset{\mathbb{Z}_2^{(t)}}{\circlearrowright}\mathrm{U}(\mathcal{H}) \twoheadrightarrow \overset{\mathbb{Z}_2^{(t)}}{\circlearrowright}\mathrm{PU}(\mathcal{H})\,. \tag{5.19}$$

(iv) The fixed locus of this action is the group of orthogonal operators, equipped with its operator topology, on the *real* Hilbert space $\mathcal{H}^{\mathbb{Z}_2^{(t)}}$ (5.7):

$$\mathrm{O}\big(\mathcal{H}^{\mathbb{Z}_2^{(t)}}\big) \simeq \big(\mathrm{U}(\mathcal{H})\big)^{\mathbb{Z}_2^{(t)}}\,. \tag{5.20}$$

(v) In turn, the quotient of the latter by the subgroup of operators acting by multiplication with real units is the infinite *projective orthogonal group* [Ro89, MMS03, §3]:

$$\mathrm{PO}\big(\mathcal{H}^{\mathbb{Z}_2^{(t)}}\big) = \big(\mathrm{PU}(\mathcal{H})\big)^{\mathbb{Z}_2^{(t)}} = \mathrm{O}\big(\mathcal{H}^{\mathbb{Z}_2^{(t)}}\big)/\{\pm 1\} \quad \in \mathrm{Grp}(\mathrm{Top})\,. \tag{5.21}$$

5.1.2 Quantum Symmetries

Remark 5.1.3. We have an isomorphism

$$\begin{array}{ccc} \mathrm{U}(\mathcal{H}) \rtimes \mathbb{Z}_2^{(t)} & \xrightarrow{\sim} & \mathrm{U}(\mathcal{H}) \sqcup \mathrm{U}_{\mathrm{anti}}(\mathcal{H}) \\ (U, \mathrm{id}) & \mapsto & U \\ (U, T) & \mapsto & U \circ T \end{array} \tag{5.22}$$

between the semidirect product of the unitary group (5.15) with $\mathbb{Z}_2^{(t)}$ (5.6) and the group of *unitary or anti-unitary* maps (cf. [Uhl16, p. 29-30]); and analogously for the projective group (5.17):

$$\begin{array}{ccc} \mathrm{PU}(\mathcal{H}) \rtimes \mathbb{Z}_2^{(t)} & \xrightarrow{\sim} & \mathrm{PU}(\mathcal{H}) \sqcup \mathrm{PU}_{\mathrm{anti}}(\mathcal{H}) \\ ([U], \mathrm{id}) & \mapsto & [U] \\ ([U], T) & \mapsto & [U \circ T]\,. \end{array} \tag{5.23}$$

This is the group of *quantum symmetries* according to Wigner's theorem (cf. [nL-WT]), in which context the operator T (5.3) is interpreted as *time-reversal symmetry*.

We next enlarge this group by what one may think of as *particle/anti-particle* symmetry (cf. [SS23b, Fact 2.3]) or "charge conjugation".

Definition 5.1.4 (Graded projective unitary group)**.**

(i) We write

$$\begin{aligned} \mathrm{U}(\mathcal{H})^{\mathrm{gr}} &:= \big(\mathrm{U}(\mathcal{H}) \times \mathrm{U}(\mathcal{H})\big) \rtimes \mathbb{Z}_2^{(p)} \\ &= \left\{ \begin{pmatrix} U_{++} & 0 \\ 0 & U_{--} \end{pmatrix}, \begin{pmatrix} 0 & U_{+-} \\ U_{-+} & 0 \end{pmatrix} \,\middle|\, U_{\bullet,\bullet} \in \mathrm{U}(\mathcal{H}) \right\} \subset \mathrm{U}(\mathcal{H}_{\mathrm{gr}}) \end{aligned} \tag{5.24}$$

for the *graded unitary group* acting in homogeneous degrees on the graded Hilbert space $\mathcal{H}_{\mathrm{gr}}$ (5.8).

(ii) The *graded projective unitary group* $\mathrm{PU}(\mathcal{H})^{\mathrm{gr}}$ (cf. [MP88, Prop. 2.2][CW08, p. 5]) is the quotient of this graded unitary group (5.24) by the diagonal subgroup $\mathrm{U}(1) \hookrightarrow \mathrm{U}(\mathcal{H}) \hookrightarrow \mathrm{U}(\mathcal{H})^{\mathrm{gr}}$, and the $\mathbb{Z}_2^{(t)}$-action (5.19) evidently extends to an automorphism action on all these graded groups:

$$\begin{array}{ccccc} \circlearrowright \mathbb{Z}_2^{(t)} & & \circlearrowright \mathbb{Z}_2^{(t)} & & \circlearrowright \mathbb{Z}_2^{(t)} \\ \mathrm{U}(1) & \hookrightarrow & \mathrm{U}(\mathcal{H})^{\mathrm{gr}} & \twoheadrightarrow & \mathrm{PU}(\mathcal{H})^{\mathrm{gr}}\,. \end{array} \tag{5.25}$$

Notation 5.1.5 (Graded quantum symmetries)**.** The further semidirect product $\mathrm{PU}(\mathcal{H})^{\mathrm{gr}} \rtimes \mathbb{Z}_2^{(t)}$ of the graded projective unitary group (5.25) with $\mathbb{Z}_2^{(t)}$ (5.6) plays a key role in the following, to be called the *group of quantum symmetries* (cf. [nL-WT][(15)]SS23-ATO):

$$\mathrm{QS} := \frac{\mathrm{U}(\mathcal{H})^2}{\mathrm{U}(1)} \rtimes \left(\mathbb{Z}_2^{(t)} \times \mathbb{Z}_2^{(c)}\right), \tag{5.26}$$

where on the right we used (5.14).

Remark 5.1.6. The group QS (5.26) sits in this system of (split) short exact sequences of topological groups:

$$\begin{array}{ccccc} \mathrm{U}(1) & = & \mathrm{U}(1) & \twoheadrightarrow & 1 \\ \downarrow & & \downarrow & & \downarrow \\ \mathrm{U}(\mathcal{H})^2 & \hookrightarrow & \mathrm{U}(\mathcal{H})^{\mathrm{gr}} \rtimes \mathbb{Z}_2^{(t)} & \twoheadrightarrow & \mathbb{Z}_2^{(c)} \times \mathbb{Z}_2^{(t)} \\ \downarrow & & \downarrow & & \| \\ \mathrm{U}(\mathcal{H})^2/\mathrm{U}(1) & \hookrightarrow & \mathrm{QS} & \underset{\iota_{\mathrm{ct}}}{\overset{p_{\mathrm{ct}}}{\rightleftarrows}} & \mathbb{Z}_2^{(c)} \times \mathbb{Z}_2^{(t)}\,. \end{array} \tag{5.27}$$

With the bottom row we also have the diagonal inclusion of the ordinary projective group:

$$\mathrm{PU}(\mathcal{H}) \;\equiv\; \frac{\mathrm{U}(\mathcal{H})}{\mathrm{U}(1)} \xhookrightarrow{\;\;\frac{\mathrm{diag}}{\mathrm{U}(1)}\;\;} \frac{\mathrm{U}(\mathcal{H})^2}{\mathrm{U}(1)} \hookrightarrow \mathrm{QS}, \qquad i_{\mathrm{pu}} : \mathrm{PU}(\mathcal{H}) \to \mathrm{QS} \tag{5.28}$$

Definition 5.1.7. A *PCT quantum symmetry* is

(i) a subgroup

$$G \subset \mathbb{Z}_2^{(t)} \times \mathbb{Z}_2^{(c)}$$

of the PCT group (5.14),

(ii) equipped with a lift $\big[\widehat{(-)}\big]$ to the group QS of graded quantum symmetries

(Ntn. 5.1.5), hence with a dashed homomorphism making this diagram commute:

$$
\begin{array}{ccc}
G & \xrightarrow{\;[\widehat{(-)}]\;} & \mathrm{QS}\,. \\
& \searrow \quad \mathbb{Z}_2^{(t)} \times \mathbb{Z}_2^{(c)} \quad \swarrow &
\end{array}
\tag{5.29}
$$

In other words, this is a graded or ungraded and unitary or anti-unitary projective representation of G where the grading and anti-unitarity is fixed by the embedding of G into the PCT group.

Here the notation for the homomorphism (5.29) means that $[-]$ is the U(1)-equivalence class of a representative $\widehat{(-)}$:

$$
\begin{array}{ccl}
G & \xrightarrow{\;[\widehat{(-)}]\;} & \mathrm{QS} \simeq \Big(\mathrm{U}(\mathcal{H})^2 \rtimes \big(\mathbb{Z}_2^{(t)} \times \mathbb{Z}_2^{(c)}\big)\Big)/\mathrm{U}(1) \\
g & \longmapsto & [\,\widehat{g}\,] \;\; \text{for } \widehat{g} \in \mathrm{U}(\mathcal{H})^2 \rtimes \big(\mathbb{Z}_2^{(t)} \times \mathbb{Z}_2^{(c)}\big).
\end{array}
\tag{5.30}
$$

Table 5.1 **Rows 1-3**: The possible cases of PCT quantum symmetries (Def. 5.1.7) according to Prop. 5.1.8.
Row 4: The spaces of Fredholm operators commuting with the given PCT quantum symmetries are classifying spaces for the 10 flavors of topological K-theory (cf. § 5.1.4).
Row 5: Traditional labels of these 10 cases in the context of the *10-fold way* classification of topological phases of matter (going back to [SRFL08, Tbl. 1], review in [CTSR16, Tbl. 1]).

PCT subgroup	$G =$	{e}	{e, P}	$\mathbb{Z}_2^{(t)} := \{\mathrm{e}, T\}$		$\mathbb{Z}_2^{(c)} := \{\mathrm{e}, C\}$		$\{\mathrm{e}, T, C, P = CT\}$			
Quantum symmetry $G \dashrightarrow^{[\widehat{(-)}]} \mathrm{QS}$ via $\mathbb{Z}_2^{(t)} \times \mathbb{Z}_2^{(c)}$	$\widehat{T}^2 =$			+1	−1			+1	+1	−1	−1
	$\widehat{C}^2 =$					+1	−1	+1	−1	+1	−1
Fred. operators commuting with $\widehat{G}$	$\mathrm{Fred}_{\mathrm{gr}}^{\widehat{G}} \simeq$	KU^0	KU^1	KO^0	KO^4	KO^2	KO^6	KO^1	KO^7	KO^5	KO^3
10-fold way class label		A	AIII	AI	AII	D	C	BDI	CI	CII	DIII

Proposition 5.1.8. *The PCT quantum symmetries* (Def. 5.1.7) *satisfy:*

(i) *If* $G = \{\mathrm{id}, P = TC\}$, *then* $[\widehat{P}]$ *has a representative* $\widehat{P}$ (5.30) *satisfying*

$$\widehat{P}^2 = \mathrm{id}\,. \tag{5.31}$$

(ii) *If* $T \in G$, *then* $[\widehat{T}]$ *has a representative* $\widehat{T}$ (5.30) *satisfying*

$$\widehat{T}^2 \in \{\pm\mathrm{id}\}\,. \tag{5.32}$$

(iii) *If* $C \in G$, *then* $[\widehat{C}]$ *has a representative* $\widehat{C}$ (5.30) *satisfying*

$$\widehat{C}^2 \in \{\pm\mathrm{id}\}\,, \tag{5.33}$$

and all these cases occur. In consequence, PCT quantum symmetries fall into 10 classes as shown in Tab. 5.1.

Proof. **(i)** The condition that $\big[\widehat{(-)}\big]$ be a group homorphism is equivalently that

$$\big[\widehat{P}\big]^2 = [\mathrm{id}] \quad\Leftrightarrow\quad \widehat{P}^2 = \omega\,\mathrm{id}\,, \;\; \text{for some } \omega \in \mathrm{U}(1). \tag{5.34}$$

But then for any choice of square root $\sqrt{\omega} \in \mathrm{U}(1)$ we have

$$\big[\widehat{P}\big] = \Big[\tfrac{1}{\sqrt{\omega}}\widehat{P}\Big], \tag{5.35}$$

and since $\widehat{P}$ is unitary (5.22) and hence in particular complex-linear, this rescaled operator has the claimed property:

$$\Big(\tfrac{1}{\sqrt{\omega}}\widehat{P}\Big)^2 = \tfrac{1}{\sqrt{\omega}}\widehat{P}\tfrac{1}{\sqrt{\omega}}\widehat{P} = \tfrac{1}{\sqrt{\omega}}\tfrac{1}{\sqrt{\omega}}\widehat{P}\widehat{P} = 1\,. \tag{5.36}$$

(ii) As in the previous case, homomorphy requires that

$$\big[\widehat{T}\big]^2 = [\mathrm{id}] \quad\Leftrightarrow\quad \widehat{T}^2 = \omega\,\mathrm{id}\,, \;\; \text{for some } \omega \in \mathrm{U}(1),$$

but since now $\widehat{T}$ is anti-unitary (5.22) and hence complex anti-linear, there is a constraint forcing ω to be real:

$$\begin{array}{ccc} \widehat{T}\widehat{T}^2 & = & \widehat{T}^2\widehat{T} \\ \| & & \| \\ \omega^*\widehat{T} & = & \omega\widehat{T} \end{array} \quad\Leftrightarrow\quad \omega \in \mathrm{U}(1) \cap \mathbb{R} = \{\pm 1\}\,.$$

Moreover, the one remaining non-trivial value, $\omega = -1$, can *not* be scaled away as in (5.35) and (5.36), since anti-linearity of $\widehat{T}$ implies that

$$\big(\pm\,\mathrm{i}\widehat{T}\,\big)^2 = (\pm\mathrm{i})\widehat{T}(\pm\mathrm{i})\widehat{T} = (\pm\mathrm{i})(\mp\mathrm{i})\widehat{T}\widehat{T} = \widehat{T}^2.$$

(iii) This case works verbatim like the previous case. □

5.1.3 Fredholm Operators

Our construction of twisted orbifold K-theory (in § 5.1.5) is based on the fact that *classifying spaces* for topological K-theory are given by space of *Fredholm operators* (the *Atiyah-Jänich theorem*, Prop. 5.1.16). Here we compile some basics on Fredholm operators that we will refer to for this discussion.

For standard review of Fredholm operators see [Mur90, §1.4][Arv02, §3.3] [DSBW23, §3]. As has become common, we will consider the space of *graded self-adjoint* Fredholm operators in the following Def. 5.1.9, since this is what lends itself naturally to the classification of the different "quantum symmetry protected" flavors of K-theory (in § 5.1.4) and to the construction of twisted equivariant K-theory (as observed by [ASe04]), and therefore eventually to the construction of twisted orbifold K-theory (in § 5.1.5).

Definition 5.1.9. **(i)** A *Fredholm operator* on $\mathcal{H}$ (5.1) is a bounded operator whose kernel and cokernel are of finite dimension:

$$\mathrm{Fred}(\mathcal{H}) := \left\{ f \in \mathcal{B}(\mathcal{H}) \;\middle|\; \begin{array}{l} \dim\big(\ker(f)\big) < \infty \\ \dim\big(\mathrm{coker}(f)\big) < \infty \end{array} \right\}. \tag{5.37}$$

We consider this space of Fredholm operators as equipped with the operator norm topology.[2]

(ii) Homeomorphically, we regard this space as that of *odd graded* but self-adjoint Fredholm operators on $\mathcal{H}_{\mathrm{gr}}$ (5.8):

$$\mathrm{Fred}_{\mathrm{gr}}(\mathcal{H}_{\mathrm{gr}}) := \left\{ F = \begin{pmatrix} 0 & f \\ f^\dagger & 0 \end{pmatrix} \in \mathcal{B}(\mathcal{H}_{\mathrm{gr}}) \,\middle|\, f \in \mathrm{Fred}(\mathcal{H}) \right\}. \tag{5.38}$$

(iii) We take this space to be pointed by:

$$F_0 := \begin{pmatrix} 0 & \mathrm{id} \\ \mathrm{id} & 0 \end{pmatrix}. \tag{5.39}$$

(iv) The kernel of a graded Fredholm operatorator F (5.38) is hence a finite-dimensional, virtual $\mathbb{C}$-vector space:

$$\ker(F) = \begin{array}{c} \ker(f) \\ \ominus \\ \ker(f^\dagger) \end{array} = \begin{array}{c} \ker(f) \\ \ominus \\ \mathrm{coker}(f), \end{array} \tag{5.40}$$

whose virtual dimension is the *Fredholm index* of f:

$$\dim\big(\ker(F)\big) = \mathrm{ind}(f) := \dim_{\mathbb{C}}\big(\ker(f)\big) - \dim_{\mathbb{C}}\big(\mathrm{coker}(f)\big). \tag{5.41}$$

(v) In this graded form (5.38), the graded quantum symmetries (Ntn. 5.1.5)

$$\left(\begin{bmatrix} U_{++} & U_{+-} \\ U_{-+} & U_{--} \end{bmatrix}, \sigma \right) \in \mathrm{PU}(\mathcal{H})^{\mathrm{gr}} \rtimes \mathbb{Z}_2^{(t)} \equiv \mathrm{QS}(\mathcal{H}) \tag{5.42}$$

have a canonical topological group action Def. 4.1.7 by conjugation:

$$\begin{array}{ccc} \mathrm{Fred}_{\mathrm{gr}} & \longrightarrow & \mathrm{Fred}_{\mathrm{gr}} \\ \begin{pmatrix} 0 & f_{+-} \\ f_{+-}^\dagger & 0 \end{pmatrix} & \longmapsto & \begin{pmatrix} U_{++} & U_{+-} \\ U_{-+} & U_{--} \end{pmatrix} \circ \begin{pmatrix} 0 & f_{+-}^{\sigma} \\ f_{+-}^{\dagger\sigma} & 0 \end{pmatrix} \circ \begin{pmatrix} U_{++} & U_{+-} \\ U_{-+} & U_{--} \end{pmatrix}^{-1}. \end{array} \tag{5.43}$$

(Here either $U_{+-}, U_{-+} = 0$ or $U_{++}, U_{--} = 0$; the square bracket denotes the $\mathrm{diag}(\mathrm{U}(1))$-equivalence class of the matrix; and F^σ equals F when $\sigma = \mathrm{e}$ and equals its complex conjugate (5.19) otherwise, cf. [Mat71b, §5.B].)

Proposition 5.1.10 (Atkinson's theorem (cf. [Mur90, Thm. 1.4.16])). *A bounded operator is Fredholm* (5.37) *iff there exists another bounded operator (a parametrix) such that their composites differ from the identity by a compact operator:*

$$f \in \mathcal{B}(\mathcal{H}) \text{ is Fredholm} \quad \Leftrightarrow \quad \exists \tilde{f} \in \mathcal{B}(\mathcal{H}) : \begin{cases} f \circ \tilde{f} - \mathrm{id} \in \mathcal{K}(\mathcal{H}), \\ \tilde{f} \circ f - \mathrm{id} \in \mathcal{K}(\mathcal{H}). \end{cases} \tag{5.44}$$

Remark 5.1.11 (Arithmetic on Fredholm operators)**.** Given a graded Fredholm operator F as in (5.38), its *charge reversal*

$$\ominus F := \begin{pmatrix} 0 & f^\dagger \\ f & 0 \end{pmatrix} \in \mathrm{Fred}_{\mathrm{gr}} \tag{5.45}$$

[2]More ambitiously, one modifies the space of graded Fredholm operators in (5.37) up to homotopy equivalence ([ASe04, §4]) such that the $\mathrm{PU}(\mathcal{H})$-action (5.43) on it becomes continuous also in the latter's strong/weak operator topology. Here we disregard this, for simplicity.

is also graded Fredholm, with negative virtual kernel:

$$\ker\big(\ominus F_P\big) = \ominus \ker\big(F_P\big)\,. \tag{5.46}$$

Given a pair $F_1, F_2 \in \mathrm{Fred}_{\mathrm{gr}}$, their *direct sum* is

$$F_1 \oplus F_2 = \begin{pmatrix} 0 & f_1 \oplus f_2 \\ f_1^\dagger \oplus f_2^\dagger & 0 \end{pmatrix} \tag{5.47}$$

acting on $\mathcal{H} \oplus \mathcal{H} \simeq \mathcal{H}$, with virtual kernel the direct sum of those of the summands:

$$\ker\big(F_1 \oplus F_2\big) = \ker(F_1) \oplus \ker(F_2)\,. \tag{5.48}$$

In particular, the *virtual difference* between a pair of Fredholm operators is the direct sum of the first with the charge reversal of the second:

$$F_1 \ominus F_2 := F_1 \oplus \big(\ominus F_2\big)\,. \tag{5.49}$$

Proposition 5.1.12 ([Jän65, Lem 4]). *For X a topological space and $F_{(-)} : X \to \mathrm{Fred}_{\mathrm{gr}}$ a continuous map, this Fredholm index* (5.41) *is continuous and hence a locally constant function* $\mathrm{ind}\big(F_{(-)}\big) : X \to \mathbb{Z}$. *The graded contributions* $\dim_{\mathbb{C}}\big(\ker(f_X)\big)$ *and* $\dim_{\mathbb{C}}\big(\mathrm{coker}(f_X)\big)$ *need not be locally constant, but if they are, then the virtual kernel bundle is a virtual topological vector bundle over X:*

$$\left.\begin{array}{r} F_X : X \xrightarrow{\;\mathrm{cntns}\;} \mathrm{Fred}_{\mathrm{gr}} \\ \dim_{\mathbb{C}}\big(\ker(f_X)\big) : X \xrightarrow{\;\mathrm{cntns}\;} \mathbb{N} \end{array}\right\} \;\Rightarrow\; \ker\big(F_X\big) \in \mathbb{C}\mathrm{Vec}^{\mathrm{gr}}_X\,. \tag{5.50}$$

Example 5.1.13. For $k \in \mathbb{N}$ and identifying $\mathcal{H} \simeq \bigoplus_{\mathbb{N}} \mathbb{C}^k$, the following graded Fredholm operator (a "left shift operator")

$$F^{\mathbb{C}^k} := \begin{pmatrix} 0 & f^{\mathbb{C}^k} \\ \big(f^{\mathbb{C}^k}\big)^\dagger & 0 \end{pmatrix}, \quad f^{\mathbb{C}^k} := \begin{pmatrix} 0 & \mathrm{id} & 0 & 0 & \cdots \\ 0 & 0 & \mathrm{id} & 0 & \cdots \\ 0 & 0 & 0 & \mathrm{id} & \cdots \\ \vdots & \vdots & \vdots & \vdots & \ddots \end{pmatrix}, \quad \begin{array}{ccc} \mathcal{H} & \xrightarrow{\;f^{\mathbb{C}^k}\;} & \mathcal{H} \\ \wr\wr & & \wr\wr \\ \mathbb{C}^k & & \mathbb{C}^k \\ \oplus & \nearrow^{\mathrm{id}} & \oplus \\ \mathbb{C}^k & & \mathbb{C}^k \\ \oplus & \nearrow^{\mathrm{id}} & \oplus \\ \mathbb{C}^k & & \mathbb{C}^k \\ \oplus & \nearrow & \oplus \\ \vdots & & \vdots \end{array}, \tag{5.51}$$

has virtual kernel (5.40) isomorphic to $\mathbb{C}^k$:

$$\ker\big(F^{\mathbb{C}^k}\big) = \mathbb{C}^k \ominus 0\,. \tag{5.52}$$

Hence a map constant on $F^{\mathbb{C}^k}$ represents, under (5.50), the trivial rank$=k$ vector bundle:

$$\Big(X \to * \xrightarrow{\;F^{\mathbb{C}^k}\;} \mathrm{Fred}_{\mathrm{gr}}\Big) \xmapsto{\;\ker_X\;} \mathbb{C}^k_X \ominus 0 \in \mathbb{C}\mathrm{Vec}^{\mathrm{gr}}_X\,. \tag{5.53}$$

We also observe the following example, which will be crucial in § 5.2 (Lem. 5.2.25).

Example 5.1.14. For $k \in \mathbb{N}$ and

$$P : \mathbb{C}^k \to \mathbb{C}^k,\;\; P^\dagger = P,\quad P \circ P = P\,, \tag{5.54}$$

a hermitian projection operator, then we obtain a graded Fredholm operator (5.37)

$$F_P := \begin{pmatrix} 0 & f_P \\ f_P^\dagger & 0 \end{pmatrix} \in \mathrm{Fred}_{\mathrm{gr}} \tag{5.55}$$

by setting

$$f_P := \begin{pmatrix} P & 1-P & 0 & 0 & 0 & \cdots \\ 0 & P & 1-P & 0 & 0 & \cdots \\ 0 & 0 & P & 1-P & 0 & \cdots \\ \vdots & \vdots & \vdots & \vdots & \vdots & \ddots \end{pmatrix}, \tag{5.56}$$

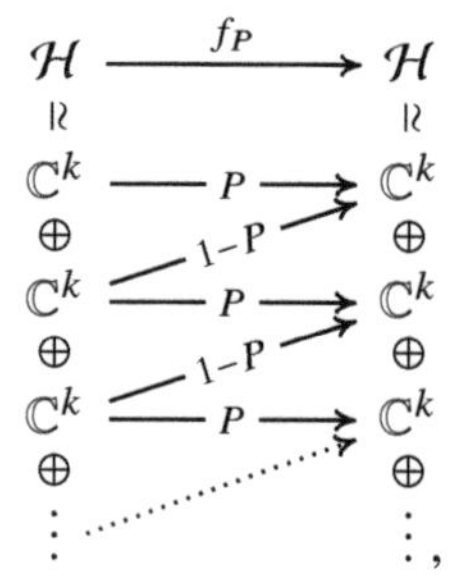

whose virtual kernel (5.40) reproduces the kernel of P:

$$\ker(F_P) = \ker(P) \ominus 0\,. \tag{5.57}$$

This construction constitutes a continuous map from Hermitian projectors to graded Fredholm operators

$$F_{(-)} : \left\{ P \in \mathcal{B}(\mathbb{C}^k) \,\middle|\, {}^{P^\dagger = P}_{P \circ P = P} \right\} \longrightarrow \mathrm{Fred}_{\mathrm{gr}}\,. \tag{5.58}$$

Proof. That the virtual kernel is as claimed in (5.57) follows by direct inspection, crucially using that P is a projector, so that P and $1-P$ have complementary kernels and images in $\mathbb{C}^k$.

Incidentally, $f_P^\dagger$ is a parametrix (5.44) for f_P (5.56): From

$$f_P^\dagger = \begin{pmatrix} P & 0 & 0 & 0 & \cdots \\ 1-P & P & 0 & 0 & \cdots \\ 0 & 1-P & P & 0 & \cdots \\ 0 & 0 & 1-P & P & \cdots \\ \vdots & \vdots & \vdots & \vdots & \ddots \end{pmatrix}, \tag{5.59}$$

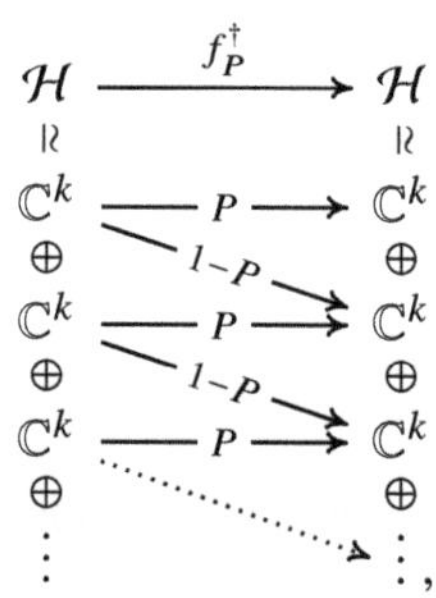

we find

$$\left.\begin{array}{l} f_P \circ f_P^\dagger - \mathrm{id} = 0 \\ f_P^\dagger \circ f_P - \mathrm{id} = \begin{pmatrix} P-1 & 0 & 0 & \cdots \\ 0 & 0 & 0 & \cdots \\ 0 & 0 & 0 & \cdots \\ \vdots & \vdots & \vdots & \ddots \end{pmatrix} \end{array}\right\} \in \mathcal{K}(\mathcal{H})\,. \tag{5.60}$$

This completes the proof. □

Example 5.1.15. In further specialization of Ex. 5.1.14, consider the projector

$$P_0 := \begin{pmatrix} 0 & 0 \\ 0 & \mathrm{id}_b \end{pmatrix} : \mathbb{C}^2_a \oplus \mathbb{C}^2_b \longrightarrow \mathbb{C}^2_a \oplus \mathbb{C}^2_b \,, \tag{5.61}$$

where the subscripts are just to name these two subspaces of $\mathbb{C}^4$, both isomorphic to $\mathbb{C}^2$. Then, under the unitary transformation which "unshuffles" the copies of these two subspaces in the infinite direct sum Hilbert space

$$\mathcal{H} := \bigoplus_{\mathbb{N}} \left(\mathbb{C}^2_a \oplus \mathbb{C}^2_b\right) \xrightarrow[\sim]{\;U\;} \underbrace{\left(\textstyle\bigoplus_{\mathbb{N}} \mathbb{C}^2_a\right)}_{\mathcal{H}_a} \oplus \underbrace{\left(\textstyle\bigoplus_{\mathbb{N}} \mathbb{C}^2_b\right)}_{\mathcal{H}_b} \,, \tag{5.62}$$

the corresponding Fredholm operator (5.56)

$$f_{P_0} = \begin{pmatrix} 0 & 0 & \mathrm{id}_a & 0 & 0 & 0 & \cdots \\ 0 & \mathrm{id}_b & 0 & 0 & 0 & 0 & \cdots \\ 0 & 0 & 0 & 0 & \mathrm{id}_a & 0 & \cdots \\ 0 & 0 & 0 & \mathrm{id}_b & 0 & 0 & \cdots \\ \vdots & \vdots & \vdots & \vdots & \vdots & \vdots & \ddots \end{pmatrix}, \qquad \begin{array}{ccc} \mathcal{H} & \xrightarrow{\;f_{P_0}\;} & \mathcal{H} \\ \wr\vert & & \wr\vert \\ \mathbb{C}^4 & \xrightarrow{\;P_0\;} & \mathbb{C}^4 \\ \oplus & \nearrow_{1-P_0} & \oplus \\ \mathbb{C}^4 & \xrightarrow{\;P_0\;} & \mathbb{C}^4 \\ \oplus & \nearrow_{1-P_0} & \oplus \\ \mathbb{C}^4 & \xrightarrow{\;P_0\;} & \mathbb{C}^4 \\ \oplus & \nearrow & \oplus \\ \vdots & & \vdots \end{array} \tag{5.63}$$

transforms into the direct sum of the left shift operator (5.51) and the identity:

$$\begin{aligned} U \circ f_{P_0} \circ U^{-1} &= \begin{pmatrix} 0 & \mathrm{id}_a & 0 & 0 & 0 & \cdots \\ 0 & 0 & \mathrm{id}_a & 0 & 0 & \cdots \\ 0 & 0 & 0 & \mathrm{id}_a & 0 & \cdots \\ 0 & 0 & 0 & 0 & \mathrm{id}_a & \cdots \\ \vdots & \vdots & \vdots & \vdots & & \end{pmatrix} \oplus \begin{pmatrix} \mathrm{id}_b & 0 & 0 & 0 & 0 & \cdots \\ 0 & \mathrm{id}_b & 0 & 0 & 0 & \cdots \\ 0 & 0 & \mathrm{id}_b & 0 & 0 & \cdots \\ 0 & 0 & 0 & \mathrm{id}_b & 0 & \cdots \\ \vdots & \vdots & \vdots & \vdots & \ddots & \end{pmatrix} \\ &= f^{\mathbb{C}^2} \oplus \mathrm{id} \;:\; \mathcal{H}_a \oplus \mathcal{H}_b \longrightarrow \mathcal{H}_a \oplus \mathcal{H}_b \,. \end{aligned} \tag{5.64}$$

This simple transformation is going to be useful in the proof of Prop. 5.2.30, via Rem. 5.2.28.

5.1.4 Classifying K-Theory

The space of Fredholm operators (5.37) and its subspaces compatible with PCT quantum symmetries (Prop. 5.1.8) turn out to be the classifying spaces (cf. Fig. 2.1) for the flavors of the generalized cohomology theory called *topological K-theory* (cf. [nL-K]). As such, one may take the following statements to be the *definition* of topological K-theory in its various flavors and degrees.

Proposition 5.1.16 (Atiyah-Jänich theorem [Jän65][Thm. A1]Atiyah1967).

(i) *The space of Fredholm operators* (5.37) *is a classifying space for complex K-theory in degree = 0,*

$$\pi_0\big\{X \dashrightarrow \mathrm{Fred}\big\} \simeq \mathrm{KU}^0(X)\,. \tag{5.65}$$

(ii) *Over compact Hausdorff spaces X, where*

$$\mathrm{KU}^0(X) \simeq K\big[\mathbb{C}\mathrm{Vec}_X\big] \tag{5.66}$$

is the Grothendieck group of complex vector bundles, this equivalence is given by passing to virtual kernel bundles (5.50) *of good representative maps* F_X^{gd} *in the homotopy class* (meaning that $\dim(\ker(f_X^{\mathrm{gd}}))$ is locally constant, see Prop. 5.1.12)*:*

$$\begin{array}{ccc} \pi_0\{X \dashrightarrow \mathrm{Fred}\} & \longrightarrow & K\big[\mathbb{C}\mathrm{Vec}_X\big] \\ [F_X] & \longmapsto & \big[\ker(F_X^{\mathrm{gd}})\big]\,. \end{array} \tag{5.67}$$

(iii) *In particular, the connected components of* Fred,

$$\pi_0\,\mathrm{Fred} \simeq \pi_0\{* \dashrightarrow \mathrm{Fred}_{\mathrm{gr}}\}\,, \tag{5.68}$$

are indexed by the integer Fredholm index (5.41)*:*

$$\begin{array}{ccc} \pi_0\,\mathrm{Fred}_{\mathrm{gr}} & \xrightarrow{\sim} & \mathbb{Z} \\ \left[F = \begin{pmatrix} 0 & f \\ f^\dagger & 0 \end{pmatrix}\right] & \longmapsto & \begin{array}{c}\dim(\ker(F)) \\ = \mathrm{ind}(f)\,.\end{array} \end{array} \tag{5.69}$$

Notation 5.1.17 (Subspaces of Fredholm operators)**.** For

(i) $\mathbb{K} \in \{\mathbb{R}, \mathbb{C}, \overline{\mathbb{C}}, \mathbb{H}\}$

(ii) $\sigma \in \{\pm 1\}$,

consider the following subspaces of that of Fredholm operators (Def. 5.1.9):

$$\begin{array}{ccccc} \mathrm{Fred}_{\mathbb{K}}^{\sigma} & \hookrightarrow & \mathrm{Fred}_{\mathbb{K}} & \hookrightarrow & \mathrm{Fred}(\mathcal{H}) \\ \text{\rotatebox{90}{:=}} & & \text{\rotatebox{90}{:=}} & & \\ \big\{f \in \mathrm{Fred}_{\mathbb{K}} \,\big|\, f^\dagger = \sigma\cdot f\big\} & \subset & \big\{f \in \mathrm{Fred}(\mathcal{H}) \,\big|\, f \text{ is } \mathbb{K}\text{-linear}\big\}\,, & & \end{array} \tag{5.70}$$

where f being $\mathbb{R}$-linear/$\mathbb{H}$-linear means that it commutes with a fixed real structure (5.3) or quaternionic structure (5.4) on $\mathcal{H}$, respectively, and where by it being "$\overline{\mathbb{C}}$-linear" we mean that it is $\mathbb{C}$-*anti*-linear (while it being $\mathbb{C}$-linear is no additional condition, of course).

Proposition 5.1.18 ([SSi69][Kar70, Thm. 1.16])**.** *The subspaces of Fredholm operators from Ntn. 5.1.17 are classifying spaces for flavors of topological K-theory as follows:*

$$\pi_0\{X \dashrightarrow \mathrm{Fred}_{\mathbb{C}}\} \simeq \mathrm{KU}^0(X) \tag{5.71a}$$

$$\pi_0\{X \dashrightarrow \mathrm{Fred}_{\mathbb{C}}^{+}\} \simeq \mathrm{KU}^1(X) \sqcup \{+,-\} \tag{5.71b}$$

$$\pi_0\{X \dashrightarrow \mathrm{Fred}_{\mathbb{R}}\} \simeq \mathrm{KO}^0(X) \tag{5.71c}$$

$$\pi_0\{X \dashrightarrow \mathrm{Fred}_{\mathbb{R}}^{+}\} \simeq \mathrm{KO}^1(X) \sqcup \{+,-\} \tag{5.71d}$$

$$\pi_0\{X \dashrightarrow \mathrm{Fred}_{\mathbb{R}}^{-}\} \simeq \mathrm{KO}^7(X) \tag{5.71e}$$

$$\pi_0\{X \dashrightarrow \mathrm{Fred}_{\overline{\mathbb{C}}}^{+}\} \simeq \mathrm{KO}^2(X) \tag{5.71f}$$

$$\pi_0\{X \dashrightarrow \mathrm{Fred}^-_{\mathbb{C}}\} \simeq \mathrm{KO}^6(X) \tag{5.71g}$$
$$\pi_0\{X \dashrightarrow \mathrm{Fred}_{\mathbb{H}}\} \simeq \mathrm{KO}^4(X) \tag{5.71h}$$
$$\pi_0\{X \dashrightarrow \mathrm{Fred}^+_{\mathbb{H}}\} \simeq \mathrm{KO}^5(X) \sqcup \{+,-\} \tag{5.71i}$$
$$\pi_0\{X \dashrightarrow \mathrm{Fred}^-_{\mathbb{H}}\} \simeq \mathrm{KO}^3(X)\,. \tag{5.71j}$$

Remark 5.1.19. **(i)** The disjoint sets $\{+,-\}$ in (5.71) witness pairs of contractible components present in each of the spaces of self-adjoint $\mathbb{R}/\mathbb{C}/\mathbb{H}$-linear Fredholm operators, corresponding to such operators whose essential spectrum (Def. 3.2.2) is purely positive or purely negative, respectively [SSi69, Thm. B][Kar70, §I].

(ii) When the self-adjoint $\mathbb{C}$-linear Fredholm operators are interpreted as tachyon field values of open strings stretching between coincident $\mathrm{D9}/\overline{D9}$-branes (cf. [Sza02, §3.1][GH10, §8]), then those with a mixed essential spectrum are interpreted as witnessing solitonic pair annihilation in a background where both $\mathrm{D9}/\overline{D9}$-branes appear with infinite multiplicity (cf. [Ho98, (3.11)][Wi01, p. 7] and § 3.2). Accordingly, the components $\pm \in \{+,-\}$ would have to be interpreted as backgrounds where an infinite number of D9-branes is compensated only by a finite number of $\overline{D9}$-branes or vice versa, respectively.

(iii) For the purpose of K-theory classification, these components are to be disregarded, but if we regard the full spaces of Fredholm operators as classifying a (slightly) non-abelian cohomology theory (cf. Fig. 2.1) variant in its own right, then $\pm \in \{+,-\}$ are valid classes in the cohomology set of any space X which reflect a pair of "collapsed" cases that one may keep track of.

We now reformulate the classical result Prop. 5.1.18 in the guise of the "10-fold way" (cf. [nL-TS]) of topological phases protected by PCT quantum symmetry (Def. 5.1.7), following [SS23b, Fact 2.12], as shown in Tab. 5.1:

Proposition 5.1.20. *The 10 subspaces of Fredholm operators* (Ntn. 5.1.17) *appearing in* (5.71), *when regarded as subspaces of* $\mathrm{Fred}_{\mathrm{gr}}(\mathcal{H}_{\mathrm{gr}})$ (5.38) *are homeomorphic to the fixed subspaces* (4.15)

$$\mathrm{Fred}^{\widehat{G}}_{\mathrm{gr}} \subset \mathrm{Fred}_{\mathrm{gr}}(\mathcal{H}_{\mathrm{gr}}) \tag{5.72}$$

for the conjugation action (5.43) *of the 10 PCT quantum symmetries* (Prop. 5.1.8 and Tab. 5.1), *respectively:*

$$\mathrm{Fred}_{\mathrm{gr}} \simeq \mathrm{Fred}_{\mathbb{C}} \tag{5.73a}$$
$$\mathrm{Fred}^{\langle \widehat{P}\rangle}_{\mathrm{gr}} \simeq \mathrm{Fred}^{\pm}_{\mathbb{C}} \quad \text{for } \widehat{P}^2 = \pm\mathrm{id} \tag{5.73b}$$
$$\mathrm{Fred}^{\langle \widehat{T}\rangle}_{\mathrm{gr}} \simeq \mathrm{Fred}_{\mathbb{R}} \quad \text{for } \widehat{T}^2 = +\mathrm{id} \tag{5.73c}$$
$$\mathrm{Fred}^{\langle \widehat{T},\widehat{C}\rangle}_{\mathrm{gr}} \simeq \mathrm{Fred}^{+}_{\mathbb{R}} \quad \text{for } \widehat{T}^2 = +\mathrm{id},\ \widehat{C}^2 = +\mathrm{id}, \tag{5.73d}$$
$$\mathrm{Fred}^{\langle \widehat{T},\widehat{C}\rangle}_{\mathrm{gr}} \simeq \mathrm{Fred}^{-}_{\mathbb{R}} \quad \text{for } \widehat{T}^2 = +\mathrm{id},\ \widehat{C}^2 = -\mathrm{id}, \tag{5.73e}$$

$$\mathrm{Fred}_{\mathrm{gr}}^{\langle\widehat{C}\rangle} \simeq \mathrm{Fred}_{\mathbb{C}}^{\pm} \quad \text{for } \widehat{C}^2 = +\mathrm{id}, \tag{5.73f}$$

$$\mathrm{Fred}_{\mathrm{gr}}^{\langle\widehat{C}\rangle} \simeq \mathrm{Fred}_{\mathbb{C}}^{-} \quad \text{for } \widehat{C}^2 = -\mathrm{id}, \tag{5.73g}$$

$$\mathrm{Fred}_{\mathrm{gr}}^{\langle\widehat{T}\rangle} \simeq \mathrm{Fred}_{\mathbb{H}} \quad \text{for } \widehat{T}^2 = -\mathrm{id}, \tag{5.73h}$$

$$\mathrm{Fred}_{\mathrm{gr}}^{\langle\widehat{T},\widehat{C}\rangle} \simeq \mathrm{Fred}_{\mathbb{H}}^{+} \quad \text{for } \widehat{T}^2 = -\mathrm{id},\ \widehat{C}^2 = +\mathrm{id}, \tag{5.73i}$$

$$\mathrm{Fred}_{\mathrm{gr}}^{\langle\widehat{T},\widehat{C}\rangle} \simeq \mathrm{Fred}_{\mathbb{H}}^{-} \quad \text{for } \widehat{T}^2 = -\mathrm{id},\ \widehat{C}^2 = -\mathrm{id}. \tag{5.73j}$$

Proof. The quantum symmetry $\widehat{P}$ with $\widehat{P}^2 = \pm\mathrm{id}$ (5.31) is represented by swapping graded summands (5.9), possibly up to a sign, so that a graded Fredholm operator F of the form (5.38) commutes with $\widehat{P}$ iff its component Fredholm operator f is (anti-)self-adjoint:

$$\begin{pmatrix} 0 & f \\ f^\dagger & 0 \end{pmatrix} \cdot \begin{pmatrix} 0 & \pm 1 \\ 1 & 0 \end{pmatrix} = \begin{pmatrix} 0 & \pm 1 \\ 1 & 0 \end{pmatrix} \cdot \begin{pmatrix} 0 & f \\ f^\dagger & 0 \end{pmatrix} \quad \Leftrightarrow \quad f^\dagger = \pm f\,. \tag{5.74}$$

This gives the homeomorphy (5.73b).

The quantum symmetry $\widehat{T}$ with $\widehat{T}^2 = \pm\mathrm{id}$ (5.32) is manifestly a real/quaternionic structure on $\mathcal{H}$, (5.3) and (5.4), whence operators commuting with it are manifestly $\mathbb{R}/\mathbb{C}$-linear. This gives the homeomorphisms (5.73c) and (5.73h).

Combining these two cases, if the quantum symmetries involve $\widehat{T}$ and $\widehat{C} = \widehat{T}\widehat{P}$ then the fixed locus is equivalent to that of $\widehat{T}$ and $\widehat{P}$ and is hence given by $\mathbb{R}/\mathbb{C}$-linear (anti-)self-adjoint operators. This gives the homeomorphisms (5.73d), (5.73e), (5.73i) and (5.73j).

Finally, if there is only $\widehat{C}$ then the condition on a Fredholm operator F to be fixed is that

$$\begin{pmatrix} 0 & f \\ f^\dagger & 0 \end{pmatrix} \cdot \begin{pmatrix} 0 & \widehat{T} \\ \widehat{T} & 0 \end{pmatrix} = \begin{pmatrix} 0 & \widehat{T} \\ \widehat{T} & 0 \end{pmatrix} \cdot \begin{pmatrix} 0 & f \\ f^\dagger & 0 \end{pmatrix} \quad \Leftrightarrow \quad f \circ \widehat{T} = \widehat{T} \circ f^\dagger\,. \tag{5.75}$$

Now to observe that f satisfying this condition is equivalent to the anti-linear Fredholm operator $f \circ \widehat{T}$ being (anti-)self-adjoint

$$\big(f \circ \widehat{T}\big)^\dagger = \widehat{T}^\dagger \circ f^\dagger = (\pm)\widehat{T} \circ f^\dagger \underset{(5.75)}{=} \pm f \circ \widehat{T}\,, \tag{5.76}$$

where in the second step we used that real structures are self-adjoint (5.3) while quaternionic structures are anti-self-adjoint (5.4). Therefore, the remaining homeomorphisms (5.73f) and (5.73g) are given by $f \mapsto f \circ \widehat{T}$. □

5.1.5 Twisted Orbifold K-Theory

We have in hand now:

(i) a general definition of (generalized nonabelian) twisted orbifold cohomology (§ 4.2.2) with coefficients in any space Y equipped with a topological group action $\Gamma \circlearrowright Y$,

(ii) a classifying space $Y = \mathrm{Fred}_{\mathrm{gr}}(\mathcal{H}_{\mathrm{gr}})$ for topological K-theory (§ 5.1.4) equipped with a topological action $\mathrm{QS} \circlearrowright \mathrm{Fred}_{\mathrm{gr}}$ (5.43) by "graded quantum symmetries"

QS (Ntn. 5.1.5), combining the graded projective unitary group with "PCT symmetries": $\mathrm{QS} = \big(\mathrm{U}(\mathcal{H})^2/\mathrm{U}(1)\big) \rtimes \big(\mathbb{Z}_2^{(t)} \times \mathbb{Z}_2^{(c)}\big)$, and hence the corresponding stacky incarnation of the universal globally equivariant QS-associated $\mathrm{Fred}_{\mathrm{gr}}$-fiber bundle (according to Ex. 4.1.45):

$$\begin{array}{c} \mathrm{QS} \backslash\!\backslash \mathrm{Fred}_{\mathrm{gr}} \\ {\scriptstyle p_{\mathrm{univ}}^{\mathrm{QS} \,\circlearrowleft\, \mathrm{Fred}_{\mathrm{gr}}}} \Big\downarrow {\scriptstyle \in \mathrm{Fib}} \\ \mathbf{B}\mathrm{QS} \end{array} \tag{5.77}$$

Therefore, we now immediately obtain a good definition of twisted orbifold K-theory (Def. 5.1.22).

In fact, we obtain a highly structured version of such: Since the fixed loci by the PCT quantum symmetries inside the classifying space $\mathrm{Fred}_{\mathrm{gr}}$ are the classifying spaces of topological K-theory in *all its flavors and degrees* (Prop. 5.1.20 and Tab. 5.1), the orbifolding of the cohomology theory represented by $\mathrm{Fred}_{\mathrm{gr}}$ not only imposes equivariance conditions on any given flavor of topological K-theory, but it also serves to "dial" which flavor and degree of K-theory is in effect where on the orbifold. Since for the T- and C-symmetries one wants (in applications to quantum physics) the corresponding orbifolding to be by orientation-reversing isotropy actions, one also speaks of *orbi-orientifolds* (or just: *orientifolds*, for short):

Definition 5.1.21. **(i)** An *orbi-orientifold* (or *real parity orbifold*, with capital "R") is an orbifold $\mathcal{X}$ (Def. 4.2.1) equipped with a map to the delooping (Ex. 4.1.12) of the PCT group (5.14):

$$\sigma : \mathcal{X} \longrightarrow \mathbf{B}\Big(\mathbb{Z}_2^{(t)} \times \mathbb{Z}_2^{(c)}\Big). \tag{5.78}$$

(ii) A *geometric K-theory twist*, or just *twist* for short, on such an orbi-orientifold is map τ to the delooping of the group of graded quantum symmetries (Ntn. 5.1.5) which lifts the PCT grading (5.78), hence a dashed map making this diagram commute:

$$\begin{array}{ccc} \tau : \mathcal{X} & \dashrightarrow & \mathbf{B}\mathrm{QS}\,. \\ & {\scriptstyle \sigma} \searrow \quad \swarrow {\scriptstyle \mathbf{B}P_{\mathrm{ct}}} & \\ & \mathbf{B}\Big(\mathbb{Z}_2^{(t)} \times \mathbb{Z}_2^{(c)}\Big) & \end{array} \tag{5.79}$$

Of course, such τ fully determines σ.

Now recall that for a topological group Γ acting continuously on a topological space X, we obtain the topological *homotopy quotient* groupoid $\Gamma \backslash\!\backslash X$ (Def. 4.1.14), canonically X-fibered over $\mathbf{B}\Gamma \equiv * /\!/ G$ (4.97). Using this for the action $\mathrm{QS} \circlearrowleft \mathrm{Fred}_{\mathrm{gr}}$ from Def. 5.1.9, we obtain the following concise and transparent definition in specialization of Def. 4.2.5:

Definition 5.1.22 (Twisted orbi-orientifold K-theory [SS26c, Ex. 6.2.5][SS23b, §2.2])**.** Given an orbi-orientifold $(\mathcal{X}, \sigma)$ with a K-theory twist τ (Def. 5.1.21) then its *τ-twisted K-cohomology* is the connected components of dashed slice maps

(Def. 4.1.46) from τ into the universal stacky QS-associated $\mathrm{Fred}_{\mathrm{gr}}$-fiber bundle (5.77), as shown on the right here:

$$
\mathrm{K}^{\tau}(\mathcal{X}) := \pi_0 \left\{ \begin{array}{ccc} & & \mathrm{QS}\backslash\!\!\backslash\mathrm{Fred}_{\mathrm{gr}} \\ & \nearrow & \downarrow \\ \mathcal{X} & \xrightarrow{\;\tau\;} & \mathbf{B}\mathrm{QS} \end{array} \right\}. \tag{5.80}
$$

Example 5.1.23 (Reduction to equivariant KU)**.** Def. 5.1.22 is a "globally equivariant" orbifold cohomology theory (in specialization of Thm. 4.2.6) in that the definition does not refer to any particular equivariance/isotropy group, and yet the construction automatically reduces to equivariant KU-theory $\mathrm{KU}_G(-)$ [AS69, p 10][ASe04, §6] – when:

(i) the domain $\mathcal{X}$ is presented as a global quotient $\mathcal{X} \simeq G\backslash\!\!\backslash X$ (Ex. 4.2.3),

(ii) the twist τ factors as a (stable, cf. [SS26c, Lem. 5.1.45]) projective G-representation ρ

$$
G \xrightarrow{\;\rho\;} \mathrm{PU}(\mathcal{H}) \underset{(5.28)}{\overset{i_{\mathrm{pu}}}{\hookrightarrow}} \mathrm{QS} \tag{5.81}
$$

in that

$$
\begin{array}{ccccccc} & & & \overset{\tau_\rho}{\frown\!\!\!\longrightarrow} & & & \\ G\backslash\!\!\backslash X & \twoheadrightarrow & \mathbf{B}G & \xrightarrow{\;\mathbf{B}\rho\;} & \mathbf{B}\mathrm{PU}(\mathcal{H}) & \xrightarrow{\;\mathbf{B}i_{\mathrm{pu}}\;} & \mathbf{B}\mathrm{QS} \\ \downarrow & \searrow^{\sigma} & & & & \swarrow_{\mathbf{B}p_{\mathrm{ct}}} & \\ * & \longrightarrow & & \mathbf{B}\Big(\mathbb{Z}_2^{(t)} \times \mathbb{Z}_2^{(c)}\Big) & & & \end{array} \tag{5.82}
$$

then

$$
\mathrm{K}^{\tau_\rho}(G\backslash\!\!\backslash X) \simeq \mathrm{KU}_G(X)\,. \tag{5.83}
$$

In particular, this shows that the orbifold K-theory is independent of the choice of global quotient presentation, a key property that is less manifest otherwise (cf. [PS10, Prop. 4.1]).

Proof. Using the general results on orbifold cohomology from chapter 4, we straightforwardly obtain that:

$$
\begin{aligned}
\mathrm{K}^{\tau_\rho}(G\backslash\!\!\backslash X) \;&\equiv\; \pi_0 \left\{ \begin{array}{ccccc} & & & & \mathrm{QS}\backslash\!\!\backslash\mathrm{Fred}_{\mathrm{gr}} \\ & & & \nearrow & \downarrow \\ G\backslash\!\!\backslash X & \twoheadrightarrow & \mathbf{B}G & \xrightarrow{\;\mathbf{B}\rho\;} & \mathbf{B}\mathrm{QS} \end{array} \right\} \\
&\underset{(4.99)}{\simeq}\; \pi_0 \left\{ \begin{array}{ccccc} & & G\backslash\!\!\backslash^{\rho}\mathrm{Fred} & \longrightarrow & \mathrm{QS}\backslash\!\!\backslash\mathrm{Fred}_{\mathrm{gr}} \\ & \nearrow & \downarrow & \lrcorner & \downarrow \\ G\backslash\!\!\backslash X & \twoheadrightarrow & \mathbf{B}G & \xrightarrow{\;\mathbf{B}\rho\;} & \mathbf{B}\mathrm{QS} \end{array} \right\} \\
&\underset{(4.107)}{\simeq}\; \pi_0 \left\{ \begin{array}{ccc} \overset{G}{\circlearrowright} & & \overset{G^{\rho}}{\circlearrowright} \\ X & \dashrightarrow & \mathrm{Fred} \end{array} \right\}.
\end{aligned} \tag{5.84}
$$

But the last line is the traditional definition of KU_G (cf. [AS69, p 10]). □

We saw around Prop. 4.1.51 that equivariance is just a special case of twisting, and indeed, in direct generalization of Ex. 5.1.23:

Example 5.1.24 (Reduction to twisted equivariant KU). When:

(i) the domain $\mathcal{X}$ is presented as a global quotient $\mathcal{X} \simeq G\backslash\!\backslash X$ (Ex. 4.2.3),

(ii) the twist τ classifies a G-equivariant $\mathrm{PU}(\mathcal{H})$-principal bundle $\mathcal{P} \xrightarrow{p} X$ in that it factors through a corresponding equivariant Čech cocycle (cf. Ex. 4.1.45, the arrows denote maps of stacks, hence notationally suppressing the inverse equivalences seen on the left of (4.100))

$$\begin{array}{ccc} G\backslash\!\backslash \mathcal{P} & \longrightarrow & \mathrm{PU}(\mathcal{H})\backslash\!\backslash \mathrm{PU}(\mathcal{H}) \\ \big\downarrow & \lrcorner & \big\downarrow {\scriptstyle p_{\mathrm{univ}}} \\ G\backslash\!\backslash X & \xrightarrow{\ \gamma\ } & \mathbf{B}\mathrm{PU}(\mathcal{H}) \end{array} \tag{5.85}$$

as

$$\begin{array}{ccccc} & & \xrightarrow{\qquad\qquad \tau_{\mathcal{P}} \qquad\qquad} & & \\ G\backslash\!\backslash X & \xrightarrow{\ \gamma\ } & \mathbf{B}\mathrm{PU}(\mathcal{H}) & \xrightarrow{\ \mathbf{B}i_{\mathrm{pu}}\ } & \mathbf{BQS} \\ \big\downarrow & \searrow^{\sigma} & & \swarrow_{\mathbf{B}p_{\mathrm{ct}}} & \\ * & \longrightarrow & \mathbf{B}\big(\mathbb{Z}_2^{(t)} \times \mathbb{Z}_2^{(c)}\big) & & \end{array} \tag{5.86}$$

then the twisted orbi-K-theory (Def. 5.1.22) reduces to the corresponding $\mathcal{P}$-twisted equivariant K-theory $\mathrm{KU}_G^{[\mathcal{P}]}(X)$ of [ASe04, §6]:

$$\mathrm{K}^{\tau_{\mathcal{P}}}\big(G\backslash\!\backslash X\big) \simeq \mathrm{KU}_G^{[P]}(X)\,. \tag{5.87}$$

Proof. As before, by the universal property of the pullback:

$$\begin{aligned} \mathrm{K}^{\tau_{\mathcal{P}}}(G\backslash\!\backslash X) &= \pi_0 \left\{ \begin{array}{ccccc} & & & & \mathrm{QS}\backslash\!\backslash \mathrm{Fred}_{\mathrm{gr}} \\ & & \nearrow & & \big\downarrow \\ G\backslash\!\backslash X & \xrightarrow{\ \gamma\ } & \mathbf{B}\mathrm{PU}(\mathcal{H}) & \xrightarrow{\ \mathbf{B}i_{\mathrm{pu}}\ } & \mathbf{BQS} \end{array} \right\} \\ &\underset{(4.99)}{\simeq} \pi_0 \left\{ \begin{array}{ccccc} & & \mathrm{PU}(\mathcal{H})\backslash\!\backslash \mathrm{Fred}_{\mathrm{gr}} & \longrightarrow & \mathrm{QS}\backslash\!\backslash \mathrm{Fred}_{\mathrm{gr}} \\ & \nearrow & \big\downarrow & \lrcorner & \big\downarrow \\ G\backslash\!\backslash X & \xrightarrow{\ \gamma\ } & \mathbf{B}\mathrm{PU}(\mathcal{H}) & \xrightarrow{\ \mathbf{B}i_{\mathrm{pu}}\ } & \mathbf{BQS} \end{array} \right\} \\ &\underset{(4.102)}{\simeq} \pi_0 \left\{ \begin{array}{ccccc} G\backslash\!\backslash(\mathcal{P} \times_{\Gamma} \mathrm{Fred}_{\mathrm{gr}}) & \longrightarrow & \mathrm{PU}(\mathcal{H})\backslash\!\backslash \mathrm{Fred}_{\mathrm{gr}} & \longrightarrow & \mathrm{QS}\backslash\!\backslash \mathrm{Fred}_{\mathrm{gr}} \\ \big\uparrow \big\downarrow & \lrcorner & \big\downarrow & \lrcorner & \big\downarrow \\ G\backslash\!\backslash X & \xrightarrow{\ \gamma\ } & \mathbf{B}\mathrm{PU}(\mathcal{H}) & \xrightarrow{\ \mathbf{B}i_{\mathrm{pu}}\ } & \mathbf{BQS} \end{array} \right\} \\ &\underset{(4.106)}{\simeq} \pi_0\, \Gamma_X\big(\mathcal{P} \otimes_{\Gamma} \mathrm{Fred}\big)^G . \end{aligned} \tag{5.88}$$

But the last expression is just the traditional definition ([ASe04, Def. 6.1]) of twisted equivariant KU^0. (The equivariant local triviality condition on $\mathcal{P}$ which is required by [ASe04] is actually implied by our stacky construction, see [SS26c, §5.2 & Ex. 6.1.2].) □

Example 5.1.25 (Reduction to KR in any degree)**.** When:

(i) the grading σ is *globally constant* over $\mathcal{X}$, in that there is a PCT-subgroup G of which $\mathcal{X}$ is a global quotient orbifold (Ex. 4.2.3) of a manifold X fixed by the action:

$$G \overset{\iota}{\hookrightarrow} \mathbb{Z}_2^{(t)} \times \mathbb{Z}_2^{(c)}\,, \quad \mathcal{X} \simeq G \backslash\!\!\backslash X \simeq X \times \mathbf{B}G\,, \tag{5.89}$$

(ii) and there is no further twist besides a PCT quantum symmetry $\big[\widehat{(-)}\big]$

$$\begin{array}{ccccc}
\mathcal{X}' \times \mathbf{B}G & \xrightarrow{\;\mathrm{pr}_2\;} & \mathbf{B}G & \xrightarrow{\;[\widehat{(-)}]\;} & \mathbf{BQS} \\
& {\scriptstyle \sigma}\searrow & \downarrow{\scriptstyle \mathbf{B}\iota} & \swarrow{\scriptstyle \mathbf{B}p_{\mathrm{ct}}} & \\
& & \mathbf{B}\big(\mathbb{Z}_2^T \times \mathbb{Z}_2^C\big)\,, & &
\end{array}
\qquad \text{(top arrow } \tau_{\widehat{G}} : \mathcal{X}' \times \mathbf{B}G \to \mathbf{BQS})$$

then the $\tau_{\widehat{G}}$-twisted orbi K-cohomology according to Def. 5.1.22 reduces to the flavor and degree of K-theory corresponding to the PCT quantum symmetry as shown in Tab. 5.1:

$$\mathrm{K}^{\tau_{\widehat{G}}}(X \times \mathrm{B}G) \simeq \left\{
\begin{array}{l|l}
\mathrm{KU}^0(X) & \widehat{G} = \mathrm{A} \\
\mathrm{KU}^1(X) \sqcup \{+,-\} & \widehat{G} = \mathrm{AIII} \\
\mathrm{KO}^0(X) & \widehat{G} = \mathrm{AI} \\
\mathrm{KO}^1(X) \sqcup \{+,-\} & \widehat{G} = \mathrm{BDI} \\
\mathrm{KO}^2(X) & \widehat{G} = \mathrm{D} \\
\mathrm{KO}^3(X) & \widehat{G} = \mathrm{BDI} \\
\mathrm{KO}^4(X) & \widehat{G} = \mathrm{AII} \\
\mathrm{KO}^5(X) \sqcup \{+,-\} & \widehat{G} = \mathrm{CII} \\
\mathrm{KO}^6(X) & \widehat{G} = \mathrm{C} \\
\mathrm{KO}^7(X) & \widehat{G} = \mathrm{CI}\,.
\end{array}\right.$$

Proof. Using the general results from chapter 4 , we obtain straightforwardly:

$$\begin{aligned}
\mathrm{K}^{\tau_{\widehat{G}}}(X \times \mathbf{B}G) \;&\equiv\; \pi_0 \left\{
\begin{array}{ccccc}
& & & & \mathrm{QS}\backslash\!\!\backslash \mathrm{Fred}_{\mathrm{gr}} \\
& & & \nearrow & \downarrow \\
X \times \mathbf{B}G & \longrightarrow & \mathbf{B}G & \xrightarrow{[\widehat{-}]} & \mathbf{BQS}
\end{array}\right\} \\
&\underset{(4.99)}{\simeq}\; \pi_0 \left\{
\begin{array}{ccccc}
& & \widehat{G}\backslash\!\!\backslash \mathrm{Fred}_{\mathrm{gr}} & \to & \mathrm{QS}\backslash\!\!\backslash \mathrm{Fred}_{\mathrm{gr}} \\
& \nearrow & \downarrow & & \downarrow \\
X \times \mathbf{B}G & \longrightarrow & \mathbf{B}G & \xrightarrow{[\widehat{-}]} & \mathbf{BQS}
\end{array}\right\} \\
&\underset{(4.110)}{\simeq}\; \pi_0 \Big\{ \; X \dashrightarrow \mathrm{Fred}^{\widehat{G}}_{\mathrm{gr}} \Big\}\,.
\end{aligned} \tag{5.90}$$

But the last line implies the claim by Prop. 5.1.20. □

These examples show that the twisted orbi-orientifold K-theory of Def. 5.1.22 unifies and mixes these aspects of K-theory: $\mathbb{R}/\mathbb{C}/\mathbb{H}$-flavors, degrees, twisting, and equivariance. Along the same lines the reader can find the sectors of twisted orbi KU^1 and KO^n, etc.

In the following subsections we focus on the KU^0-sector, twisted/orbifolded by $\Gamma \in \{\mathrm{U}(2), \mathrm{Sp}(2)\}$.

5.2 Unstable Orientation

In preparation for constructing the four/ten-dimensional $\mathrm{U}(2)/\mathrm{Sp}(2)$-equivariant orientation of orbi-KU^0, § 5.2, here we discuss the canonical $\mathrm{U}(2)/\mathrm{Sp}(2)$-equivariant $\mathbb{C}/\mathbb{H}$-line bundle on $\mathbb{C}P^1/\mathbb{H}P^1$, and the equivariant trivialization of its pullback along the $\mathbb{C}/\mathbb{H}$-Hopf fibration. We use quaternionic 2-component spinor calculus (following [FSS22, §2.1]) in a way that lends itself to constructing the required Fredholm operator families.

This subsection uses well-known constructions in quaternionic algebra, and the result will not be surprising to experts. But since it does not seem properly citable, and in order to establish the notation needed in § 5.2, we spell out the pleasant construction explicitly.

5.2.1 Quaternion algebra

Definition 5.2.1 (cf. [Zha97][MGS14, §1]). We write:

- $\mathbb{H}$ for the real vector space of *quaternions*,
- with their real associative product $(\ )\ (\ )$,
- which is a real star-algebra via conjugation $(-)^*$,
- and equipped with a norm $|-|$ given by $|q|^2 = qq^*$.
- $\mathbb{H}_{\mathrm{in}} \subset \mathbb{H}$ for the real subspace of *imaginary quaternions*, $q^* = -q$,
- for which we choose, as usual, an orthonormal linear basis $\mathbf{i}, \mathbf{j}, \mathbf{k} \in \mathbb{H}_{\mathrm{im}}$ such that
$$\mathbf{i} \cdot \mathbf{j} = \mathbf{k}, \tag{5.91}$$
- which in fact generates $\mathbb{H}$ subject to the further relations
$$\mathbf{i}\,\mathbf{j} = -\mathbf{j}\,\mathbf{i}, \quad \mathbf{i}^2 = \mathbf{j}^2 = -1, \tag{5.92}$$
- and exhibits a star-algebra inclusion of the complex numbers:
$$\begin{array}{ccc} \mathbb{C} & \hookrightarrow & \mathbb{H} \\ \mathrm{i} & \longmapsto & \mathbf{i}\,. \end{array} \tag{5.93}$$
- $\mathcal{B}(\mathbb{H}^2)$ for the real algebra of 2×2 matrices with quaternion entries,

- with star-operation

$$\begin{pmatrix} a & b \\ c & d \end{pmatrix}^{\dagger} := \begin{pmatrix} a^* & c^* \\ b^* & d^* \end{pmatrix}. \tag{5.94}$$

- $\mathrm{Sp}(1) = S(\mathbb{H})$ for the unit norm quaternions, with their group structure under quaternion multiplication.

Example 5.2.2 (cf. [MGS14, §1.27]). The *Pauli matrices* define a homomorphism of real star-algebras from the quaternions (Def. 5.2.1) to the linear operators on $\mathbb{C}^2$:

$$\mathbb{H} \xrightarrow{\ \gamma\ } \mathcal{B}(\mathbb{C}^2)\,, \tag{5.95}$$

given by

$$1 \longmapsto \begin{bmatrix} 1 & 0 \\ 0 & 1 \end{bmatrix}, \quad \mathbf{i} \longmapsto \mathrm{i}\begin{bmatrix} 1 & 0 \\ 0 & -1 \end{bmatrix}, \quad \mathbf{j} \longmapsto \mathrm{i}\begin{bmatrix} 0 & 1 \\ 1 & 0 \end{bmatrix}, \quad \mathbf{k} \longmapsto \mathrm{i}\begin{bmatrix} 0 & \mathrm{i} \\ -\mathrm{i} & 0 \end{bmatrix}.$$

One immediately checks that for a general quaternion $x := x_0 + x_1\mathbf{i} + x_2\mathbf{j} + x_3\mathbf{k} \in \mathbb{H}$, with $(x_i \in \mathbb{R})_{i=0}^3$, one has $\det(\gamma_x) = \sum_{i=0}^3 (x_i)^2 = \gamma_x \gamma_x^\dagger$. This implies that under the matrix representation (5.95) the unit norm quaternions are identified with the special unitary 2×2 matrices:

$$S(\mathbb{H}) \;=\; \mathrm{Sp}(1) \xrightarrow[\sim]{\ \gamma\ } \mathrm{SU}(2) \subset \mathcal{B}(\mathbb{C}^2)\,. \tag{5.96}$$

Definition 5.2.3 (cf. [CDL00, p 11][VB20, (2.4)]). The *group of unimodular quaternionic* 2×2 *matrices* is

$$\mathrm{SL}(\mathbb{H}^2) := \left\{ A \in \mathcal{B}(\mathbb{H}^2) \;\middle|\; \det_D(A) = 1 \right\}, \tag{5.97}$$

where

$$\det_D \begin{pmatrix} a & b \\ c & d \end{pmatrix} = \begin{cases} 0 & \text{if } \; b = c = d = 0 \\ \left|ad - aca^{-1}b\right| & \text{if } \; a \neq 0 \\ \left|da - dbd^{-1}c\right| & \text{if } \; d \neq 0 \\ \left|bdb^{-1}a - bc\right| & \text{if } \; b \neq 0 \\ \left|cac^{-1}d - cb\right| & \text{if } \; c \neq 0\,. \end{cases} \tag{5.98}$$

Definition 5.2.4 (cf. [Zha97, p. 28]). The *group of unitary quaternionic* $2{\times}2$ *matrices* (also: *compact symplectic group*) is:

$$\mathrm{Sp}(2) \equiv \mathrm{U}(\mathbb{H}^2) := \left\{ U \in \mathcal{B}(\mathbb{H}^2) \;\middle|\; U \cdot U^\dagger = 1 \right\}. \tag{5.99}$$

Remark 5.2.5. The condition in (5.99) is indeed sufficient, because for $B \in \mathcal{B}(\mathbb{H}^2)$, we have (cf. [Zha97, Prop. 4.1]):

$$B \cdot B^\dagger = 1 \quad \Leftrightarrow \quad B^\dagger \cdot B = 1\,. \tag{5.100}$$

Remark 5.2.6. Applying the star-algebra homomorphism γ (5.95) entrywise yields a star-algebra homomorphism

$$\begin{array}{ccc} \mathcal{B}(\mathbb{H}^2) & \xrightarrow{\ \gamma\ } & \mathcal{B}(\mathbb{C}^4) \\ \begin{pmatrix} a & b \\ c & d \end{pmatrix} & \longmapsto & \begin{pmatrix} \gamma_a & \gamma_b \\ \gamma_c & \gamma_d \end{pmatrix}, \end{array} \tag{5.101}$$

where on the right the 2x2 matrix of complex 2x2 matrices is canonically regarded as a 4x4 matrix. In particular, (5.101) gives a complex-unitary representation of the quaternionic unitary group (5.99):

$$\overset{\mathrm{Sp}(2)}{\underset{\mathbb{C}^4}{\circlearrowright}} \;:\; \mathrm{Sp}(2) \xrightarrow{\;\gamma\;} \mathrm{SU}(4)\,. \tag{5.102}$$

Lemma 5.2.7 (cf. [CDL00, Cor. 6.4]). *Every unitary matrix according to* (5.99) *is unimodular according to* (5.97)*:*

$$\mathrm{Sp}(2) \subset \mathrm{SL}(\mathbb{H}^2)\,. \tag{5.103}$$

Proof. The characteristic properties of $\det_D$ (5.98) are (cf. [CDL00, Thm. 5.1, & Cor. 6.4]), for all $A, B, \in \mathrm{Mat}_{2\times 2}(\mathbb{H})$

$$\begin{aligned} \det_D(A) &\in \mathbb{R}_{\geq 0} \\ \det_D(1) &= 1 \\ \det_D(A \cdot B) &= \det_D(A) \cdot \det_D(B) \\ \det_D\big(A^\dagger\big) &= \det_D(A)\,. \end{aligned} \tag{5.104}$$

Thereby, $U \cdot U^\dagger = 1$ implies

$$\begin{aligned} 1 &= \det_D\big(U \cdot U^\dagger\big) \\ &= \det_D(U) \cdot \det_D\big(U^\dagger\big) \\ &= \big(\det_D(U)\big)^2\,, \end{aligned} \tag{5.105}$$

and the only solution to that in $\mathbb{R}_{\geq 0}$ is $\det_D(U) = 1$. □

Remark 5.2.8. Lem. 5.2.7 entails that the quaternionic unitary group (5.99) plays the role of a *special* unitary group, certainly in its following appearance in Prop. 5.2.9. In view of this, the notation "Sp(2)" is more suggestive than "U(2, ℍ)", with its (coincidental but fortunate) alliteration to "SU(2)".

5.2.2 The Tautological Line Bundle

What drives the following construction is this fact:

Proposition 5.2.9 ([KT82], streamlined review in [BH09][VB20][FSS21c, §3.2]). *The following isomorphism of quadratic real vector spaces, from 6D Minkowski spacetime with signature* $\eta := \mathrm{diag}(-1, +1, \cdots, +1)$ *to the space of hermitian* 2×2 *quaternionic matrices with quadratic form being minus the ordinary determinant, intertwines the canonical* $\mathrm{Spin}(1,5) \to \mathrm{SO}(1,5)$*-action on the left with the conjugation action*

$$\begin{array}{ccc} \mathrm{SL}(\mathbb{H}^2) & \xrightarrow{\;\mathrm{conj}\;} & \mathrm{Aut}_{\mathbb{R}}\Big(\big\{A \in \mathcal{B}(\mathbb{H}^2)\,\big|\,A^\dagger = A\big\}\Big) \\ A & \longmapsto & A \cdot (-) \cdot A^\dagger \end{array} \tag{5.106}$$

of $\mathrm{SL}(\mathbb{H}^2)$ (Def. 5.2.3) *on the right:*

$$
\begin{array}{ccc}
\mathrm{Spin}(1,5) \circlearrowright_{\mathrm{can}} \left(\mathbb{R}^{1,5}, \eta\right) & \xrightarrow{\sim} & \mathrm{SL}(\mathbb{H}^2) \circlearrowright_{(5.106)} \left(\left\{A \in \mathcal{B}(\mathbb{H}^2) \,\middle|\, A^\dagger = A\right\}, -\det\right) \\
\begin{bmatrix} x^0 \\ x^1 \\ \vdots \\ x^5 \end{bmatrix} & \longmapsto & \begin{pmatrix} x^0 - x^1 & x^2 + \mathbf{i}x^3 + \mathbf{j}x^4 + \mathbf{k}x^5 \\ x^2 - \mathbf{i}x^3 - \mathbf{j}x^4 - \mathbf{k}x^5 & x^0 + x^1 \end{pmatrix},
\end{array}
\tag{5.107}
$$

and analogously so for 4D Minkowski spacetime and complex matrices, by restriction along the inclusion $\mathbb{C} \hookrightarrow \mathbb{H}$ (5.93)*:*

$$
\begin{array}{ccc}
\mathrm{Spin}(1,3) \circlearrowright_{\mathrm{can}} \left(\mathbb{R}^{1,3}, \eta\right) & \xrightarrow{\sim} & \mathrm{SL}(\mathbb{C}^2) \circlearrowright \left(\left\{A \in \mathcal{B}(\mathbb{C}^2) \,\middle|\, A^\dagger = A\right\}, -\det\right) \\
\begin{bmatrix} x^0 \\ x^1 \\ \vdots \\ x^3 \end{bmatrix} & \longmapsto & \begin{pmatrix} x^0 - x^1 & x^2 + \mathrm{i}x^3 \\ x^2 - \mathrm{i}x^3 & x^0 + x^1 \end{pmatrix}.
\end{array}
\tag{5.108}
$$

Corollary 5.2.10. *Under the identification of Prop. 5.2.9, the subspace* $\mathbb{R}^5 \subset \mathbb{R}^{1,5}$ *with its Euclidean norm* g*, and furthermore (pointed)* $S^4 \subset \mathbb{R}^5$*, are identified with those Hermitian matrices that are traceless and traceless unitary, respectively, and the corresponding subgroups* $\mathrm{Spin}(4) \subset \mathrm{Spin}(5) \subset \mathrm{Spin}(1,5)$ *are identified with* $\mathrm{Sp}(1) \times \mathrm{Sp}(1) \subset \mathrm{Sp}(2) \subset \mathrm{SL}(\mathbb{H}^2)$ (5.103)*, as follows:*

$$
\begin{array}{ccc}
\mathrm{Spin}(1,5) \circlearrowright_{\mathrm{can}} \left(\mathbb{R}^{1,5}, \eta\right) & \xrightarrow{\sim} & \mathrm{SL}(\mathbb{H}^2) \circlearrowright_{\mathrm{conj}} \left(\left\{A \in \mathcal{B}(\mathbb{H}^2) \,\middle|\, A^\dagger = A\right\}, -\det\right) \\
\uparrow & & \uparrow \\
\mathrm{Spin}(5) \circlearrowright_{\mathrm{can}} \left(\mathbb{R}^{5}, g\right) & \xrightarrow{\sim} & \mathrm{Sp}(2) \circlearrowright_{\mathrm{conj}} \left(\left\{A \in \mathcal{B}(\mathbb{H}^2) \,\middle|\, \substack{A^\dagger = A \\ \mathrm{tr}(A) = 0}\right\}, -\det\right) \\
\uparrow & & \uparrow \\
\mathrm{Spin}(5) \circlearrowright_{\mathrm{can}} S^4, & \xrightarrow{\sim} & \mathrm{Sp}(2) \circlearrowright_{\mathrm{conj}} \left\{A \in \mathcal{B}(\mathbb{H}^2) \,\middle|\, \substack{A^\dagger = A \\ \mathrm{tr}(A) = 0 \\ A \cdot A^\dagger = 1}\right\} \\
\uparrow & & \uparrow \\
\mathrm{Spin}(4) \circlearrowright_{\mathrm{can}} S^4_{\mathrm{nth}}, & \xrightarrow{\sim} & \mathrm{Sp}(1)^2 \circlearrowright_{\mathrm{conj}} \left\{A \in \mathcal{B}(\mathbb{H}^2) \,\middle|\, \substack{A^\dagger = A \\ \mathrm{tr}(A) = 0 \\ A \cdot A^\dagger = 1}\right\}_{\left(\begin{smallmatrix} 1 & 0 \\ 0 & -1 \end{smallmatrix}\right)}.
\end{array}
\tag{5.109}
$$

Proof. From the component expression in (5.107), it is manifest that the condition $x^0 = 0$, characterizing the subset $\mathbb{R}^5 \subset \mathbb{R}^{1,5}$, corresponds to vanishing trace on the Hermitian matrices. To see that it is precisely $\mathrm{Sp}(2) = \mathrm{U}(\mathbb{H}^2) \subset \mathrm{SL}(\mathbb{H}^2)$ which preserves this tracelessness condition, we note that this is equivalent to this subgroup preserving the orthogonal complement of the traceless Hermitian matrices. Again by the component expression in (5.107), this orthogonal complement is spanned by the identity matrix 1. Therefore, $G \in \mathrm{SL}(\mathbb{H}^2)$ preserves the traceless matrices iff $G \cdot 1 \cdot G^\dagger = 1$, hence iff G is unitary (5.99).

Further from the component expression in (5.107) one sees that

$$\left.\begin{smallmatrix} A^\dagger = A \\ \mathrm{tr}(A)=0 \end{smallmatrix}\right\} \quad \Rightarrow \quad AA^\dagger = -\det(A)\,, \tag{5.110}$$

which shows that the unitary traceless Hermitian matrices form the unit sphere inside $\mathbb{R}^5$. Finally, it is readily seen that the subgroup of Sp(2) whose conjugation action fixes the base point $\left(\begin{smallmatrix} 1 & 0 \\ 0 & -1 \end{smallmatrix}\right)$ among these matrices is the diagonal

$$\mathrm{Sp}(1) \times \mathrm{Sp}(1) \simeq \begin{pmatrix} \mathrm{Sp}(1) & 0 \\ 0 & \mathrm{Sp}(1) \end{pmatrix}.$$

This concludes the proof. □

Remark 5.2.11. The direct analog of Cor. 5.2.10 over the complex numbers gives

$$\mathrm{Spin}(3) \circlearrowright_{\mathrm{can}} S^2 \xrightarrow{\quad\sim\quad} \mathrm{SU}(2) \circlearrowright_{\mathrm{conj}} \left\{ A \in \mathcal{B}(\mathbb{C}^2) \,\middle|\, \begin{smallmatrix} A^\dagger = A \\ \mathrm{tr}(A)=0 \\ A\cdot A^\dagger = 1 \end{smallmatrix} \right\}. \tag{5.111}$$

However, in line with the fact that $\mathrm{Sp}(2) = \mathrm{U}(\mathbb{H}^2)$ is actually the quaternionic unitary group (5.99), and since unity of the determinant no longer plays a role when we conjugating matrices A constrained to have $\det(A) = -1$, the SU(2)-action on the right of (5.111) evidently extends to a $\mathrm{U}(2) \simeq \mathrm{Spin}^c(3)$-action:

$$\mathrm{Spin}^c(3) \circlearrowright_{\mathrm{can}} S^2 \xrightarrow{\quad\sim\quad} \mathrm{U}(2) \circlearrowright_{\mathrm{conj}} \left\{ A \in \mathcal{B}(\mathbb{C}^2) \,\middle|\, \begin{smallmatrix} A^\dagger = A \\ \mathrm{tr}(A)=0 \\ A\cdot A^\dagger = 1 \end{smallmatrix} \right\}. \tag{5.112}$$

Definition 5.2.12 (cf. [Bro, §5.3]). For $\mathbb{K} \in \{\mathbb{R}, \mathbb{C}, \mathbb{H}\}$, the $\mathbb{K}$-*projective space* of dimension $n \in \mathbb{N}$ is the space of *right* $\mathbb{K}$-lines in $\mathbb{K}^{n+1}$,

$$\mathbb{K}P^n = \left\{ v \cdot \mathbb{K} \subset \mathbb{K}^{n+1} \,\middle|\, v \in \mathbb{K}^{n+1} \setminus \{0\} \right\}, \tag{5.113}$$

traditionally denoted

$$\left[v_1 : v_2 : \cdots : v_{n+1}\right] := v \cdot \mathbb{K}\,, \quad \text{for } v = (v_1, \cdots, v_{n+1}) \in \mathbb{K}^{n+1} \setminus \{0\}\,, \tag{5.114}$$

with standard injections

$$\begin{array}{ccc} \mathbb{K}P^n & \hookrightarrow & \mathbb{K}P^{n+1} \\ \left[v_1 : \cdots : v_{n+1}\right] & \mapsto & \left[v_1 : \cdots : v_{n+1} : 0\right], \end{array} \tag{5.115}$$

canonical fibrations

$$\begin{array}{ccc} \mathbb{R}P^{2n+1} & \twoheadrightarrow & \mathbb{C}P^n \\ v \cdot \mathbb{R} & \mapsto & v \cdot \mathbb{C}\,, \end{array} \qquad \begin{array}{ccc} \mathbb{C}P^{2n+1} & \twoheadrightarrow & \mathbb{H}P^n \\ v \cdot \mathbb{C} & \mapsto & v \cdot \mathbb{H}\,, \end{array} \tag{5.116}$$

and their composites

$$\mathbb{R}P^{2n} \hookrightarrow \mathbb{R}P^{2n+1} \twoheadrightarrow \mathbb{C}P^n\,, \quad \mathbb{C}P^{2n} \hookrightarrow \mathbb{C}P^{2n+1} \twoheadrightarrow \mathbb{H}P^n. \tag{5.117}$$

(where the composite $\mathbb{R}P^{2n} \to \mathbb{C}P^n$ is labelled $f^{\mathbb{R}}_{\mathbb{C}}$ and the composite $\mathbb{C}P^{2n} \to \mathbb{H}P^n$ is labelled $f^{\mathbb{C}}_{\mathbb{H}}$)

These spaces carry a canonical action of the unitary group $\mathrm{U}(n+1, \mathbb{K})$ (5.99) by left multiplication:

$$
\begin{array}{ccc}
\mathrm{O}(n+1) \times \mathbb{R}P^n & \xrightarrow{\text{mult}} & \mathbb{R}P^n \\
\mathrm{U}(n+1) \times \mathbb{C}P^n & \xrightarrow{\text{mult}} & \mathbb{C}P^n \\
\mathrm{Sp}(n+1) \times \mathbb{H}P^n & \xrightarrow{\text{mult}} & \mathbb{H}P^n \\
(G, v \cdot \mathbb{K}) & \longmapsto & G \cdot v \cdot \mathbb{K}\,.
\end{array}
\tag{5.118}
$$

Corollary 5.2.13. *We have an identification of the two-four-sphere with the* $\mathbb{C}/\mathbb{H}$*-projective line* (5.113) *which intertwines the canonical* $\mathrm{Spin}^c(3)/\mathrm{Spin}(5)$*-action on* S^2/S^4 *with the multiplication action* (5.118) *of* $\mathrm{U}(2)/\mathrm{Sp}(2)$ *on* $\mathbb{C}P^1/\mathbb{H}P^1$ (5.113)*:*

$$
\begin{array}{ccc}
\mathrm{Spin}^c(3) \circlearrowleft_{\mathrm{can}} S^2 & \xrightarrow{\sim} & \mathrm{U}(2) \circlearrowleft_{\mathrm{mult}} \mathbb{C}P^1, \\
\mathrm{Spin}(5) \circlearrowleft_{\mathrm{can}} S^4 & \xrightarrow{\sim} & \mathrm{Sp}(2) \circlearrowleft_{\mathrm{mult}} \mathbb{H}P^1.
\end{array}
\tag{5.119}
$$

Proof. We spell out the argument over $\mathbb{H}$: By Cor. 5.2.10, the 4-sphere is identified with traceless unitary hermitian matrices. These are, furthermore, identified, first with rank=1 projection operators on $\mathbb{H}^2$, and then with lines in $\mathbb{H}^2$, like this:

$$
\begin{array}{ccccc}
S^4 \underset{(5.109)}{\overset{\sim}{\longleftrightarrow}} \left\{ A \in \mathcal{B}(\mathbb{H}^2) \,\middle|\, \begin{smallmatrix} A^\dagger = A \\ \mathrm{tr}(A)=0 \\ A \cdot A = 1 \end{smallmatrix} \right\} & \overset{\sim}{\longleftrightarrow} & \left\{ P \in \mathcal{B}(\mathbb{H}^2) \,\middle|\, \begin{smallmatrix} P^\dagger = P \\ \mathrm{tr}(P)=1 \\ P \cdot P = P \end{smallmatrix} \right\} & \overset{\sim}{\longleftrightarrow} & \mathbb{H}P^1 \\
A & \mapsto & P_A := \tfrac{1}{2}(1 - A) & \mapsto & \ker(P_A) \\
A_P := (1 - 2P) & \leftarrow\!\shortmid & P,\; P_v := 1 - \frac{v \cdot v^\dagger}{|v|^2} & \leftarrow\!\shortmid & v \cdot \mathbb{H}\,.
\end{array}
\tag{5.120}
$$

Now, Cor. 5.2.10 also shows that the Spin(5)-action on S^4 becomes the Sp(2)-conjugation action (5.106) on the matrices A. Under the above bijection (5.120), this translates first to the same kind of conjugation action on projection operators P and then to left multiplication on $\mathbb{H}$-lines, as claimed, since:

$$
\ker\big(G \cdot P \cdot G^\dagger\big) \simeq G \cdot \ker(P)\,.
\tag{5.121}
$$

□

The argument over $\mathbb{C}$ is verbatim the same, now using the U(2)-action from (5.112).

Definition 5.2.14. The kernel bundle of the family of the $\mathbb{C}P^1/\mathbb{H}P^1$-parameterized projection operators (5.120) is the *tautological* $\mathbb{C}/\mathbb{H}$*-line bundle* over $\mathbb{C}P^1/\mathbb{H}P^1$:

$$
\begin{array}{ccc}
\mathcal{L}^{\mathrm{taut}}_{\mathbb{K}P^1} := \left\{ (P, v) \,\middle|\, \begin{smallmatrix} P \in \mathbb{K}P^1 \\ v \in \ker(P) \end{smallmatrix} \right\} & \hookrightarrow & \mathbb{K}P^1 \times \mathbb{K}^2\,. \\
& \searrow \quad \mathbb{H}P^1 \quad \swarrow &
\end{array}
\tag{5.122}
$$

Next we are after the canonical trivialization of this tautological line bundle after pullback along the *Hopf fibration*.

5.2.3 The Hopf Fibration

Definition 5.2.15. The $\mathbb{R}/\mathbb{C}/\mathbb{H}$-*Hopf fibration* is the map which sends unit norm elements in $\mathbb{R}^2/\mathbb{C}^2/\mathbb{H}^2$ (Def. 5.2.1) to the $\mathbb{R}/\mathbb{C}/\mathbb{H}$-lines which they span (5.113):

$$\begin{array}{ccccc}
S^1 \simeq S(\mathbb{R}^2) \ni v & & S^3 \simeq S(\mathbb{C}^2) \ni v & & S^7 \simeq S(\mathbb{H}^2) \ni v \\
\big\downarrow{\scriptstyle p_{\mathbb{R}}} \quad \big\downarrow \quad \big\downarrow & & \big\downarrow{\scriptstyle p_{\mathbb{C}}} \quad \big\downarrow \quad \big\downarrow & & \big\downarrow{\scriptstyle p_{\mathbb{H}}} \quad \big\downarrow \quad \big\downarrow \\
S^1 \simeq \mathbb{R}P^1 \ni v\cdot\mathbb{R}, & & S^2 \simeq \mathbb{C}P^1 \ni v\cdot\mathbb{C}, & & S^4 \underset{(5.119)}{\simeq} \mathbb{H}P^1 \ni v\cdot\mathbb{H}.
\end{array} \tag{5.123}$$

Remark 5.2.16. By incrementally forming lines along the sequence of inclusions $\mathbb{R} \subset \mathbb{C} \subset \mathbb{H}$, the $\mathbb{C}/\mathbb{H}$-Hopf fibration (5.123) canonically factors through further relevant fibrations:

$$\begin{array}{ccccc}
S(\mathbb{C}^2) & \twoheadrightarrow & \mathbb{R}P^3 & \xrightarrow{t_{\mathbb{C}}} & \mathbb{C}P^1 \\
v & \longmapsto & v\cdot\mathbb{R} & \longmapsto & v\cdot\mathbb{C},
\end{array} \qquad (\text{composite } p_{\mathbb{C}}) \tag{5.124}$$

and

$$\begin{array}{ccccccc}
S(\mathbb{H}^2) & \twoheadrightarrow & \mathbb{R}P^7 & \twoheadrightarrow & \mathbb{C}P^3 & \xrightarrow{t_{\mathbb{H}}} & \mathbb{H}P^1 \\
v & \longmapsto & v\cdot\mathbb{R} & \longmapsto & v\cdot\mathbb{C} & \longmapsto & v\cdot\mathbb{H}.
\end{array} \qquad (\text{composite } p_{\mathbb{H}}) \tag{5.125}$$

The map $t_{\mathbb{C}} : \mathbb{R}P^3 \twoheadrightarrow \mathbb{C}P^1$ (5.124) plays a key role in § 3.1.4, while $t_{\mathbb{H}} : \mathbb{C}P^3 \twoheadrightarrow \mathbb{H}P^1$ (5.125), also known as the *twistor fibration* (cf. [FSS22, §2]), plays a key role in more refined variants of flux quantization on M5-branes, discussed elsewhere (cf. [SS25e][SS25f][FSS23, §12][SS25g]).

Lemma 5.2.17. *The* $\mathbb{C}/\mathbb{H}$*-Hopf fibration* (5.123) *is equivariant with respect to*

(i) *the canonical* $\mathrm{U}(2)/\mathrm{Sp}(2)$*-action on* $S(\mathbb{C}^2)/S(\mathbb{H}^2)$*;*

(ii) *the multiplication action* (5.118) *on* $\mathbb{C}P^1/\mathbb{H}P^1$*:*

$$\begin{array}{ccccccc}
\mathrm{Spin}^c(3) \curvearrowright S^3 \simeq S(\mathbb{C}^2) \curvearrowleft \mathrm{U}(2) & & \mathrm{Spin}(5) \curvearrowright S^7 \simeq S(\mathbb{H}^2) \curvearrowleft \mathrm{Sp}(2) \\
\big\downarrow{\scriptstyle p_{\mathbb{C}}} \qquad \big\downarrow{\scriptstyle v \,\mapsto\, v\cdot\mathbb{C}} & & \big\downarrow{\scriptstyle p_{\mathbb{H}}} \qquad \big\downarrow{\scriptstyle v \,\mapsto\, v\cdot\mathbb{H}} \\
\mathrm{Spin}^c(3) \curvearrowright S^2 \simeq \mathbb{C}P^1 \curvearrowleft \mathrm{U}(2) & & \mathrm{Spin}(5) \curvearrowright S^4 \simeq \mathbb{H}P^1 \curvearrowleft \mathrm{Sp}(2).
\end{array} \tag{5.126}$$

This equivariance extends canonically to the factorizations of Rem. 5.2.16.

Proof. Evidently, for $v \in \mathbb{H}^2$ and $G \in \mathrm{Sp}(2)$, the assignments

$$\begin{array}{ccc}
v & \longmapsto & G\cdot v \\
\big\downarrow & & \big\downarrow \\
v\cdot\mathbb{H} & \longmapsto & G\cdot v\cdot\mathbb{H}
\end{array} \tag{5.127}$$

commute. Analogously so over $\mathbb{C}$. □

Lemma 5.2.18. *The cell attachments* (Ex. 4.1.3) *along the Hopf fibrations* (Def. 5.2.15) *and their factorizations* (Rem. 5.2.16) *are projective spaces* (Def. 5.2.12) *as follows:*

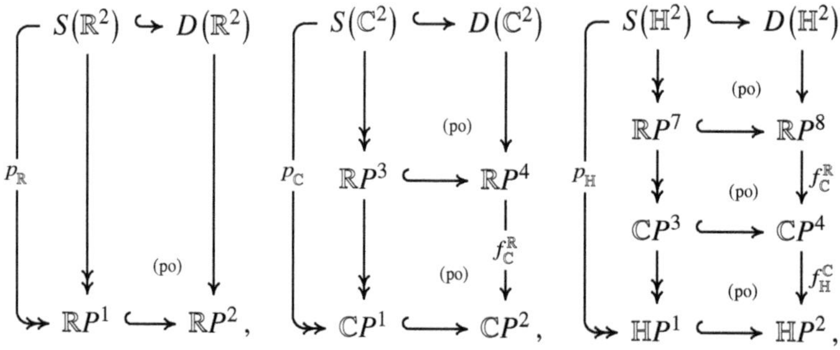

(5.128)

where the top right maps are given by

$$\begin{array}{ccc} D(\mathbb{R}^n) & \longrightarrow & \mathbb{R}P^n \\ (v_1, \cdots, v_n) & \mapsto & \left[v_1 : \cdots : v_n : \sqrt{1 - |\vec{v}|^2}\right]. \end{array} \tag{5.129}$$

Proof. By the pasting law (Prop. 4.1.4), the claim follows from the general and standard statement (cf. [Mas91, §IX.3.6]) that the following is a pushout for all $n \in \mathbb{N}$ and $\mathbb{K} \in \{\mathbb{R}, \mathbb{C}, \mathbb{H}\}$:

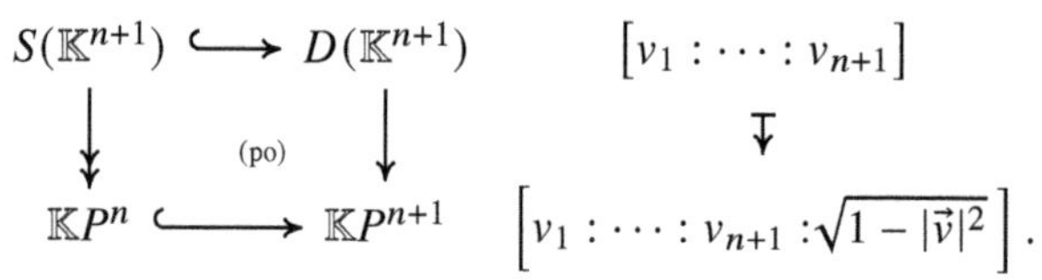

(5.130)

Namely, this expresses the fact that an element $[v_1 : \cdots : v_{n+1} : v_{n+2}] \in \mathbb{K}P^{n+1}$:

- either has $v_{n+2} = 0$, in which case it is in the image of $\mathbb{K}P^n \hookrightarrow \mathbb{K}P^{n+1}$ and equivalently in the image of $S(\mathbb{K}^{n+1}) \hookrightarrow D(\mathbb{K}^{n+1}) \to \mathbb{K}P^{n+1}$,
- or else it equals (setting $v'_n := v_n / v_{n+2}$ for $n \in \{1, \cdots, n+1\}$):

$$\left[v'_1 : \cdots : v'_{n+1} : 1\right] = \left[\frac{v'_1}{\sqrt{|\vec{v}'|^2+1}} : \cdots : \frac{v'_{n+1}}{\sqrt{|\vec{v}'|^2+1}} : \frac{1}{\sqrt{|\vec{v}'|^2+1}} = \sqrt{1 - \left|\frac{\vec{v}'}{\sqrt{|\vec{v}'|^2+1}}\right|^2}\right]$$

and hence is in the image of $D(\mathbb{K}^{n+1}) \to \mathbb{K}P^{n+1}$. □

Lemma 5.2.19. *The* $\mathbb{R}/\mathbb{C}/\mathbb{H}$*-Hopf fibrations* (Def. 5.2.15) *are equivalently the following coset coprojections:*

$$
\begin{array}{ccccccc}
S^1 & \simeq & S(\mathbb{R}^2) & \simeq & \dfrac{\mathrm{O}(2)}{\mathrm{O}(1)_{\mathrm{stb}}} & := & \mathrm{O}(2)\Big/\begin{pmatrix} 1 & 0 \\ 0 & \mathrm{O}(1) \end{pmatrix} \\
\Big\downarrow{\scriptstyle p_{\mathbb{R}}} & & \Big\downarrow {\scriptstyle v \mapsto v\cdot\mathbb{R}} & & \Big\downarrow & & \\
S^1 & \simeq & \mathbb{R}P^1 & \simeq & \dfrac{\mathrm{O}(2)}{\mathrm{O}(1)^2} & := & \mathrm{O}(2)\Big/\begin{pmatrix} \mathrm{O}(1) & 0 \\ 0 & \mathrm{O}(1) \end{pmatrix},
\end{array}
\tag{5.131a}
$$

$$
\begin{array}{ccccccc}
S^3 & \simeq & S(\mathbb{C}^2) & \simeq & \dfrac{\mathrm{U}(2)}{\mathrm{U}(1)_{\mathrm{stb}}} & := & \mathrm{U}(2)\Big/\begin{pmatrix} 1 & 0 \\ 0 & \mathrm{U}(1) \end{pmatrix} \\
\Big\downarrow{\scriptstyle p_{\mathbb{C}}} & & \Big\downarrow {\scriptstyle v \mapsto v\cdot\mathbb{C}} & & \Big\downarrow & & \\
S^2 & \simeq & \mathbb{C}P^1 & \simeq & \dfrac{\mathrm{Sp}(2)}{\mathrm{U}(1)^2} & := & \mathrm{U}(2)\Big/\begin{pmatrix} \mathrm{U}(1) & 0 \\ 0 & \mathrm{U}(1) \end{pmatrix},
\end{array}
\tag{5.131b}
$$

$$
\begin{array}{ccccccc}
S^7 & \simeq & S(\mathbb{H}^2) & \simeq & \dfrac{\mathrm{Sp}(2)}{\mathrm{Sp}(1)_{\mathrm{stb}}} & := & \mathrm{Sp}(2)\Big/\begin{pmatrix} 1 & 0 \\ 0 & \mathrm{Sp}(1) \end{pmatrix} \\
\Big\downarrow{\scriptstyle p_{\mathbb{H}}} & & \Big\downarrow {\scriptstyle v \mapsto v\cdot\mathbb{H}} & & \Big\downarrow & & \\
S^4 & \simeq & \mathbb{H}P^1 & \simeq & \dfrac{\mathrm{Sp}(2)}{\mathrm{Sp}(1)^2} & := & \mathrm{Sp}(2)\Big/\begin{pmatrix} \mathrm{Sp}(1) & 0 \\ 0 & \mathrm{Sp}(1) \end{pmatrix}.
\end{array}
\tag{5.131c}
$$

Proof. We indicate the argument over $\mathbb{H}$, the other two cases are directly analogous: By Lem. 5.2.17, we have a transitive $\mathrm{Sp}(2)$-action, exhibiting the 7-sphere as the $\mathrm{Sp}(2)$-translations of its base point

$$
\mathrm{Sp}(2)\cdot\begin{pmatrix} 1 \\ 0 \end{pmatrix} \simeq S(\mathbb{H}^2)\,. \tag{5.132}
$$

Now the stabilizer group of the base point is manifestly

$$
\mathrm{Stab}_{\mathrm{Sp}(2)}\begin{pmatrix} 1 \\ 0 \end{pmatrix} = \begin{pmatrix} 0 & 0 \\ 0 & \mathrm{Sp}(1) \end{pmatrix} =: \mathrm{Sp}(1)_{\mathrm{stb}} \subset \mathrm{Sp}(2)\,, \tag{5.133}
$$

while the fiber of the base point $[1:0] \in \mathbb{H}P^1$ is manifestly

$$
\mathrm{fib}(p_{\mathbb{H}}) = \begin{pmatrix} \mathrm{Sp}(1) \\ 0 \end{pmatrix} = \begin{pmatrix} 1 \\ 0 \end{pmatrix}\cdot\begin{pmatrix} \mathrm{Sp}(1) & 0 \\ 0 & 1 \end{pmatrix}. \tag{5.134}
$$

This implies the claim by the *orbit-stabilizer theorem* for Lie group actions (cf. [War83, Thm. 3.62][Lee12, Thm. 21.18]). □

Over $\mathbb{C}$ we obtain from this yet another incarnation of the Hopf fibration, which is useful to make explicit (cf. [Lyo03, (3)]):

Corollary 5.2.20. *The* $\mathbb{C}$*-Hopf fibration is equivalently given by the conjugation action of* $\mathrm{SU}(2)$ *on the matrix representing the basepoint of* S^2 (5.109)*:*

$$
\begin{array}{ccc}
S^3 & \overset{\sim}{\longleftrightarrow} & \mathrm{SU}(2) \\
\Big\downarrow{\scriptstyle p_{\mathbb{C}}} & & \Big\downarrow \begin{array}{c} g \\ \downarrow \\ g\cdot\begin{pmatrix} +1 & 0 \\ 0 & -1 \end{pmatrix}\cdot g^{-1} \end{array} \\
S^2 & \overset{\sim}{\longleftrightarrow} & \left\{ A \in \mathcal{B}(\mathbb{C}^2) \,\middle|\, \begin{array}{c} A^\dagger = A \\ \mathrm{tr}(A)=0 \\ A\cdot A = 1 \end{array} \right\}.
\end{array}
\tag{5.135}
$$

Proof. First to observe that we have a homeomorphism of underlying spaces:

$$
\begin{array}{ccc}
\frac{\mathrm{U}(2)}{\mathrm{U}(1)_{\mathrm{stb}}} & \overset{\sim}{\longleftarrow} & \mathrm{SU}(2) \\
g\cdot \mathrm{U}(1)_{\mathrm{stb}} & \longleftarrow\!\shortmid & g\,,
\end{array}
\tag{5.136}
$$

conversely reflecting the fact that every coset in $\mathrm{U}(2)/\mathrm{U}(1)_{\mathrm{stb}}$ contains precisely one special unitary matrix.

Next to note that under the isomorphisms of (5.120), one incarnation of the $\mathbb{C}$-Hopf fibration is as shown on the left and middle of the following diagram:

$$
\begin{array}{ccc}
S(\mathbb{C}^2) & U\cdot v & g\cdot\begin{pmatrix}1\\0\end{pmatrix} \\
\Big\downarrow & \Big\downarrow & \Big\downarrow \\
\left\{ A \in \mathcal{B}(\mathbb{C}^2) \,\middle|\, \begin{array}{c} A^\dagger = A \\ \mathrm{tr}(A)=0 \\ A\cdot A = 1 \end{array} \right\} & U\cdot\big(2v\cdot v^\dagger - 1\big)\cdot U^\dagger & g\cdot\begin{pmatrix} +1 & 0 \\ 0 & -1 \end{pmatrix}\cdot g^{-1}.
\end{array}
\tag{5.137}
$$

Combining these two statements gives that $g \in \mathrm{SU}(2)$ maps as shown on the right, which is the claim to be proven. □

In § 5.2 we will crucially use the following curious incarnation of the Hopf fibration in geometric homotopy theory, which may be less widely appreciated:

Corollary 5.2.21. *After passage to homotopy* $\mathrm{Sp}(2)$*-quotients, the quaternionic Hopf fibration is equivalent to the delooping of the inclusion of the stabilizer subgroup* $\mathrm{Sp}(1)_{\mathrm{stb}}$ *into its product with the active* $\mathrm{Sp}(1)$*, as shown on the right here:*

$$
\begin{array}{ccccccc}
\mathrm{Spin}(5)\backslash\!\!\backslash S^7 & \simeq & \mathrm{Sp}(2)\backslash\!\!\backslash S(\mathbb{H}^2) & \simeq & \mathrm{Sp}(2)\backslash\!\!\backslash \frac{\mathrm{Sp}(2)}{\mathrm{Sp}(1)_{\mathrm{stb}}} & \simeq & \mathrm{Sp}(1)_{\mathrm{stb}}\backslash\!\!\backslash * \\
{\scriptstyle \mathrm{Spin}(5)\backslash\!\!\backslash p_{\mathbb{H}}}\Big\downarrow & & \Big\downarrow & & \Big\downarrow & & \Big\downarrow \\
\mathrm{Spin}(5)\backslash\!\!\backslash S^4 & \simeq & \mathrm{Sp}(2)\backslash\!\!\backslash \mathbb{H}P^1 & \simeq & \mathrm{Sp}(2)\backslash\!\!\backslash \frac{\mathrm{Sp}(2)}{\mathrm{Sp}(1)^2} & \simeq & \mathrm{Sp}(1)^2\backslash\!\!\backslash *\,.
\end{array}
\tag{5.138}
$$

Proof. By the general equivalence $\mathbf{B}H \simeq G\backslash\!\!\backslash G/H$ (4.67), for subgroups $H \subset G$. □

5.2.4 Trivialization along Hopf Fibration

For completeness, we close this discussion by highlighting the traditional trivialization of the pullback of the tautological line bundle along the Hopf fibration (but in § 5.2 we instead use Cor. 5.2.21 for a more powerful argument):

Proposition 5.2.22. *The unit sphere bundle of the tautological line bundle* (5.122) *on* $\mathbb{C}P^1/\mathbb{H}P^1$ *is isomorphic to the Hopf fibration* (5.123)*:*

$$\begin{array}{ccc} S\big(\mathcal{L}^{\mathrm{taut}}_{\mathbb{H}P^1}\big) & \xrightarrow{\;\sim\;} & S^7 \\ \downarrow & & \big\downarrow{\scriptstyle p_{\mathbb{H}}} \\ \mathbb{H}P^1 & \xrightarrow{\;\sim\;} & S^4. \end{array} \tag{5.139}$$

Proof. Unwinding the definitions, this is again essentially a tautology:

$$\begin{array}{ccccc} S(\mathcal{H}_{\mathbb{H}P^1}) & & \xrightarrow{\qquad\sim\qquad} & & S(\mathbb{H}^2) \\ \Big\downarrow & (P, v) & \rightleftarrows & v & \Big\downarrow \\ & \downarrow & & \downarrow & \\ & P & = & 1 - \frac{v v^\dagger}{|v|^2} & \\ \mathbb{H}P^1 & & =\!=\!=\!=\!=\!= & & \mathbb{H}P^1, \end{array} \tag{5.140}$$

using here that every $v \in \mathbb{H}^2$ is in the kernel of a unique self-adjoint rank=1 projector (5.120), and observing that the condition $|v| = 1$ is the same on both sides. □

Proposition 5.2.23. *The pullback of the tautological* $\mathbb{H}$*-line bundle* (5.122) *along the* $\mathbb{H}$*-Hopf fibration* (5.123) *trivializes* $\mathrm{Sp}(2)$*-equivariantly:*

$$\begin{array}{ccc} \overset{\mathrm{Sp}(2)}{\circlearrowright} & & \overset{\mathrm{Sp}(2)}{\circlearrowright} \\ S^7 \times \mathbb{H} & \xrightarrow{\qquad\sim\qquad} & p_{\mathbb{H}}^* \mathcal{L}^{\mathrm{taut}}_{\mathbb{H}P^1} \,. \\ & \searrow \quad S^7 \quad \swarrow & \end{array} \tag{5.141}$$

Proof. In components, the bundle isomorphism may be given as

$$\begin{array}{ccc} \overbrace{\left\{ (u, q) \;\middle|\; \begin{smallmatrix} u \in \mathbb{H}^2 \\ q \in \mathbb{H} \\ |u| = 1 \end{smallmatrix} \right\}}^{S(\mathbb{H}^2) \times \mathbb{H}} & \xleftrightarrow{\;\sim\;} & \overbrace{\left\{ (u, P, v) \;\middle|\; \begin{smallmatrix} P \in \mathbb{H}P^1 \\ u, v \in \ker(P) \\ |u| = 1 \end{smallmatrix} \right\}}^{p_{\mathbb{H}}^* \mathfrak{L}_{\mathbb{H}P^1}} \\ (u, q) & \longmapsto & \big(u, 1 - u u^\dagger, u \cdot q\big) \\ \big(u, v/u\big) & \longleftarrow\!\!\!\shortmid & (u, P, v)\,, \end{array} \tag{5.142}$$

whose equivariance is manifest. □

Remark 5.2.24. In terms of classifying maps, the trivialization of Thm. 5.2.22 is a homotopy of this form

$$\begin{array}{ccc} S^7 & \longrightarrow & * \\ \big\downarrow{\scriptstyle p_{\mathbb{C}}} & \overset{\sim}{\swarrow\!\!\!\!=} & \big\downarrow \\ \mathbb{H}P^1 & \xrightarrow{\;\vdash \mathcal{L}^{\mathbb{H}P^1}\;} & \mathbb{H}P^\infty. \end{array} \tag{5.143}$$

This shows that postcomposition with a quaternion orientation of complex K-theory, $\mathbb{H}P^\infty \to \mathrm{KU}$, gives a homotopy h_3^{KU} according to Fig. 2.6.

Next we turn to constructing this in detail and equivariantly (cf. Prop. 5.2.30).

Based on the above construction of the Sp(2)-equivariant tautological quaternionic line bundle $\mathcal{L}^{\mathrm{taut}}_{\mathbb{H}P^1}$ (Def. 5.2.14) over $S^4 \simeq \mathbb{H}P^1$, we now explicitly construct its incarnation in the orbifold K-theory of Def. 5.1.22 for trivial PCT symmetry (hence in the KU^0-sector, by Ex. 5.1.25) and exhibit the trivialization of its pullback along the quaternionic Hopf fibration. This explicitly exhibits the four/ten-dimensional equivariant orientation of KU^0 in the manner explained around Fig. 2.6.

5.2.5 The equivariant unit

In fact, the tautological realization of $\mathcal{L}^{\mathrm{taut}}_{\mathbb{H}P^1}$ as the kernel bundle of the projectors which are the points of $\mathbb{H}P^1$ (Def. 5.2.14) induces a similarly tautological incarnation of its reduced K-theory class as the virtual kernel of parameterized Fredholm operators (5.67), by use of Ex. 5.1.14:

Lemma 5.2.25. *The complex vector bundle underlying the tautological* $\mathbb{H}$*-line bundle over* $\mathbb{H}P^1$ (Def. 5.2.14) *is the virtual kernel bundle* (5.50) *of the following* $\mathbb{H}P^1$*-parameterized graded Fredholm operator:*

$$
\begin{array}{ccccccc}
\mathbb{H}P^1 & \underset{(5.120)}{\xrightarrow{\sim}} & \left\{ P \in \mathcal{B}(\mathbb{H}^2) \,\middle|\, \substack{P^\dagger = P \\ P \circ P = P \\ \mathrm{tr}_{\mathbb{H}}(P) = 1} \right\} & \underset{(5.101)}{\xrightarrow{\gamma}} & \left\{ P \in \mathcal{B}(\mathbb{C}^4) \,\middle|\, \substack{P^\dagger = P \\ P \circ P = P} \right\} & \underset{(5.58)}{\xrightarrow{F_{(-)}}} & \mathrm{Fred}_{\mathrm{gr}} \\
P & \mapsto & P & \mapsto & \gamma(P) & \mapsto & F_{\gamma(P)}\,,
\end{array}
\qquad (5.144)
$$

(the composite map $\mathbb{H}P^1 \to \mathrm{Fred}_{\mathrm{gr}}$ being labelled $F^{\mathrm{taut}}_{\mathbb{H}P^1}$)

in that

$$
\ker\big(F^{\mathrm{taut}}_{\mathbb{H}P^1}\big) = \gamma(\mathcal{L}^{\mathrm{taut}}_{\mathbb{H}P^1}) \ominus 0\,. \qquad (5.145)
$$

Proof. By Ex. 5.1.14, the virtual kernel of $F^{\mathrm{taut}}_P \equiv F_{\gamma(P)}$ is that of $\gamma(P)$, by Rem. 5.2.6 this is the underlying complex vector space of the kernel of P, and that, by Def. 5.2.14, is the fiber of the tautological $\mathbb{H}$-line bundle over P. □

Example 5.2.26. The virtual difference (5.49) of the above (5.144) and the constant $\mathbb{H}P^1$-parametrized Fredholm operator $P \mapsto F_{\mathbb{C}^2}$ (5.53)

$$
\mathbb{H}P^1 \xrightarrow{F^{\mathrm{taut}}_{\mathbb{H}P^1} \ominus F^{\mathbb{C}^2}_{\mathbb{H}P^1}} \mathrm{Fred}_{\mathrm{gr}} \qquad (5.146)
$$

has as virtual kernel bundle the "reduced" version of the (underlying complex vector bundle) of the tautological $\mathbb{H}$-line bundle,

$$
\ker\big(F^{\mathrm{taut}}_{\mathbb{H}P^1} \ominus F^{\mathbb{C}^2}_{\mathbb{H}P^1}\big) = \gamma(\mathcal{L}^{\mathrm{taut}}_{\mathbb{H}P^1}) \ominus \mathbb{C}^2_{\mathbb{H}P^1}, \qquad (5.147)
$$

of vanishing virtual dimension, $\dim\Big(\ker\big(F^{\mathrm{taut}}_{\mathbb{H}P^1} \ominus F^{\mathbb{C}^2}_{\mathbb{H}P^1}\big)\Big) = 0\,.$

Notation 5.2.27. Let

$$
\mathrm{Sp}(2) \xrightarrow{\oplus_{\mathbb{N}} \gamma} \mathrm{U}(\mathcal{H}) \twoheadrightarrow \mathrm{PU}(\mathcal{H}) \hookrightarrow \mathrm{PU}(\mathcal{H})^{\mathrm{gr}} \hookrightarrow \mathrm{Aut}\big(\mathrm{Fred}_{\mathrm{gr}}\big) \qquad (5.148)
$$

be (the projective image of) $\mathbb{N}$ direct summands of the 4-dimensional irrep (5.102), realizing the Hilbert space $\mathcal{H}$ (5.1) as the unitary Sp(2)-representation which is the $\mathbb{N}$-indexed direct sum of these irreps (5.102):[3]

$$\overset{\mathrm{Sp}(2)}{\overset{\circlearrowright}{\mathcal{H}}} := \bigoplus_{\mathbb{N}} \overset{\mathrm{Sp}(2)}{\overset{\circlearrowright}{\mathbb{C}^4}} . \tag{5.149}$$

Remark 5.2.28. We will be concerned with the restriction of this Sp(2)-representation (5.149) to a representation of the stabilizer subgroup (5.133),

$$\mathrm{Sp}(1)_{\mathrm{stb}} \xhookrightarrow{\iota_{\mathrm{stb}}} \mathrm{Sp}(2)\,, \tag{5.150}$$

in which form it becomes equivariantly unitarily equivalent, as in (5.62), to the direct sum of an infinite-dimensional trivial representation with an $\mathbb{N}$-indexed sum of copies of the defining representation $\mathbb{C}^2_{\mathrm{def}}$ of $\mathrm{Sp}(1) \simeq \mathrm{SU}(2)$ (5.96):

$$\begin{aligned} \overset{\mathrm{Sp}(1)_{\mathrm{stb}}}{\overset{\circlearrowright}{\mathcal{H}}} &= \bigoplus_{\mathbb{N}} \big(\overset{\mathrm{Sp}(1)_{\mathrm{stb}}}{\overset{\circlearrowright}{\mathbb{C}^4}} \big) = \bigoplus_{\mathbb{N}} \big(\mathbb{C}^2_{\mathrm{triv}} \oplus \overset{\mathrm{Sp}(1)_{\mathrm{stb}}}{\overset{\circlearrowright}{\mathbb{C}^2_{\mathrm{def}}}} \big) \\ &\xrightarrow[\sim]{U} \underbrace{\Big(\bigoplus_{\mathbb{N}} \mathbb{C}^2_{\mathrm{triv}} \Big)}_{\mathcal{H}_{\mathrm{triv}}} \oplus \underbrace{\Big(\bigoplus_{\mathbb{N}} \overset{\mathrm{Sp}(1)_{\mathrm{stb}}}{\overset{\circlearrowright}{\mathbb{C}^2_{\mathrm{def}}}} \Big)}_{\mathcal{H}_{\mathrm{def}}} . \end{aligned} \tag{5.151}$$

Under this transformation, the Fredholm operator F_{P_0} transforms into (5.64)

$$U \circ F_{P_0} \circ U^{-1} = F^{\mathbb{C}^2_{\mathrm{triv}}} \oplus F^{0 \cdot \mathbb{C}^2_{\mathrm{def}}} . \tag{5.152}$$

Lemma 5.2.29. *The tautological $\mathbb{H}P^1$-parameterized Fredholm operator* (5.144) *is* Sp(2)*-equivariant with respect to the canonical action on* $\mathbb{H}P^1$ (5.126) *and the action on* $\mathrm{Fred}_{\mathrm{gr}}$ *from* (5.148)*:*

$$\overset{\mathrm{Sp}(2)}{\overset{\circlearrowright}{\mathbb{H}P^1}} \xrightarrow{\;F^{\mathrm{taut}}_{\mathbb{H}P^1}\;} \overset{\mathrm{Sp}(2)}{\overset{\circlearrowright}{\mathrm{Fred}_{\mathrm{gr}}}}\,, \tag{5.153}$$

and hence descends to a map of homotopy quotient stacks (4.107) *of this form:*

$$\begin{array}{ccc} \mathrm{Sp}(2)\backslash\!\!\backslash \mathbb{H}P^1 & \xrightarrow{\;\mathrm{Sp}(2)\backslash\!\!\backslash F^{\mathrm{taut}}_{\mathbb{H}P^1}\;} & \mathrm{Sp}(2)\backslash\!\!\backslash \mathrm{Fred}_{\mathrm{gr}} . \\ & \searrow \quad \mathbf{B}\mathrm{Sp}(2) \quad \swarrow & \end{array} \tag{5.154}$$

Proof. The first two maps in (5.144) are tautologically equivariant, as around (5.121). The third map in (5.144) is clearly equivariant by the block matrix form (5.56) of F_P

[3] Generally one would take $\mathcal{H}$ to be a *stable representation* of Sp(2), namely, the $\mathbb{N}$-fold direct sum of the direct sum of all the complex irreducible representations of Sp(2). But for the present purpose, and not to overburden the notation unnecessarily, we may stick with (5.148).

being compatible with the direct sum of representations (5.149):

$$
\begin{aligned}
&\big(\oplus_{\mathbb{N}}\gamma(G)\big)\begin{pmatrix} P & 0 & 0 & \cdots \\ 1-P & P & 0 & \cdots \\ 0 & 1-P & P & \cdots \\ 0 & 0 & 1-P & \cdots \\ \vdots & \vdots & \vdots & \ddots \end{pmatrix}\big(\oplus_{\mathbb{N}}\gamma(G)^{-1}\big) \\
&= \begin{pmatrix} \gamma(G)P\gamma(G)^{-1} & 0 & 0 & \cdots \\ 1-\gamma(G)P\gamma(G)^{-1} & \gamma(G)P\gamma(G)^{-1} & 0 & \cdots \\ 0 & 1-\gamma(G)P\gamma(G)^{-1} & \gamma(G)P\gamma(G)^{-1} & \cdots \\ 0 & 0 & 1-\gamma(G)P\gamma(G)^{-1} & \cdots \\ \vdots & \vdots & \vdots & \ddots \end{pmatrix}.
\end{aligned}
\tag{5.155}
$$

This establishes the claim. □

5.2.6 The orienting homotopy

With this in hand, we have the following Fredholm operator analogue of Prop. 5.2.23, cf. Rem. 5.2.24, which realizes the phenomenon of Figs. 2.6 and 2.7 for $E = \mathrm{KU}$:

Proposition 5.2.30. *The pullback of the reduced version* (5.146) *of the tautological* $\mathbb{C}P^1/\mathbb{H}P^1$*-parameterized Fredholm operator* (5.144) *along the* $\mathbb{C}/\mathbb{H}$*-Hopf fibration* (5.123) *trivializes* $\mathrm{U}(2)/\mathrm{Sp}(2)$*-equivariantly, in that we have an equivariant homotopy* (4.21) *of this form:*

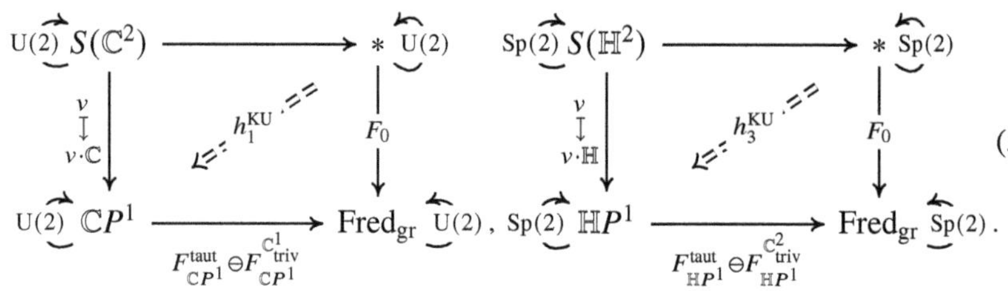

(5.156)

Proof. We spell out the proof over $\mathbb{H}$; for $\mathbb{C}$ it is verbatim the same up to the evident substitutions.

After equivalent passage to Sp(2)-homotopy quotients (4.107) and under the equivalence of Cor. 5.2.21, we are reduced to exhibiting **B**Sp(2)-sliced homotopy (4.21) as indicated by the dashed arrow in this diagram:

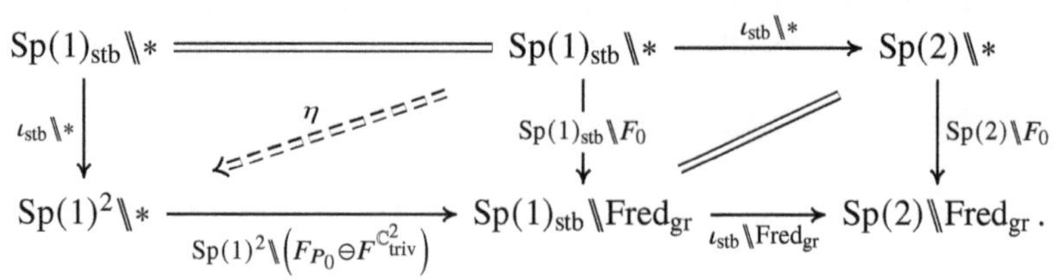

(5.157)

Here, in identifying the bottom map as shown, we have used that restriction along the equivalence $\mathrm{Sp}(1)^1\backslash\!\!\backslash * \xrightarrow{\sim} \mathrm{Sp}(2)\backslash\!\!\backslash \mathrm{Sp}(2)/\mathrm{Sp}(1)^1 \simeq \mathrm{Sp}(2)\backslash\!\!\backslash \mathbb{H}P^1$ means (4.67) to (restrict the isotropy action along $\mathrm{Sp}(1)^2 \hookrightarrow \mathrm{Sp}(2)$ and) evaluate on the neutral coset $e \cdot \mathrm{Sp}(1)^2 \in \mathrm{Sp}(2)/\mathrm{Sp}(1)^2$, hence equivalently (by Lem. 5.2.19) on the base point $P_0 \in \mathbb{H}P^1$.

However, precisely because the left map comes from inclusion of the stabilizer subgroup, its composite with the bottom map produces the situation (5.152) discussed in Rem. 5.2.28, whence the above diagram is equivalently of this form:

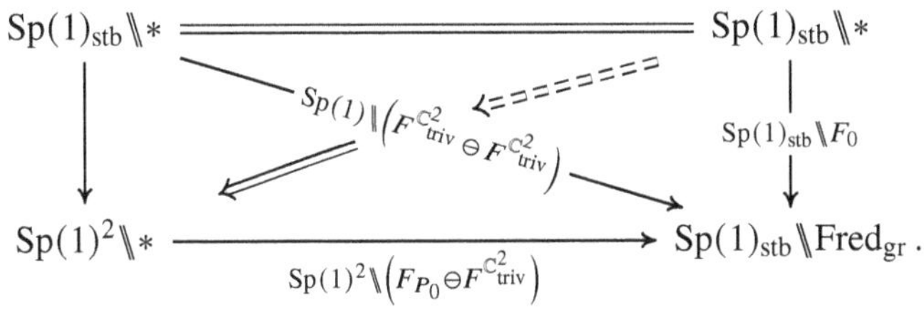

(5.158)

But now the maps of the top right triangle both pick Fredholm operators that are nontrivial only on the first summand of $\mathcal{H} \simeq \mathcal{H}_{\mathrm{triv}} \oplus \mathcal{H}_{\mathrm{def}}$ (5.151), hence on which the group action is trivial. Therefore, the remaining dashed homotopy is obtained from any plain homotopy (no equivariance constraint) in

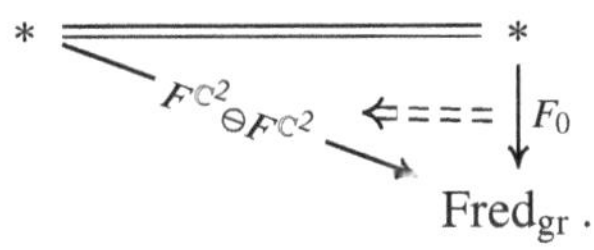

(5.159)

This does exist, by (5.69), since both these Fredholm operators have vanishing index. □

Remark 5.2.31. In summary, by Prop. 5.2.30, we have constructed a homotopy of topological groupoids:

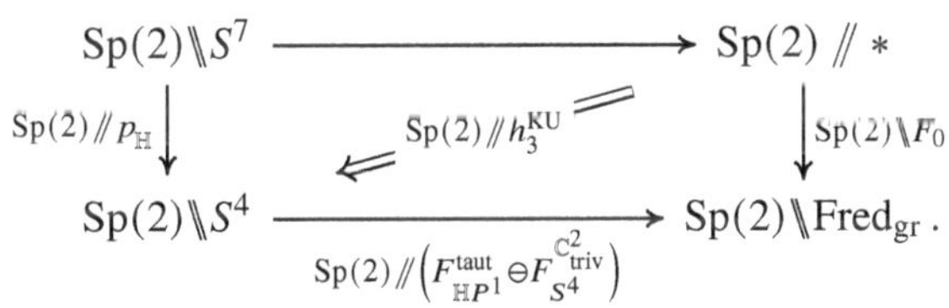

(5.160)

This is the announced unstable equivariant quaternionic orientation in topological K-theory, according to Fig. 2.7 (bottom row).

The construction for the complex orientation is obtained essentially verbatim the same way by restricting all of the above discussion along the inclusion $\mathbb{C} \hookrightarrow \mathbb{H}$.

5.2.6.1 Full Equivariance

We note here that the equivariance group of the quaternionic Hopf fibration is larger than the group Sp(2) from Lem. 5.2.17. We discuss to which extent the above results generalize accordingly – the trivialization of the tautological K-class does not generalize (Rem. 5.2.39).

Definition 5.2.32 (Central product groups [Gr69][FSS20, Def. 2.11]). For $n \in \mathbb{N}$, write

$$\begin{aligned} \mathrm{Sp}(n).\mathrm{Sp}(1) &:= \big(\mathrm{Sp}(n) \times \mathrm{Sp}(1)\big) \big/ \Big\{ \big(\mathrm{diag}(q,q), q\big) \,\big|\, q \in Z\big(\mathrm{Sp}(1)\big) \Big\} \\ &\simeq \big(\mathrm{Sp}(n) \times \mathrm{Sp}(1)\big) \big/ (-I_n, -1)\,. \end{aligned} \tag{5.161}$$

Remark 5.2.33. Since U(1) and O(1) are abelian, the analogous central product groups in these cases are isomorphic to the plain orthogonal/unitary groups:

$$\begin{aligned} \mathrm{O}(n).\mathrm{O}(1) &\simeq \mathrm{O}(n) \\ \mathrm{U}(n).\mathrm{U}(1) &\simeq \mathrm{U}(n)\,, \end{aligned}$$

which is the reason why the following equivariance enhancement occurs for the $\mathbb{H}$-but not for the $\mathbb{R}$/$\mathbb{C}$-Hopf fibrations.

In enhancement of Lem. 5.2.17 we have:

Lemma 5.2.34 (Full equivariance of $\mathbb{H}$-Hopf fibration ([GWZ86, Prop. 4.1], cf. [FSS20, Prop. 2.20])). *The $\mathbb{H}$-Hopf fibration* (5.123) *is equivariant with respect to the central product group* Sp(2).Sp(1) (5.161) *acting*

1. *on S^7 by:*

$$\begin{array}{ccc} \mathrm{Sp}(2).\mathrm{Sp}(1) \times S(\mathbb{H}^2) & \longrightarrow & S(\mathbb{H}^2) \\ \big([G,q], v\big) & \mapsto & G \cdot v \cdot \bar{q}\,, \end{array} \tag{5.162}$$

2. *on S^4 by:*

$$\begin{array}{ccc} \mathrm{Sp}(2).\mathrm{Sp}(1) \times \mathbb{H}P^1 & \longrightarrow & \mathbb{H}P^1 \\ \big([G,q], [v]\big) & \mapsto & [G \cdot v \cdot \bar{q}] = [G \cdot v]\,, \end{array} \tag{5.163}$$

in that:

$$\begin{array}{ccc} \mathrm{Spin}(5).\mathrm{Spin}(3) \circlearrowright S^7 & \simeq & S(\mathbb{H}^2) \circlearrowleft \mathrm{Sp}(2).\mathrm{Sp}(1) \\ \Big\downarrow{\scriptstyle p_{\mathbb{H}}} & & \Big\downarrow{\scriptstyle v \,\mapsto\, v\cdot\mathbb{H}} \\ \mathrm{Spin}(5).\mathrm{Spin}(3) \circlearrowright S^4 & \simeq & \mathbb{H}P^1 \circlearrowleft \mathrm{Sp}(2).\mathrm{Sp}(1)\,. \end{array} \tag{5.164}$$

This larger equivariance no longer extends to the factorizations of Rem. 5.2.16.

Proof. For $v \in \mathbb{H}^2$ and $G \in \mathrm{Sp}(2)$ and $q \in \mathrm{Sp}(1) \subset \mathbb{H}$, the assignments

$$\begin{array}{ccc} v & \longmapsto & G \cdot v \cdot \bar{q} \\ \Big\downarrow & & \Big\downarrow \\ & & G \cdot v \cdot \bar{q} \cdot \mathbb{H} \\ & & \big\| \\ v \cdot \mathbb{H} & \longmapsto & G \cdot v \cdot \mathbb{H} \cdot \bar{q} \end{array} \tag{5.165}$$

still commute. □

Thus, in enhancement of Lem. 5.2.19 we have:

Lemma 5.2.35. *The quaternionic Hopf fibration is equivalently the following coset coprojection:*

$$
\begin{array}{ccccc}
S(\mathbb{H}^2) & \simeq & \dfrac{\mathrm{Sp}(2).\mathrm{Sp}(1)}{\mathrm{Sp}(1)_{\mathrm{stb}}.\mathrm{Sp}(1)_{\mathrm{stb}'}} & := & \dfrac{\mathrm{Sp}(2).\mathrm{Sp}(1)}{\left\{ \left[\begin{pmatrix} q' & 0 \\ 0 & q \end{pmatrix}, q' \right] \,\middle|\, q, q' \in \mathrm{Sp}(1) \right\}} \\
\Big\downarrow {\scriptstyle p_{\mathbb{H}}} & & & & \Big\downarrow \\
\mathbb{H}P^1 & \simeq & \dfrac{\mathrm{Sp}(2).\mathrm{Sp}(1)}{\big(\mathrm{Sp}(1)_{\mathrm{stb}} \times \mathrm{Sp}(1)_{\mathrm{act}}\big).\mathrm{Sp}(1)_{\mathrm{stb}'}} & := & \dfrac{\mathrm{Sp}(2).\mathrm{Sp}(1)}{\left\{ \left[\begin{pmatrix} p & 0 \\ 0 & q \end{pmatrix}, q' \right] \,\middle|\, q, q', p \in \mathrm{Sp}(1) \right\}} .
\end{array}
\tag{5.166}
$$

Proof. The action (5.162) of $\mathrm{Sp}(2).\mathrm{Sp}(1)$ on S^7 is still transitive

$$
\mathrm{Sp}(2).\mathrm{Sp}(1) \cdot \begin{pmatrix} 1 \\ 0 \end{pmatrix} \simeq S(\mathbb{H}^2)
$$

and the stabilizer of the base point is evidently the subgroup $\mathrm{Sp}(1)_{\mathrm{stb}}.\mathrm{Sp}(1)_{\mathrm{stb}'}$ given in (5.166), whence the claim for the top row again follows by the orbit-stabilizer theorem. □

Hence in enhancement of Cor. 5.2.21 we have:

Corollary 5.2.36 ([FSS20, Prop. 2.22]). *After passage to homotopy* $\mathrm{Sp}(2).\mathrm{Sp}(1)$*-quotients, the quaternionic Hopf fibration is equivalently identified as shown on the right here:*

$$
\begin{array}{ccc}
\mathrm{Sp}(2).\mathrm{Sp}(1) \backslash\!\!\backslash S(\mathbb{H}^2) & \simeq & \mathbf{B}\big(\mathrm{Sp}(1)_{\mathrm{stb}}.\mathrm{Sp}(1)_{\mathrm{stb}'}\big) \\
\Big\downarrow {\scriptstyle \mathrm{Sp}(2).\mathrm{Sp}(1) \backslash\!\!\backslash p_{\mathbb{H}}} & & \Big\downarrow {\scriptstyle \mathbf{B}\left([q,q'] \mapsto [q,q',q']\right)} \\
\mathrm{Sp}(2).\mathrm{Sp}(1) \backslash\!\!\backslash \mathbb{H}P^1 & \simeq & \mathbf{B}\big(\big(\mathrm{Sp}(1)_{\mathrm{stb}} \times \mathrm{Sp}(1)_{\mathrm{act}}\big).\mathrm{Sp}(1)_{\mathrm{stb}'}\big) .
\end{array}
\tag{5.167}
$$

Now in generalization of Ntn. 5.2.27 we set:

Notation 5.2.37. We have an action of $\mathrm{Sp}(2).\mathrm{Sp}(1)$ on graded Fredholm operators given by

$$
\begin{array}{ccccccc}
\mathrm{Sp}(2).\mathrm{Sp}(1) & \longrightarrow & \mathrm{PU}(\mathcal{H}) & \hookrightarrow & \mathrm{PU}(\mathcal{H})^{\mathrm{gr}} & \hookrightarrow & \mathrm{Aut}\big(\mathrm{Fred}_{\mathrm{gr}}\big) \\
{[G,q]} & \mapsto & \big[\oplus_{\mathbb{N}} \gamma(G)\big] , & & & &
\end{array}
\tag{5.168}
$$

which is well-defined since the sign ambiguity in the representative G is absorbed by the equivalence relation (5.17) for the projective unitary group.

This way we may enhance Lem. 5.2.29 to:

Lemma 5.2.38. *The tautological* $\mathbb{H}P^1$*-parameterized Fredholm operator* (5.144) *is* $\mathrm{Sp}(2).\mathrm{Sp}(1)$*-equivariant with respect to the action* (5.163) *on* $\mathbb{H}P^1$ *and the action on* $\mathrm{Fred}_{\mathrm{gr}}$ *from* (5.168)*:*

$$\begin{array}{ccc} \mathrm{Sp}(2).\mathrm{Sp}(1) & & \mathrm{Sp}(2).\mathrm{Sp}(1) \\ \circlearrowright & & \circlearrowright \\ \mathbb{H}P^1 & \xrightarrow{\;F^{\mathrm{taut}}_{\mathbb{H}P^1}\;} & \mathrm{Fred}_{\mathrm{gr}}\,, \end{array} \tag{5.169}$$

and hence descends to a map of homotopy quotient stacks (4.107) *of this form:*

$$\begin{array}{ccc} \mathrm{Sp}(2).\mathrm{Sp}(1)\backslash\!\!\backslash \mathbb{H}P^1 & \xrightarrow{\;\mathrm{Sp}(2).\mathrm{Sp}(1)\backslash\!\!\backslash F^{\mathrm{taut}}_{\mathbb{H}P^1}\;} & \mathrm{Sp}(2).\mathrm{Sp}(1)\backslash\!\!\backslash \mathrm{Fred}_{\mathrm{gr}}\,. \\ & \searrow \qquad \swarrow & \\ & \mathbf{B}\big(\mathrm{Sp}(2).\mathrm{Sp}(1)\big) & \end{array} \tag{5.170}$$

Proof. By the nature of the action (5.168), where the extra datum q just drops out, the proof of Lem. 5.2.29 still goes through essentially verbatim. □

Remark 5.2.39. While all these generalizations so far have been straightforward, the generalization of Prop. 5.2.30 fails: The tautological $\mathbb{H}P^1$-parameterized Fredholm operator, which trivializes Sp(2)-equivariantly according to Prop. 5.2.30, does *not* trivialize Sp(2).Sp(1)-equivariantly. This is clear from the proof of Prop. 5.2.30, which relies on the absence in (5.158) of the second factor, $\mathrm{Sp}(1)_{\mathrm{stb}'}$, appearing now in (5.167).

Part III

Transition to Higher Toposes

In passing from the previous part II to the following part IV, we will shift gears and pass from ordinary topology (and be it the "higher topology" of topological groupoids) to *higher topos theory*, specifically to $(\infty, 1)$-*topos theory* in the sense of Grothendieck-Rezk-Lurie, hence to *geometric homotopy theory*.

Even though higher topos theory has matured in the last decade as a topic in itself, there is arguably still room left in motivating it and connecting it to what it naturally wants to be (like all category theory): Not an end in itself but a backdrop on which to develop (geometric) theory: A *τόπος* (Greek: "place") is where geometry may *take place*.

Similarly, higher topoi provide a substrate for higher geometry, such as groupoids, orbifolds, and higher étale stacks. While this is well understood by experts, in general, in this brief intermediate part of the book, we will informally motivate the approach of the following parts to the non-expert reader, and maybe highlight a point or two that the expert may also find noteworthy.

For similar exposition see also [Sc25].

6

Geometric Homotopy

6.1 Generalized Spaces

6.1.1 Convenient categories of spaces

From the point of view of geometry, a *topos* is a "really convenient" category of geometric spaces, in the technical sense going back to [St67]:

Here a *convenient category of spaces* is one where many *universal constructions* exist, meaning that one can conveniently proceed making intended constructions of geometric spaces without worrying whether the result may fail to actually exist, while being guaranteed that it enjoys all the intended defining universal property.

For example, in the category of plain topological spaces, the mapping spaces $\mathrm{Map}(X, -)$ in general do not exist with their universal property of being right adjoint to the product spaces $X \times (-)$ (4.10), while in the category of D-topological spaces (Def. 4.1.5) they do. In this sense, D-topological spaces form a *more convenient category*, namely here: a Cartesian closed category.

This issue of Cartesian closure becomes much more pronounced as richer geometric structure is considered. Notably, as we now want to pass from topological spaces to smooth manifolds, even though these have underlying D-topological spaces so that mapping spaces $\mathrm{Map}(X, Y)$ between smooth manifolds X, Y exist as D-topological spaces, these mapping spaces may still not admit the structure of smooth manifolds themselves: Namely, only when X is compact does $\mathrm{Map}(X, Y)$ admit the structure of a (generally infinite-dimensional) Fréchet smooth manifold.

Similarly, while categories of (D-)topological spaces are closed under forming general limits and colimits of diagrams of spaces, this is far from true for the category of smooth manifolds (even if we are admitting infinite-dimensional Fréchet smooth manifolds). For example, already the fiber product of a pair of smooth maps between smooth manifolds exists as a smooth manifold only when these two maps are "transverse" to each other.

These problems are resolved when regarding smooth manifolds among more general smooth spaces forming more convenient ambient categories: *topoi*.

6.1.2 Gros Topoi of Probeable Spaces

Smooth manifolds, of course, are a globalization of the idea of *Cartesian spaces* $\mathbb{R}^n$ (for $n \in \mathbb{N}$) with smooth functions between them: A smooth manifold is defined to

DOI: 10.1201/9781003675969-6

be a topological space covered by open subsets identified with Cartesian spaces, such that on overlaps these are glued together via smooth functions between Cartesian spaces.[1] We may recognize that there are two stages in this classical definition: First the idea that Cartesian spaces are what serves to "chart out" smooth manifolds, and second that these charts are actually open subsets.

The idea of the topos-theoretic generalization of the notion of smooth manifolds follows essentially by dropping the second of these two conditions.

6.1.2.1 The Topos of Smooth Sets

We want to say that a *smooth set* X is anything that can be determined by charting out Cartesian spaces inside it. But to avoid the usual connotation that these charts need to be open subsets, we will speak of "plots" instead of "charts": A *smooth set* X is to be anything that can be determined by *probing* it with Cartesian spaces $\mathbb{R}^n$, namely by considering the system of would-be smooth maps from $\mathbb{R}^n$ to X, to be called the *plots* of X by the probe space $\mathbb{R}^n$. We may naturally make this precise as follows:

To identify a smooth set X we demand:

- for each Cartesian space $\mathbb{R}^n$, a set $\mathrm{Plt}(\mathbb{R}^n, X)$, to be thought of as the set of ways of smoothly mapping $\mathbb{R}^n$ into the space X (which is going to be bootstrapped into existence thereby),
- for each smooth function ϕ between Cartesian spaces, a map of sets of *plots* reflecting the precomposition ϕ^* of plots with ϕ:

$$\begin{array}{ccc} \mathbb{R}^{n_1} & \mapsto & \mathrm{Plt}(\mathbb{R}^{n_1}, X) \\ \uparrow{\scriptstyle \phi} & \mapsto & {\scriptstyle \phi^*}\downarrow \\ \mathbb{R}^{n_2} & \mapsto & \mathrm{Plt}(\mathbb{R}^{n_2}, X)\,, \end{array} \tag{6.1}$$

  such that composition and identity maps are respected.

So far, this makes the system $\mathrm{Plt}(-, X)$ of plots (and hence the space X to be defined thereby) a *presheaf* on the category CrtSp of Cartesian spaces.

There is just one more natural demand, expressing the "cohesion" among elements of X, namely, that plots from larger probes may be obtained by gluing plots from smaller probes; we demand:

- for each differentially good open cover

$$\left\{ U_i \simeq \mathbb{R}^n \overset{\iota_i}{\hookrightarrow} \mathbb{R}^n \right\}_{i \in I} \tag{6.2}$$

  a bijection between global plots and systems of plots glued from the patches:

$$\begin{array}{ccc} \mathrm{Plt}(\mathbb{R}^n, X) & \xrightarrow{\sim} & \left\{ \big(\phi_i \in \mathrm{Plt}(U_i, X)\big)_{i \in I} \,\middle|\, \underset{U_i \cap U_j}{\forall} \big(\iota_{ij}^* \phi_i = \iota_{ij}^* \phi_j\big) \right\} \\ \phi & \mapsto & \big(\iota_i^* \phi\big)_{i \in I} \end{array} \tag{6.3}$$

[1] This informal description refers to *differentiably good open covers* of smooth manifolds, which one may assume without restriction, by Lem. 4.1.31.

This last condition makes the system $\mathrm{Plt}(-, X)$, and hence the smooth set X defined thereby, a *sheaf* on the *site* CrtSp.

Proceeding in this vein of understanding generalized spaces via the systems of their sets of probes, a *smooth map* $f : X \longrightarrow Y$ between such smooth sets should be reflected for each probe space $\mathbb{R}^n$ by the rule f_* for how it turns the plots of X into plots of Y by postcomposition, compatible with the precomposition operation (6.1) by maps of plots:

$$\begin{array}{cccc} & X & \xrightarrow{\;\;f\;\;} & Y \\ \mathbb{R}^{n_1} & \mathrm{Plt}(\mathbb{R}^{n_1}, X) & \xrightarrow{f_*} & \mathrm{Plt}(\mathbb{R}^{n_1}, Y) \\ \uparrow{\scriptstyle \phi} & \downarrow{\scriptstyle \phi^*} & & \downarrow{\scriptstyle \phi^*} \\ \mathbb{R}^{n_2} & \mathrm{Plt}(\mathbb{R}^{n_2}, X) & \xrightarrow{f_*} & \mathrm{Plt}(\mathbb{R}^{n_2}, Y)\,. \end{array} \tag{6.4}$$

But this just means that maps of smooth sets are reflected in *natural transformations* between their systems of sets of plots, which are just the morphisms between these sheaves of sets.

Hence, we conclude that generalized smooth spaces in the guise of *smooth sets*, determined by how they may be *probed* by Cartesian space, form the *category of sheaves* over the latter:

$$\mathrm{SmthSet} := \mathrm{Sh}(\mathrm{CrtSp})\,. \tag{6.5}$$

Such categories of sheaves are (Grothendieck-)*topoi*; hence, here we have the *topos of smooth sets*. (Identified in this form in [Sc13, §1.2.2, Def. 1.2.16, 1.3.58][KS17, Def. 2.1], earlier discussion includes [MW07, §A.1], review with application to field theory in [Sc17b][GS25][IM25].)

For example, ordinary smooth manifolds are faithfully regarded as smooth sets by taking their sets of plots to be the usual sets $C^\infty(-,-)$ of smooth maps into them:

$$\begin{array}{ccc} \mathrm{SmthMfd} & \hookrightarrow & \mathrm{SmthSet} \\ X & \mapsto & C^\infty(-, X)\,. \end{array} \tag{6.6}$$

This applies actually in the generality of infinite-dimensional Fréchet smooth manifolds.

In particular, the Cartesian probe spaces $\mathbb{R}^n$, that the whole construction is bootstrapped become smooth sets this way. This raises a potential conceptual inconsistency: Because now there are two different notions of smooth maps from $\mathbb{R}^n$ into any $X \in \mathrm{SmthSet}$: On the one hand there are the *predefined* plots in $\mathrm{Plt}(\mathbb{R}^n, X)$, and on the other hand there are now the *resulting* maps of smooth sets in $\mathrm{Hom}_{\mathrm{SmthSet}}(\mathbb{R}^n, X)$. Remarkably, that these two notions coincide and hence that the bootstrap definition of smooth sets is consistent and thereby complete, is the statement of the notorious *Yoneda lemma*, asserting that these two notions of maps are in natural bijection:

$$\mathrm{Hom}_{\mathrm{SmthSet}}(\mathbb{R}^n, X) \simeq \mathrm{Plt}(\mathbb{R}^n, X)\,. \tag{6.7}$$

The construction of more general smooth sets is typically straightforward. For example the *product* $X \times Y$ of a pair of smooth sets should clearly have as plots just the pairs of individual plots of X and Y, and this already defines it:

$$\mathrm{Plt}(-, X \times Y) \simeq \mathrm{Plt}(-, X) \times \mathrm{Plt}(-, Y)\,. \tag{6.8}$$

Similarly, their *mapping space* $\mathrm{Map}(X,Y)$ should clearly be such that its $\mathbb{R}^n$-plots are just the smoothly $\mathbb{R}^n$-parameterized smooth maps, and again, this already defines it as a smooth set:

$$\mathrm{Plt}\big(\mathbb{R}^n, \mathrm{Map}(X,Y)\big) \simeq \mathrm{Hom}_{\mathrm{SmthSet}}\big(\mathbb{R}^n \times X, Y\big) \tag{6.9}$$

and entails the desired universal property,

$$\mathrm{Map}\big(Z, \mathrm{Map}(X,Y)\big) \simeq \mathrm{Map}\big(Z \times X, Y\big)\,, \tag{6.10}$$

naturally in $X, Y, Z \in \mathrm{SmthSet}$.

Remarkably, when $\mathrm{Map}(X,Y)$ exists as a traditional Fréchet smooth manifold (namely, when X and Y are smooth manifolds with X compact), then it matches this simple construction as a smooth set under the embedding (6.6).

Beyond such examples at the boundary of what can be described by infinite-dimensional smooth manifolds, there are very natural smooth sets that are substantially not like smooth manifolds. For example, for each $p \geq 1$ there is the smooth *moduli space for smooth p-forms* $\mathbf{\Omega}^p_{\mathrm{dR}} \in \mathrm{SmthSet}$, and again its construction is straightforward: The characteristic universal property of such a moduli space is that maps into it are in natural bijection to differential p-form on the domain space,

$$\mathrm{Hom}_{\mathrm{SmthSet}}\big(X, \mathbf{\Omega}^p_{\mathrm{dR}}\big) \simeq \Omega^p_{\mathrm{dR}}(X)\,, \tag{6.11}$$

and so this already defines its plots and thereby the space itself:

$$\mathrm{Plt}\big(\mathbb{R}^n, \mathbf{\Omega}^p_{\mathrm{dR}}\big) := \Omega^p_{\mathrm{dR}}\big(\mathbb{R}^n\big)\,. \tag{6.12}$$

In fact, this defines conversely also what we mean by a differential p-form ω_p on a general smooth set X, namely, equivalently a map $\vdash \omega_p : X \longrightarrow \mathbf{\Omega}^p_{\mathrm{dR}}$. In particular, there is a *universal* p-form on the moduli space itself, given by the identity map, and any other p-form is uniquely the pullback of that universal one, as befits a moduli space.

Such statements are formidable to approximate with traditional tools of smooth manifolds, while here among smooth sets they reduce to tautologies.

In the following we incrementally pass from Cartesian spaces to more general probe spaces (cf. Tab. 6.1) and with these to more general generalized geometric spaces probeable by these probes.

6.1.2.2 Cohesion of Smooth Sets

The category (6.5) of smooth sets is a *gros topos* (namely a *cohesive topos*, as we will discuss, due to [Sc13] following [La07]):

First, in the informal sense, it is to be regarded as a *category of spaces* — in contrast to the *petits topoi* that most expositions of the subject focus on, such as the categories $\mathrm{Sh}(X)$ of sheaves on (the open subsets of) a fixed topological space X, which are in themselves nothing but re-incarnations of that single space X.

More formally, gros topoi of the kind we are concerned with are characterized as being *local and strongly and locally connected* when regarded as spaces, meaning that their "petit aspect" is essentially trivial in that as petit topoi they are essentially point-like. What this means is the following, for the example of smooth sets:

Table 6.1 A sequence of increasingly rich generalized abstract coordinate charts is to serve as “probes” which witness the structure of generalized global geometric spaces. The theory of (higher, *gros*) topoi may be understood as making this precise: Probes form a (higher) *site*, and global spaces form the (higher) *sheaf topos* on such a site. (Table adapted from [Sc25].)

	Cartesian space	infinitesimal halo	super space	higher morphism	orbi-singularity	neg-dim sphere
probe	$\mathbb{R}^n \;\times$	$\mathbb{D}^m_k \;\times$	$\mathbb{R}^{0\vert q} \;\times$	$\Delta^r \;\times$	$\underset{\curlyvee}{G} \;\times$	$\mathbb{S}^{-d}$
geometry	differential topology	differential geometry	super geometry	homotopy theory	proper equivariance	stable homotopy
physics	fields	variations	fermions	gauge symmetry	orbi-singularities	quantum

As every topos, SmthSet is connected to the *base topos* Set of plain sets by an essentially unique pair of adjoint functors

$$\mathrm{SmthSet} \underset{\mathrm{Pnts}}{\overset{\mathrm{Disc}}{\underset{\longrightarrow}{\overset{\longleftarrow}{\bot}}}} \mathrm{Set} \tag{6.13}$$

such that Disc preserves finite limits (whence Pnts is called a *geometric morphism*). Here:

- the right adjoint takes a smooth set to its *set of points*, namely, the set of plots by the point probe $\mathbb{R}^0$:

$$\mathrm{Pnts} : X \mapsto \mathrm{Plt}(\mathbb{R}^0, X)\,, \tag{6.14}$$

- the left adjoint regards a set as equipped with the *discrete* smooth structure, all whose plots are constant maps:

$$\mathrm{Disc} : S \mapsto \mathrm{Plt}\big(-, \mathrm{Disc}(S)\big) := S\,. \tag{6.15}$$

But in fact there is more structure along these lines:

- First, Disc (besides preserving all colimits as every left adjoint does) evidently preserves not only finite limits but all limits and, as such, has itself a left adjoint to be called Shp: This is given by forming sets of what should be regarded as *connected components* $\pi_0(-)$ of a smooth set (those parts whose substructure cannot be resolved by maps to a discrete space):

$$\mathrm{Shp} : X \mapsto \pi_0(X)\,. \tag{6.16}$$

While there is no further left adjoint in that Shp does not preserve all limits, it does preserve all products:

$$\mathrm{Shp}(X \times Y) \simeq \pi_0(X) \times \pi_0(Y)\,. \tag{6.17}$$

- Secondly, Pnts clearly preserves all limits, as every right adjoint does, but also all colimits and, as such, has a further right adjoint to be called Chtc. This functor regards a set as equipped with its *chaotic* or *codiscrete* smooth structure, which regards every map of points as being smooth:

$$\mathrm{Chtc} : S \mapsto \mathrm{Plt}\big(-; \mathrm{Chtc}(S)\big) := \mathrm{Hom}_{\mathrm{Set}}(-, S)\,. \tag{6.18}$$

- Moreover, Disc is fully faithful, clearly, whence equivalently so is Chtc.

In summary, the default geometric morphism (6.13) extends to the following highly structured situation:

$$\mathrm{SmthSet}\;\begin{array}{c}\xleftarrow{\quad\mathrm{Shp}\quad}\!\!\!\times\\ \perp\\ \xleftarrow{\quad\mathrm{Disc}\quad}\\ \perp\\ \xrightarrow{\quad\mathrm{Pnts}\quad}\\ \perp\\ \xleftarrow{\quad\mathrm{Chtc}\quad}\end{array}\;\mathrm{Set} \tag{6.19}$$

A topos carrying such a system of adjoint functors is called a *cohesive topos* since, as we just saw in this example of smooth sets, the adjoints reflect how the elements of the spaces in the topos do or do not hang together ("cohere", like molecules in a fluid).

6.1.2.3 The Quasitopos of Diffeological Spaces

More attention in the past has been devoted to the subcategory of *concrete* smooth sets $X \in \mathrm{SmthSet}$, namely those that are supported on their plain set of points (6.14),

$$\mathrm{Pnts}(X) := \mathrm{Plt}(\mathbb{R}^0, X) \in \mathrm{Set}$$

in that their smooth plots are all among the maps of sets of points to this underlying set, in that there are natural injections

$$\mathrm{Plt}\big(\mathbb{R}^n, X\big) \hookrightarrow \mathrm{Hom}_{\mathrm{Set}}\big(\mathbb{R}^n, \mathrm{Pnts}(X)\big)\,. \tag{6.20}$$

These concrete smooth sets are known as *diffeological spaces* (due to [Ch77][So80], developed in [IZ13]).

For instance, ordinary manifolds are concrete as smooth sets, and the products (6.8) and mapping spaces (6.9) of concrete smooth sets are themselves concrete.

Also D-topological spaces (cf. § 4.1.2) become diffeological spaces, simply by taking their would-be smooth plots to be all continuous maps from the Cartesian probe spaces into them

$$\begin{array}{ccc} \mathrm{TopSpc} & \hookrightarrow & \mathrm{DfflSpc} \\ X & \mapsto & \mathrm{Plt}(-, X) := C^0(-, X)\,. \end{array} \tag{6.21}$$

In contrast, the moduli spaces of $p \geq 1$-forms (6.12) are *not* concrete: They have a single underlying point (the 0-form),

$$\mathrm{Pnts}\Big(\mathbf{\Omega}^{p\geq 1}_{\mathrm{dR}}\Big) \simeq *\,,$$

and yet have an infinitude of plots over probes $\mathbb{R}^n$ from dimension $n \geq p$ on.

Accordingly, diffeological spaces form the proper full subcategory of smooth sets on the concrete ones. This subcategory is not quite a topos but it is close (being Cartesian closed and bicomplete), called a *quasitopos*. This "fairly convenient" quasitopos of diffeological spaces (cf. [BH08]) is hence intermediate between the traditional categories of (D-)topological spaces/smooth manifolds and the more convenient topos of smooth sets:

$$\begin{array}{l} \mathrm{TopSpc} \searrow \\ \qquad\qquad \mathrm{DfflSpc} \hookrightarrow \mathrm{SmthSet}\,. \\ \mathrm{SmthMfd} \nearrow \end{array} \tag{6.22}$$

6.1.2.4 The Topos of Haloed Smooth Sets

We have seen above, for the example of Cartesian probe spaces, that topos theory serves as a tool that produces for a given small category of geometric probe spaces (a *site*) a good large category of generalized geometric spaces that are characterized by the system of ways in which these probe spaces may be plotted out inside them. Therefore, it is now straightforward to produce good notions of yet further generalized geometric spaces simply by defining more general probe spaces.

For the purpose of differential geometry, it will be most useful to consider probe spaces that are Cartesian spaces but possibly with further *infinitesimal extension*.

Indeed, it is a truism that many traditional definitions in contemporary rigorous differential geometry aim to capture a notion of *infinitesimals* without explicitly saying so. For instance, the concept of a *tangent vector* $v \in T_x X$, formulated in terms of equivalence classes of smooth curves through a given point x of a manifold X, clearly tries to indirectly describe the *infinitesimal curves* through that point without, however, having a primitive way of saying this.

More concretely, where a smooth curve through X is a smooth map $\gamma : \mathbb{R}^1 \longrightarrow X$ taking 0 to x, one envisions that the tangent vector it describes is the restriction of γ to a neighborhood $\mathbb{D}^1_1 \subset \mathbb{R}^1$ of $0 \in \mathbb{R}^1$ which is "so tiny" that the numbers $\epsilon \in \mathbb{D}^1_1$ it contains are "so small" that their square (which must be far smaller still) is in fact literally indistinguishable from zero: $\epsilon^2 = 0$. If that existed, then the algebra of smooth functions on $\mathbb{D}^1_1$ would be the polynomial algebra generated by the coordinate function ϵ quotiented by the ideal generated by ϵ^2, as follows:[2]

$$C^\infty(\mathbb{D}^1_1) \simeq \mathbb{R}[\epsilon]/(\epsilon^2) \simeq_{\mathbb{R}} \left\{ f = f(0) + f'(0) \cdot \epsilon \,\middle|\, f(0), f'(0) \in \mathbb{R} \right\} \tag{6.23}$$

hence with product operation given by

$$\begin{aligned} f \cdot g &= \big(f(0) + f'(0) \cdot \epsilon\big) \cdot \big(g(0) + g'(0) \cdot \epsilon\big) \\ &= f(0) \cdot g(0) + \big(f(0) \cdot g'(0) + f'(0) \cdot g(0)\big) \cdot \epsilon\,. \end{aligned} \tag{6.24}$$

Such algebraic manipulations with would-be infinitesimal quantities go back to P. de Fermat in the 1630s, were popularized by G. W. Leibniz's development of modern calculus in the 1680s, and are said to have tacitly been used by S. Lie when laying the foundations of modern differential geometry (cf. [Mer15, §3.3]).

But of course, in the category of smooth manifolds (and even in that of smooth sets), this infinitesimal subobject $\mathbb{D}^1_1 \subset \mathbb{R}^1$ does not exist, as the only real number which squares to zero is zero itself.

Historically this was a stumbling block for a long time, but with our understanding of topos theory as producing generalized spaces it becomes a triviality to solve: If we want a theory of generalized smooth spaces that includes infinitesimal neighborhoods, then we simply need to write down a small site with of probe spaces including the archetypical examples of infinitesimals as in (6.23)!

It becomes obvious how to do that once we note that the functor which sends smooth manifolds to their commutative real algebras of smooth functions is fully

[2] The elements of (6.23) were called "dual numbers" by W. Clifford (1873), which however is clearly not a useful terminology.

faithful ("Milnor's exercise", cf. [KMS93, §35]):

$$\begin{array}{ccc} \mathrm{SmthMfd} & \hookrightarrow & \mathrm{CAlg}_{\mathbb{R}}^{\mathrm{op}} \\ X & \mapsto & C^\infty(X)\,. \end{array} \tag{6.25}$$

In particular, also the subcategory of Cartesian spaces is equivalently a full subcategory of formal duals of commutative real algebras:

$$\begin{array}{ccc} \mathrm{CartSp} & \hookrightarrow & \mathrm{CAlg}_{\mathbb{R}}^{\mathrm{op}} \\ \mathbb{R}^n & \mapsto & C^\infty(\mathbb{R}^n)\,. \end{array} \tag{6.26}$$

But seen in this dual algebraic incarnation, we may readily implement the desired infinitesimal neighborhoods simply by adjoining their would-be algebras of smooth functions (6.23). We may say that $\mathbb{D}^1_1$ *is* the object of $\mathrm{CAlg}_{\mathbb{R}}^{\mathrm{op}}$ corresponding to its algebra of functions (6.23).

Generally, we may declare that *haloed* (or *infinitesimally thickened*) Cartesian spaces $\mathbb{R}^n \times \mathbb{D}$ are the objects of $\mathrm{CAlg}_{\mathbb{R}}^{\mathrm{op}}$ which correspond to their intended algebra of smooth functions, being the $\mathbb{R}$-tensor product of smooth functions on $\mathbb{R}^n$ with any $\mathbb{R}$-algebra with a maximal and nilpotent ideal W, thus defining the full category

$$\begin{array}{ccccc} \mathrm{CrtSp} & \overset{i}{\hookrightarrow} & \mathrm{HldCrtSp} & \xhookrightarrow{C^\infty(-)} & \mathrm{Calg}_{\mathbb{R}}^{\mathrm{op}} \\ & & \mathbb{R}^n \times \mathbb{D} & \mapsto & C^\infty(\mathbb{R}^n) \otimes_{\mathbb{R}} C^\infty(\mathbb{D}) \end{array} \tag{6.27}$$

on all commutative $\mathbb{R}$-algebras as on the right, for which there exists $n, k \in \mathbb{N}$ and linear isomorphisms

$$C^\infty(\mathbb{D}) \simeq_{\mathbb{R}} \big(\mathbb{R} \oplus W\big) \quad \text{with} \quad \dim(W) < 0 \text{ and } W^{k+1} = 0\,. \tag{6.28}$$

One says that the minimal such k in each case is the *order of infinitesimality* of the infinitesimal space $\mathbb{D}$.

For example, the *infinitesimal disk* of dimension m and order k, $\mathbb{D}^m_k$, corresponds to the polynomial ring on m generators quotiented by the ideal generated by the monomials of any $k+1$ of these generators:

$$C^\infty(\mathbb{D}^m_k) := \mathbb{R}\big[\epsilon_1, \cdots, \epsilon_m\big] \big/ \big((\epsilon_{i_1} \cdots \epsilon_{i_{k+1}})_{i \in \{1,\cdots,m\}^{k+1}}\big)\,. \tag{6.29}$$

Now, to pass from these simple local coordinate expressions for haloed Cartesian spaces to a full-blown theory of generalized smooth spaces with infinitesimals, we simply need to declare how to glue these probe spaces and then pass to the sheaf topos over their site.

But there is no room for nontrivial gluing in infinitesimal directions, so the covering families in HldCrtSpc should be of the simple form

$$\Big\{U_i \times \mathbb{D} \simeq \mathbb{R}^n \times \mathbb{D} \xhookrightarrow{\iota_i \times \mathrm{id}_{\mathbb{D}}} \mathbb{R}^n \times \mathbb{D}\Big\}_{i \in I} \tag{6.30}$$

for $(\iota_i)_{i \in I}$ a differentiably good open cover of the finitely extended parts as previously in (6.2).

With respect to this notion of gluing and in evident generalization of the topos of smooth sets (6.5), we thus obtain the topos of *haloed* (or *infinitesimally thickened*) *smooth sets*:

$$\mathrm{HldSmthSet} := \mathrm{Sh}(\mathrm{HldCrtSp})\,. \tag{6.31}$$

The canonical inclusion of sites i (6.27) induces (by left Kan extension) a full inclusion of the topos of smooth sets and hence also of quasitopos of diffeological spaces and of the ordinary category of smooth manifolds (6.22):

$$\mathrm{SmthMfd} \hookrightarrow \mathrm{DfflSpc} \hookrightarrow \mathrm{SmthSet} \hookrightarrow \mathrm{HldSmthSet} \tag{6.32}$$

This topos (6.31) is equivalently (cf. [GS26, §4.1]) known as the "Cahiers topos" due to E. Dubuc [Du79a], originally introduced as a *well-adapted model* for the axioms of *synthetic differential geometry* due to W. Lawvere (cf. [La96]).

As such, the origins of the discussion of the topos of haloed smooth sets are about as old as that of the mere quasi-topos of diffeological spaces, while already subsuming and improving on it. That diffeological spaces have received broader public acclaim as a convenient category for differential geometry in the past is a quirk of scholarly history, probably owing less to a difference in the complexity of definitions (which is negligible) than to the subconscious prejudice of mathematicians that spaces need to be defined as underlying sets of points equipped with extra structure instead of operationally as entities chartable by probe spaces more general than points. This deserves to change.

6.1.2.5 The Topos of Super Haloed Smooth Sets

Continuing in this vein, it is now a pleasantly small step to proceed further to generalized *superspaces*. Despite the conceptually awkward (but long and entirely standard) terminology of *super*-geometry alluding to the physics concept of "supersymmetry" (cf. [CCF11]), it is an open secret that this is really the *fermionic* geometry that in field theory is needed to describe configuration spaces of any physical system containing fermions (like electrons and quarks, cf. [Sc17b]). The key characteristic of these fermionic fields is that as maps θ (sections) on spacetime, they are skew-commuting among each other:

$$\theta_i \cdot \theta_j = -\theta_j \cdot \theta_i\,. \tag{6.33}$$

Incidentally, this means that such fermionic elements are first-order infinitesimals in the sense of § 6.1.2.4, in that

$$\theta_i^2 = 0\,. \tag{6.34}$$

But due to the sign rule (6.33), fermionic fields do not exist in the topos of haloed smooth sets, and certainly not in traditional differential geometry.

Of course, the success of the physics literature in dealing with a notion of geometry that does not (or did not) even exist in the textbooks rests on the willingness to disregard an elusive globally defined notion of generalized space in favor of focusing on algebraic manipulations with objects that look like coordinate functions on charts. After all, it is straightforward to recognize a proper algebraic home for the sign rule (6.33): This just defines the *Grassmann algebras* (cf. [Ber87, §1]) introduced way back in [Gr1844]:

$$C^\infty\big(\mathbb{D}^{0|q}\big) := \wedge^\bullet(\mathbb{R}^q)^* \in \mathrm{Alg}_{\mathbb{R}}\,. \tag{6.35}$$

On the left we are notationally indicating that these are to be thought of as the smooth functions on a first-order infinitesimal disk like (6.29), whose q infinitesimal directions are "odd graded" (or just *odd*, for short), meaning that their coordinate functions $(\theta_i)_{i=1}^q$ skew-commute among each other (6.33).

While historically it was a challenge to understand how to formulate a differential geometry of *supermanifolds* whose local coordinate functions contain such Grassmann algebras (6.35), in our picture of topoi of probable space it is immediate to go much further even and define the topos of *super smooth sets* that subsumes not only supermanifolds but also previously unheard-of generalizations to *super-diffeological* spaces and super-moduli spaces of differential super-forms, etc.

Namely, in evident generalization of the previous discussion, we declare *super haloed Cartesian spaces* to form the full subcategory of formal duals of real algebras on the tensor products of the following form:

$$\begin{array}{ccccc} \mathrm{HldCrtSp} & \hookrightarrow & \mathrm{SprHldCrtSp} & \hookrightarrow & \mathrm{Alg}_{\mathbb{R}}^{\mathrm{op}} \\ & & \mathbb{R}^n \times \mathbb{D} \times \mathbb{D}^{0|q} & \mapsto & C^\infty(\mathbb{R}^n) \otimes_{\mathbb{R}} C^\infty(\mathbb{D}) \otimes_{\mathbb{R}} C^\infty(\mathbb{R}^{0|q}), \end{array} \tag{6.36}$$

where on the right we are using the notation introduced above in (6.35) and (6.28).

Since this only further generalizes the infinitesimal extension of these probe spaces, the gluing of such super haloed Cartesian spaces is still governed by the gluing of their underlying Cartesian spaces, so that in direct generalization of (6.30), we take the covering families in their category to be

$$\left\{ U_i \times \mathbb{D} \times \mathbb{D}^{0|q} \simeq \mathbb{R}^n \times \mathbb{D} \times \mathbb{D}^{0|q} \xhookrightarrow{\iota_i \times \mathrm{id}_{\mathbb{D}} \times \mathrm{id}_{\mathbb{D}^{0|q}}} \mathbb{R}^n \times \mathbb{D} \times \mathbb{D}^{0|q} \right\}_{i \in I} \tag{6.37}$$

for $(\iota_i)_{i\in I}$ a differentiably good open cover of the underlying ordinary Cartesian space.

With this in hand, we right away find the topos of *super haloed smooth sets*, being a convenient category for differential supergeometry:

$$\mathrm{SprHldSmthSet} := \mathrm{Sh}(\mathrm{SprHldCrtSp}), \tag{6.38}$$

and faithfully subsuming all the previous categories discussed:

$$\cdots \hookrightarrow \mathrm{SmthSet} \hookrightarrow \mathrm{HldSmthSet} \hookrightarrow \mathrm{SprHldSmthSet}. \tag{6.39}$$

6.1.2.6 The 2-Topos of Smooth Groupoids

Besides fermionic fields (matter fields) with their curious supergeometric nature (§ 6.1.2.5), the universe consists of bosonic *gauge fields* (force fields). Remarkably, their peculiar nature is a notion of *identity* more subtle than observed for other fields. Namely, gauge fields A, A' may be nominally different while still being (physically) equivalent, witnessed by *gauge transformations* λ relating them

$$A \xrightarrow[\sim]{\lambda} A'.$$

These gauge transformations compose in an associative, unital, and invertible way.

For example, electromagnetic field configurations on a Cartesian space $\mathbb{R}^n$ are represented by differential 1-forms $A \in \Omega^1_{\mathrm{dR}}(\mathbb{R}^n)$, but their moduli space is *not* just the moduli space $\mathbf{\Omega}^1_{\mathrm{dR}}$ (6.12) of these plain 1-forms, since there are gauge equivalences between these 1-forms witnessed by de Rham coboundaries.

This means that the plots of the correct moduli space of electromagnetic fields — to be denoted $\mathbf{B}\mathrm{U}(1)_{\mathrm{conn}}$ — do not form plain sets but form *groupoids* (topologically discrete groupoids, cf. Def. 4.1.8) whose morphisms are the gauge transformations, namely from the action groupoid of the set of 1-forms acted on by the set of U(1)-valued smooth functions acting by abelian gauge transformations:

$$\begin{aligned} \mathrm{Plt}\big(\mathbb{R}^n, \mathbf{B}\mathrm{U}(1)_{\mathrm{conn}}\big) &:= \Omega^1_{\mathrm{dR}}(\mathbb{R}^n) /\!\!/ C^\infty\big(\mathbb{R}^n, \mathrm{U}(1)\big) \\ &= \left\{ \begin{array}{ccc} & A_1 & \\ {}^{\lambda_{01}}\nearrow & & \searrow^{\lambda_{12}} \\ A_0 & \xrightarrow[=\lambda_{12}\circ\lambda_{01}]{\lambda_{02}} & A_2 \end{array} \;\middle|\; \begin{array}{c} A_i \in \Omega^1_{\mathrm{dR}}(\mathbb{R}^n) \\ \lambda_{ij} \in \Omega^0_{\mathrm{dR}}(\mathbb{R}^n) \\ A_j = A_i + \mathrm{d}\ln\lambda_{ij} \end{array} \right\}. \end{aligned} \tag{6.40}$$

As the probe space varies, pullback of differential forms makes the system of these groupoids of plots again form a presheaf on the category of Cartesian probe spaces, but now with values not in sets but in (topologically discrete) groupoids:

$$\mathrm{Plt}(-;\mathbf{B}\mathrm{U}(1)_{\mathrm{conn}}) : \mathrm{CrtSpc}^{\mathrm{op}} \longrightarrow \mathrm{Grpd}\,. \tag{6.41}$$

In the spirit of the previous discussion of generalized spaces characterized by systems of ways that probe spaces and chart them out, we think of this as the system of plots of a single *smooth groupoid* $\mathbf{B}\mathrm{U}(1)_{\mathrm{conn}}$ as it is being probed by Cartesian spaces.

Such smooth groupoids generalize (D-)topological groupoids (§ 4.1.5) in higher generalization of how smooth sets generalize (D-)topological spaces (6.22). Given a (D-)topological groupoid $\mathcal{X}$ (Def. 4.1.8), its groupoids of plots probed by $\mathbb{R}^n$ is the underlying discrete groupoid of maps $U \to \mathcal{X}$, from (4.47).

For such systems of *higher plots* there is again a notion of gluing, generalizing (6.3), where the equalities over the double intersection of plots, appearing on the right of (6.3), are replaced by gauge transformations λ_{ij}, the equation $\lambda_{ik} = \lambda_{jk} \circ \lambda_{ij}$ is required on the triple intersection of plots, and the bijection of sets in (6.3) is generalized to an equivalence of groupoids (Def. 4.1.23).

Presheaves of groupoids satisfying this higher gluing condition are traditionally called *stacks* (for better or worse) but may also be called $(2,1)$-*sheaves* or 2-*sheaves*, for short.

Accordingly, we find that (systems of plots of) *smooth groupoids* — generalizing the smooth sets from § 6.1.2.1 by admitting gauge transformations between their plots by the Cartesian probe spaces — should form the 2-category of 2-sheaves on CrtSp.

Now, just as we saw for topological groupoids in Def. 4.1.24, the correct maps between such smooth groupoids are not just the plain morphisms of presheaves of groupoids, but are *Morita morphisms*: We say that

(i) A natural transformation $\mathcal{X}' \xrightarrow{\sim} \mathcal{X}$ between (systems of groupoids of plots of) smooth groupoids is an *equivalence* if on all stalks — as in (4.47) — it is an equivalence of groupoids of germs of plots.

(ii) A *Morita morphism* between smooth groupoids is a span of natural transformations (of groupoids of plots) where the left leg is such an equivalence:

$$\mathcal{X} \xleftarrow{\;\sim\;} \widehat{\mathcal{X}} \xrightarrow{\;F\;} \mathcal{Y}\,. \tag{6.42}$$

The groupoid-enriched category formed by sheaves of groupoids with such Morita morphisms between them — a (2, 1)-*category* — is accordingly a *(2,1)-topos* $\mathrm{Sh}_2(-)$ or *2-topos*, for short. [3]

Accordingly, we find that *smooth groupoids* — generalizing the smooth sets from § 6.1.2.1 by admitting gauge transformations between their plots by the Cartesian probe spaces — are to form the 2-topos

$$\mathrm{SmthGrpd} := \mathrm{Sh}_2(\mathrm{CrtSp})\,. \tag{6.43}$$

Since regarding sets as groupoids with trivial morphisms is fully faithful, this 2-topos faithfully contains the topos of smooth sets (§ 6.1.2.1) and with it the quasi-topos of diffeological spaces (§ 6.1.2.3), etc.:

$$\begin{array}{l} \mathrm{TopSpc} \hookrightarrow \\ \qquad\qquad\qquad \mathrm{DfflSpc} \hookrightarrow \mathrm{SmthSet} \longrightarrow \mathrm{SmthGrpd}\,. \\ \mathrm{SmthMfd} \hookrightarrow \end{array} \tag{6.44}$$

In the directly analogous manner, we similarly obtain 2-topoi of *haloed smooth groupoids* in higher generalization of § 6.1.2.4 and *super haloed smooth groupoids* in generalization of § 6.1.2.5.

6.1.2.7 The ∞-Topos of Smooth ∞-Groupoids

But the notion of gauge equivalence goes further, applying not just to field configurations themselves but also to the gauge transformations between them: For *higher* gauge field configurations B, B', a pair of gauge transformations $\lambda, \lambda' B \longrightarrow B'$ between them may in turn itself be nominally different and yet equivalent as witnessed by *higher gauge transformations*:

$$B \underset{\lambda'}{\overset{\lambda}{\rightrightarrows}} B'' \quad (\Downarrow \mu) \tag{6.45}$$

In particular, the composition of ordinary gauge transformations is thus determined only up to such higher gauge transformations μ, in general,

$$\begin{array}{ccccc} & & B_1 & & \\ & {}^{\lambda_{01}}\nearrow & \Downarrow \mu_{012} & \searrow^{\lambda_{12}} & \\ B_0 & & \xrightarrow[\lambda_{02}]{} & & B_2\,, \end{array} \tag{6.46}$$

whence the associativity law for this composition in general only holds up to yet higher gauge transformations ν:

$$\begin{array}{ccc} B_1 & \xrightarrow{\lambda_{12}} & B_2 \\ {\scriptstyle\lambda_{01}}\uparrow & \overset{\mu_{012}}{\searrow}\ \nearrow_{\lambda_{02}}\ \underset{\mu_{023}}{\searrow} & \downarrow{\scriptstyle\lambda_{23}} \\ B_0 & \xrightarrow{\lambda_{03}} & B_3 \end{array} \quad \overset{\nu_{0123}}{\Rrightarrow} \quad \begin{array}{ccc} B_1 & \xrightarrow{\lambda_{12}} & B_2 \\ {\scriptstyle\lambda_{01}}\uparrow & \overset{\mu_{123}}{\swarrow}\ \searrow^{\lambda_{13}}\ \underset{\mu_{013}}{\swarrow} & \downarrow{\scriptstyle\lambda_{23}} \\ B_0 & \xrightarrow{\lambda_{03}} & B_3\,, \end{array} \tag{6.47}$$

[3] More general 2-topoi specifically called $(2, 2)$-*topoi* are 2-categories of sheaves of small categories (instead of just of groupoids), but these we will not (need to) consider here.

which may be thought of as filling a tetrahedral diagram:

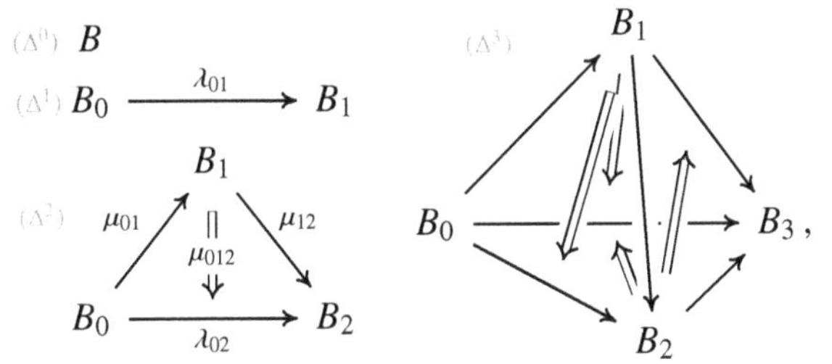

(6.48)

The shapes of these diagrams are known as cellular *simplices* Δ^n, higher-dimensional generalizations of the cellular triangle Δ^2. With the evident cellular maps between them (sending vertices to vertices, edges to edges, and generally k-cells to k-cells), these abstract shape diagrams form a category known as the *simplex category*, cleverly denoted Δ. The simplex category, hence, serves as a good category of *probes* detecting higher gauge transformations:

The *higher groupoid* (∞*-groupoid*) $\mathcal{G}$ of configurations of a given higher gauge field species, with its system of higher gauge transformations, is for each $k \in \mathbb{N}$, hence for each k-simplex $\Delta^n \in \Delta$, a set of order=k higher gauge transformations (which for $k = 0$ we understand to be the fields themselves), and for each cellular map $\Delta^k \xrightarrow{f} \Delta^{k'}$ a converse map between the corresponding sets of gauge transformations:

$$\begin{array}{ccc} \Delta^k & \mapsto & \mathrm{Plt}(\Delta^k, \mathcal{G}) \\ \uparrow f & \mapsto & f^* \downarrow \\ \Delta^{k'} & \mapsto & \mathrm{Plt}(\Delta^{k'}, \mathcal{G})\,. \end{array} \tag{6.49}$$

In particular:

(i) when f is injective on vertices, then it is the inclusion of a part of the boundary of the n-simplex and hence f^* is the map that sends n fold gauge transformations to the corresponding piece of the lower-rank gauge transformations that they act between;

(ii) when f is surjective on vertices, then it collapses some of the cells, and hence f^* is the map that regards k-fold gauge transformations as correspondingly trivial higher k'-fold gauge transformations on them.

These assignments (6.49) clearly ought to respect composition and identity maps, whence we find that higher groupoids of higher gauge transformations are encoded by (pre)sheaves on Δ of sets of n-fold transformations which they form, called *simplicial sets* (cf. Fig. 6.1):

$$\mathrm{sSet} := \mathrm{Sh}(\Delta)\,. \tag{6.50}$$

For example, groupoids $\mathcal{G}$ are re-incarnated as simplicial sets called their *simplicial nerves*

$$N\mathcal{G} : \Delta^k \mapsto \left\{ \begin{array}{c} \text{sequences of } k \text{ composable} \\ \text{morphism in } \mathcal{G} \end{array} \right\}, \tag{6.51}$$

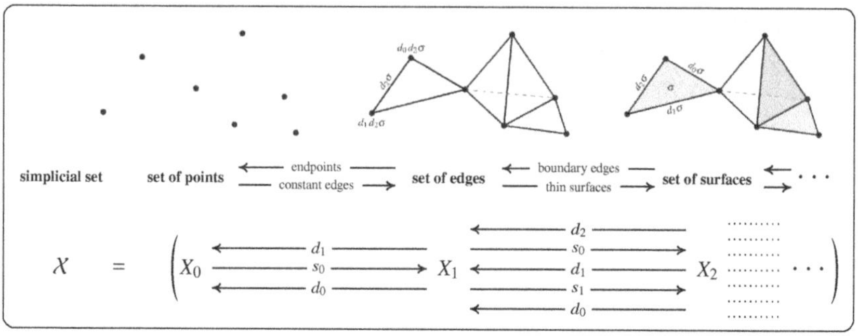

Figure 6.1 Simplicial sets (6.50) are encodings of structures consisting of $k+1$-*cells* ($k+1$-fold transformations) between k-cells for all $k \in \mathbb{N}$.

where the probes Δ^0 and Δ^1 detect the objects and the morphisms of the groupoid, the probe Δ^2 detects the composition law, and all higher probes $\Delta^{\geq 3}$ witness the associativity of this composition.

For another example: For X a topological space, there is the simplicial set $\int X$ (its *singular simplicial complex*) whose n-fold transformations are the continuous images of the *topological simplices*

$$\Delta^k_{\mathrm{Top}} := \left\{ \vec{x} \in \mathbb{R}^{k+1}_{\geq 0} \,\middle|\, \textstyle\sum_i x_i = 1 \right\} \subset \mathbb{R}^{k+1} \tag{6.52}$$

in X, hence:

$$\int X \,:\, \Delta^k \mapsto \mathrm{Hom}_{\mathrm{Top}}\big(\Delta^k_{\mathrm{Top}}, X\big)\,. \tag{6.53}$$

This is known as the *fundamental* or *path ∞-groupoid* of X, which we shall also call the *shape* of X.

Conversely, given a simplicial set $\mathcal{S} \in \mathrm{sSet}$, one recovers a topological space by regarding each of its abstract $k+1$ simplices as a topological $k+1$-simplex (6.52), glued along common face k-simplices, called the *topological realization* of $\mathcal{S}$ and denoted by the following coend formula

$$|\mathcal{S}| := \int^{\Delta^k \in \Delta} \mathcal{S}_k \times \Delta^k_{\mathrm{Top}} \quad \in \ \mathrm{TopSpc}\,. \tag{6.54}$$

These operations come with a canonical comparison map

$$\left|\int X\right| \xrightarrow{\ \sim\ } X \tag{6.55}$$

which is a *weakly homotopy equivalence* (in that it induces isomorphisms on all homotopy groups), reflecting the fact that every simplicial set $\mathcal{S}$ may be understood as presenting the ∞-groupoid $\int|\mathcal{S}|$ whose higher transformations are those of $\mathcal{S}$ with all inverses adjoined that may have been missing.

This simplicial way of describing higher groupoids of higher transformations is the same kind of idea of charting out a complex object (here: an ∞-groupoid) by simple probe spaces (here: the cellular simplices) that we have seen above for generalized geometric spaces, and hence, it seamlessly combines with the previous logic:

A *smooth ∞-groupoid* $\mathbf{X}$ is represented by, equivalently:

(i) a simplicial set of higher gauge transformations for each Cartesian probe space $\mathbb{R}^n$, hence a *simplicial presheaf* in

$$\mathrm{sPSh}(\mathrm{CrtSpc}) := \mathrm{Func}(\mathrm{CrtSpc}^{\mathrm{op}}, \mathrm{sSet})$$

(ii) a *smooth set* of k-fold transformations for each Δ^k-probe,

(iii) a set of plots for each formal product of a Cartesian space with a cellular simplex:

$$\begin{array}{ccc} (\mathrm{CrtSpc} \times \Delta)^{\mathrm{op}} & \longrightarrow & \mathrm{Set} \\ \mathbb{R}^n \times \Delta^k & \mapsto & \mathrm{Plt}(\mathbb{R}^n \times \Delta^k, \mathbf{X}) \,. \end{array} \tag{6.56}$$

For example, the smooth groupoid $\mathbf{B}\mathrm{U}(1)_{\mathrm{conn}}$ has a set of probes by $\mathbb{R}^n \times \Delta^k$, the set of k-simplices in the nerve (6.51) of the groupoid of U(1)-gauge fields on $\mathbb{R}^n$ (6.40):

$$\mathrm{Plt}\big(\mathbb{R}^n \times \Delta^k, \mathbf{B}\mathrm{U}(1)_{\mathrm{conn}}\big) = N\Big(\Omega^1_{\mathrm{dR}}(\mathbb{R}^n) /\!\!/ C^\infty\big(\mathbb{R}^n, \mathrm{U}(1)\big)\Big)_k \,. \tag{6.57}$$

In conclusion, this shows that in order to describe spaces whose elements may be related by higher gauge transformations, we adjoin formal products with cellular simplices to whatever probe spaces that we already have for detecting the geometric structure. The cellular simplices are the probe spaces for higher transformations.

For instance, in order to describe *super haloed smooth ∞-groupoids* we take the probe spaces to have all of: a Cartesian space factor $\mathbb{R}^n$, a bosonic infinitesimal factor $\mathbb{D}$, a super-infinitesimal factor $\mathbb{D}^{0|q}$ and a cellular simplex factor Δ^k, with the systems of plots, hence being presheaves of the form

$$\begin{array}{ccc} (\mathrm{SprHldCrtSp} \times \Delta)^{\mathrm{op}} & \longrightarrow & \mathrm{Set} \\ \mathbb{R}^n \times \mathbb{D} \times \mathbb{D}^{0|q} \times \Delta^k & \mapsto & \mathrm{Plt}\big(\mathbb{R}^n \times \mathbb{D} \times \mathbb{D}^{0|q} \times \Delta^k, \mathbf{X}\big) \end{array} \tag{6.58}$$

hence equivalently (singling out the higher transformation aspect) *simplicial presheaves* of the form

$$\begin{array}{ccc} \mathrm{SprHldCrtSp}^{\mathrm{op}} & \longrightarrow & \mathrm{sSet} \\ \mathbb{R}^n \times \mathbb{D} \times \mathbb{D}^{0|q} & \mapsto & \Big(\Delta^k \mapsto \mathrm{Plt}\big(\mathbb{R}^n \times \mathbb{D} \times \mathbb{D}^{0|q} \times \Delta^k, \mathbf{X}\big)\Big) \end{array} \tag{6.59}$$

forming the category

$$\mathrm{sPhv}(\mathrm{SprHldCrtSp}) := \mathrm{Func}\big(\mathrm{SprHldCrtSp}^{\mathrm{op}}, \mathrm{sSet}\big)$$

What remains is just to think again about the correct notion of equivalence between these systems of probes. This now needs to combine the *local* geometric notion of equivalence which previously we captured by passage to germs of plots, with the global notion of homotopical equivalence of systems of higher gauge transformations.

In view of (6.55) one says that a map between simplicial presheaves is an *equivalence*

$$F : \mathbf{X} \xrightarrow{\;\sim\;} \mathbf{Y} \tag{6.60}$$

if it is a *local weak homotopy equivalence*, meaning that on all simplicial sets of germs $\mathbb{G}^n$ of Cartesian probe spaces (Ex. 4.1.17), it induces a weak homotopy equivalence (isomorphism on all homotopy groups) under topological realization:

$$\big|\mathrm{Plt}\big(\mathbb{G}^n \times \mathbb{D} \times \mathbb{D}^{0|q}, \mathbf{X}\big)\big| \xrightarrow[F_*]{\sim} \big|\mathrm{Plt}\big(\mathbb{G}^n \times \mathbb{D} \times \mathbb{D}^{0|q}, \mathbf{Y}\big)\big| \tag{6.61}$$

This expresses exactly the idea that F witnesses the ∞-groupoids of plots of $\mathbf{X}$ and $\mathbf{Y}$ by small probes to be equivalent, and hence witnesses $\mathbf{X}$ to be operationally equivalent to $\mathbf{Y}$.

Then the ∞-*topos* of (super haloed) smooth ∞-groupoids ([Sc13][SS26c]), denoted

$$\begin{aligned} \mathrm{Grpd}_\infty &:= \mathrm{Sh}_\infty(*) \\ \mathrm{SmthGrpd}_\infty &:= \mathrm{Sh}_\infty(\mathrm{CrtSpc}) \\ \mathrm{HldSmthGrpd}_\infty &:= \mathrm{Sh}_\infty(\mathrm{HldCrtSp}) \\ \mathrm{SprHldSmthGrpd}_\infty &:= \mathrm{Sh}_\infty(\mathrm{SprHldCrtSp}), \end{aligned} \tag{6.62}$$

is the category of simplicial presheaves, regarded (via *simplicial localization*) with a simplicial set of morphisms between any pair of objects, such that the equivalences (6.60) universally become actual homotopy equivalences. One way to make precise what this means is model category theory cf. § 8.1.5 (see also [FSS23, §1]).

6.2 Geometric Modalities

While convenient categories of spaces, as established above, are indeed convenient in that they contain very general spaces realizing all desired universal constructions, in practice one needs to identify among these generalized spaces their more specialized versions (such as actual manifolds among haloed smooth spaces), and then one needs to generalize traditional geometric constructions (such as the definition of Riemannian metrics, etc.) from these standard spaces to all of the generalized spaces.

To accomplish such vast generalization, one needs to extract the conceptual essence of classical structures, independent of their traditional concrete implementation, in a way that may be formulated for all generalized spaces of which we only know that they have certain *qualitative aspects* ("synthetic mathematics", cf. [Bau21]).

The formalism for speaking about *quality* in mathematics is that of *modal operators* (cf. § 1.1.9), originating in the *modal logic* that goes back to the ancients, where they express *modes of being true* (for instance, being *possibly* true or being *necessarily* true, cf. [SS25c, §2.2]).

Beyond mere logic, in category/type theory, modal operators express *modes of being*, such as (as we will see, following [Sc13]) *being discrete*, *being reduced*, or *being bosonic*.

We set up these modal operators in chapter 9 and then use them to formulate Cartan geometry internal to (super) haloed smooth ∞-groupoids in § 10.2.

7

Equivariant Homotopy

The basic idea of equivariant homotopy theory is simple, but its phenomenology is rich and its theoretical development surprisingly subtle. We first survey the basic ideas (in § 7.1) and then explain the need for the perspective of "orbi-singular cohesion" (in § 7.2).

7.1 Equivariance & Orbits

After geometric and higher homotopy structures on generalized spaces, we now turn to the idea of *equivariant* and *orbi-singular* structure.

At first sight it appears obvious: For G a topological group and $G \curvearrowright X$, $G \curvearrowright Y$ a pair of its continuous actions on topological spaces X, Y (as in § 4.1.3), we are to consider as maps the equivariant maps

$$\underset{X}{\overset{G}{\circlearrowright}} \xrightarrow{\quad f \quad} \underset{Y}{\overset{G}{\circlearrowright}}$$

and as homotopies the *equivariant homotopies*, namely the G-equivariant continuous maps of the form shown on the right here:

$$\underset{X}{\overset{G}{\circlearrowright}} \underset{g}{\overset{f}{\underset{\longrightarrow}{\overset{\longrightarrow}{\Downarrow \mu}}}} \underset{Y}{\overset{G}{\circlearrowright}} \quad\Leftrightarrow\quad \begin{array}{ccc} \overset{G}{\circlearrowright} X & & \\ \big\downarrow{\scriptstyle (\mathrm{id},\{0\})} & \overset{f}{\searrow} & \\ G \circlearrowright X \times [0,1] & \dashrightarrow{\scriptstyle \mu} & \overset{G}{\circlearrowright} Y, \\ \big\uparrow{\scriptstyle (\mathrm{id},\{1\})} & \underset{g}{\nearrow} & \\ \underset{G}{\circlearrowleft} X & & \end{array} \tag{7.1}$$

where the G-action on the interval $[0, 1]$ is taken to be trivial.

But this situation becomes more subtle as we next turn to regarding these topological G-spaces as mere stand-ins for their fundamental ∞-groupoids $\int X$(6.53). This transition to *equivariant homotopy theory* is hence fundamentally a topic of cohesive

DOI: 10.1201/9781003675969-7

homotopy theory: The cohesive structure of topological spaces is used to present, under the shape modality, a notion of equivariant homotopy types.

The key question then concerns the identification of these equivariant homotopy types, hence the question of when the shape of X *with its* G*-action* is equivalent to the shape of Y with its G-action. Technically this is the question for the generalization of the notion of *weak homotopy equivalence* between topological spaces to the generality of G-spaces.

Namely, in the non-equivariant base case, when X and Y admit the structure of CW-complexes, then *Whitehead's theorem* says that a continuous map between them is a homotopy equivalence (4.22) iff it is a *weak homotopy equivalence* in that it induces isomorphisms on all homotopy groups. Since, moreover, every topological space is weak homotopy equivalent to a CW-complex, by (6.55), this shows that the identification of the shapes of topological spaces is equivalently given by regarding these spaces up to weak homotopy equivalence.

In order to generalize this situation to the equivariant case, one hence needs to ask for the analogous notion of *equivariant weak homotopy equivalence* which makes the analogous statement true for G-spaces. The answer to this is given by *Elmendorf's theorem* (recalled as Prop. 8.1.27): This says that a map between G-CW complexes is a G-homotopy equivalence in the sense of (7.1) iff it is an ordinary weak homotopy equivalence *on all* H*-fixed loci* for all closed subgroups $H \subset G$.

Focusing now on finite equivariance groups G, for simplicity, this leads to the following picture of *equivariant homotopy types*: First, for each subgroup $H \subset G$, we have the H-fixed subspace $X^H \subset X$, which we may think of as the space of G-equivariant maps from the coset G-space G/H to X:

$$\mathrm{Map}(G/H, X)^G \simeq X^H \simeq \{x \in X \,|\, \forall_{h \in H} : h \cdot x = x\} \tag{7.2}$$

and for each G-equivariant map between such coset spaces, we have the induced converse map of fixed loci

$$\begin{array}{ccc} G \circlearrowright G/H & \mapsto & X^H \\ \phi \uparrow & \mapsto & \downarrow \phi^* \\ G \circlearrowright G/H' & \mapsto & X^{H'} . \end{array} \tag{7.3}$$

This construction of systems of fixed loci exhibits the G-space $G \circlearrowright X$ as a presheaf of topological spaces on the sub-category of G-spaces of the form G/H for $H \subset G$. But, up to isomorphism, these G-spaces are exactly the possible *orbits* of points under any G-action, whence one speaks of the *category of* G*-orbits* or just of the G*-orbit category*, for short:

$$G\mathrm{Orb} := \left\{ \overset{\circlearrowright G}{G/H} \to \overset{\circlearrowright G}{G/H'} \,\middle|\, H \subset G \right\}. \tag{7.4}$$

Curiously, Elmendorf's theorem, hence, indicates that G-orbits G/H play the role of *probe spaces* that detect the equivariant or orbi-singular nature of G-spaces (in analogy to how we have seen that Cartesian spaces $\mathbb{R}^n$ probe any smooth structure that may be present additionally), in that the fixed loci X^H play the role of the (topological spaces of) plots of X by the probe space G/H:

$$
\begin{array}{ccc}
G\mathrm{TopSp} & \longrightarrow & \mathrm{PSh}\big(G\mathrm{Orb}, \mathrm{TopSp}\big) \\
G \circlearrowleft X & \mapsto & \big(G/H \mapsto \mathrm{Map}(G/H, X)^G \simeq X^H\big)\,.
\end{array}
\tag{7.5}
$$

Namely, Elmendorf's theorem says that under this perspective the G-equivariant homotopy equivalences (between G-CW-complexes) on the left correspond to the canonical equivalences in the presheaf category on the right, namely to the *probe-wise* weak homotopy equivalences.

But from § 6.1.2.7 we know that topological spaces regarded up to weak homotopy equivalence are a presentation of the ∞-topos of (their fundamental) ∞-groupoids, whence one sees that the ∞-topos of *G-equivariant ∞-groupoids* is just the corresponding ∞-presheaf category:

$$
G\mathrm{Grpd}_\infty := \mathrm{PSh}_\infty(G\mathrm{Orb})\,. \tag{7.6}
$$

This remarkable insight of Elmendorf's theorem, that, in our perspective, *G-equivariant ∞-groupoids are generalized ∞-groupoids probeable by G-orbits*, thus ties in neatly with the above discussion of geometric ∞-groupoids as those chartable by suitable geometric probe spaces. This allows for an immediate combination of these notions:

An ∞-groupoid that is equipped with both geometric structure and G-orbi-singular structure will hence be probeable by formal products of geometric probes and G-orbits, witnessing that each H-fixed locus exhibits the given generalized geometric structure. This way we obtain the ∞-topoi of *G-orbi singular cohesive* generalized spaces in G-equivariant generalization of (6.62):

$$
\begin{aligned}
G\mathrm{Grpd}_\infty &:= \mathrm{Sh}_\infty\big(G\mathrm{Orb}\big) \\
G\mathrm{SmthGrpd}_\infty &:= \mathrm{Sh}_\infty\big(\mathrm{CrtSpc} \times G\mathrm{Orb}\big) \\
G\mathrm{HldSmthGrpd}_\infty &:= \mathrm{Sh}_\infty\big(\mathrm{HldCrtSp} \times G\mathrm{Orb}\big) \\
G\mathrm{SprHldSmthGrpd}_\infty &:= \mathrm{Sh}_\infty\big(\mathrm{SprHldCrtSp} \times G\mathrm{Orb}\big)\,.
\end{aligned}
\tag{7.7}
$$

These are hence very convenient categories for generalized orbi-singular cohesive geometry over a fixed equivariance group G. However, in applications, one often does not want to fix the equivariance group G but work "globally" by allowing all possible equivariance groups (from a given class, say finite groups).

To this end consider the full sub-∞-category of ∞-groupoids

$$
\begin{array}{ccccc}
\mathrm{Snglrt} & \xrightarrow{\ \sim\ } & \Big\{\mathbf{B}K \,\Big|\, K \in \mathrm{Grp}^{\mathrm{fin}}\Big\} & \hookrightarrow & \mathrm{Grpd}_\infty \\
\overset{K}{\curlyvee} & \mapsto & \mathbf{B}K & &
\end{array}
\tag{7.8}
$$

(cf. Def. 9.2.1) whose objects are the delooping groupoids (4.35) of finite groups and whose hom-groupoids consist of the functors and their natural transformations between these. Recalling that such a delooping groupoid is a point with G worth of automorphisms:

$$
\overset{K}{\curlyvee} \simeq \Big\{ \bullet \circlearrowleft^{K} \Big\}
$$

we may understand it as the archetype of a K-orbi singularity (which is what the new notation "$\rightthreetimes^{K}$" is trying to visualize), hence as the *probe space* detecting concrete K-orbi singularities, whence we regard Snglrt (7.8) as the *category of abstract orbi-singularities*.

Elsewhere, this category (7.8) is called the *global orbit category*. That term is less suggestive, but is justified by the remarkable fact that the actual G-orbit categories GOrb(7.4) are close to being "local slices" of Snglrt: they are 0-truncations $\tau_0(-)$ (hence the full subcategories on the faithful functors) of the slice over $\mathbf{B}G$:

$$\mathrm{Snglrt}_{/\mathbf{B}G} \underset{\hookleftarrow}{\overset{[-]_0}{\underset{\perp}{\longrightarrow}}} \tau_0\big(\mathrm{Snglrt}_{/\mathbf{B}G}\big) \simeq G\mathrm{Orb} \tag{7.9}$$

But this means that the generalized spaces probeable by abstract orbi-singularities, namely the ∞-topos over Snglrt, are a good category of generalized *orbi-singular ∞-groupoids*, whose slices over the abstract singularities $\mathbf{B}G$ are cohesive over the G-singular ∞-topos (7.7) (this observation is due to [Re14], cf. [SS26c, Prop. 4.3.38]).

In conclusion, this shows that Snglrt (7.8) serves as a good category of probe spaces for the definition of generalized *singular-geometric* spaces (geometric spaces which may exhibit orbi-singularities). And hence, in the "globally equivariant" generalization of (7.7), we finally arrive at the *singular-cohesive* ∞-topoi that will be of concern in the following:

$$\begin{aligned} \mathrm{SnglrGrpd}_\infty &:= \mathrm{Sh}_\infty\big(\mathrm{Snglrt}\big) \\ \mathrm{SnglrSmthGrpd}_\infty &:= \mathrm{Sh}_\infty\big(\mathrm{CrtSpc} \times \mathrm{Snglrt}\big) \\ \mathrm{SnglrHldSmthGrpd}_\infty &:= \mathrm{Sh}_\infty\big(\mathrm{HldCrtSp} \times \mathrm{Snglrt}\big) \\ \mathrm{SnglrSprHldSmthGrpd}_\infty &:= \mathrm{Sh}_\infty\big(\mathrm{SprHldCrtSp} \times \mathrm{Snglrt}\big)\,. \end{aligned} \tag{7.10}$$

7.2 Differential Equivariance

Above, in seeing how to bring (differential) geometry into the discussion of orbifold cohomology via topological groupoids and thus in order to promote the latter to a theory of *geometric orbifold cohomology*, we have produced ever larger and ever more "convenient" higher categories of generalized spaces characterized as being probeable by a whole sequence of kinds of test spaces, first a hierarchy of geometric test spaces and then a class of abstract orbi-singularities (cf. Tab. 6.1).

It remains to highlight the deeper reason for bringing in these orbi-singular probes. Namely, orbifolds in their incarnation as Lie groupoids (Def. 4.2.1) and hence as smooth groupoids (§ 6.1.2.6) do faithfully exist already in the cohesive ∞-topos $\mathrm{SmthGrpd}_\infty$ and its super haloed siblings (6.62), but now they are"re-incarnated", also faithfully, in their alternative *orbi-singular* incarnation in $\mathrm{SnglrSmoothGrpd}_\infty$ and *its* super haloed siblings (6.62).

Here, in this last section of the transitional part, we indicate why this alternative orbi-singular incarnation of orbifolds is necessary for the definition of their (generalized) *differential cohomology*, hence for the definition of (generalized) *differential orbi-cohomology* and hence in particular also for (generalized) *equivariant differential cohomology*. (See also analogous comments in [SS26c, §6.2.1].)

The point is simply that given an ∞-topos $\mathbf{H}$ we may and want to think of its Hom-∞-groupoids $\mathbf{H}(X, A) \in \mathrm{Grpd}_\infty$ (for $X, A \in \mathbf{H}$) as the *cocycle ∞-groupoids*, and hence of their 0-truncation τ_0 (cf. § 8.1.15) as the intrinsic (generalized nonabelian) *cohomology* of X with coefficients in A (cf. § 1.1.11):

$$H^0(X; A) := \tau_0 \mathbf{H}(X, A) . \tag{7.11}$$

This extremely simple-looking definition is extremely powerful, witnessing (cf. [NSS12a, p. 4]) ∞-topoi as ∞-categorifications of (nonabelian) cohomology sets (cf. [SS25g]).

But note that it is the geometrically discrete hom ∞-groupoids that appear here, the underlying point ∞-groupoid of the internal mapping stack $\mathrm{Map}(X, A) \in \mathbf{H}$, where all notions of coboundary are encoded by the 1-morphisms:

$$\mathbf{H}(X, A) \simeq \mathrm{PntsMap}(X, A) ,$$

hence

$$H^0(X; A) \simeq \tau_0 \mathrm{PntsMap}(X, A) . \tag{7.12}$$

But when $\mathbf{H}$ is cohesive, there is another canonical way to extract a geometrically discrete ∞-groupoid from the mapping stack, namely by forming its shape $\int(-)$. The 0-truncation of the shape of a cohesive ∞-groupoid is a generalization of the geometric notion of the set of *path-connected components* π_0:

$$\pi_0 \mathrm{Map}(X, A) := \tau_0 \mathrm{ShpMap}(X, A) \tag{7.13}$$

This is analogous to but crucially different from (7.12), in general, hence (7.13) is not actually a notion of cohomology sets intrinsic to an ∞-topos, or at least not generally and not manifestly so.

But it is this latter form (7.13) in which (twisted) orbifold cohomology appeared (4.122) in § 4.2.2! Therefore, Def. 4.2.5, there is not yet a definition of orbifold cohomology which is *manifestly intrinsic* to an ambient ∞-topos, in the sense of (7.11).

It is this issue which is solved (via [SS26c, Thm. 6.2.3]) by generalizing from the cohesive ∞-topos $\mathrm{SmthGrpd}_\infty$ (and its siblings) to the *singular cohesive* ∞-topos $\mathrm{SnglrSmthGrpd}_\infty$ (and its siblings), where orbifolds $\mathcal{X}$ are re-incarnated in their *orbi-singular* version $\curlyvee \mathcal{X}$ (discussed in § 9.2)!

In short, the answer to "Of which class of ∞-topoi is geometric orbifold cohomology the *intrinsic* notion of cohomology?" is *singular-cohesive ∞-topoi*. And that is the reason why we develop geometric orbifold cohomology in this abstractly appropriate context in this book.

Finally, a key use case where it matters to be dealing with the *intrinsic* cohomology of ∞-topoi is *differential cohomology*. While here we will not further dwell on this point but be content with laying foundations for it, the basic idea is this:

Differential (generalized nonabelian) cohomology arises by pulling back the *character map* on geometrically discrete cocycle ∞-groupoids, $\mathrm{Grpd}_\infty(\int X, \mathcal{A})$, to L_∞-algebra-valued differential forms (this is the content of [FSS23]). In order to equivariantize this one needs to be able to apply this same construction *probe-wise*, pulling back to probe-wise defined systems of L_∞-algebra-valued differential forms (see [SS25g]). But this is exactly what works in singular cohesive ∞-topoi with respect to its abstract orbi-singularity probes discussed in § 7.1.

Part IV

Cohesive Homotopy Theory

8

Higher Geometry

We recall basics of higher geometry in the form of *geometric homotopy theory*, namely *higher topos theory*, in §8.1 and lay out in §8.2 the *internal formulation*, in ∞-toposes, of group actions and the classification of fiber bundles (more on which in [SS26c]).

8.1 Topos theory

Here we briefly record basics of ∞-topos theory [TV05][Lu09a][Re10] (review is in [Re19], exposition with an eye toward applications in differential geometry and physics is in [FSS14][Sc25]). This is to set up our notation and to highlight some less widely used aspects that we need further below.

8.1.1 Categories

We make free use of the language and the basic facts of category theory and homotopy theory (see [GJ99][Rie14][Ri20]) as well as of ∞-category theory (see [Joy08a][Joy08a][Lu09a][Rie14][Ci19]).

(i) We write Cat_∞ for the ("very large") ∞-category of (large) ∞-categories [Re98][Be05][Lu09a, Ch. 3], though we only use this for declaring ∞-categories. Inside Cat_∞, there is the sequence of full sub-∞-categories (Def. 8.1.1) of n-categories (i.e., $(n, 1)$-categories) as well as of n-groupoids (see Def. 8.1.28) for all $n \in \mathbb{N}$, denoted thus:

$$
\begin{array}{ccccccccc}
\mathrm{Sets} & \hookrightarrow & \mathrm{Cat}_1 & \hookrightarrow & \mathrm{Cat}_2 & \hookrightarrow & \cdots & \hookrightarrow & \mathrm{Cat}_\infty \\
\| & & \uparrow & & \uparrow & & & & \uparrow \dashv \downarrow \mathrm{Core} \\
\mathrm{Sets} & \hookrightarrow & \mathrm{Grpd}_1 & \hookrightarrow & \mathrm{Grpd}_2 & \hookrightarrow & \cdots & \hookrightarrow & \mathrm{Grpd}_\infty
\end{array}
\tag{8.1}
$$

(ii) Here $\mathrm{Core}(C)$ denotes the maximal ∞-groupoid inside an ∞-category C.

(iii) For $C \in \mathrm{Cat}_\infty$ and for $X, Y \in C$ a pair of objects, we write

$$
C(X,Y) \;:=\; \mathrm{Hom}_C(X,Y) \;\in\; \mathrm{Grpd}_\infty \tag{8.2}
$$

for the *hom-∞-groupoid*, i.e., the ∞-groupoid of morphisms between them, and higher homotopies between these (see [Lu09a, 1.2.2][DS09]). This is well-defined, up to equivalence of ∞-groupoids, independently of which model for

DOI: 10.1201/9781003675969-8

∞-categories is used, since these are all equivalent to each other [Be06][Be14]. We have no need to specify any particular model for ∞-categories (except for the construction of examples, in §8.1.5).

Definition 8.1.1 (Fully faithful functor [Lu09a, 1.2.10]). For $C, \mathcal{D} \in \mathrm{Cat}_\infty$ (8.1), a functor $F : C \longrightarrow \mathcal{D}$ is called *fully faithful*, to be denoted

$$C \xhookrightarrow{\;F\;} \mathcal{D}, \tag{8.3}$$

if it is an equivalence on all hom-∞-groupoids (8.2):

$$\underset{X,Y \in C}{\forall} \quad C(X, Y) \xrightarrow[\simeq]{F_{X,Y}} \mathcal{D}\big(F(X), F(Y)\big). \tag{8.4}$$

In this case we also say that (8.3) exhibits a *full sub-∞-category inclusion.*

8.1.2 Topology

The category of Δ-*generated* or *D-topological* spaces (Rem. 8.1.3) is both a convenient foundation for homotopy theory (Prop. 8.1.4) and pivotal for our key example context (Example 9.1.19):

Definition 8.1.2 (Topological spaces). We write

$$\mathrm{CWCmplx} \hookrightarrow \mathrm{DTopSpc} \hookrightarrow \mathrm{TopSpc} \quad \in \mathrm{Cat}_1 \tag{8.5}$$

for (from right to left):

(i) the category of all topological spaces with continuous functions between them;
(ii) the full subcategory of *D-topological spaces* (§ 4.1.2),
(iii) the further full subcategory on those that admit the structure of a CW-cell complex (Ex. 4.1.3).

Remark 8.1.3 (D-topological is Δ-generated).

(i) Since the topological n-simplex Δ^n_{top} is a retract of the Euclidean space $\mathbb{R}^n$, the condition on $X \in \mathrm{TopSpc}$ of being D-topological (Def. 4.1.2) is equivalent to being Δ-*generated*, in that the open subsets of X are precisely those whose pre-images under all continuous functions of the form $\Delta^n_{\mathrm{top}} \to X$ are open.
(ii) The concept of Δ-generated spaces is due to [Sm][Dug03] and independently due to [SYH10], where they are called *numerically generated.*
(iii) We say *D-topological* to better bring out their conceptual role, in view of Prop. 8.1.7.

Proposition 8.1.4 (D-topological spaces are convenient). *The category of* DTopSpc *(Def. 4.1.2) is a convenient category of topological spaces in the sense of [St67] in that it:*

(i) *contains all CW-complexes* (8.5) *[SYH10, Cor. 4.4];*
(ii) *has all small limits and colimits [SYH10, Prop. 3.4];*
(iii) *is locally presentable [FR07, Cor. 3.7];*

(iv) *is Cartesian closed [SYH10, Cor. 4.6]: the mapping space between* $X, Y \in$ DTopSpc *is the reflection* (8.12) *of the internal mapping space* **Map** (8.82) *of* DiffSp *[SYH10, Prop. 4.7]:*

$$\mathrm{Map}(X,Y) \;=\; \mathrm{Dtplg}\Big(\mathbf{Map}\big(\mathrm{Cdfflg}(X),\, \mathrm{Cdfflg}(Y)\big)\Big). \tag{8.6}$$

8.1.3 Differential topology

D-topological spaces lend themselves to differential topology via their joint (co-) reflection (Prop. 8.1.7) both into all topological spaces and into diffeological spaces (Def. 8.1.6):

Definition 8.1.5 (Cartesian spaces)**.** We write

$$\mathrm{CrtSpc} \hookrightarrow \mathrm{SmthMfd} \quad \in \mathrm{Cat}_1 \tag{8.7}$$

for the category whose objects are the natural numbers $n \in \mathbb{N}$, thought of as representing the Cartesian spaces $\mathbb{R}^n$, and whose morphisms are the *smooth* functions between these. We regard this category as equipped with the coverage (Grothendieck pre-topology) whose covers are the differentially good open covers (i.e., such that all non-empty finite intersections of patches are *diffeomorphic* to a Cartesian space [FStS12, 6.3.9]).

Definition 8.1.6 (Diffeological spaces)**.**

(i) The category of *diffeological spaces* ([So80][So84][IZ85], see [BH08][IZ13]) is the full subcategory of sheaves on CrtSpc (Def. 8.1.5)

$$\mathrm{DiffSp} \hookrightarrow \mathrm{Shv}(\mathrm{CrtSpc}) \tag{8.8}$$

on those $X \in \mathrm{Shv}(\mathrm{CrtSpc})$ which are *concrete sheaves* [Du79b] supported on their *underlying set*

$$X_s \;:=\; \mathrm{Shv}(\mathrm{SmthMfd})(*, X) \tag{8.9}$$

in that the canonical function

$$X(U) \hookrightarrow (U_s, X_s) \tag{8.10}$$

is an injection, for all $U \in \mathrm{CrtSpc}$, with U_s denoting their underlying set U.

(ii) We call

$$X(U) \underset{\text{Prop. 8.1.54}}{\simeq} \mathrm{DiffSp}(U, X) \;\in \tag{8.11}$$

the set of U-*plots* of the diffeological space X.

Proposition 8.1.7 (Topological/diffeological adjunction)**.**
(i) *There is an adjunction [SYH10, Prop. 3.1]*

$$\mathrm{TopSpc} \underset{\mathrm{Cdfflg}}{\overset{\mathrm{Dtplg}}{\underset{\longrightarrow}{\overset{\longleftarrow}{\bot}}}} \mathrm{DiffSp} \tag{8.12}$$

between the categories of topological spaces (Def. 8.1.2) and of diffeological spaces (Def. 8.1.6), where

- *the right adjoint* Cdfflg *sends a topological space to the same underlying set equipped with the* **topological diffeology** *whose plots* (8.11) *are precisely the continuous functions;*
- *the left adjoint* Dtplg *sends a diffeological space to the same underlying set equipped with the* **diffeological topology** *("D-topology" [IZ13, 2.38][CSW13]), which is the final topology with respect to all plots* (8.11)*, hence such that a subset is open precisely if its pre-image under all plots is open.*

(ii) *The fixed points* $X \in$ TopSpc *of this adjunction are the D-topological spaces (Remark 8.1.3)*

$$X \text{ is D-topological} \quad \Leftrightarrow \quad \mathrm{Dtplg}\big(\mathrm{Cdfflg}(X)\big) \xrightarrow[\simeq]{\epsilon_X} X\,. \tag{8.13}$$

(iii) *The adjunction is idempotent [SYH10, Lemma 3.3], hence factors through the category of D-topological spaces, exhibiting them as a co-reflective subcategory of* TopSpc *and a reflective subcategory of* DiffSp*:*

$$\mathrm{TopSpc} \underset{\mathrm{Cdfflg}}{\overset{\hookleftarrow}{\underset{\bot}{\longrightarrow}}} \mathrm{DTopSpc} \underset{}{\overset{\mathrm{Dtplg}}{\underset{\bot}{\hookrightarrow}}} \mathrm{DiffSp}\,. \tag{8.14}$$

The following Prop. 8.1.8 is due to [Har13, Thm. 3.3].

Proposition 8.1.8 (Model structure on D-topological spaces)**.**

(i) *The standard cell inclusions define a cofibrantly generated model category structure on* DTopSpc *(Def. 4.1.2).*

(ii) *With respect to this model structure and the standard model structure on* TopSpc*, the co-reflection* (8.12) *becomes a Quillen equivalence:*

$$\mathrm{TopSpc} \underset{\mathrm{Cdfflg}}{\overset{\hookleftarrow}{\underset{\simeq_{\mathrm{Quillen}}}{\longrightarrow}}} \mathrm{DTopSpc}\,. \tag{8.15}$$

8.1.4 Differential geometry

Definition 8.1.9 (Smooth Manifolds)**.** We write

$$\mathrm{SmthMfd} \in \mathrm{Cat} \tag{8.16}$$

for the category of finite-dimensional paracompact smooth manifolds with smooth functions between them. We regard this as a site with the Grothendieck topology of open covers.

Proposition 8.1.10 (Cartesian spaces are dense in the site of manifolds)**.** *With respect to the coverages in Def. 8.1.9 and Def. 8.1.5, the inclusion* $\mathrm{CrtSpc} \overset{i}{\hookrightarrow} \mathrm{SmthMfd}$ *is a dense sub-site in that it induces an equivalence of categories of sheaves*

$$\mathrm{Shv(CrtSpc)} \underset{i_*}{\overset{i^*}{\underset{\simeq}{\rightleftarrows}}} \mathrm{Shv(SmthMfd)}\,. \tag{8.17}$$

Proposition 8.1.11 (Smooth manifolds inside diffeological spaces). *Every* $X \in$ SmthMfd (8.16) *becomes a diffeological space (Def. 8.1.6) on its underlying set by taking its plots* (8.11) *of shape* $U \in$ CrtSpc *to be the ordinary smooth functions:*

$$X(U) := \mathrm{SmthMfd}(U, X)\,. \tag{8.18}$$

More generally, every possibly infinite-dimensional Fréchet manifold (e.g., [KS17, 2.2]) becomes a diffeological space this way. Moreover, this constitutes fully faithful embeddings (Def. 8.1.1) into the category of Diffeological spaces [Lo94, Thm. 3.1.1]:

$$\underset{\text{finite-dimensional}}{\mathrm{SmthMfd}} \hookrightarrow \underset{\substack{\text{possibly}\\ \text{infinite-dimensional}}}{\mathrm{FréSmthMfd}} \hookrightarrow \mathrm{DiffSp}\,. \tag{8.19}$$

8.1.5 Model toposes

We recall some basics of model categories (e.g. [GJ99, 2]) of simplicial presheaves ([Jar87][Jar96][Jar15]) as presentations of ∞-toposes ([Lu09a, A.2, A.3]).

Model categories of simplicial presheaves.

Definition 8.1.12 (Model category of simplicial presheaves).
Let C be a site. We write

(i)

$$\mathrm{sPShv}(C)_{\mathrm{loc}} \;\in\; \mathrm{HomotopicalCategories} \tag{8.20}$$

for the category of simplicial presheaves on C, regarded as a homotopical category with weak equivalences, the local weak homotopy equivalences of simplicial sets.

(ii)

$$\mathrm{sPShv}(C)_{\substack{\mathrm{inj}/\\ \mathrm{proj}},\mathrm{loc}} \;\in\; \mathrm{ModelCategories} \tag{8.21}$$

for the same category regarded as either the corresponding injective or projective model category.

(iii)

$$\mathrm{sPShv} \xrightarrow{\;\ell\;} L_{\mathrm{lwhe}}\mathrm{sPShv}_{\mathrm{loc}} \;=:\; \mathbf{H} \tag{8.22}$$

for the corresponding simplicial localization.

Lemma 8.1.13 (Cofibrancy in projective model structure [Du01, Cor. 9.4]). *Let* C *be a site. For a simplicial presheaf* $X_\bullet \in \mathrm{sPShv}(C)_{\mathrm{proj,loc}}$ *in the projective model structure* (8.21) *to be cofibrant, it is sufficient that* $X_\bullet$ *is degreewise*

(i) *a coproduct of representables, such that*

(ii) *the degenerate cells split off as a direct summand.*

Lemma 8.1.14 (Simplicial presheaf represents its own hocolim [DHI04, 2.1][Sc13, 2.3.21]). *Let* C *be a site and* $X_\bullet \in \mathrm{sPShv}(C)$ *a simplicial presheaf (Def. 8.1.12). Then its image under simplicial localization* (8.22) *is equivalently the simplicial homotopy colimit over the images of its component presheaves:*

$$\ell(X_\bullet) \;\simeq\; \underset{\longrightarrow}{\lim}(\ell X)_\bullet \;\in\; \mathbf{H}\,.$$

8.1.6 Topological mapping stacks

Example 8.1.15 (Model category presentation of smooth ∞-groupoids)**.** Let C = CrtSpc (Def. 8.1.5). Then the simplicial localization (8.22) of $\mathrm{sPShv}(C)_{\mathrm{loc}}$ (8.21) is $\mathrm{SmthGrpd}_\infty$ (Ex. 9.1.19):

$$L_{\mathrm{lwhe}}\mathrm{sPShv}(\mathrm{CrtSpc})_{\mathrm{loc}} \simeq \mathrm{SmthGrpd}_\infty\,.$$

Lemma 8.1.16 (Mapping stack from delooping of discrete group to topological stack)**.** *In* $\mathrm{SmthGrpd}_\infty$ *(Ex. 9.1.19) consider*

- *a finite group embedded via* (9.154)

$$G \in \mathrm{Grp} \xrightarrow{\mathrm{Disc}} \mathrm{Grp}\big(\mathrm{SmthGrpd}_\infty\big), \tag{8.23}$$

- *a topological groupoid, embedded via* (9.53)

$$\begin{array}{ccc} \mathrm{TopGrpd} & \xrightarrow{\mathrm{Cdfflg}} & \mathrm{SmthGrpd}_\infty \\ \mathcal{X}_{\mathrm{top}} & \mapsto & \mathcal{X}_{\cup} \end{array} \tag{8.24}$$

Then the mapping stack (8.82) *formed in* $\mathrm{SmthGrpd}_\infty$ *is the degree-wise image under* Cdfflg (8.12) *of the topological groupoid representing the mapping stack of topological groupoids (which exists by [No10] since G is finite, hence compact):*

$$\mathbf{Map}\big(\mathbf{B}G\,,\,\mathcal{X}_{\cup}\big) \simeq \mathrm{Cdfflg}\,\mathbf{Map}\big(\mathbf{B}G\,,\,\mathcal{X}_{\mathrm{top}}\big). \tag{8.25}$$

Proof. Since (by Ex. 9.1.19)

$$\mathrm{SmthGrpd}_\infty \simeq \mathrm{Shv}_\infty(\mathrm{CrtSpc}) \underset{\longrightarrow}{\overset{L}{\hookleftarrow}}\!\!\!\perp\; \mathrm{PShv}_\infty(\mathrm{CrtSpc})$$

it is sufficient to show that we have an equivalence of ∞-presheaves of the form

$$\mathbb{R}^n \longmapsto \begin{array}{l} \mathrm{PShv}_\infty\big(\mathrm{CrtSpc}\big)\big(\mathbb{R}^n \times \mathbf{B}G\,,\,\mathcal{X}_{\cup}\big) \\ \simeq\; \mathrm{PShv}_\infty\big(\mathrm{CrtSpc}\big)\big(\mathbb{R}^n\,,\,\mathrm{Cdfflg}\,\mathbf{Map}(\mathbf{B}G, \mathcal{X}_{\mathrm{top}}).\big) \end{array} \tag{8.26}$$

By Ex. 8.1.15, we may model this in the global projective model structure on simplicial presheaves over CrtSpc:

$$\mathrm{sPShv}(\mathrm{CrtSpc}) \xrightarrow{\ \ell\ } L_{\mathrm{lwhe}}\mathrm{sPShv}(\mathrm{CrtSpc})_{\mathrm{proj}} \simeq \mathrm{Shv}_\infty(\mathrm{CrtSpc}) \tag{8.27}$$

by the following models (Lemma 8.1.14):

(a) A model under ℓ (8.27) of the Cartesian product $\mathbb{R}^n \times \mathbf{B}G$ with the delooping $\mathbf{B}G \simeq \varinjlim G^{\times_\bullet}$ (8.154), is given by the simplicial presheaf

$$\mathbb{R}^n \times G^{\times_\bullet} \in \mathrm{sPShv}(\mathrm{CrtSpc})_{\mathrm{proj}}\,. \tag{8.28}$$

(b) A model under ℓ (8.27) for the image (8.24) of a topological groupoid $\mathcal{X}_{\mathrm{top}}$ is given by its nerve regarded as a simplicial presheaf, componentwise via (9.52)

$$N_\bullet(\mathcal{X}_{\mathrm{top}}) \in \mathrm{sPShv}(\mathrm{CrtSpc})_{\mathrm{proj}}\,. \tag{8.29}$$

Moreover,

- The object (8.28) is projectively cofibrant, by Lemma 8.1.13, as is its Cartesian product with a k simplex $\Delta[k]$.

- The object (8.29) is projectively fibrant (objectwise a Kan complex) by the groupoid property of $\mathcal{X}_{\mathrm{top}}$.

Therefore, to get (8.26), it is, in turn, sufficient to exhibit for $\mathbb{R}^n \in \mathrm{CrtSpc}$ a natural isomorphism of simplicial sets of the form

$$\begin{aligned} &\int_{[k]\in\Delta} \mathrm{PShv}\big(\mathbb{R}^n \times \big(G^{\times_k} \times \Delta(k,\bullet)\big)\,,\, \mathrm{Cdfflg}(N_k(\mathcal{X}_{\mathrm{top}}))\big) \\ &\simeq\; \mathrm{PShv}\left(\mathbb{R}^n\,,\, \mathrm{Cdfflg}\left(\int_{[k]\in\Delta} N_k(\mathcal{X}_{\mathrm{top}})^{(G^{\times_k}\times\Delta(k,\bullet))}\right)\right)\,, \end{aligned} \tag{8.30}$$

where the *end* $\int_{[k]\in\Delta}(-)$ expresses the limit that computes the morphism of simplicial sets as a subset of the product of the function spaces of components. We obtain this as the following composite of natural isomorphisms:

$$\begin{aligned} &\int_{[k]\in\Delta} \mathrm{PShv}\Big(\mathbb{R}^n \times \big(G^{\times_k} \times \Delta(k,\bullet)\big)\,,\, \mathrm{Cdfflg}\big(N_k(\mathcal{X}_{\mathrm{top}})\big)\Big) \\ &\simeq \int_{[k]\in\Delta} \mathrm{PShv}\left(\mathbb{R}^n\,,\, \big(\mathrm{Cdfflg}(N_k(\mathcal{X}_{\mathrm{top}}))\big)^{(G^{\times_k}\times\Delta(k,\bullet))}\right) \\ &\simeq \int_{[k]\in\Delta} \mathrm{PShv}\left(\mathbb{R}^n\,,\, \mathrm{Cdfflg}\left(\big(N_k(\mathcal{X}_{\mathrm{top}})\big)^{(G^{\times_k}\times\Delta(k,\bullet))}\right)\right) \\ &\simeq \mathrm{PShv}\left(\mathbb{R}^n\,,\, \int_{[k]\in\Delta} \mathrm{Cdfflg}\left(\big(N_k(\mathcal{X}_{\mathrm{top}})\big)^{(G^{\times_k}\times\Delta(k,\bullet))}\right)\right) \\ &\simeq \mathrm{PShv}\left(\mathbb{R}^n\,,\, \mathrm{Cdfflg}\left(\int_{[k]\in\Delta} \big(N_k(\mathcal{X}_{\mathrm{top}})\big)^{(G^{\times_k}\times\Delta(k,\bullet))}\right)\right). \end{aligned}$$

Here the first step is the definition of function spaces $(-)^{(-)}$, the second step uses that Cdfflg, being a right adjoint, preserves products (Prop. 8.1.42). The third step uses that the Hom-functor preserves limits (hence ends) in its second argument, while the fourth step uses that Cdfflg, being a right adjoint, preserves limits (hence ends), again by Prop. 8.1.42. □

8.1.7 Smooth ∞-Groupoids

We write

$$\mathrm{CartSp} = \left\{ \begin{array}{ccc} & \mathbb{R}^{n_1} & \\ {}^{f}\nearrow & & \searrow^{g} \\ \mathbb{R}^{n_0} & \xrightarrow{\;g\circ f\;} & \mathbb{R}^{n_2} \end{array} \right\} \tag{8.31}$$

for the category whose objects are the Cartesian spaces $\mathbb{R}^n$, $n \in \mathbb{N}$ (4.1), and whose morphisms are the *smooth* functions between these. We regard this as a site with respect to the coverage (Grothendieck pre-topology) of differentially good open covers (Def. 4.1.30). The sheaves on this site we call *smooth sets* ([Sc13, §1.2.2][KS17, Def. 2.1][SS26c, Ntn. 4.3.15]); these faithfully subsume (cf. [SS26c, Prop. 4.3.19]) smooth manifolds, D-topological spaces (Def. 4.1.5), and diffeological spaces:

$$\begin{array}{ccccc} \mathrm{DTopSp} & \searrow & & & \\ & & \mathrm{DiflSp} & \hookrightarrow & \mathrm{SmthSet} := \mathrm{Sh}(\mathrm{CartSp})\,. \\ \mathrm{SmthMfd} & \nearrow & & & \end{array} \tag{8.32}$$

The higher groupoidal/homotopy theoretic version of this sheaf topos is the (hypercomplete) ∞-sheaf topos over CartSp, whose objects we call *smooth ∞-groupoids* ([Sc13, §4.4][SS26c, Ntn. 4.3.27], or *smooth ∞-stacks*) and which faithfully subsume the Morita (stack) theory of D-topological groupoids (Def. 4.1.7), Lie groupoids, and diffeological groupoids (Rem. 4.1.9):

$$\begin{array}{l} \mathrm{DTopGrpd} \hookrightarrow \\ \qquad\qquad\qquad \mathrm{DiflGrpd} \hookrightarrow \mathrm{SmthGrpd}_\infty := \mathrm{Sh}_\infty(\mathrm{CartSp})\,. \\ \mathrm{LieGrpd} \hookrightarrow \end{array} \tag{8.33}$$

More concretely, $\mathrm{SmthGrpd}_\infty$ is equivalently the ∞-category presented by the projective model structure on simplicial presheaves [nL-MSP] over CartSp (8.31), left Bousfield-localized at the class W of stalkwise simplicial weak equivalences (cf. [FSS23, Ex. 1.20]):

$$\mathrm{SmthGrpd}_\infty \simeq L^W \mathrm{PSh}(\mathrm{CartSp} \times \Delta)\,. \tag{8.34}$$

A D-topological groupoid is represented here by (cf. [SS26c, Ntn. 2.2.24]) the simplicial presheaf which to $\mathbb{R}^n$ assigns the *simplicial nerve* of its underlying discrete groupoid $\flat\mathrm{Map}(\mathbb{R}^n, \mathcal{X})$ of $\mathbb{R}^n$-plots (Exs. 4.1.15 and 4.1.17), hence to $\mathbb{R}^n \times \Delta^k$ the set of $\mathbb{R}^n$-plots of its space of k-tuples of sequentially composable morphisms:

$$\begin{array}{ccc} \mathrm{DTopGrpd} & \xhookrightarrow{\;\;N\;\;} & \mathrm{PSh}(\mathrm{CartSp} \times \Delta) \\ \mathcal{X} & \longmapsto & \Big(\mathbb{R}^n \times \Delta^k \mapsto \flat\mathrm{Map}\big(\mathbb{R}^n, \mathrm{Mor}(\mathcal{X})_s\!\times_t^k\big)\Big), \end{array} \tag{8.35}$$

and this constitutes a fully faithful embedding of 1-categories with naive (not Morita) morphisms on both sides. Under this embedding (8.35):

- The internal hom is given by the topological functor groupoid Def. 4.1.22.
- The weak equivalences in W (8.34) are just the equivalences of topological groupoids according to Def. 4.1.23.
- The global fibrations are just the global fibrations of topological groupoids according to Def. 4.1.41.

Moreover, the key point for our purpose then is that:

Fact 8.1.17. *With topological groupoids regarded among simplicial presheaves via* (8.35)*:*

- *Their (correctly "derived") mapping stack is computed* (by general model category theory, cf. [FSS23, Ex. 1.10]) *as the functor groupoid*
  - *out of a global projective cofibrant resolution,*
  - *into a local projective fibrant resolution.*
- *The nerves of*
  - *Dugger-cofibrant topological groupoids according to Def. 4.1.38 are global projective cofibrant* (cf. [Du01, Cor. 9.4][FSS23, Prop. 1.23]),
  - *delooping groupoids* $\mathbf{B}\Gamma$ *and with them the action groupoids* $G \backslash\!\!\backslash \Gamma$ *(Def. 4.1.14) are local projective fibrant* (by [Pav22, Prop. 4.13][SS26c, Lem. 4.3.30])

over CartSp.

Together, these imply that the (sliced) mapping stacks according to Defs. 4.1.39 and 4.1.46 represent, under (8.35), the derived internal hom — which proves Lems. 4.1.40 and 4.1.54 and thereby ultimately the final Thm. 4.2.6 about Morita invariance of twisted orbifold cohomology.

8.1.8 Equivariant homotopy

For reference, we recall some basics of unstable equivariant homotopy theory (see [May96][Blu17]). We focus here on finite groups, for simplicity and since this is what we need in the main text (Remark 9.2.20), but all statements in the following, notably Elmendorf's theorem (Prop. 8.1.27), generalize to compact Lie groups.

Definition 8.1.18 (Topological G-spaces)**.** Let $G \in \mathrm{Grp}^{\mathrm{fin}}$ be a finite group.

(i) We write

$$G\mathrm{DTopSpc} \hookrightarrow G\mathrm{TopSpc} \;\in\; \mathrm{Cat} \tag{8.36}$$

for the categories whose objects

$$G \circlearrowright X \;:=\; \big(X,\, G \times X \xrightarrow{\rho} X\big) \tag{8.37}$$

are topological spaces X (as in Def. 8.1.2) or specifically D-topological spaces (as in Def. 4.1.2), respectively, equipped with continuous left G-actions ρ, and whose morphisms are the G-equivariant continuous functions:

$$\begin{aligned} & G\mathrm{TopSpc}\big(G \circlearrowright X_1,\; G \circlearrowright X_2\big) \\ & := \left\{ X_1 \xrightarrow[\text{continuous}]{f} X_2 \;\middle|\; \begin{array}{ccc} G \times X_1 & \xrightarrow{\rho_1} & X_1 \\ {\scriptstyle f}\downarrow & & \downarrow{\scriptstyle f} \\ G \times X_2 & \xrightarrow[\rho_2]{} & X_2 \end{array} \right\}. \end{aligned} \tag{8.38}$$

(ii) For $G \circlearrowright X_1$ a (D-)topological G-space and $H \overset{\iota}{\hookrightarrow} G$ a subgroup, we write

$$X^H \;:=\; \Big\{ x \in X \;\big|\; \underset{h \in H \subset G}{\forall}\; \rho(h,x) = x \Big\} \tag{8.39}$$

for the topological subspace of H-fixed points (which, if X is D-topological, is itself again D-topological, by Prop. 8.1.4).

(iii) For $G \circlearrowright X_1$ and $G \circlearrowright X_2$ two (D-)topological G-spaces, the mapping space (8.6) between their underlying (D-)topological spaces canonically becomes a G-space via the conjugation action and the corresponding fixed point space (8.39)

$$\mathrm{Map}(X_1, X_2)^G \hookrightarrow \mathrm{Map}(X_1, X_2) \tag{8.40}$$

is the subspace on the G-equivariant functions (8.38).

Example 8.1.19 (G-cells)**.** For $G \in \mathrm{Grp}_{\mathrm{fin}}$, $H \subset G$ a subgroup and $n \in \mathbb{N}$, we have the G-spaces (Def. 8.1.18)

$$(G/H) \times D^n \;,\; (G/H) \times S^{n-1} \;\in\; G\mathrm{DTopSpc}$$

being the product spaces of the discrete orbit spaces with the standard topological unit disk and unit circle, respectively, the latter equipped with the trivial G-action. The boundary inclusions $\partial D^n = S^{n-1} \overset{\iota_n}{\hookrightarrow} D^n$ induce G-equivariant maps

$$\iota_{n,H} : \quad (G/H) \times S^{n-1} \xhookrightarrow{(\mathrm{id}, \iota_n)} (G/H) \times D^n \tag{8.41}$$

for all $n \in \mathbb{N}$, $H \subset G$.

Definition 8.1.20 (G-CW-complexes).

(i) A *G-CW-complex* X is a D-topological G-space (Def. 8.1.18) which is equipped with the realization as a colimit

$$X \simeq \underset{\longrightarrow_n}{\lim} X_n \quad \in G\mathrm{DTopSpc}$$

over a sequence

$$X_{-1} \longrightarrow X_0 \longrightarrow X_1 \longrightarrow X_2 \longrightarrow \cdots \quad \in G\mathrm{DTopSpc}\,,$$

where $X_{-1} = \varnothing$ and where each $X_n \to X_{n-1}$ is given by a set of attachments of G-cells along (8.41), hence by a pushout of the form:

$$\begin{array}{ccc} \coprod_{\substack{H \subset G \\ i \in I_n}} G/H \times S^{n-1} & \longrightarrow & X_{n-1} \\ {\scriptstyle (\iota_{n,H})_{n,H}} \big\downarrow & \text{(po)} & \big\downarrow \\ \coprod_{\substack{H \subset G \\ i \in I_n}} G/H \times D^n & \longrightarrow & X_n \end{array}$$

(ii) Write

$$G\mathrm{Sets} \hookrightarrow G\mathrm{CWCmplx} \hookrightarrow G\mathrm{DTopSpc} \tag{8.42}$$

for the full subcategories on those D-topological G-spaces which admit the structure of G-CW-complexes.

Definition 8.1.21 (Homotopy theory of D-topological G-spaces). The *homotopy theory of topological G-spaces* is the ∞-category

$$G\mathrm{Grpd}_\infty \in \mathrm{Cat}_\infty \tag{8.43}$$

which has the same objects as GCWCmplx (Def. 8.1.20), and with ∞-groupoids the topological shapes (Def. 8.1.29) of the mapping spaces (8.40) of G-equivariant maps:

$$G\mathrm{Grpd}_\infty(G \circlearrowright X_1, G \circlearrowright X_2) := \mathrm{Shp}_{\mathrm{Top}}\left(\mathrm{Map}(X_1, X_2)^G\right). \tag{8.44}$$

Definition 8.1.22 (Shape of G-topological spaces). **(i)** We write

$$\mathrm{Shp}_{G\mathrm{Top}} : G\mathrm{CWCmplx} \longrightarrow G\mathrm{Grpd}_\infty \tag{8.45}$$

for the canonical ∞-functor (topologically enriched functor) from the 1-category of G-CW-complexes (Def. 8.1.20) to the ∞-category of G-∞-groupoids (Def. 8.1.21), which is the identity on objects and which on Hom-spaces is the continuous map given by the identity function from the discrete set of G-equivariant maps (8.38) to the topological space of G-equivariant maps (8.44).

(ii) For any choice of G-CW-approximation functor

$$G\mathrm{TopSpc} \xrightarrow{(-)_{\mathrm{cof}}} G\mathrm{CWComplex}$$

we get the corresponding shape functor on all of GTopSpc (Def. 8.1.18) and hence on GDTopSpc, which we denote by the same symbol:

$$\mathrm{Shp}_{G\mathrm{Top}} \;:\; G\mathrm{TopSpc} \xrightarrow{(-)_{\mathrm{cof}}} G\mathrm{CWCmplx} \xrightarrow{\mathrm{Shp}_{G\mathrm{top}}} G\mathrm{Grpd}_{\infty}\,. \tag{8.46}$$

Definition 8.1.23 (Proper G-equivariant generalized cohomology of topological G-spaces)**.** For $G \in \mathrm{Grp}^{\mathrm{fin}}$, we say that the *proper G-equivariant cohomology* of a topological G-space (Def. 8.1.18) $X \in G\mathrm{TopSpc}$ with coefficients in a (pointed) G-∞-groupoid (Def. 8.1.21), $A \in G\mathrm{Grpd}_{\infty}$, is

$$H_G^{-n}(X, A) \;:=\; \pi_n\Big(G\mathrm{Grpd}\big(\mathrm{Shp}_{G\mathrm{Top}}(X)\,,\, A\big)\Big),$$

where on the right we have the nth homotopy group (at the given basepoint) of the hom-∞-groupoid (8.44) from the G-topological shape of X (8.46) to A.

8.1.9 Elmendorf's theorem

Definition 8.1.24 (Orbit of action of a finite group)**.** Let G be a finite group. If $G \circlearrowright S$ is a set equipped with an action by G, then an *orbit* of G in S is a subset of points $\{g(s) | g \in G\} \subset S$ obtained from any single point $s \in S$ by acting on it with all elements of G.

Definition 8.1.25 (Orbit category of a finite group)**.** The *category of G-orbits* or *orbit category of G*

$$G\mathrm{Orb} \hookrightarrow G\mathrm{Sets} \in \mathrm{Cat}$$

is the category whose objects correspond to subgroup inclusions $H \overset{\iota}{\hookrightarrow} G$ and whose morphisms are G-equivariant functions, hence morphisms of G-sets (8.42), between the corresponding coset spaces $G/H_1 \longrightarrow G/H_2$.

Example 8.1.26 (Systems of fixed point spaces)**.** Consider a topological space equipped with a G-action $G \circlearrowright X \in G\mathrm{DTopSpc}$ (Def. 8.1.18) and $H \subset G$ a subgroup. Then a G-equivariant function $G/H \xrightarrow{f} X$ from the corresponding G-orbit (Def. 8.1.25) is determined by its image $f\big([e]\big) \in X$ of the class of the neutral element, and that image has to be fixed by the action of $H \subset G$ of X. Therefore, the corresponding G-equivariant mapping spaces (8.40)

$$\mathrm{Map}\big(G/H, X\big)^G \simeq X^H := \Big\{x \in X \,|\, \underset{h \in H \subset G}{\forall} \big(h(x) = x\big)\Big\} \subset X$$

are the topological subspaces of H-fixed points inside X (8.39). By functoriality of the mapping space construction, these fixed point spaces are exhibited as arranging into a topological presheaf on the G-orbit category (Def. 8.1.25):

$$X^{(-)} \;:\; G\mathrm{Orb}^{\mathrm{op}} \xrightarrow{\mathrm{Map}(-,X)^G} \mathrm{TopSpc}$$

Proposition 8.1.27 (Elmendorf's theorem [El83][DwKa84, §1.2, 1.7 & Thm. 3.1], see [Blu17, Thm. 1.3.6 and 1.3.8]). *Let G be a finite group. The functor which sends a G-space $G \curvearrowright X$ (Def. 8.1.18) to its system of H-fixed point spaces (Ex. 8.1.26) constitutes an equivalence of ∞-categories*

$$\begin{array}{ccc} G\mathrm{Grpd}_\infty & \xrightarrow{\;\simeq\;} & \mathrm{Shv}_\infty(G\mathrm{Orb}) \\ G \curvearrowright X & \longmapsto & X^{(-)} = \mathrm{Map}(-, X)^G . \end{array} \tag{8.47}$$

8.1.10 Homotopy theory

Definition 8.1.28 (∞-Groupoids).

(i) We write

$$\mathrm{Grpd}_\infty \in \mathrm{Cat}_\infty \tag{8.48}$$

for the ∞-category which is presented by the topologically enriched category whose objects are the CW-complexes (8.5) and whose hom-spaces are the mapping spaces (8.6).

(ii) The full sub-∞-category (Def. 8.1.1) on the homotopy n-types is that of *n-groupoids*

$$\mathrm{Grpd}_n \hookrightarrow \mathrm{Grpd}_\infty . \tag{8.49}$$

Definition 8.1.29 (Topological shape).

(i) We write

$$\mathrm{Shp}_{\mathrm{Top}} \;:\; \mathrm{CWCmplx} \longrightarrow \mathrm{Grpd}_\infty \tag{8.50}$$

for the ∞-functor from the 1-category of CW-complexes (8.5) to the ∞-category of ∞-groupoids (Def. 8.1.28) which, as a topologically enriched functor, is the identity on objects and is on hom-spaces the continuous map given by the identity function from the discrete set of continuous maps to the mapping space (8.6).

(ii) For any choice of CW-approximation functor

$$\mathrm{TopSpc} \xrightarrow{\;(-)_{\mathrm{cof}}\;} \mathrm{CWCmplx} \tag{8.51}$$

we get the corresponding functor on all topological spaces (Def. 8.1.2), hence on D-topological spaces (Def. 4.1.2) which we denote by the same symbol:

$$\mathrm{Shp}_{\mathrm{Top}} \;:\; \mathrm{TopSpc} \xrightarrow{\;(-)_{\mathrm{cof}}\;} \mathrm{CWCmplx} \xrightarrow{\;\mathrm{Shp}_{\mathrm{Top}}\;} \mathrm{Grpd}_\infty . \tag{8.52}$$

Example 8.1.30 (Delooping groupoids). For $G \in \mathrm{Grp}^{\mathrm{fin}}$, consider the groupoid with a single object $*$, and with G as its set of morphisms, whose composition is given by the product in the group:

$$\begin{array}{ccccc} & & * & & \\ & {}^{g_1}\nearrow & & \searrow^{g_2} & \\ * & & \xrightarrow[\;g_2 \cdot g_1\;]{} & & * \end{array} \tag{8.53}$$

This groupoid is the topological shape (8.1.29) of the Eilenberg-MacLane space $K(G,1)$ as well as (since G is assumed to be finite) the classifying space BG. More intrinsically, this groupoid is, equivalently, the homotopy quotient of the point by the trivial G-action:

$$*/\!\!/ G \;\in\; \mathrm{Grpd}_1 \hookrightarrow \mathrm{Grpd}_\infty\,. \tag{8.54}$$

More generally:

Example 8.1.31 (Action groupoids). For $G \in \mathrm{Grp}^{\mathrm{fin}}$, a finite group, and for $G \circlearrowright X$ a set equipped with a G-action

$$\begin{array}{ccc} G \times X & \xrightarrow{\;\;\rho\;\;} & X \\ (g,x) & \longmapsto & g \cdot x\,, \end{array} \tag{8.55}$$

the corresponding *action groupoid* has as objects the elements of X, and its morphisms and their composition are given as follows:

$$x \xrightarrow{g_1} g_1 \cdot x \xrightarrow{g_2} g_2 \cdot g_1 \cdot x\,, \qquad x \xrightarrow{g_2 \cdot g_1} g_2 \cdot g_1 \cdot x \tag{8.56}$$

This action groupoid is a model for the homotopy quotient of X by its G-action

$$X /\!\!/ G \;\in\; \mathrm{Grpd}_1 \hookrightarrow \mathrm{Grpd}_\infty\,. \tag{8.57}$$

The following elementary example plays a pivotal role in later constructions (Lem 10.1.7):

Example 8.1.32 (Hom-groupoid into action groupoid). Let $G \in \mathrm{Grp}^{\mathrm{fin}}$, $X \in$ equipped with a G-action (8.55), hence with action groupoid/homotopy quotient $X/\!\!/G \;\in\; \mathrm{Grpd}_1$ (Example 8.1.31). Let $K \in \mathrm{Grp}^{\mathrm{fin}}$ be any finite group, with $*/\!\!/K \;\in\; \mathrm{Grpd}_1$ its delooping groupoid (Example 8.1.30). Then the hom-groupoid (functor groupoid) of morphisms (functors) $*/\!\!/K \longrightarrow X/\!\!/G$ is, equivalently, the action groupoid of G acting on the set of pairs consisting of a group homomorphism $\phi : K \to G$ and a point in X fixed by the image of ϕ:

$$\mathrm{Grpd}_1\big(*/\!\!/K\,,\; X/\!\!/G\big) \;\simeq\; \Bigg(\coprod_{\phi \in \mathrm{Grp}(K,G)} X^{\phi(K)}\Bigg) /\!\!/ G\,. \tag{8.58}$$

Here

- $\phi(K) \subset G$ denotes the subgroup of G which is image of the group homomorphism $\phi : K \to G$;
- $X^{\phi(K)} = \Big\{ x \in X \;\Big|\; \underset{h \in \phi(K)}{\forall}\; h \cdot x = x \Big\}$ denotes the $\phi(K)$-fixed-point set in X;
- the G-action by which the homotopy quotient is taken is the conjugation action on ϕ, hence $g \cdot \phi \;:=\; \mathrm{Ad}_g \circ \phi$, and the given G-action on $x \in X$.

This follows by direct unwinding of the definition of functors and of natural transformations between the groupoids (8.53) and (8.56).

Definition 8.1.33 (Simplicial-topological shape). Let

$$X_\bullet \;:\; \Delta^{\mathrm{op}} \longrightarrow \mathrm{TopSpc} \tag{8.59}$$

be a simplicial topological space, for instance, the nerve of a topological groupoid. Then we say that its *simplicial-topological shape* is the homotopy colimit (Prop. 8.1.52) of its degreewise topological shape (Def. 8.1.29):

$$\mathrm{Shp}_{\mathrm{sTop}}(X_\bullet) \;:=\; \underset{\longrightarrow}{\lim}\big(\mathrm{Shp}_{\mathrm{Top}}(X)\big)_\bullet \;\in\; \mathrm{Grpd}_\infty\,. \tag{8.60}$$

The following Prop. 8.1.34 appears as [Wa18, 4.3,. 4.4]:

Proposition 8.1.34 (Simplicial-topological shape of degreewise cofibrant spaces is fat geometric realization)**.** *If $X_\bullet$ is a simplicial topological space* (8.59) *which degreewise admits the structure of a retract of a cell complex (for instance: degreewise a CW-complex* (8.5)*), then its simplicial topological shape* (8.1.33) *is equivalent to its fat geometric realization* $\|-\|$ [Se74] (see [HG07, 2.3])*:*

$$\underset{\substack{\text{Degreewise cofibrant} \\ \text{simplicial topological spaces}}}{X_\bullet \in (\mathrm{TopSpc}_{\mathrm{cof}})^{\Delta^{\mathrm{op}}}} \quad \Rightarrow \quad \underset{\substack{\text{simplicial} \\ \text{topological shape}}}{\mathrm{Shp}_{\mathrm{sTop}}(X_\bullet)} \simeq \underset{\substack{\text{fat geometric} \\ \text{realization}}}{\|X_\bullet\|}\,. \tag{8.61}$$

Definition 8.1.35 (Diffeological simplices)**.**

(i) We write

$$\begin{array}{ccc} \Delta & \xrightarrow{\Delta^\bullet_{\mathrm{smth}}} & \mathrm{DiffSp} \\ [n] & \longmapsto & \left\{ \vec{x} \in \mathbb{R}^{n+1} \mid \sum_i x^i = 1 \right\} \end{array} \tag{8.62}$$

for the *diffeological extended simplices*, hence for the simplicial object in diffeological spaces (Def. 8.1.6) (in fact, in smooth manifolds, under Prop. 8.1.11) which in degree n is the extended n-simplex in $\mathbb{R}^{n+1}$, regarded with its sub-diffeology, and whose face and degeneracy maps are the standard ones (see [CW14, Def. 4.3][BEBP19, p. 1]).

(ii) The induced nerve/realization construction is a pair of adjoint functors (Def. 8.1.40)

$$\mathrm{DiffSp} \underset{\mathrm{Sing}_{\mathrm{diff}}}{\overset{|-|_{\mathrm{diff}}}{\underset{\bot}{\rightleftarrows}}} \mathrm{SimplicialSets} \tag{8.63}$$

between the categories of simplicial sets and of diffeological spaces (Def. 8.1.6), where the right adjoint $\mathrm{Sing}_{\mathrm{diff}}$ sends $X \in \mathrm{DiffSp}$ to its *smooth singular simplicial set*

$$\mathrm{Sing}_{\mathrm{diff}}(X)_\bullet := \mathrm{DiffSp}(\Delta^\bullet_{\mathrm{diff}}, X). \tag{8.64}$$

The following Prop. 8.1.36 is due to [CW14, Prop. 4.14]:

Proposition 8.1.36 (Diffeological singular simplicial set of continuous diffeology)**.**

(i) *For all $X_{\mathrm{top}} \in \mathrm{TopSpc}$ there is a weak homotopy equivalence between the diffeological singular simplicial set (Def. 8.1.35) of its continuous diffeology (Def. 8.1.7) and its ordinary singular simplicial set:*

$$\mathrm{Sing}(X_{\mathrm{top}}) \simeq_{\mathrm{whe}} \mathrm{Sing}_{\mathrm{diff}}(\mathrm{Cdfflg}(X_{\mathrm{top}}))\,. \tag{8.65}$$

(ii) *Equivalently this means, in the terminology to be introduced in a moment, that the topological shape* (8.1.29) *of topological spaces is equivalent to the cohesive shape (Def. 9.1.1) of their incarnation as continuous-diffeological spaces (see Example 9.1.19):*

$$\mathrm{Shp}_{\mathrm{Top}}(X_{\mathrm{top}}) \simeq \mathrm{Shp}(\mathrm{Cdfflg}(X_{\mathrm{top}})) \quad \in \mathrm{Grpd}_\infty\,. \tag{8.66}$$

8.1.11 Universal constructions

All diagrams we consider now are homotopy-coherent, even if we do not notationally indicate the higher cells, unless some are to be highlighted. Similarly, all universal constructions we consider now are ∞-categorical, even if this is not further pronounced by the terminology. In particular, we say "colimit" $\varinjlim$ for "homotopy colimit", "limit" $\varprojlim$ for "homotopy limit" (see Prop. 8.1.52), "Cartesian square" for "homotopy Cartesian square", etc.:

Notation 8.1.37 (Cartesian squares). We say a square in an ∞-category is *Cartesian*, to be denoted

$$\begin{array}{ccc} X \times_B Y & \longrightarrow & Y \\ {\scriptstyle f^* g}\downarrow & \text{(pb)} & \downarrow{\scriptstyle g} \\ X & \xrightarrow{\;\;f\;\;} & B \end{array} \tag{8.67}$$

if it is a limit cone over the diagram consisting of f and g. We also say this is the *pullback square* of g along f.

Example 8.1.38 (Pullback of equivalence is equivalence). Let $C \in \mathrm{Cat}_\infty$. Then a square in C whose right vertical morphism is an equivalence is Cartesian (Ntn. 8.1.37) precisely if the left vertical morphism is also an equivalence:

$$\begin{array}{ccc} A & \longrightarrow & B \\ \downarrow & \text{(pb)} & \downarrow{\scriptstyle \simeq} \\ C & \longrightarrow & D \end{array} \quad\Leftrightarrow\quad \begin{array}{c} A \\ \downarrow{\scriptstyle \sim} \\ C \end{array} \tag{8.68}$$

hence precisely if $C \to D$ is equivalent to $A \to B$ in C^{Δ^1}.

Proposition 8.1.39 (Pasting law [Lu09a, Lemma 4.4.2.1]). *In any ∞-category, consider a diagram of the form*

$$\begin{array}{ccccc} A & \longrightarrow & B & \longrightarrow & C \\ \downarrow & \swarrow & \downarrow & \swarrow & \downarrow \\ D & \longrightarrow & E & \longrightarrow & F \end{array} \tag{8.69}$$

such that the right square is Cartesian (Ntn. 8.1.37). Then the left square is Cartesian if and only if the total rectangle is Cartesian.

Definition 8.1.40 (Adjoint ∞-functors [Lu09a, 5.2.2.7, 5.2.2.8][RV13, 4.4.4]). Let $C, \mathcal{D} \in \mathrm{Cat}_\infty$ (8.1) and $L : C \longleftrightarrow \mathcal{D} : R$ two functors between them, back and forth. This is an *adjoint pair* with L *left adjoint* and R *right adjoint*, to be denoted $(L \dashv R)$:

$$\mathcal{D} \underset{R}{\overset{L}{\underset{\bot}{\rightleftarrows}}} C \tag{8.70}$$

if there is a natural equivalence of hom-∞-groupoids (8.2) of the form

$$\mathcal{D}\big(L(-),\, -\big) \;\simeq\; C\big(-,\, R(-)\big) \tag{8.71}$$

(This is unique when it exists [Lu09a, Prop. 5.2.1.3, 5.2.6.2]). In this case, one says:

(i) The *adjunction unit* is the natural transformation

$$X \xrightarrow{\eta_X} R \circ L(X) \tag{8.72}$$

which is the (pre-)image under (8.71) of the identity on $R(X)$.

(ii) The *adjunction co-unit* is the natural transformation

$$L \circ R(X) \xrightarrow{\epsilon_X} X \tag{8.73}$$

which is the image under (8.71) of the identity on $L(X)$.

As in the classical situation of 1-category theory, it follows that:

Proposition 8.1.41 (Triangle identities). *Let* $\mathcal{D} \underset{R}{\overset{L}{\rightleftarrows}} C$ *be a pair of adjoint ∞-functors (Def. 8.1.40). Then their adjunction unit* η *(8.72) and counit* ϵ *(8.73) satisfy the following natural equivalences:*

(i) *for all* $c \in C$,

$$\begin{array}{ccc} & L \circ R \circ L(c) & \\ {}^{L(\eta_c)}\nearrow & & \searrow^{\epsilon_{L(c)}} \\ L(c) & = = = = = & L(c)\,; \end{array} \tag{8.74}$$

(ii) *for all* $d \in \mathcal{D}$,

$$\begin{array}{ccc} & R \circ L \circ R(d) & \\ {}^{\eta_{R(d)}}\nearrow & & \searrow^{R(\epsilon_d)} \\ R(d) & = = = = = & R(d)\,. \end{array} \tag{8.75}$$

Proposition 8.1.42 (Right/left adjoints preserve limits/colimits [Lu09a, 5.2.3.5]). *Let* $\mathcal{D} \underset{R}{\overset{L}{\rightleftarrows}} C$ *be a pair of adjoint ∞-functors (Def. 8.1.40) and let* $I \in \mathrm{Cat}_\infty$.

(i) *If* $X_\bullet : I \to \mathcal{D}$ *is a diagram whose limit exists, then this limit is preserved by the right adjoint R:*

$$R(\underleftarrow{\lim}\, X_\bullet) \simeq \underleftarrow{\lim}\, RX_\bullet\,. \tag{8.76}$$

(ii) *If* $X_\bullet : I \to C$ *is a diagram whose colimit exists, then this colimit is preserved by the left adjoint L:*

$$L(\underrightarrow{\lim}\, X_\bullet) \simeq \underrightarrow{\lim}\, LX_\bullet\,. \tag{8.77}$$

Conversely:

Proposition 8.1.43 (Adjoint ∞-functor theorem [Lu09a, 5.5.2.9]). *Let* $C_{1,2} \in \mathrm{Cat}_\infty$ *be presentable (e.g., ∞-toposes, Def. 8.1.46), then an ∞-functor* $C_1 \to C_2$ *is a:*

(i) *right adjoint (i.e., has a left adjoint, Def. 8.1.40) precisely if it preserves limits (8.76);*

(ii) *left adjoint (i.e., has a right adjoint, Def. 8.1.40) precisely if it preserves colimits (8.77).*

Proposition 8.1.44 (Fully faithful adjoints [Lu09a, 5.2.7.4]). *For adjoint ∞-functors (Def. 8.1.40)* $\mathcal{D} \underset{R}{\overset{L}{\rightleftarrows}} C$,

(i) *L is fully faithful* $\mathcal{D} \overset{L}{\hookleftarrow} C$ *(Def. 8.1.1) iff the adjunction unit* η *(8.72) is an equivalence:* $\mathrm{id} \underset{\simeq}{\overset{\eta}{\to}} R \circ L$;

(ii) *R is fully faithful* $\mathcal{D} \overset{R}{\hookrightarrow} \mathcal{C}$ *(Def. 8.1.1) iff the adjunction counit* ϵ (8.73) *is an equivalence* $L \circ R \xrightarrow[\simeq]{\epsilon} \mathrm{id}$.

Proposition 8.1.45 (Idempotent Monads and Comonads)**.** *For* $\mathcal{D} \underset{R}{\overset{L}{\rightleftarrows}} \mathcal{C}$ *a pair of adjoint ∞-functors (Def. 8.1.40):*

(i) *If R is fully faithful (Def. 8.1.1) then* $\bigcirc := R \circ L$ *is idempotent, exhibited by the* $\bigcirc$*-image of the adjunction unit* η (8.72)*:*

$$\bigcirc(c) \xrightarrow[\simeq]{\bigcirc(\eta_{L(c)})} \bigcirc \circ \bigcirc(c)\,. \tag{8.78}$$

(ii) *If L is fully faithful (Def. 8.1.1) then* $\square := L \circ R$ *is idempotent, exhibited by the* $\square$*-image of the adjunction counit* ϵ (8.73)*:*

$$\square \circ \square(d) \xrightarrow[\simeq]{\square(\epsilon_{R(d)})} \square(d)\,. \tag{8.79}$$

Proof. Consider case **(i)**, the other case is formally dual. Since R is fully faithful, by assumption, the condition that $\bigcirc(\eta_{L(c)}) := R \circ L(\eta_{L(c)})$ is an equivalence is equivalent to $L(\eta_{L(c)})$ being an equivalence. But, by the triangle identity (Prop. 8.1.41), we have that the composite $\epsilon_{L(L(c))} \circ L(\eta_{L(c)})$ is an equivalence, while by Prop. 8.1.44 the counit ϵ is a natural equivalence. By cancellation, this implies that $L(\eta_{L(c)})$ is an equivalence. □

8.1.12 ∞-Toposes

For our purposes, we take the following characterization to be the definition of ∞-toposes. This is due to Rezk and Lurie [Lu09a, 6.1.6.8]; we follow the presentation in [NSS12a, Prop. 2.2]:

Definition 8.1.46 (∞-Topos)**.** An ∞-topos $\mathbf{H}$ is a presentable ∞-category with the following properties:

(i) **Universal colimits.** For all morphisms $f : X \longrightarrow B$ and all small diagrams $A : I \longrightarrow \mathbf{H}_{/B}$, there is an equivalence:

$$\varinjlim_i f^* A_i \simeq f^*\big(\varinjlim_i A_i\big) \tag{8.80}$$

between the pullback (8.67) of the colimit and the colimit over the pullbacks of its components.

(ii) **Univalent universes.** For every sufficiently large regular cardinal κ, there exists a morphism $\widehat{\mathrm{Objects}}_\kappa \longrightarrow \mathrm{Objects}_\kappa$ in $\mathbf{H}$, such that for every object $X \in \mathbf{H}$, pullback (8.67) along morphisms $X \longrightarrow \mathrm{Objects}_\kappa$ constitutes an equivalence

$$\begin{array}{ccc} \mathrm{Core}(\mathbf{H}_{/_\kappa X}) & \simeq & \mathbf{H}(X, \mathrm{Objects}_\kappa) \\ E & \longmapsto & \vdash E \end{array} \qquad \begin{array}{ccc} E & \longrightarrow & \widehat{\mathrm{Objects}}_\kappa \\ \downarrow & \text{(pb)} & \downarrow \\ X & \xrightarrow[\vdash E]{} & \mathrm{Objects}_\kappa \end{array} \tag{8.81}$$

between the ∞-groupoid core (8.1) of bundles (Ntn. 8.1.61) which are κ-small over X, and the hom-∞-groupoid (8.2) of morphisms from X to the *object classifier* $\mathrm{Objects}_\kappa$.

Example 8.1.47 (Internal mapping space in an ∞-topos)**.** Let $\mathbf{H}$ be an ∞-topos (Def. 8.1.46) and $X \in \mathbf{H}$ an object. As a special case of universality of colimits (8.80), we have that the functor $X \times (-)$ of Cartesian product with X preserves all colimits. Hence, by the adjoint ∞-functor theorem (Prop. 8.1.43), this functor has a right adjoint, to be denoted $\mathbf{Map}(X, -)$, the *internal hom-* or *internal mapping space-* or *mapping stack-*functor:

$$\mathbf{H} \underset{\underset{\text{internal mapping space}}{\mathbf{Map}(X,-)}}{\overset{X\times(-)}{\underset{\longrightarrow}{\overset{\longleftarrow}{\bot}}}} \mathbf{H}\,. \tag{8.82}$$

By adjointness, the probes of the internal mapping space over any $U \in \mathbf{H}$ are given by

$$\mathbf{H}\big(U, \mathbf{Map}(X,Y)\big) \;\simeq\; \mathbf{H}\big(U \times X\,,\, Y\big). \tag{8.83}$$

Proposition 8.1.48 (Colimits and equifibered transformations [Lu09a, 6.1.3.9(4)][Re10, 6.5])**.** *Let $\mathbf{H}$ be an ∞-topos (Def. 8.1.46), $\mathcal{I}$ a small ∞-category, $X_\bullet, Y_\bullet : \mathcal{I} \to \mathbf{H}$ two $\mathcal{I}$-shaped diagrams.*

(i) *If $X_\bullet \overset{f_\bullet}{\Rightarrow} Y_\bullet$ is a natural transformation which is equifibered [Re10, p. 9] in that its value on all morphisms $i_1 \xrightarrow{\phi} i_2$ in $\mathcal{Y}$ is a Cartesian square (Ntn. 8.1.37), then the value of $\underset{\longrightarrow}{\lim} f_\bullet$ on all colimit component morphisms is also Cartesian:*

$$\underset{i_1 \xrightarrow{\phi} i_2}{\forall}\;\begin{array}{ccc} X_{i_1} & \xrightarrow{f_{i_1}} & Y_{i_1} \\ {\scriptstyle X_\phi}\downarrow & \text{(pb)} & \downarrow{\scriptstyle Y_\phi} \\ X_{i_2} & \xrightarrow[f_{i_2}]{} & Y_{i_2}\end{array} \;\Rightarrow\; \underset{i}{\forall}\;\begin{array}{ccc} X_i & \xrightarrow{f_i} & Y_{i_1} \\ {\scriptstyle q_{X_i}}\downarrow & \text{(pb)} & \downarrow{\scriptstyle q_{Y_i}} \\ \underset{\longrightarrow}{\lim} X_\bullet & \xrightarrow[\underset{\longrightarrow}{\lim} f_\bullet]{} & \underset{\longrightarrow}{\lim} Y_\bullet \end{array}. \tag{8.84}$$

(ii) *Let $X_\bullet^{\triangleright} : \mathcal{I}^{\triangleright} \to \mathbf{H}$ be a cocone under $X_\bullet$, with tip $\mathcal{X} \in \mathbf{H}$, and let $Y_\bullet^{\triangleright} : \mathcal{I}^{\triangleright} \to \mathbf{H}$ denote the colimiting cocone under $Y_\bullet$ with tip $\underset{\longrightarrow}{\lim} Y_\bullet$. If $X_\bullet^{\triangleright} \overset{f_\bullet^{\triangleright}}{\Rightarrow} Y_\bullet^{\triangleright}$ is a natural transformation of cocone diagrams which is equifibered, then $X_\bullet^{\triangleright}$ is a colimiting cocone:*

$$\underset{i_1 \xrightarrow{\phi} i_2}{\forall}\;\begin{array}{ccc} X_{i_1} & \xrightarrow{f_{i_1}} & Y_{i_1} \\ {\scriptstyle X_\phi}\downarrow & \text{(pb)} & \downarrow{\scriptstyle Y_\phi} \\ X_{i_2} & \xrightarrow[f_{i_2}]{} & Y_{i_2}\end{array} \;\text{ and }\; \underset{i}{\forall}\;\begin{array}{ccc} X_i & \xrightarrow{f_i} & Y_{i_1} \\ {\scriptstyle q_{X_i}}\downarrow & \text{(pb)} & \downarrow{\scriptstyle q_{Y_i}} \\ \mathcal{X} & \xrightarrow[\underset{\longrightarrow}{\lim} f_\bullet]{} & \underset{\longrightarrow}{\lim} Y_\bullet \end{array} \tag{8.85}$$

$$\Rightarrow \quad \mathcal{X} \simeq \underset{\longrightarrow}{\lim} X_\bullet\,.$$

Example 8.1.49 (Initial object in ∞-topos is empty object [Re19, p. 16])**.** Let $\mathbf{H}$ be an ∞-topos (Def. 8.1.46). Applying the implication (8.85) in Prop. 8.1.48 to the colimit over the empty diagram, which is the initial object, shows that any object with a morphism to the initial object is itself equivalent to the initial object. Hence if we write

$$\varnothing \in \mathbf{H} \qquad \text{s.t.} \qquad \underset{X \in \mathbf{H}}{\forall} \big(\mathbf{H}(\varnothing, X) \simeq * \big) \tag{8.86}$$

for the initial object, this means that

$$X \xrightarrow{\ \exists\ } \varnothing \quad \Rightarrow \quad X \simeq \varnothing . \tag{8.87}$$

Proposition 8.1.50 (Tensoring of ∞-toposes over ∞-groupoids). *Let* $\mathbf{H}$ *be an ∞-topos (Def. 8.1.46) with an inverse base geometric morphism (Prop. 8.1.59) denoted* $\Delta : \mathrm{Grpd}_\infty \longrightarrow \mathbf{H}$. *Then, for* $S \in \mathrm{Grpd}_\infty$ *and* $X, Y \in \mathbf{H}$, *there is a natural equivalence of ∞-groupoids*

$$\mathbf{H}\big(\Delta(S) \times X\,,\, Y\big) \simeq \mathrm{Grpd}_\infty\big(S\,,\, \mathbf{H}(X,Y)\big). \tag{8.88}$$

Proof. By [Lu09a, Cor. 4.4.4.9] we have, for $S \in \mathrm{Grpd}_\infty \hookrightarrow \mathrm{Cat}_\infty$ and $X, Y \in \mathbf{H}$, natural equivalences

$$\underset{\overrightarrow{S}}{\lim}\, \mathrm{const}_* \simeq S \qquad \text{and} \qquad \mathbf{H}\Big(\underset{\overrightarrow{S}}{\lim}\, \mathrm{const}_X\,,\, Y\Big) \simeq \mathrm{Grpd}_\infty\big(S\,,\, \mathbf{H}(X,Y)\big). \tag{8.89}$$

This implies the statement in the form (8.88) by using **(a)** that Δ preserves all colimits as well as finite limits (Prop. 8.1.59) and **(b)** that Cartesian products may be taken inside colimits, as a special case of (8.80):

$$\begin{aligned} \mathbf{H}\big(\Delta(S) \times X\,,\, Y\big) &\simeq \mathbf{H}\big(\Delta\big(\underset{\overrightarrow{S}}{\lim}\, *\big) \times X\,,\, Y\big) \simeq \mathbf{H}\Big(\big(\underset{\overrightarrow{S}}{\lim}\, \underbrace{\Delta(*)}_{\simeq *}\big) \times X\,,\, Y\Big) \\ &\simeq \mathbf{H}\Big(\big(\underset{\overrightarrow{S}}{\lim}\, \underbrace{(* \times X)}_{\simeq X}\big)\,,\, Y\Big) \simeq \mathrm{Grpd}_\infty\big(S\,,\, \mathbf{H}(X,Y)\big). \end{aligned} \tag{8.90}$$

The composite equivalence is (8.88). □

8.1.13 Sheaves

Notation 8.1.51 (∞-Presheaves). For C a small ∞-category, we write

$$\mathrm{PShv}_\infty(C) := \mathrm{Func}_\infty\big(C^{\mathrm{op}}\,,\, \mathrm{Grpd}_\infty\big) \tag{8.91}$$

for the ∞-category of ∞-presheaves on C. More generally, if $\mathbf{H}$ is any ∞-topos (Def. 8.1.46) we also write

$$\mathrm{PShv}_\infty\big(C, \mathbf{H}\big) := \mathrm{Func}_\infty\big(C^{\mathrm{op}}\,,\, \mathbf{H}\big). \tag{8.92}$$

Proposition 8.1.52 (Limits and colimits in an ∞-topos [Lu09a, Lem. 4.2.4.3]). *Let* $\mathbf{H}$ *be an ∞-topos (Def. 8.1.46) and* C *a small ∞-category. Then the ∞-functor which sends an object in* $\mathbf{H}$ *to the* $\mathbf{H}$*-valued presheaf* (8.92) *constant on this object has a right- and a left-adjoint (Def. 8.1.40), given by the limit and colimit construction, respectively:*

$$\mathrm{Func}_\infty\big(C\,,\, \mathbf{H}\big) \; \underset{\underset{\overleftarrow{\lim}}{\longrightarrow}}{\overset{\overset{\overrightarrow{\lim}}{\longrightarrow}}{\underset{\perp}{\overset{\perp}{\longleftarrow \mathrm{const} \longrightarrow}}}} \; \mathbf{H}\,. \tag{8.93}$$

Proposition 8.1.53 (∞-Yoneda embedding [Lu09a, Lemma 5.5.2.1]). *Let C be an ∞-category. Then the ∞-functor from C to its ∞-presheaves* (8.91) *which assigns representable presheaves*

$$\begin{array}{rcl} C & \overset{y}{\hookrightarrow} & \mathrm{PShv}_\infty(C) \\ c & \longmapsto & C(-,c) \end{array} \tag{8.94}$$

is fully faithful (Def. 8.1.1).

Proposition 8.1.54 (∞-Yoneda lemma [Lu09a, Lemma 5.5.2.1]). *Let C be an ∞-category. Then for $X \in \mathrm{PShv}_\infty(C)$* (8.91) *and $c \in C$, there is a natural equivalence*

$$\mathrm{PShv}_\infty\big(y(c), X\big) \;\simeq\; X(c)\,, \tag{8.95}$$

where y is the Yoneda embedding (8.94) *from Prop. 8.1.53.*

Proposition 8.1.55 ((Co-)Limits of presheaves are computed objectwise [Lu09a, Cor. 5.1.2.3]). *Let $\mathbf{H}$ be an ∞-topos, let C and $\mathcal{D}$ be small ∞-categories, and let*

$$I \;:\; \mathcal{D} \longrightarrow \mathrm{PShv}_\infty(C, \mathbf{H}) \tag{8.96}$$

be a diagram of $\mathbf{H}$-valued ∞-presheaves over C. Then the limit and colimit over I exist and are given objectwise over $c \in C$ by the limit and colimit of the components in Grpd_∞*:*

$$\begin{aligned} \big(\underrightarrow{\lim}\, I\big) &\;:\; c \longmapsto \big(\underrightarrow{\lim}\, I_c\big)\,, \\ \big(\underleftarrow{\lim}\, I\big) &\;:\; c \longmapsto \big(\underleftarrow{\lim}\, I_c\big)\,. \end{aligned}$$

Lemma 8.1.56 (Colimit of representable functor is contractible). *Let C be a small ∞-category, and consider an ∞-functor $yC : C^{\mathrm{op}} \longrightarrow \mathrm{Grpd}_\infty$ to the ∞-category of ∞-groupoids* (8.48)*, which is representable, hence which is in the essential image of the ∞-Yoneda embedding* (8.94)*. Then the colimit of this functor is contractible:*

$$\underset{C}{\underrightarrow{\lim}}(yC) \;\simeq\; *\,. \tag{8.97}$$

Proof. The terminal $* \in \mathrm{Grpd}_\infty$ is characterized by the fact that for $S \in \mathrm{Grpd}_\infty$ there is a natural equivalence

$$S \;\simeq\; \mathrm{Grpd}_\infty(*,\, S)\,. \tag{8.98}$$

Hence, it is sufficient to see that $\underrightarrow{\lim}\,(yC)$ satisfies the same property. But we have the following sequence of natural equivalences:

$$\begin{aligned} \mathrm{Grpd}_\infty\Big(\underrightarrow{\lim}\,(yC)\,,\; S\Big) &\;\simeq\; \mathrm{Func}_\infty\big(C^{op}\big)\big(yC\,,\, \mathrm{const}\big) \\ &\;\simeq\; (\mathrm{const}\, S)(C) \;\simeq\; S\,. \end{aligned} \tag{8.99}$$

Here the first step is the adjunction (8.93), while the second step is the ∞-Yoneda lemma (Prop. 8.1.54). □

Proposition 8.1.57 (Topos is accessibly lex reflective in presheaves [Lu09a, 6.1.0.6]). *Let $\mathbf{H}$ be an ∞-topos (Def. 8.1.46).*

(i) *Then there exists an ∞-site for $\mathbf{H}$, namely a small $C \in \mathrm{Cat}_\infty$ equipped with a pair of adjoint ∞-functors (Def. 8.1.40) between $\mathbf{H}$ and $\mathrm{PShv}_\infty(C)$ (Ntn. 8.1.51):*

$$\mathbf{H} \underset{\longleftarrow}{\overset{L}{\underset{\bot}{\hookleftarrow}}} \mathrm{PShv}_\infty(C) \tag{8.100}$$

such that **(a)** *the right adjoint is accessible and fully faithful (Def. 8.1.1) and* **(b)** *the left adjoint preserves finite limits (in addition to preserving all colimits, by Prop. 8.1.42).*

(ii) *Conversely, any such accessibly embedded lex reflective sub-∞-category of an ∞-category of ∞-presheaves is an ∞-topos.*

Definition 8.1.58 (Sheaf ∞-topos [Lu09a, 6.2]). An ∞-topos **H** (Def. 8.1.46) is called an *∞-category of ∞-sheaves* or *of ∞-stacks*, or just a *sheaf topos* for short, to be denoted

$$\mathbf{H} \simeq \mathrm{Shv}_\infty(C) \tag{8.101}$$

if there exists a *site* C, namely a small $C \in \mathrm{Cat}_\infty$ with a reflection $(L\mathrm{const} \dashv \Gamma)$ (8.100) as in Prop. 8.1.57, such that Lconst exhibits localization at a set

$$\underbrace{\left\{U \hookrightarrow y(c)\right\}}_{\text{covering sieves}} \subset \coprod_{c \in C} \mathrm{SubObjects}\big(y(c)\big) \tag{8.102}$$

of monomorphisms (Def. 8.1.75) into representable presheaves (8.94).

Proposition 8.1.59 (Base geometric morphism [Lu09a, 6.3.4.1]). *Let* **H** *be an ∞-topos (Def. 8.1.46). There is an essentially unique pair of adjoint ∞-functors (Def. 8.1.40) between* **H** *and* Grpd$_\infty$ *(Def. 8.1.28)*

$$\mathbf{H} \underset{\Gamma}{\overset{L\mathrm{const}}{\underset{\longrightarrow}{\overset{\longleftarrow}{\bot}}}} \mathrm{Grpd}_\infty \tag{8.103}$$

such that the left adjoint Lconst *preserves finite limits (in addition to preserving all colimits, by Prop. 8.1.42).*

Example 8.1.60 (Base geometric morphism via site). Let **H** be an ∞-topos (Def. 8.1.46) and C a site (Prop. 8.1.57). Then the composite of pairs of adjoint ∞-functors (Def. 8.1.40)

$$\mathbf{H} \underset{}{\overset{L}{\underset{\hookrightarrow}{\overset{\longleftarrow}{\bot}}}} \mathrm{PShv}_\infty(C) \underset{\lim\limits_{\longleftarrow}}{\overset{\mathrm{const}}{\underset{\longrightarrow}{\overset{\longleftarrow}{\bot}}}} \mathrm{Grpd}_\infty \tag{8.104}$$

of **(a)** the reflection into presheaves over the site (Prop. 8.1.57) with **(b)** the limit-construction on presheaves (Prop. 8.1.52) is such that the composite left adjoint Lconst preserves finite limits (since L does by Prop. 8.1.57 and const does by Prop. 8.1.42 with Prop. 8.1.52). Hence, by the essential uniqueness of Prop. 8.1.59, the composite (8.104) is a factorization of the base geometric morphism of **H**.

8.1.14 Bundles

Notation 8.1.61 (Bundles and slicing). Let **H** an ∞-topos (Def. 8.1.46) and $X \in \mathbf{H}$ an object. We write:

(i) $(X, p) \in \mathbf{H}_{/X}$ for objects in the slice ∞-category of **H** over X, corresponding to morphisms p to X in $\mathbf{H}$ (*bundles* over X):

$$\begin{array}{c} E \\ \downarrow{\scriptstyle p} \\ X \end{array} \tag{8.105}$$

(ii) $(f, \alpha) \in \mathbf{H}_{/X}\big((E_1, p_1), (E_2, p_2)\big)$ for morphisms in the slice ∞-category, corresponding to diagrams in $\mathbf{H}$ of the form

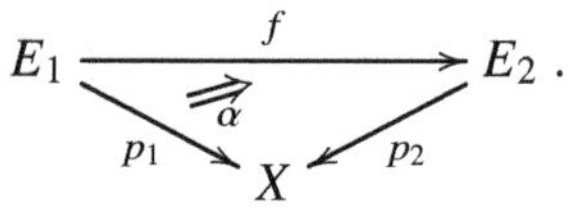

(8.106)

Proposition 8.1.62 (Slice ∞-topos [Lu09a, Prop. 6.3.5.1 (1)]). *Let $\mathbf{H}$ be an ∞-topos (Def. 8.1.46) and $X \in \mathbf{H}$ an object. Then the slice ∞-category $\mathbf{H}_{/X}$ (Ntn. 8.1.61) is also an ∞-topos.*

Example 8.1.63 (Iterated slice ∞-topos). Let $\mathbf{H}$ be an ∞-topos (Def. 8.1.46), $X \in \mathbf{H}$ and $(Y, p) \in \mathbf{H}_{/X}$ an object in the slice, hence (Ntn. 8.1.61) a morphism $Y \xrightarrow{p} X$. Then $\mathbf{H}_{/X}$ is itself an ∞-topos, by Prop. 8.1.62, and we may slice again to obtain the iterated slice ∞-topos

$$\big(\mathbf{H}_{/X}\big)_{/(Y,p)} \in \mathrm{Cat}_\infty \,. \tag{8.107}$$

(i) An object in (8.107) is a diagram in $\mathbf{H}$ of this form:

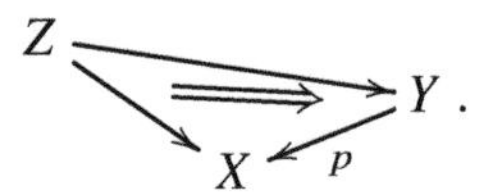

(8.108)

(ii) A morphism in (8.107) is a diagram in $\mathbf{H}$ of this form (which is furthermore filled by a 3-morphism that we Ntn.ally suppress, for readability):

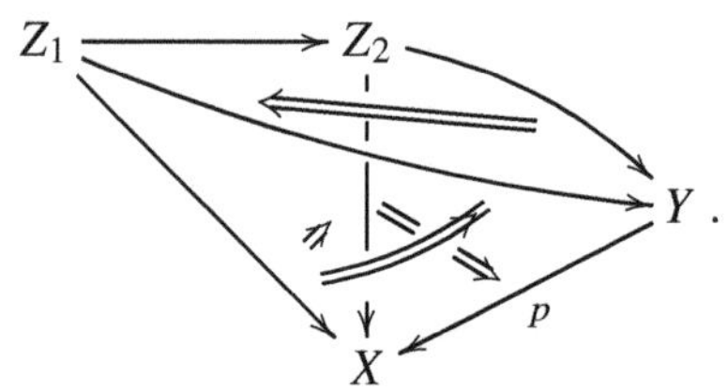

(8.109)

Proposition 8.1.64 (Hom-∞-groupoids in slices [Lu09a, Prop. 5.5.5.12]). *Let $\mathbf{H}$ be an ∞-topos (Def. 8.1.46) and $B \in \mathbf{H}$ an object. Then for $(X_1, p_1), (X_2, p_2) \in \mathbf{H}_{/B}$, two objects in the slice over B (Prop. 8.1.62), the hom-∞-groupoid between them is given by the following homotopy fiber-product of hom-∞-groupoids of $\mathbf{H}$:*

$$\mathbf{H}_{/B}\big((X_1, p_1), (X_2, p_2)\big) \simeq \{p_1\} \underset{\mathbf{H}(X_1, B)}{\times} \mathbf{H}(X_1, X_2) \tag{8.110}$$

hence by the ∞-groupoid given by the following Cartesian square (Ntn. 8.1.37):

$$\begin{array}{ccc} \mathbf{H}_{/B}\big((X_1, p_1), (X_2, p_2)\big) & \longrightarrow & \mathbf{H}(X_1, X_2) \\ \big\downarrow & \text{(pb)} & \big\downarrow {\scriptstyle p_2 \circ (-)} \\ * & \underset{\vdash p_1}{\longrightarrow} & \mathbf{H}(X_1, B) \,. \end{array} \tag{8.111}$$

Proposition 8.1.65 (Base change [Lu09a, HTT 6.3.5]). *Let* $\mathbf{H}$ *be an* ∞*-topos (Def. 8.1.46). Then for every morphism* $X \xrightarrow{f} Y$ *in* $\mathbf{H}$ *there is an induced base change adjoint triple (Def. 8.1.40) between the corresponding slice* ∞*-toposes (Prop. 8.1.62):*

$$\mathbf{H}_{/X} \underset{f_*}{\overset{f_!}{\underset{\bot}{\overset{\bot}{\rightleftarrows}}}} \mathbf{H}_{/Y} \quad (f_! \dashv f^* \dashv f_*) \tag{8.112}$$

where, in $\mathbf{H}$*,* $f_!$ *is given by postcomposition with* f *while* f^* *is given by pullback along* f*.*

Example 8.1.66 (Bundle morphisms covering base morphisms). For $\mathbf{H}$ an ∞-topos (Def. 8.1.46), the system of all its slice ∞-toposes (Prop. 8.1.62)

$$\begin{array}{ccc} \mathbf{H}^{\mathrm{op}} & \xrightarrow{\mathbf{H}_{/(-)}} & \mathrm{Cat}_\infty \\ X & \longmapsto & \mathbf{H}_{/X} \end{array} \tag{8.113}$$

related via contravariant base change (8.112) arranges into the "arrow ∞-topos" [Lu09a, 2.4.7.12]

$$\mathrm{Bundles}(\mathbf{H}) := \int_X \mathbf{H}_{/X} \simeq \mathbf{H}^{\Delta[1]}, \tag{8.114}$$

which, in view of Ntn. 8.1.61, may be thought of as the ∞-category of bundles in $\mathbf{H}$, but now with bundle morphisms allowed to cover non-trivial base morphisms.

Example 8.1.67 (Spectral bundles and tangent ∞-topos). Let $\mathbf{H}$ be an ∞-topos (Def. 8.1.46). Instead of the system (8.113) of its plain slices, consider the corresponding system of *stabilized* slices (stabilized under the suspension/looping adjunction on pointed objects, e.g., [Lu07, 1.4]):

$$\begin{array}{ccc} \mathbf{H}^{\mathrm{op}} & \xrightarrow{\mathrm{Stab}(\mathbf{H}_{/(-)})} & \mathrm{Cat}_\infty \\ X & \longmapsto & \mathrm{Stab}(\mathbf{H}_{/X}) . \end{array} \tag{8.115}$$

The resulting total ∞-category

$$\mathrm{SpectralBundles}(\mathbf{H}) := \int_X \mathrm{Stab}(\mathbf{H}_{/X}), \tag{8.116}$$

is that of *bundles of spectra* in $\mathbf{H}$ (parametrized spectrum objects). Remarkably, this is itself an ∞-topos [Joy08a, 35.5][Lu17, 6.1.1.11], also called the *tangent* ∞*-topos* $T\mathbf{H}$ of $\mathbf{H}$ (e.g., [Lu07][BM19]).

Example 8.1.68 (Base change along terminal morphism). Let $\mathbf{H}$ be an ∞-topos (Def. 8.1.46) and $X \in \mathbf{H}$ be any object. With $\mathbf{H} \simeq \mathbf{H}_{/*}$ regarded as its own slice (Prop. 8.1.62) over the terminal object, base change (Prop. 8.1.65) along the terminal morphism $X \to *$ is of the form

$$\mathbf{H}_{/X} \underset{\bot}{\overset{\mathrm{dom}}{\underset{\bot}{\rightleftarrows}}} \mathbf{H} \quad (\text{middle: } X \times (-)) \tag{8.117}$$

where **(a)** the top functor sends a morphism $Y \to X$ to its domain object Y, and **(b)** the middle functor is Cartesian product with X. In particular, it follows that:

(i) The base geometric morphism (Prop. 8.1.40) of the slice ∞-topos $\mathbf{H}_{/X}$ (Prop. 8.1.62) is given by

$$\big(\Delta \dashv \Gamma\big) \;\simeq\; \big((X \to *)^* \dashv (X \to *)_*\big) \tag{8.118}$$

(since $(X \to *)^*$ is a left adjoint that also preserves finite limits, as it is also a right adjoint, Prop. 8.1.42).

(ii) The forgetful functor $\mathrm{dom} : \mathbf{H}_{/X} \to \mathbf{H}$ is a left adjoint $(X \to *)_!$ and hence preserves all colimits (Prop. 8.1.42).

While dom (8.117) does not preserve all limits, it does preserve fiber products:

Proposition 8.1.69 (Fiber products in slice ∞-toposes). *Let* $\mathbf{H}$ *be an* ∞*-topos (Def. 8.1.46),* $B \in \mathbf{H}$, $\mathbf{H}_{/B}$ *the slice* ∞*-topos (Prop. 8.1.62), and* $\mathbf{H}_{/B} \xrightarrow{\mathrm{dom}} \mathbf{X}$ *its forgetful functor* (8.117) *from Example 8.1.68.*

(i) *Given a cospan* $(Y, \phi_Y) \to (X, \phi_X) \leftarrow (Z, \phi_Z)$ *in* $\mathbf{H}_{/B}$*, the underlying object of its fiber product is the fiber product of its underlying objects:*

$$\mathrm{dom}\Big((Y,\phi_Y) \underset{(X,\phi_X)}{\times} (Z,\phi_Z)\Big) \;\simeq\; Y \underset{X}{\times} Z\,. \tag{8.119}$$

(ii) *In particular, since* (X, id_X) *is the terminal object in* $\mathbf{H}_{/X}$*, the plain product in the slice is*

$$(Y,\phi_Y) \times (Z,\phi_Z) \;=\; (Y,\phi_Y) \underset{(X,\mathrm{id}_X)}{\times} (Z,\phi_Z)\,, \tag{8.120}$$

we have that the product in $\mathbf{H}_{/X}$ *is given by the fiber product over* X *in* $\mathbf{H}$*:*

$$\mathrm{dom}\Big((Y,\phi_Y) \times (Z,\phi_Z)\Big) \;\simeq\; Y \underset{X}{\times} Z\,. \tag{8.121}$$

Proof. Generally, limits in $\mathbf{H}_{/X}$ are given by limits in $\mathbf{H}$ over the underlying co-cone diagram. Specifically: for $Y : I \to \mathbf{H}$ we have $\mathrm{dom}\big(\varprojlim Y_\bullet\big) \simeq \varprojlim\, \big(Y/X\big)_\bullet$. With this, the claim follows via [Lu09a, Prop. 4.1.1.8] from the fact that the canonical inclusion of diagram categories

$$\{\, y \to b \leftarrow z \,\} \hookrightarrow \left\{ \begin{array}{ccccc} y & \longrightarrow & b & \longleftarrow & z \\ & \searrow & \downarrow & \swarrow & \\ & & t & & \end{array} \right\} \tag{8.122}$$

is an initial functor (i.e., under $(-)^{\mathrm{op}}$ it is a final functor), as one finds by direct inspection from [Lu09a, Prop. 4.1.3.1]. □

Proposition 8.1.70 (Terminal right base change of bare ∞-groupoids). *In the base* ∞*-topos* $\mathbf{H} = \mathrm{Grpd}_\infty$ *(8.48), the right base change along the terminal morphism (Example 8.1.68) of an object* $X \in \mathrm{Grpd}_\infty$ *is given by the hom-*∞*-groupoid out of* X*, regarded as the terminal object in the slice:*

$$(X \to *)_* \;\simeq\; \mathbf{H}_{/X}(X, -) \;:\; \big(\mathrm{Grpd}_\infty\big)_{/X} \longrightarrow \mathrm{Grpd}_\infty\,. \tag{8.123}$$

Proof. We have the following chain of natural equivalences:

$$\begin{aligned} \mathrm{Grpd}_\infty\big(A, (\mathrm{Grpd}_\infty)_{/X}(X, B)\big) &\simeq (\mathrm{Grpd}_\infty)_{/X}\big(\Delta(A) \times_X X, B\big) \\ &\simeq (\mathrm{Grpd}_\infty)_{/X}\big(\Delta(A), B\big) \\ &\simeq (\mathrm{Grpd}_\infty)_{/X}\big((X \to *)^*(A), B\big). \end{aligned} \tag{8.124}$$

Here the first step observes that the slice $(\mathrm{Grpd}_\infty)_{/X}$ is itself an ∞-topos by Prop. 8.1.62, so that the tensoring equivalence of Prop. 8.1.50 applies. The second step uses the fact that X is regarded as the terminal object in its own slice so that forming Cartesian product with it is equivalently the identity operation. The last step observes that for the slice ∞-topos $\Delta \simeq (X \to *)^*$ (8.118) by Example 8.1.68. In summary, the total equivalence of (8.124) is the hom-equivalence that characterizes $\mathbf{H}_{/X}(X, -)$ as a right adjoint to $(X \to *)^*$. □

Proposition 8.1.71 (Conservative base change along effective epi [NSS12a, 3.15]). *Let $\mathbf{H}$ be an ∞-topos (Def. 8.1.46). For $Y \twoheadrightarrow X$ an effective epimorphism (Def. 8.1.80) in $\mathbf{H}$, the induced base change (Prop. 8.1.65)*

$$\mathbf{H}_{/X} \xrightarrow{\;i^*\;} \mathbf{H}_{/Y} \tag{8.125}$$

is a conservative ∞-functor, meaning that a morphism $f \in \mathbf{H}_{/X}$ is an equivalence if its base change $i^*(f)$ in $\mathbf{H}_{/Y}$ is an equivalence.

Proposition 8.1.72 (Colimits of classifying maps are classifying maps of colimits). *Let $\mathbf{H}$ be an ∞-topos (Def. 8.1.46), $\mathcal{I}$ a small ∞-category, $X_\bullet : \mathcal{I} \to \mathbf{H}$ a diagram, and $(\vdash E)_\bullet : X_\bullet \to \mathrm{const}_{\mathrm{Objects}_\kappa}$ a transformation to the diagram constant on the object classifier* (8.81), *thus classifying a diagram $E_\bullet : \mathcal{I} \to \mathbf{H}$ of bundles over $X_\bullet$. Then the colimit of $(\vdash E)_\bullet$ formed in the slice $\mathbf{H}_{/\mathrm{Objects}_\kappa}$ (Prop. 8.1.62) is the colimit of $X_\bullet$ equipped with the classifying map for the colimit of $E_\bullet$:*

$$\varinjlim\, (\vdash E)_\bullet \;\sim\; \vdash\big(\varinjlim E_\bullet\big)\,. \tag{8.126}$$

Proof. Since underlying the colimit $\varinjlim(\vdash E)_\bullet$ in the slice ∞-topos $\mathbf{H}_{/\mathrm{Objects}_\kappa}$ is the colimit $\varinjlim X_\bullet$ in $\mathbf{H}$ (by Example 8.1.68), we are dealing with a situation as shown in the diagram on the right (where a simplicial diagram shape is shown just for definiteness of illustration). We need to demonstrate that the front square in this diagram is Cartesian. Observe that

(a) the vertical squares over each $\vdash E_i$ are Cartesian by assumption, whence

(b) also the solid vertical squares over each $X_i \twoheadrightarrow X_j$ are Cartesian, by the pasting law (Prop. 8.1.39).

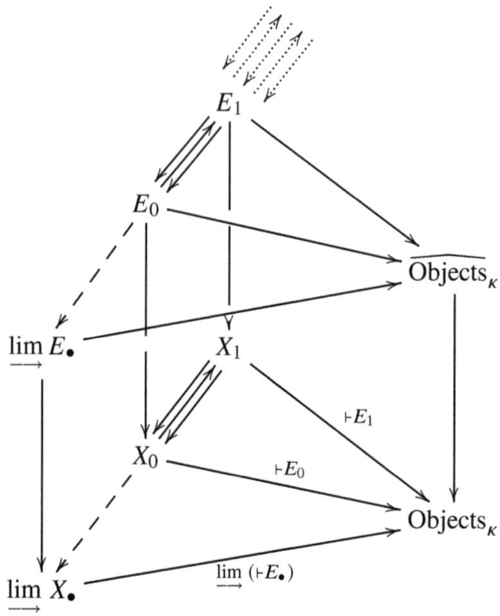

(8.127)

This means that the assumption of Prop. 8.1.48 is satisfied for the left part of the diagram (regarded as a transformation of diagrams from top to bottom), implying that the dashed square is Cartesian.

This implies, together with **(a)**, that the front square is Cartesian, again by the pasting law (Prop. 8.1.39). □

8.1.15 *n*-Truncation

Definition 8.1.73 (*n*-truncated objects [Lu09a, Def. 5.5.6.1]).
Let $n \in \{-2, -1, 0, 1, 2, \cdots\}$.

(i) An ∞-groupoid is called *n-truncated* for $n \geq 0$ if all its homotopy groups of degree $> n$ are trivial. It is called (-1)*-truncated* if it is either empty or contractible and *(-2)-truncated* if it is (non-empty and) contractible.

(ii) Let C be an ∞-category. Then an object $X \in C$ is *n-truncated* if for all objects $U \in C$ the hom-∞-groupoid $C(U, X)$ is n-truncated, in the above sense.

Definition 8.1.74 (*n*-truncated morphisms [Lu09a, Def. 5.5.6.8]).
Let $n \in \{-2, -1, 0, 1, 2, \cdots\}$.

(i) A morphism of ∞-groupoids is called *n-truncated* if all its homotopy fibers are n-truncated ∞-groupoids according to Def. 8.1.73.

(ii) Let C be an ∞-category. A morphism $X \xrightarrow{f} Y$ in C is called *n-truncated* if for all objects $U \in C$ the induced morphism of hom-∞-groupoids $C(U, X) \xrightarrow{C(U,f)} C(U, Y)$ is n-truncated in the above sense.

Definition 8.1.75 (Monomorphisms). A (-1)-truncated morphism (Def. 8.1.74) is also called a *monomorphism*, to be denoted

$$X \hookrightarrow Y\,. \tag{8.128}$$

Proposition 8.1.76 (Monomorphisms are preserved by pushout [Re19, p. 21]). *Let* **H** *be an ∞-topos (Def. 8.1.46). Then the class of monomorphisms in* **H** (Def. 8.1.75) *is closed under* **(i)** *pullback and* **(ii)** *composition.*

Definition 8.1.77 (Poset of subobjects). Let **H** be an ∞-topos and $X \in \mathbf{H}$ any object. Then the *poset of subobjects* of X is the sub-∞-category (Def. 8.1.78) of (-1)-truncated objects of the slice over X:

$$\mathrm{SubObjects}(X) \hookrightarrow \mathbf{H}_{/X} \tag{8.129}$$

whose objects are equivalently the monomorphisms (Def. 8.1.75) $U \hookrightarrow X$.

Proposition 8.1.78 (n-Truncation modality [Lu09a, 5.5.6.18]). *For* **H** *an ∞-topos* (Def. 8.1.46)*, and for* $n \in \{-1, 0, 1, 2, \cdots\}$*, its full sub-∞-category* (Def. 8.1.1) *of n-truncated objects (Def. 8.1.73) is reflective, in that the inclusion functor has a left adjoint (Def. 8.1.40):*

$$\underset{\infty\text{-topos}}{\mathbf{H}} \underset{i_n}{\overset{\tau_n}{\underset{\longleftarrow}{\overset{\longrightarrow}{\bot}}}} \underset{\substack{\text{sub-}\infty\text{-category} \\ \text{of } n\text{-truncated objects}}}{\mathbf{H}_{\leq n}}\,. \tag{8.130}$$

We denote the induced n-truncation modality (1.18) *by:*

$$\underset{\text{“}n\text{-truncated”}}{\boldsymbol{\tau}_n} := i_n \circ \tau_n \;:\; \mathbf{H} \longrightarrow \mathbf{H}\,. \tag{8.131}$$

Remark 8.1.79 (Reflection of n-truncated objects). The reflection (8.130) entails that maps into an n-truncated object factor essentially uniquely through the n-truncation of their domain:

$$Y \text{ is } n\text{-truncated} \quad \vdash \quad \mathbf{H}(\boldsymbol{\tau}_n X, Y) \xrightarrow[\sim]{(\eta_X^{\tau_n})^*} \mathbf{H}(X, Y)\,. \tag{8.132}$$

Definition 8.1.80 (Effective epimorphisms [Lu09a, Cor. 6.2.3.5]). Let **H** be an ∞-topos. A morphism in **H** is called an *effective epimorphism*, to be denoted

$$Y \overset{f}{\twoheadrightarrow} Z \tag{8.133}$$

if, when regarded as an object of the slice over X (Prop. 8.1.62), its (-1)-truncation (Prop. 8.1.78) is the terminal object

$$\tau_{(-1)}(f) \simeq * \quad \in \mathbf{H}_{/X}\,. \tag{8.134}$$

We write

$$\mathrm{EffEpi}(\mathbf{H}) \subset \mathbf{H}^{(0 \to 1)} \in \mathrm{Cat}_\infty \tag{8.135}$$

for the full sub-∞-category (Def. 8.1.1) of the arrow-category of **H** on those that are effective epimorphisms.

Definition 8.1.81 (n-Connected morphisms [Lu09a, Prop. 6.5.1.12]). Let **H** be an ∞-topos (Def. 8.1.46) and $n \in \{-1, 0, 1, 2 \cdots\}$. Then a morphism $X \xrightarrow{f} Y$ in **H**

is called *n-connected* if, regarded as an object in the slice over X (Prop. 8.1.62), its n-truncation (Def. 8.1.78) is the terminal object:

$$Y \xrightarrow{\;f\;} X \quad \text{is } n\text{-truncated} \qquad \Leftrightarrow \qquad \tau_n(f) \;\simeq\; * \;\in \mathbf{H}_{/X}\,. \tag{8.136}$$

Hence the (-1)-connected morphisms are equivalently the effective epimorphisms (Def. 8.1.80).

Lemma 8.1.82 (Effective epimorphisms are preserved by pullback [Lu09a, 6.2.3.15]). *Let* $\mathbf{H}$ *be an* ∞*-topos (Def. 8.1.46). Then the class of effective epimorphisms in* $\mathbf{H}$ *(Def. 8.1.80) is closed under* **(i)** *pullback and* **(ii)** *composition.*

8.1.16 n-Image factorization

Proposition 8.1.83 (Connected/truncated factorization system [Lu09a, Ex. 5.2.8.16][Re10, Prop. 5.8]). *Let* $\mathbf{H}$ *be an* ∞*-topos. Then, for all* $n \in \{-1, 0, 1, 2, \cdots\}$, *the pair of classes of n-connected/n-truncated morphisms* (Def. 8.1.81, Def. 8.1.74) *forms an orthogonal factorization system in that:*

(i) *every morphism* f *in* $\mathbf{H}$ *factors essentially uniquely as*

$$\begin{array}{ccccc} X & & \xrightarrow{\quad f \quad} & & Y \\ & {\scriptstyle n\text{-connected}}\searrow & & \nearrow{\scriptstyle n\text{-truncated}} & \\ & & \mathrm{im}_n(f) & & \end{array} \tag{8.137}$$

(ii) *every commuting square as follows has an essentially unique dashed lift:*

$$\begin{array}{ccc} X & \longrightarrow & A \\ {\scriptstyle n\text{-connected}}\downarrow & \nearrow & \downarrow{\scriptstyle n\text{-truncated}} \\ Y & \longrightarrow & B\,. \end{array} \tag{8.138}$$

Remark 8.1.84 (Functoriality of n-image factorization). In consequence, the n-factorization of Def. 8.137 is *functorial* in that for a homotopy-commuting square there is an essentially unique compatible induced map (shown dashed below) between the n-images of the vertical morphisms:

$$f\left(\begin{array}{ccc} X & \longrightarrow & A \\ \downarrow{\scriptstyle n\text{-con}} & & \downarrow{\scriptstyle n\text{-con}} \\ \mathrm{im}_n(f) & \overset{\exists !}{\dashrightarrow} & \mathrm{im}_n(g) \\ \downarrow{\scriptstyle n\text{-trunc}} & & \downarrow{\scriptstyle n\text{-trnc}} \\ Y & \longrightarrow & B \end{array}\right)g \tag{8.139}$$

Namely, the dashed map arises as the essentially unique lift (8.138) of the top left map against the bottom right map.

Example 8.1.85 (Epi/mono factorization). For $n = -1$, the connected/truncated factorization system (Prop. 8.1.83) has as left class the effective epimorphisms (Def.

8.1.80) and as right class the monomorphisms (Def. 8.1.75). Hence, with the Ntn. from (8.133) and (8.128):

(i) The (-1)-image factorization (8.137) reads:

$$\begin{array}{ccc} X & \xrightarrow{\quad f \quad} & Y\,. \\ & \searrow\!\!\!\!\searrow \quad \nearrow\!\!\!\!\hookleftarrow & \\ & \mathrm{im}_{-1}(f) & \end{array} \tag{8.140}$$

(ii) The lifting property (8.138) for $n = -1$ reads:

$$\begin{array}{ccc} X & \longrightarrow & A \\ \downarrow\!\!\downarrow & \nearrow & \downarrow \\ Y & \longrightarrow & B\,. \end{array} \tag{8.141}$$

8.1.17 Groupoids and Stacks

Definition 8.1.86 (Groupoids internal to an ∞-topos [Lu09a, 6.1.2.7])**.** Let $\mathbf{H}$ be an ∞-topos (Def. 8.1.46).

(i) A *groupoid in* $\mathbf{H}$ is a simplicial diagram

$$X_\bullet : \Delta^{\mathrm{op}} \longrightarrow \mathbf{H} \tag{8.142}$$

which satisfies the *groupoidal Segal condition*: For all $n \in \mathbb{N}$ and for all partitions of the set of $n+1$ elements by two subsets that share a unique element, the corresponding image under $X_\bullet$ is a Cartesian square (Ntn. 8.1.37):

$$\begin{array}{ccccc} & \{0,1,\cdots,n\} & & & \\ S_1 \hookrightarrow & \text{(po)} & \hookleftarrow S_2 & & \\ \nwarrow & * & \nearrow & & \end{array} \;\overset{X_\bullet}{\longmapsto}\; \begin{array}{ccc} & X_n & \\ X_{|S_1|-1} \swarrow & \text{(pb)} & \searrow X_{|S_2|-1}\,. \\ \searrow & X_0 & \swarrow \end{array} \tag{8.143}$$

(ii) We write

$$\mathrm{Grpd}(\mathbf{H}) \hookrightarrow \mathbf{H}^{(\Delta^{\mathrm{op}})} \quad \in \mathrm{Cat}_\infty \tag{8.144}$$

for the full sub-∞-category of that of simplicial diagrams in $\mathbf{H}$ on those that are groupoids.

Example 8.1.87 (Nerves)**.** Let $\mathbf{H}$ be an ∞-topos (Def. 8.1.46) and $X \xrightarrow{f} \mathcal{X}$ a morphism in $\mathbf{H}$. Its *nerve* is the simplicial diagram of its iterated homotopy fiber products:

$$\begin{array}{cccc} \mathrm{Nerve}_\bullet(f) : & \Delta^{\mathrm{op}} & \longrightarrow & \mathbf{H} \\ & [n] & \longmapsto & \underbrace{X \times_{\mathcal{X}} X \times_{\mathcal{X}} \cdots \times_{\mathcal{X}} X}_{n \text{ factors}}\,, \end{array} \tag{8.145}$$

with face maps, the projections, and degeneracy maps, the diagonals. This is evidently a groupoid object according to Def. 8.1.86:

$$\mathrm{Nerve}_\bullet(f) \in \mathrm{Grpd}(\mathbf{H})\,. \tag{8.146}$$

Proposition 8.1.88 (Groupoids equivalent to stacks with atlases [Lu09a, 6.2.3.5]). *Let* **H** *be an* ∞*-topos (Def. 8.1.46). Then the* ∞*-functor sending* $X_\bullet \in \mathrm{Grpd}(\mathbf{H})$ *(Def. 8.1.86) to the* X_0*-component of its colimiting cocone*

(i) *lands in effective epimorphisms* (8.135) *and*

(ii) *constitutes an equivalence of* ∞*-categories whose inverse is given by the construction of nerves (Example 8.1.87):*

$$\begin{array}{ccc} \mathrm{Grpd}(\mathbf{H}) & \xrightarrow{\ \simeq\ } & \mathrm{EffEpi}(\mathbf{H}) \\ X_\bullet & \longmapsto & (X_0 \twoheadrightarrow \varinjlim X_\bullet) \\ \mathrm{Nerve}_\bullet(a) & \longleftarrow & (X \overset{a}{\twoheadrightarrow} \mathcal{X}\,) \end{array} \tag{8.147}$$

$$\begin{array}{ccccc} \vdots & & \vdots & & \\ X_0 \times_{\mathcal{X}} X_0 & \simeq & X_1 & & \text{“groupoid”} \\ {\scriptstyle \mathrm{pr}_1}\downarrow\uparrow\downarrow{\scriptstyle \mathrm{pr}_2} & & {\scriptstyle s}\downarrow\uparrow{\scriptstyle e}\downarrow{\scriptstyle t} & & \\ X_0 & = & X_0 & & \\ {\scriptstyle a}\downarrow & & \downarrow & & \text{“atlas”} \\ \mathcal{X} & \simeq & \varinjlim X_\bullet & & \text{“stack”} \end{array} \tag{8.148}$$

Remark 8.1.89 (Internal groupoids with prescribed properties). Often one considers $X_\bullet \in \mathrm{Grpd}(\mathbf{H})$ (Def. 8.1.86) whose simplicial component diagram (8.142) is inside a chosen sub-∞-category of **H**. Key examples are étale groupoids (Def. 9.1.37) and V-étale groupoids (Remark 10.2.3).

Remark 8.1.90 (Morita morphisms of groupoids). A morphism between stacks $\mathcal{X} := \varinjlim X_\bullet$ underlying groupoids $X_\bullet$ (according to Prop. 8.1.88) *without* (i.e., disregarding) the corresponding atlas is also known as a *Morita morphism* (in particular, a *Morita equivalence* if it is an equivalence), or a *Hilsum-Skandalis morphism* [HS87][Pr89], or a *groupoid bibundle* [Bl07][Nu13, Prop. 2.2.34] between the corresponding groupoids:

$$\begin{array}{lccccccl} & \mathrm{Grpd}(\mathbf{H}) & \xrightarrow{\ \simeq\ } & \mathrm{EffEpi}(\mathbf{H}) & \xrightarrow{\ \mathrm{codom}\ } & \mathbf{H} & & \\ \text{groupoid} & X_\bullet & \longmapsto & (X_0 \twoheadrightarrow \mathcal{X}) & \longmapsto & \mathcal{X} & & \text{“stack”} \\ & & & & & {\scriptstyle f}\downarrow & & \text{morphism of underlying stacks = “Morita morphism” of groupoids} \\ \text{groupoid} & Y_\bullet & \longmapsto & (Y_0 \twoheadrightarrow \mathcal{Y}) & \longmapsto & \mathcal{Y} & & \text{“stack”} \end{array} \tag{8.149}$$

Hence whether or not there is a conceptual distinction between "geometric groupoids" and "stacks" depends on whether morphisms of groupoids are taken to be their plain morphisms or their Morita morphisms. In practice, one is typically interested in the latter case. Indeed, the groupoid atlas of a stack, whose preservation restricts Morita morphisms to plain morphisms of groupoids, by Prop. 8.1.88, is, in practice, typically required to exist with a certain property but not required to be preserved by morphisms (this is so notably for V-étale groupoids, Remark 10.2.3). In particular, the

$\mathrm{SmthGrpd}_\infty$ of Example 9.1.19 and the $\mathrm{JetsOfSmoothGroupoids}_\infty$ of Example 9.1.26 are ∞-groupoids with Morita morphisms understood, hence could also be called *(jets of) smooth ∞-stacks*.

Proposition 8.1.91 (Equifibered morphisms of groupoids)**.** *Let* $\mathbf{H}$ *be an ∞-topos (Def. 8.1.46) and* $X_\bullet, Y_\bullet \in \mathrm{Grpd}(\mathbf{H})$ *(Def. 8.1.86). Then, under the equivalence* (8.147) *between groupoids and their stacks with atlases (Prop. 8.1.88), we have that equifibered morphisms of groupoids correspond to Cartesian squares between their atlases:*

$$X_\bullet \overset{f_\bullet}{\Longrightarrow} Y_\bullet \quad \text{such that} \quad \underset{[n_1]\xrightarrow{\phi}[n_2]}{\forall} \quad \begin{array}{ccc} X_{n_1} & \xrightarrow{\;f_{n_1}\;} & Y_{n_1} \\ {\scriptstyle X_\phi}\big\downarrow & {\scriptstyle (\mathrm{pb})} & \big\downarrow{\scriptstyle Y_\phi} \\ X_{n_2} & \xrightarrow[\;f_{n_2}\;]{} & Y_{n_2} \end{array}$$

$$\Leftrightarrow \quad \begin{array}{ccc} X_0 & \xrightarrow{\;f_0\;} & Y_0 \\ {\scriptstyle a_X}\big\downarrow & {\scriptstyle (\mathrm{pb})} & \big\downarrow{\scriptstyle a_Y} \\ \mathcal{X} & \xrightarrow[\;\varinjlim f_\bullet\;]{} & \mathcal{Y} \end{array} . \tag{8.150}$$

Proof. From right to left this follows by the pasting law (Prop. 8.1.39), while from left to right this is Prop. 8.1.48. □

8.2 Galois theory

We discuss here the *internal* formulation in ∞-toposes of the theory of *groups*, *group actions*, and *fiber bundles*, following [NSS12a][SSS12] (see [FSS14] for exposition). Externally, these concepts are known as *grouplike A_∞-algebras* or, equivalently, *grouplike E_1-algebras* (here, in ∞-stacks) and as their *A_∞-modules*, etc., and are traditionally presented by simplicial techniques [May72][Lu17]. But internally the theory becomes finitary and elementary, with all concepts emerging naturally from pastings of a few Cartesian squares.

Accordingly, much of the following constructions may readily be expressed fully formally in homotopy type theory [BvDR18] (see p. 10). Thus, the following elegant characterizations of

- groups (Prop. 8.2.1),
- group actions (Prop. 8.2.6),
- principal bundles (Prop. 8.2.15),
- fiber bundles (Prop. 8.2.19),

in an ∞-topos $\mathbf{H}$ may be taken to be the *definition* of these notions for all purposes of internal constructions.

8.2.1 Groups

The following characterization of group ∞-stacks (Prop. 8.2.1) is the time-honored *May recognition theorem* [May72] generalized from Grpd_∞ to general ∞-toposes [Lu09a, 7.2.2.11][Lu17, 6.2.6.15]:

Proposition 8.2.1 (A_∞-Group recognition theorem [NSS12a, Thm. 2.19]). *Let* $\mathbf{H}$ *be an* ∞*-topos (Def. 8.1.46). Then the operation of sending an* ∞*-group* G *to the homotopy quotient of its action on a point constitutes an equivalence of* ∞*-categories:*

$$\begin{array}{ccc} \mathrm{Grp}(\mathbf{H}) & \underset{\mathbf{B}}{\overset{\Omega}{\underset{\longrightarrow}{\overset{\longleftarrow}{\simeq}}}} & \mathbf{H}^{*/}_{\geq 1} \\ G & \longmapsto & * /\!\!/ G \end{array} \tag{8.151}$$

between the ∞*-category of* ∞*-group objects and the* ∞*-category of pointed and connected objects in* $\mathbf{H}$*. The inverse equivalence is given by forming the loop space object*

$$\begin{array}{ccc} G \simeq \Omega \mathbf{B} G & \longrightarrow & * \\ \downarrow & \text{(pb)} & \downarrow \\ * & \longrightarrow & \mathbf{B}G \end{array} \tag{8.152}$$

Example 8.2.2 (Point in delooping is an effective epi)**.** For $G \in \mathrm{Grp}(\mathbf{H})$, the morphism that exhibits its delooping as a pointed object (Prop. 8.2.1)

$$* \twoheadrightarrow \mathbf{B}G\,, \tag{8.153}$$

is an effective epimorphism (Def. 8.1.78). Thus, Prop. 8.1.88 says here that:

(i) Groups in $\mathbf{H}$ are, equivalently, the groupoids in $\mathbf{H}$ (Def. 8.1.86) that admit an atlas by the point.

(ii) With (8.152), we have

$$\mathbf{B}G \;\simeq\; \underrightarrow{\lim} G^{\times_\bullet} \;\in\; \mathbf{H}\,. \tag{8.154}$$

Example 8.2.3 (Neutral element)**.** Let $\mathbf{H}$ be an ∞-topos. Given a group $G \in \mathrm{Grp}(\mathbf{H})$ in the form of a pointed connected object $* \to \mathbf{B}G$, according to Prop. 8.2.1, its *neutral element* $* \xrightarrow{e} G$ is the diagonal morphism into the defining homotopy fiber product (8.152); hence, the canonical morphism induced by the universal property of the homotopy fiber product from the equivalence with itself of the point inclusion into $\mathbf{B}G$ (8.153).

$$\begin{array}{ccccc} & & * & & \\ & \swarrow\!\!= & \downarrow{\scriptstyle e} & =\!\!\searrow & \\ & & G & & \\ * & \swarrow & \text{(pb)} & \searrow & * \\ & \searrow & & \swarrow & \\ & & \mathbf{B}G & & \end{array} \tag{8.155}$$

Example 8.2.4 (Group division/shear map)**.** Let $\mathbf{H}$ be an ∞-topos. Given a group $G \in \mathrm{Grp}(\mathbf{H})$ in the form of a pointed connected object $* \longrightarrow \mathbf{B}G$, according to Prop. 8.2.1, the group division operation

$$G \times G \xrightarrow{(-)\cdot(-)^{-1}} G \tag{8.156}$$

is exhibited by the universal morphism shown dashed in the following diagram:

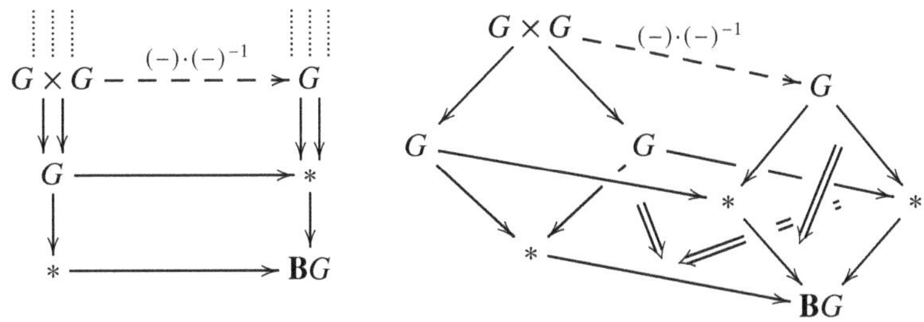

(8.157)

On the left, we are showing this as part of a morphism of Čech nerve augmented simplicial diagrams. On the right, the situation is shown in more detail: Here the right and the two bottom squares are all the looping relation (8.152), while the left square exhibits the plain product of G with itself. With this, the universal property of the right square implies the essentially unique dashed morphism making the total diagram homotopy-commute. Note the following:

- **(i)** The two top squares are also Cartesian: This follows from the pasting law (Prop. 8.1.39) using, for the top front square, that the left and right and the bottom rear squares are Cartesian; and similarly for the top rear square.
- **(ii)** The total homotopy filling the top and the right faces in (8.157) is, by commutativity, equivalent to the total homotopy filling the left and the bottom faces. But, in performing the composition this way, the direction of one of the two bottom homotopies gets reversed. This is why this construction gives the division map $(-)\cdot(-)^{-1}$ (shear map) instead of the plain group product.

Proposition 8.2.5 (Mayer-Vietoris sequence [Sc13, Prop. 3.6.142]). *Let* $\mathbf{H}$ *be an* ∞*-topos (Def. 8.1.46),* $G \in \mathrm{Grp}(\mathbf{H})$ *(Prop. 8.2.1), and* $(X, f), (Y, g) \in \mathbf{H}_{/G}$ *two objects in the slice (Prop. 8.1.62) over the underlying object of* G*. Then their homotopy fiber product*

$$\begin{array}{ccc} X \underset{G}{\times} Y & \xrightarrow{\ \mathrm{pr}_Y\ } & Y \\ {\scriptstyle \mathrm{pr}_X}\downarrow & \text{(pb)} & \downarrow{\scriptstyle g} \\ X & \xrightarrow[\ f\ ]{} & G \end{array} \tag{8.158}$$

is equivalently exhibited by the following Mayer-Vietoris homotopy fiber sequence

$$\begin{array}{ccccc} X \times_G Y & \longrightarrow & & \longrightarrow & * \\ {\scriptstyle (\mathrm{pr}_X, \mathrm{pr}_Y)}\downarrow & & \text{(pb)} & & \downarrow \\ X \times Y & \xrightarrow{\ (f,g)\ } & G \times G & \xrightarrow{\ (-)\cdot(-)^{-1}\ } & G, \end{array} \qquad X \times Y \xrightarrow{\ f\cdot g^{-1}\ } G \tag{8.159}$$

where the morphism on the bottom right is the group division map (8.157).

8.2.2 Group actions

Proposition 8.2.6 (Group actions [NSS12a, 4.1]). *Let* $\mathbf{H}$ *an* ∞*-topos and* $G \in \mathrm{Grp}(\mathbf{H})$ *(Prop. 8.2.1).*

(i) *An action* (X, ρ) *of* G *is an object* $X \in \mathbf{H}$ *and homotopy fiber sequence in* $\mathbf{H}$ *of the form*

$$\begin{array}{ccc} X & \xrightarrow{\mathrm{fib}(\rho)} & X /\!\!/ G \\ & & \downarrow{\scriptstyle \rho} \\ & & \mathbf{B}G\,, \end{array} \tag{8.160}$$

where $\mathbf{B}G$ *is the delooping of* G (8.2.1).

(ii) *The object* $X /\!\!/ G$ *appearing in* (8.160) *is, equivalently, the homotopy quotient of the action of* G *on* X*:*

$$X /\!\!/ G \simeq \underset{\longrightarrow}{\lim}\Big(\quad \cdots \quad X \times G \times G \;\substack{\longrightarrow\\ \longleftarrow\\ \longrightarrow \\ \longleftarrow \\ \longrightarrow}\; X \times G \;\substack{\longrightarrow\\ \longleftarrow\\ \longrightarrow}\; X \Big). \tag{8.161}$$

(iii) *Hence the* ∞*-category of* G*-actions is, equivalently, the slice* ∞*-topos (Prop. 8.1.62) of* $\mathbf{H}$ *over* $\mathbf{B}G$*:*

$$G\mathrm{Act}(\mathbf{H}) \simeq \mathbf{H}_{/\mathbf{B}G} \quad \in \mathrm{Cat}_\infty\,. \tag{8.162}$$

We record the following immediate but important aspect of this characterization:

Lemma 8.2.7 (Homotopy quotient maps are effective epimorphisms). *Let* $\mathbf{H}$ *be an* ∞*-topos,* $G \in \mathrm{Grp}(\mathbf{H})$ *(Prop. 8.2.1), and* $(X, \rho) \in G\mathrm{Act}(\mathbf{H})$ *(Prop. 8.2.6). Then the quotient morphism from* X *to its homotopy quotient* (8.161) *is an effective epimorphism (Def. 8.1.80):*

$$X \xrightarrow{\mathrm{fib}(\rho)} \!\!\!\!\to X /\!\!/ G\,. \tag{8.163}$$

Proof. By (8.160) in Prop. 8.2.6, the quotient map sits in a homotopy pullback square of the form

$$\begin{array}{ccc} X & \xrightarrow{\mathrm{fib}(\rho)} & X /\!\!/ G \\ \downarrow & \text{(pb)} & \downarrow{\scriptstyle \rho} \\ * & \longrightarrow & \mathbf{B}G \end{array} \tag{8.164}$$

The bottom morphism is an effective epimorphism (Example 8.2.2). Since these are preserved by pullback (Lemma 8.1.82), the claim follows. □

Example 8.2.8 (Left multiplication action). Let $\mathbf{H}$ be an ∞-topos (Def. 8.1.46) and $G \in \mathrm{Grp}(\mathbf{H})$ (Prop. 8.2.1). The defining looping relation (8.152) exhibits, by comparison with (8.160), an action of G on itself:

$$\begin{array}{ccc} G & \xrightarrow{\mathrm{fib}(\rho_\ell)} & * \\ & & \downarrow{\scriptstyle \rho_\ell} \\ & & \mathbf{B}G\,. \end{array} \tag{8.165}$$

This is the *left multiplication action* with $G /\!\!/ G \simeq *$.

Example 8.2.9 (Adjoint action)**.** Let $\mathbf{H}$ be an ∞-topos (Def. 8.1.46) and $G \in \mathrm{Grp}(\mathbf{H})$ (Prop. 8.2.1). Then the free loop space object $\mathcal{L}\mathbf{B}G$ of the delooping $\mathbf{B}G$ (8.151), defined by the Cartesian square

$$\begin{array}{ccc} \mathcal{L}\mathbf{B}G & \longrightarrow & \mathbf{B}G \\ {\scriptstyle \rho_{\mathrm{ad}}}\big\downarrow & \text{(pb)} & \big\downarrow{\scriptstyle \Delta} \\ \mathbf{B}G & \xrightarrow[\Delta]{} & \mathbf{B}G \times \mathbf{B}G \end{array} \tag{8.166}$$

sits in a homotopy fiber sequence of the form

$$\begin{array}{ccc} G & \xrightarrow{\mathrm{fib}(\rho_{\mathrm{ad}})} & \mathcal{L}\mathbf{B}G \\ & & \big\downarrow{\scriptstyle \rho_{\mathrm{ad}}} \\ & & \mathbf{B}G\,. \end{array} \tag{8.167}$$

By comparison with (8.160), this exhibits an action of G on itself. This is the *adjoint action* with $G /\!\!/_{\mathrm{ad}} G \;\simeq\; \mathcal{L}\mathbf{B}G$.

Definition 8.2.10 (Equivariant maps)**.** By the functoriality/universality of the homotopy fiber construction in (8.160) and using the equivalence (8.162), we have the ∞-functor that assigns the underlying objects of the G-actions in Def. 8.2.6:

$$G\mathrm{Act}(\mathbf{H}) \simeq \mathbf{H}_{/\mathbf{B}G} \xrightarrow{\mathrm{fib}} \mathbf{H}\,. \tag{8.168}$$

With two G-actions (X_i, ρ_i) given, we say that a morphism $X_1 \to X_2 \;\in \mathbf{H}$ between their underlying objects *is equivariant* if it lifts through this functor, hence if it is the image of a morphism $(X_1, \rho_1) \to (X_2, \rho_2) \;\in \mathrm{GAct}(\mathbf{H})$.

Example 8.2.11 (Group division is equivariant under diagonal left and adjoint action)**.** Let $\mathbf{H}$ be an ∞-topos (Def. 8.1.46) and $G \in \mathrm{Grp}(\mathbf{H})$ (Prop. 8.2.1). Then the group division operation (Example 8.2.4) is equivariant (Def. 8.2.10) with respect to the diagonal left multiplication action ρ_ℓ (Example 8.2.8) on its domain and the adjoint action ρ_{ad} (Example 8.2.9) on its codomain:

$$(G, \rho_\ell) \times (G, \rho_\ell) \xrightarrow{(-)\cdot(-)^{-1}} (G, \rho_{\mathrm{ad}}) \quad \in G\mathrm{Act}(\mathbf{H})\,. \tag{8.169}$$

Proof. Observe the following pasting of Cartesian squares:

$$\begin{array}{ccccc} G \times G & \xrightarrow{(-)\cdot(-)^{-1}} & G & \longrightarrow & * \\ {\scriptstyle (-)^{-1}\cdot(-)\circ\sigma}\big\downarrow & & \big\downarrow & & \big\downarrow \\ G & \longrightarrow & \mathcal{L}\mathbf{B}G & \longrightarrow & \mathbf{B}G \\ \big\downarrow & & \big\downarrow & & \big\downarrow{\scriptstyle \Delta} \\ * & \longrightarrow & \mathbf{B}G & \xrightarrow{\Delta} & \mathbf{B}G \times \mathbf{B}G\,. \end{array} \tag{8.170}$$

The middle horizontal composite, regarded as a morphism in the slice over $\mathbf{B}G$ and hence as a morphism of G-actions (8.160), gives (8.169). □

Proposition 8.2.12 (Restricted and induced group actions)**.** *Let* $\mathbf{H}$ *be an* ∞*-topos. Then, for* $\phi : H \to G$ *a morphism in* $\mathrm{Grp}(\mathbf{H})$ *(Prop. 8.2.1), there is a triple of adjoint*

∞-functors (Def. 8.1.40) between the corresponding ∞-categories of group actions (Prop. 8.2.6)

$$
H\mathrm{Act}(\mathbf{H})
\begin{array}{c}
\xrightarrow{\text{“left-induced”}\;\mathbf{B}\phi_!} \\
\perp \\
\xleftarrow{\mathbf{B}\phi^*} \\
\perp \\
\xrightarrow[\text{“right-induced”}]{\mathbf{B}\phi_*}
\end{array}
G\mathrm{Act}(\mathbf{H})
\tag{8.171}
$$

such that $\mathbf{B}\phi^*$ preserves the object being acted on (“restricted action”).

Proof. By (8.162) in Prop. 8.2.6, an adjoint triple (Def. 8.1.40) of the form (8.171) is given by base change (Prop. 8.1.65) of homotopy quotients (8.161) along the delooped morphism $\mathbf{B}\phi$ (Prop. 8.2.1). This means that $\mathbf{B}\phi^*$ is given by sending the homotopy fiber sequence (8.160) corresponding to a G-action to the following homotopy pullback (Prop. 8.2.1):

$$
\begin{array}{ccccc}
X & \xrightarrow{\mathrm{fib}(\phi^*\rho)} & X /\!\!/ H & \longrightarrow & X /\!\!/ G \\
 & & {\scriptstyle \phi^*\rho}\downarrow & \text{(pb)} & \downarrow{\scriptstyle \rho} \\
 & & \mathbf{B}H & \xrightarrow[\mathbf{B}\phi]{} & \mathbf{B}G\,.
\end{array}
\qquad (X \xrightarrow{\mathrm{fib}(\phi)} X /\!\!/ G)
\tag{8.172}
$$

That this preserves the object X being acted on, as indicated, follows by the pasting law (Prop. 8.1.39). □

Definition 8.2.13 (Automorphism group)**.** Let $\mathbf{H}$ be an ∞-topos and $F \in \mathbf{H}$ an object. Then the *automorphism group* $\mathrm{Aut}(F) \in \mathrm{Grp}(\mathbf{H})$ of F is the looping (Prop. 8.2.1) of the (-1)-image (8.137) of the classifying map (8.81) of F:

$$
* \xrightarrow{\text{(-1)-conn.}} \mathbf{B}\mathrm{Aut}(F) \xhookrightarrow{\text{(-1)-trunc.}} \mathrm{Objects}_\kappa\,.
\qquad (* \xrightarrow[\vdash F]{} \mathrm{Objects}_\kappa)
\tag{8.173}
$$

The canonical action of this group (Prop. 8.2.6) on F is exhibited, via (8.160), by the left square of the following pasting composite of Cartesian squares:

$$
\begin{array}{ccccc}
F & \xrightarrow{\mathrm{fib}(\rho_{\mathrm{Aut}})} & F /\!\!/ \mathrm{Aut}(F) & \longrightarrow & \widehat{\mathrm{Objects}}_\kappa \\
\downarrow & \text{(pb)} & \downarrow{\scriptstyle \rho_{\mathrm{Aut}}} & \text{(pb)} & \downarrow \\
* & \longrightarrow & \mathbf{B}\mathrm{Aut}(F) & \hookrightarrow & \mathrm{Objects}_\kappa\,,
\end{array}
\qquad (* \xrightarrow[\vdash F]{} \mathrm{Objects}_\kappa)
\tag{8.174}
$$

where we use the pasting law (Prop. 8.1.39) to identify F as the homotopy fiber of ρ_{Aut}.

Proposition 8.2.14 (Automorphism group is universal)**.** *Let* $\mathbf{H}$ *be an ∞-topos, $G \in$* $\mathrm{Grp}(\mathbf{H})$ *(Prop. 8.2.1), and* $(X, \rho) \in G\mathrm{Act}(\mathbf{H})$ *(Def. 8.2.6). Then there is a group* *homomorphism from G to the automorphism group (Def. 8.2.13)*

$$G \xrightarrow{\;i_\rho\;} \mathrm{Aut}(X)\,, \tag{8.175}$$

such that the action ρ is the restricted action (Prop. 8.2.12) along i_ρ of the canonical automorphism action (8.174), *i.e., such that there is a Cartesian square of this form:*

$$\begin{array}{ccc} X /\!\!/ G & \longrightarrow & X /\!\!/ \mathrm{Aut}(X) \\ {\scriptstyle \rho}\big\downarrow & \text{(pb)} & \big\downarrow{\scriptstyle \rho_{\mathrm{Aut}}} \\ \mathbf{B}G & \underset{\mathbf{B}i_\rho}{\longrightarrow} & \mathbf{B}\mathrm{Aut}(X)\,. \end{array} \tag{8.176}$$

Proof. Let κ be a regular cardinal such that X is κ-small, and consider the following solid diagram of classifying maps (8.81) for ρ, ρ_{Aut} and for X:

$$\begin{array}{ccccccc} X & & \longrightarrow & & X /\!\!/ \mathrm{Aut}(X) & & \\ & \searrow & & \nearrow{\scriptstyle (\text{dashed})} & & \searrow & \\ \big\downarrow & & X /\!\!/ G & & \longrightarrow & & \widehat{\mathrm{Objects}_\kappa} \\ & & & & \big\downarrow & & \\ \ast & & \longrightarrow & & \mathbf{B}\mathrm{Aut}(X) & & \big\downarrow \\ & \searrow{\scriptstyle (\text{-1})\text{-connected}} & \big\downarrow & \nearrow{\scriptstyle \mathbf{B}i_\rho\,(\text{dashed})} & & \searrow{\scriptstyle (\text{-1})\text{-truncated}} & \\ & & \mathbf{B}G & & \underset{\vdash \rho}{\longrightarrow} & & \mathrm{Objects}_\kappa \end{array} \tag{8.177}$$

Here the bottom square homotopy-commutes by the essential uniqueness of the classifying map $\vdash X$ (8.81). Hence the dashed lift exists essentially uniquely (8.138) by the connected/truncated factorization system (Prop. 8.1.83). □

8.2.3 Principal bundles

Proposition 8.2.15 (Principal bundles [NSS12a, Thm. 3.17]). *Let $\mathbf{H}$ be an ∞-topos, $X \in \mathbf{H}$, and $G \in \mathrm{Grp}(\mathbf{H})$ (Prop. 8.2.1). Then G-principal ∞-bundles $P \to X$ over X are, equivalently, given by classifying maps $\vdash P : X \to \mathbf{B}G$. Forming their homotopy fibers*

$$\begin{array}{ccc} P & & \\ {\scriptstyle \mathrm{fib}(\vdash P)}\big\downarrow & & \\ X & \underset{\vdash P}{\longrightarrow} & \mathbf{B}G \end{array} \tag{8.178}$$

constitutes an equivalence of ∞-groupoids:

$$\begin{array}{ccc} G\mathrm{Bundles}_X(\mathbf{H}) & \xleftarrow[\simeq]{\;\;\mathrm{fib}\;\;} & \mathbf{H}(X, \mathbf{B}G)\,. \\ P & \longmapsto & \vdash P \end{array} \tag{8.179}$$

Remark 8.2.16 (Principal base spaces are homotopy quotients). Comparison of the abstract characterization of **(i)** group actions (Prop. 8.2.6) and **(ii)** principal bundles (Prop. 8.2.15) reveals that these are about one and the same abstract concept, just viewed from two different perspectives: In an ∞-topos, every G-principal bundle is a G-action whose homotopy quotient is the given base space; and, conversely, every G-action is that of a principal bundle over its homotopy quotient:

$$\begin{array}{ccccc} \text{Principal G-bundle} & P \circlearrowright G & & \text{G-action} \\ & \downarrow & & \\ \text{Base space} & X \simeq & P /\!\!/ G & \text{Homotopy quotient} \end{array} \tag{8.180}$$

Notice (see [NSS12a, 3.1] for exposition) that it is the higher geometry inside an ∞-topos that makes this work.

Definition 8.2.17 (Atiyah groupoid)**.** Let $\mathbf{H}$ be an ∞-topos (Def. 8.1.46), $X \in \mathbf{H}$, $G \in \mathrm{Grp}(\mathbf{H})$ (Prop. 8.2.1), and $P \in G\mathrm{Bundles}_X$ (Prop. 8.2.15). Then the *Atiyah groupoid* of P is the groupoid $\mathrm{At}_\bullet(P) \in \mathrm{Grpd}(\mathbf{H})$ (Def. 8.1.86) whose corresponding stack with atlas (via Prop. 9.1.38) is the (-1)-image projection (Example 8.1.85) of the bundle's classifying map $\vdash P$ (8.179):

$$X \twoheadrightarrow \mathcal{A}t(P) \hookrightarrow \mathbf{B}G\,, \qquad X \xrightarrow{\vdash P} \mathbf{B}G \tag{8.181}$$

8.2.4 Fiber bundles

Definition 8.2.18 (Fiber bundle)**.** Let $\mathbf{H}$ be an ∞-topos (Def. 8.1.46).

(i) A morphism $Y \xrightarrow{p} X$ in $\mathbf{H}$ is a *fiber bundle* with *typical fiber* $F \in \mathbf{H}$ if there exists an effective epimorphism $U \overset{i}{\twoheadrightarrow} X$ (Def. 8.1.80) and a Cartesian square (Ntn. 8.1.37) of the form

$$\begin{array}{ccc} U \times F & \longrightarrow & Y \\ \downarrow & \text{(pb)} & \downarrow p \\ U & \underset{i}{\twoheadrightarrow} & X\,. \end{array} \tag{8.182}$$

(ii) We write

$$F\mathrm{FiberBundles}_X(\mathbf{H}) \subset \mathrm{Core}\big(\mathbf{H}_{/X}\big) \in \mathrm{Grpd}_\infty \tag{8.183}$$

for the full ∞-groupoid of the core (8.1) of the slice $\mathbf{H}_{/X}$ over X (Prop. 8.1.62) on the F-fiber bundles.

Proposition 8.2.19 (Classification of fiber bundles [NSS12a, Prop. 4.10])**.** *Let $\mathbf{H}$ be an ∞-topos (Def. 8.1.46) and $X, F \in \mathbf{H}$. Then fiber bundles over X (Def. 8.2.18) with typical fiber F are equivalent to morphisms $X \longrightarrow \mathbf{B}\mathrm{Aut}(F)$ from X to the delooping (Prop. 8.2.1) of the automorphism group (Def. 8.2.13) of F:*

$$\begin{array}{ccc} F\mathrm{FiberBundles}_X(\mathbf{H}) & \xrightarrow{\simeq} & \mathbf{H}\big(X\,,\,\mathbf{B}\mathrm{Aut}(F)\big)\,. \\ E & \longmapsto & \vdash E \end{array} \tag{8.184}$$

Proof. Let κ be a regular cardinal such that F is κ-small. Then, by assumption, we have the following solid diagram of classifying maps (8.81):

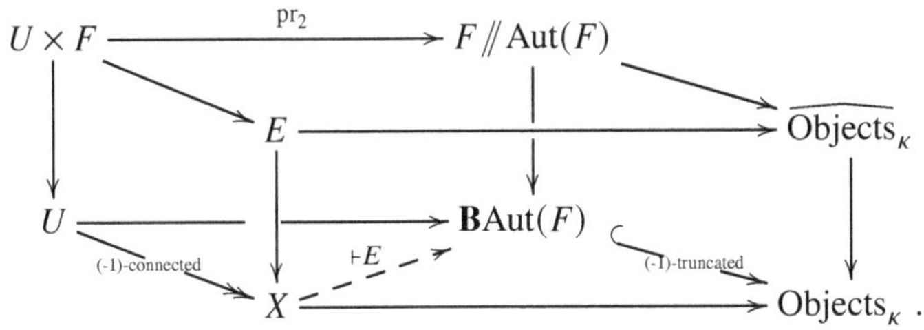

(8.185)

Now the (-1)-connected/(-1)-truncated factorization system (Prop. 8.1.83) implies that the dashed morphism exists essentially uniquely (8.138).

It just remains to see that this assignment is independent of the choice of U: For $U' \twoheadrightarrow X$ any other effective epimorphism with $(\vdash E)'$ the associated classifying map as above, observe that the fiber product $U \times_X U' \twoheadrightarrow X$ is again an effective epimorphism, since the class of effective epimorphisms is closed under pullbacks as well as under composition (Lemma 8.1.82). Therefore, $\vdash E$ and $(\vdash E)'$ are jointly lifts in a diagram as above but with $U \times_X U'$ in the top left. Hence, by the essential uniqueness of lifts in the connected/truncated orthogonal factorization system, they are equivalent, $(\vdash E) \simeq (\vdash E)'$, in an essentially unique way. □

Notation 8.2.20 (Associated bundles). We say that

(i) the morphism $\vdash E$ in (8.184) is the *classifying map* of E and

(ii) that E is *associated* to the $\mathrm{Aut}(F)$-principal bundle which is classified by $\vdash E$ according to Prop. 8.2.15.

Remark 8.2.21 (Twisted cohomology in slice ∞-toposes). Prop. 8.2.19 implies (together with the universal property of the pullback) that sections σ of A-fiber bundles E over some X are, equivalently, lifts c of the classifying map $c := \vdash E$ (8.184) through ρ_{Aut} (8.174):

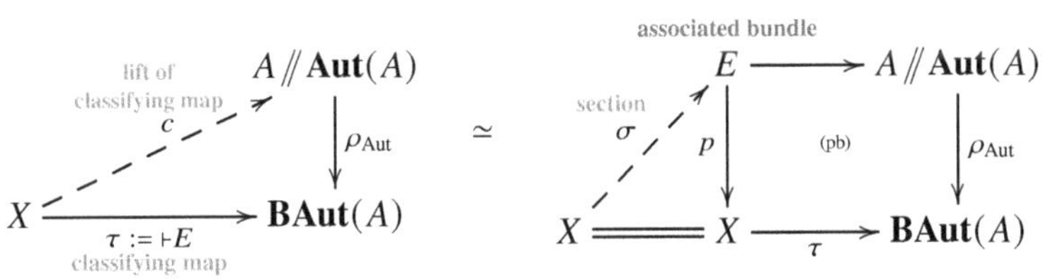

(8.186)

(i) If A is regarded here as a coefficient object for A-cohomology (1.21), then such a section σ is a locally A-valued cocycle, which is "twisted" over X according to the classifying map τ. Hence such a σ is a cocycle in (non-abelian) τ-*twisted cohomology* [NSS12a, 4.2]. But the left-hand side of (8.186) is, equivalently, a morphism (8.106) in the slice ∞-topos (Prop. 8.1.62) $\mathbf{H}_{/\mathbf{BAut}(A)}$. It follows that *twisted cohomology is the intrinsic cohomology* (1.21) *of slice ∞-toposes*:

$$
\underset{\substack{\tau\text{-twisted}\\ \text{cohomology}}}{H^{\tau}\big(X\,,\,A\big)} \;:=\; \pi_0\,\mathbf{H}_{/\mathbf{BAut}(A)}\Big(\big(X,\tau\big)\,,\,\big(A /\!\!/ \mathbf{Aut}(A), \rho_{\mathrm{Aut}}\big)\Big)
\;\simeq\; \left\{
\begin{array}{ccc}
X & \overset{\text{cocycle}}{\underset{c}{\dashrightarrow}} & A /\!\!/ \mathbf{Aut}(A) \\
& \Longleftarrow & \\
{\scriptstyle \tau}\searrow & & \swarrow{\scriptstyle \rho_{\mathrm{Aut}}} \\
& \mathbf{BAut}(A) &
\end{array}
\right\}_{/\sim}
\tag{8.187}
$$

(ii) By the universality of $\mathbf{Aut}(A)$ (Prop. 8.2.14), this holds for slicing over *any* pointed connected object $\mathbf{B}G$ (8.151).

(iii) If the base object is not connected, the intrinsic cohomology of its slice may be thought of as a mixture of twisted and parametrized cohomology. We encounter an example of this in Def. 11.2.1.

Remark 8.2.22 (Twisted cohomology as global sections)**.** The ∞-groupoid of sections of the associated bundle $E := \tau^*(A /\!\!/ G) \xrightarrow{p} X$ in (8.186), is equivalently its image $\Gamma_X(E)$ under the base geometric morphism (Prop. 8.1.59)

$$
\mathbf{H}_{/X} \underset{\Gamma_X}{\overset{\Delta_X}{\underset{\longrightarrow}{\overset{\longleftarrow}{\bot}}}} \mathrm{Grpd}_{\infty}
\tag{8.188}
$$

of the slice ∞-topos $\mathbf{H}_X$ (Prop. 8.1.62), in that (by Prop. 8.1.50) $\Gamma_X(E) \simeq \mathbf{H}_X\big(\mathrm{id}_X\,,\,p\big)$. Hence the τ-twisted cohomology (8.187) of X is equivalently the set of connected components of the ∞-groupoid of global sections:

$$
\mathbf{H}^{\tau}\big(X;\,A\big) \;\simeq\; \pi_0\,\Gamma_X\big(\tau^*(A /\!\!/ G)\big)\,.
\tag{8.189}
$$

Remark 8.2.23 (Twisted abelian cohomology in tangent ∞-toposes)**.** Let $\mathbf{H}$ be an ∞-topos (Def. 8.1.46).

(i) Notice that the intrinsic cohomology (1.21) of Bundles($\mathbf{H}$) (Example 8.1.66) is still twisted cohomology as in Remark 8.2.21, just up to a change in perspective: now the twisting τ is encoded not in the domain object, but in the cocycles on these (a morphism of the form $\mathrm{id}_X \to \rho_{\mathrm{Aut}}$ in Bundles($\mathbf{H}$) is still manifestly given by the diagrams in (8.186)).

(ii) Therefore, similarly, the intrinsic cohomology (1.21) in the tangent ∞-topos SpectralBundles($\mathbf{H}$) (Example 8.1.67) is twisted cohomology with local coefficients being spectra [Sc13, 4.1][ABGHR14][GS19a][GS19b], hence is *twisted abelian cohomology*.

(iii) In the case that $\mathbf{H} = \mathrm{Grpd}_{\infty}$, the base tangent ∞-topos

$$
T\mathrm{Grpd}_{\infty} \;=\; \mathrm{SpectralBundles}\big(\mathrm{Grpd}_{\infty}\big)
\tag{8.190}
$$

is the topic of traditional parametrized stable homotopy theory [Jam95][MSi06] [ABGHR14, §2][BM19], and its intrinsic cohomology theory (1.21) is traditional twisted generalized cohomology [Do05][ABG10].

8.2.5 Fixed points and fixed loci

Definition 8.2.24 (Fixed points and fixed loci)**.** Let $\mathbf{H}$ be an ∞-topos, $G \in \mathrm{Grp}(\mathbf{H})$ (Prop. 8.2.1) and $(X,\rho) \in G\mathrm{Act}(\mathbf{H})$ (Prop. 8.2.6).

(i) A *fixed point* of (X,ρ) is an element $* \xrightarrow{x} X$ induced from a section $x /\!\!/ G$ of ρ in (8.160), as shown on the right (where we are using the pasting law, Prop. 8.1.39, and Example 8.1.38 to identify the top square as Cartesian).

$$\begin{array}{ccc} * & \longrightarrow & \mathbf{B}G \\ {\scriptstyle x}\downarrow & \text{(pb)} & \downarrow{\scriptstyle x/\!\!/G} \\ X & \xrightarrow{\mathrm{fib}(\rho)} & X /\!\!/ G \\ \downarrow & \text{(pb)} & \downarrow{\scriptstyle \rho} \\ * & \longrightarrow & \mathbf{B}G\,. \end{array} \qquad (\text{with } \mathbf{B}G = \mathbf{B}G \text{ on the right}) \tag{8.191}$$

(ii) The *G-fixed locus* of (X,ρ) is the object

$$X^G \;:=\; \mathbf{B}(G \to *)_*\big((X,\rho)\big) \quad\in\quad 1\mathrm{Action}(\mathbf{H}) \;\simeq\; \mathbf{H}\,, \tag{8.192}$$

that is right induced (Prop. 8.2.12) along the unique morphism to the trivial group.

Example 8.2.25 (Global points of fixed loci are homotopy fixed points)**.** The global points of a homotopy-fixed locus X^G (8.192) are indeed, equivalently, the fixed points (8.191). By the adjunction (8.171), we have the hom-equivalence (8.71)

$$\big(* \longrightarrow X^G = \mathbf{B}(G \to 1)_*(X,\rho)\big) \quad\longleftrightarrow\quad \big(\mathbf{B}(G \to 1)^*(*) \longrightarrow (X,\rho)\big) \tag{8.193}$$

and, by Prop. 8.2.6, the latter morphisms are equivalent to homotopy-commuting diagrams of the form

$$\begin{array}{ccc} \mathbf{B}G & \xrightarrow{\quad x/\!\!/G \quad} & X /\!\!/ G\,. \\ & {\scriptstyle \mathbf{B}(G\to 1)^*(*)}\searrow \;\simeq\; \swarrow{\scriptstyle \rho} & \\ & \mathbf{B}G & \end{array} \tag{8.194}$$

This is just the type of diagram charactcrizing homotopy fixed points, as seen vertically on the right in (8.191).

Example 8.2.26 (Fixed loci in ∞-groupoids)**.** Consider $\mathbf{H} := \mathrm{Grpd}_\infty$, $G \in \mathrm{Grp}(\mathrm{Grpd}_\infty)$ and $(X,\rho) \in G\mathrm{Act}\big(\mathrm{Grpd}_\infty\big)$. Then the G-fixed locus (Def. 8.2.24) is given (due to Prop. 8.1.70) by

$$X^G \;\simeq\; \mathbf{H}_{/*/\!\!/G}\big(*/\!\!/G\,,\, X/\!\!/G\big) \quad \in \mathrm{Grpd}_\infty\,. \tag{8.195}$$

Definition 8.2.27 (Pointed-automorphism group)**.** Let $\mathbf{H}$ be an ∞-topos and $\big(X^{\swarrow *}\big) \in \mathbf{H}^{*/} \hookrightarrow \mathbf{H}^{\Delta[1]}$ a pointed object in $\mathbf{H}$, equivalently regarded as an object in the ∞-topos $\mathbf{H}^{\Delta[1]}$ of ∞-functors from $\Delta[1] := \{0 \to 1\}$ to $\mathbf{H}$. Noticing the evaluation functors (8.93)

$$\mathbf{H}^{\Delta[1]} \;\; \begin{array}{c} \xrightarrow{\;\mathrm{ev}_1\;} \\ {\scriptstyle \top} \\ \xleftarrow{\;\mathrm{const}\;} \\ {\scriptstyle \bot} \\ \xrightarrow{\;\mathrm{ev}_0\;} \end{array} \;\; \mathbf{H} \tag{8.196}$$

and that these preserve all ∞-limits and ∞-colimits (by Prop. 8.1.55), hence all group objects and their deloopings (by (8.160) in Prop. 8.2.6), we say that the *pointed-automorphism group* of X is the image under ev_0 of its automorphism group, according to Def. 8.2.13, formed in the arrow ∞-topos $\mathbf{H}^{\Delta[1]}$:

$$\mathrm{Aut}_*(X) := \mathrm{ev}_0\left(\mathbf{Aut}\big(X^{\nearrow^{*}}\big)\right) \in \mathrm{Grp}(\mathbf{H}). \tag{8.197}$$

This pointed-automorphism group comes with a canonical pointed action on X as follows: From the defining factorization (8.173)

$$\mathrm{const}(*) \twoheadrightarrow \mathbf{B}\mathrm{Aut}\big(X^{\nearrow^{*}}\big) \hookrightarrow \mathrm{Objects}_\kappa \tag{8.198}$$

(where now $\mathrm{Objects}_\kappa$ denotes the κ-small object classifier (8.81) of $\mathbf{H}^{\Delta}[1]$), and using again that the evaluation functors (8.196) preserves ∞-limits, hence in particular homotopy pullbacks, it follows that the front and rear faces of the following diagram are Cartesian (Ntn. 8.1.37)

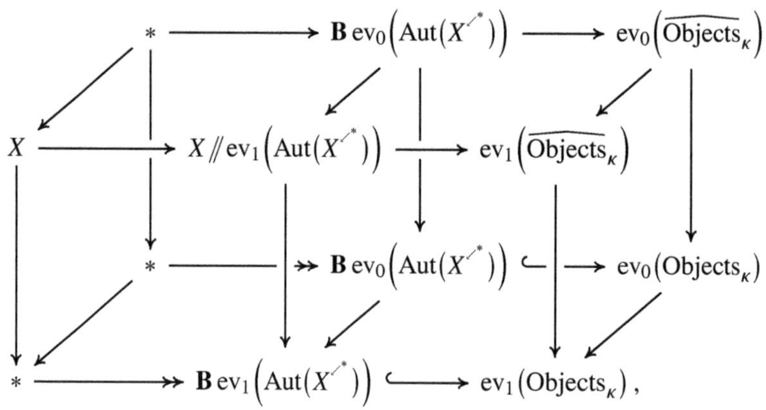

(8.199)

so that pullback along the bottom diagonal morphisms shows that the pointed automorphism ∞-group (8.197) sits in a diagram in $\mathbf{H}$ of the following form:

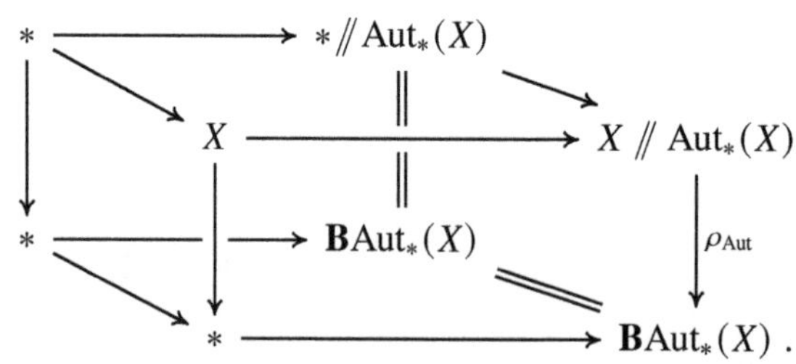

(8.200)

Here the Cartesian front face exhibits the action of the pointed-automorphism group of X on X, and the Cartesian rear face exhibits its trivial action on the base point. With this and noticing that also the bottom face is Cartesian (by Example 8.1.38), the pasting law (Prop. 8.1.39) implies that also the top square is Cartesian, exhibiting the given base point as a homotopy fixed point (Def. 8.2.24) of the pointed-automorphism action.

Definition 8.2.28 (Group-automorphism group)**.** Let $\mathbf{H}$ be an ∞-topos and $G \in \mathrm{Grp}(\mathbf{H})$ (Prop. 8.2.1). Then the group of group-automorphisms of G is the group of pointed-automorphisms (Def. 8.2.27) of its delooping $\mathbf{B}G$ (8.151):

$$\mathrm{Aut}_{\mathrm{Grp}}(G) := \mathrm{Aut}_*(\mathbf{B}G) \quad \in \mathrm{Grp}(\mathbf{H}). \tag{8.201}$$

Proposition 8.2.29 (Canonical action of group-automorphism group)**.** *Let* $\mathbf{H}$ *be an* ∞*-topos and* $G \in \mathrm{Grp}(\mathbf{H})$ *(Prop. 8.2.1). The group-automorphism group of* G *(Def. 8.2.28) has a canonical action (Prop. 8.2.6)*

$$\big(G, \rho_{\mathrm{Aut}_{\mathrm{Grp}}}\big) \in \mathrm{Aut}_{\mathrm{Grp}}(G)\mathrm{Act}(\mathbf{H}) \tag{8.202}$$

on the underlying object $G \in \mathbf{H}$*, which is such that*

(i) *The neutral element* $* \xrightarrow{e} G$ *(Example 8.2.3) is a fixed point of the action (Def. 8.2.24).*

(ii) *The homotopy quotient* $G \sslash \mathrm{Aut}_{\mathrm{Grp}}(G)$ *carries the structure of a group object* (8.151) *in the slice* (8.162)

$$G \sslash \mathrm{Aut}_{\mathrm{Grp}}(G) \in \mathrm{Grp}\big(\mathbf{H}_{/\mathbf{B}\mathrm{Aut}_{\mathrm{Grp}}}\big), \tag{8.203}$$

whose delooping (8.151) *is the homotopy quotient of the defining action* (8.201) *on the delooping* $\mathbf{B}G$ *of* G*:*

$$\mathbf{B}\big(G \sslash \mathrm{Aut}_{\mathrm{Grp}}(G)\big) \simeq (\mathbf{B}G) \sslash \mathrm{Aut}_{\mathrm{Grp}}(G). \tag{8.204}$$

Proof. First consider item **(ii)**: Write $G \sslash \mathrm{Aut}_{\mathrm{Grp}}(G)$ for the homotopy fiber product in the following pullback square

$$\begin{array}{ccc} G \sslash \mathrm{Aut}_{\mathrm{Grp}}(G) & \longrightarrow & * \sslash \mathrm{Aut}_{\mathrm{Grp}}(G) \\ \big\downarrow & \text{(pb)} & \big\downarrow \\ * \sslash \mathrm{Aut}_{\mathrm{Grp}}(G) & \longrightarrow & (\mathbf{B}G) \sslash \mathrm{Aut}_{\mathrm{Grp}}(G). \end{array} \tag{8.205}$$

Since this is the looping (8.152) in the slice (Prop. 8.1.62):

$$G \sslash \mathrm{Aut}_{\mathrm{Grp}}(G) \simeq \Omega\big((\mathbf{B}G) \sslash \mathrm{Aut}_{\mathrm{Grp}}(G)\big) \quad \in \mathbf{H}_{/\mathbf{B}\mathrm{Aut}_{\mathrm{Grp}}(G)}, \tag{8.206}$$

the looping/delooping equivalence (8.151) implies the claim (8.204) as soon as we show (in view of Prop. 8.2.6) that the homotopy fiber of the left morphism in (8.205) is indeed G, in that it makes the total solid rear rectangle of the following diagram be

Cartesian:

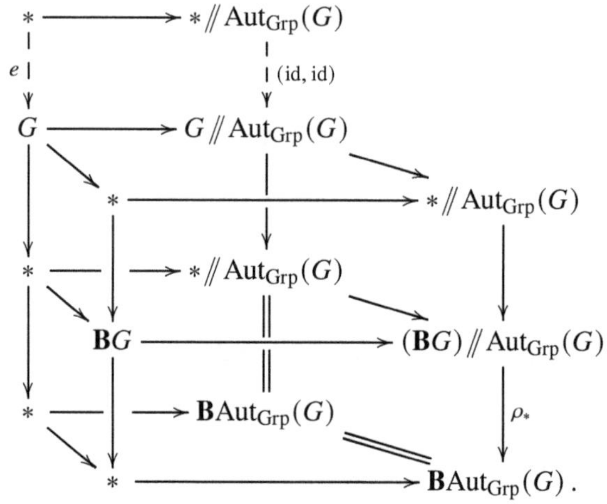

(8.207)

Here:

- the bottom part is the diagram (8.200) (for $X = \mathbf{B}G$) which exhibits the pointed-automorphism action on $\mathbf{B}G$;
- the top front square is Cartesian and exhibits the base point being a homotopy-fixed point, as in (8.200);
- the top left square is Cartesian and exhibits the looping/delooping relation (8.152);
- the top right square is (8.205) and hence Cartesian by definition.

Hence, the pasting law (Prop. 8.1.39) implies that also the solid top rear square is Cartesian.

Finally, to see item **(i)**: Observe that there is the dashed morphism shown in the top right of (8.207), this being the diagonal morphism induced from the Cartesian property of the top right square, by the above. This means, by construction, that the total vertical morphism on the right is an equivalence. Now define the dashed top square to be a pullback square. Then, by the pasting law (Prop. 8.1.39), the pullback object in the top left of the dashed square is equivalently the pullback of the total rear diagram, hence the pullback of an equivalence to a point, hence it is itself equivalent to the point, as shown. Since the point is terminal, the top left dashed morphism is thus a cone over the Cartesian square on the top left. By the universal property of the homotopy fiber product, this means that the top left dashed morphism must be the neutral element (Example 8.2.3). The top dashed square hence exhibits this as a homotopy fixed point. □

Proposition 8.2.30 (Group division is equivariant under group-automorphisms)**.** *Let* $\mathbf{H}$ *be an* ∞*-topos and* $G \in \mathrm{Grp}(\mathbf{H})$ *(Prop. 8.2.1). Then the group division morphism* $G \times G \xrightarrow{(-)\cdot(-)^{-1}} G$ *(Example 8.2.4) is equivariant (Def. 8.2.10) with respect to the canonical group-automorphism action (Prop. 8.2.29) of the group-automorphism* *group* $\mathrm{Aut}_{\mathrm{Grp}}(G)$ *(Def. 8.2.28) acting on all three copies of* G*:*

$$(G, \rho_{\mathrm{Aut}_{\mathrm{Grp}}}) \times (G, \rho_{\mathrm{Aut}_{\mathrm{Grp}}}) \xrightarrow{(-)\cdot(-)^{-1}} (G, \rho_{\mathrm{Aut}_{\mathrm{Grp}}}) \quad \in \mathrm{Aut}_{\mathrm{Grp}}(G)\mathrm{Act}(\mathbf{H}) \,. \tag{8.208}$$

Proof. By (8.157) the group division morphism is a universal morphism induced from pasting of copies of the looping square (8.152). Thus the claim follows by Prop. 8.2.29. □

9

Singular Geometry

Here, we establish "synthetic" axiomatic foundations of a geometric homotopy theory of orbifolds, unifying:

(i) §9.1 – the cohesive geometric homotopy theory due to [SSS12][Sc19], which reflects the *geometric aspects* of orbifolds;

(ii) §9.2 – the cohesive global-equivariant homotopy theory due to [Re14], understood as reflecting the *singular aspects* of orbifolds, as in Figure D.

This is to provide, in §10, a general abstract theory of geometric aspects of orbi-singular spaces and of étale ∞-stacks.

The general idea of such *synthetic modal higher geometry* is that, first, the traditional geometric objects of interest (here: orbifolds) are regarded within ambient ∞-toposes where all universal constructions on these objects, that one may wish to consider, are guaranteed to exist (in contrast to more traditional categories of these geometric objects). But then, second, one needs novel formulations of traditional geometric notions to apply in the larger generality of the objects of these ∞-topoi – and these definitions are to be formulated in terms of the modal operators (here: singular-cohesive modalities) on the ∞-topos.

9.1 Geometry

We discuss in turn:

§9.1.1: *Synthetic Differential Topology via Cohesive Modality*

§9.1.2: *Synthetic Differential Geometry via Elastic Modality*

§9.1.3: *Synthetic Supergeometry via Solid Modality*

For more introductory lecture notes along these lines see also [Sc18].

9.1.1 Differential Topology

We present a formulation of differential topology internal to ∞-toposes which we call *cohesive* [Sc13]. In ∞-categorical generalization of [La94][La07], this involves an

DOI: 10.1201/9781003675969-9

abstract *shape* operation $\int$ that relates higher geometric spaces to their bare underlying homotopy type.

Definition 9.1.1 (Cohesive ∞-topos)**.** **(i)** An ∞-topos $\mathbf{H}$ (Def. 8.1.46) is called *cohesive* if its base geometric morphism (Prop. 8.1.59), to be denoted $\mathrm{Pnts} : \mathbf{H} \twoheadrightarrow \mathrm{Grpd}_\infty$, is part of an adjoint quadruple of ∞-functors (Def. 8.1.40)

$$
\begin{array}{cccccc}
\text{"shape"} & & \xleftarrow{\ } \mathrm{Shp} \longrightarrow & & \\
& & \bot & & \\
\text{"discrete"} & & \longleftarrow \mathrm{Disc} \hookrightarrow & & \\
& & \bot & & \\
\text{"points"} & \mathbf{H} & \longrightarrow \mathrm{Pnts} \longrightarrow & \mathbf{B} & \\
& & \bot & & \\
\text{"chaotic"} & & \longleftarrow \mathrm{Chtc} \hookrightarrow & & \\
& \substack{\text{cohesive} \\ \infty\text{-topos}} & & \substack{\text{discrete} \\ \text{sub-topos}} &
\end{array}
\tag{9.1}
$$

such that **(a)** Disc and Chtc are fully faithful (Def. 8.1.1), and **(b)** such that Shp preserves finite products.

(ii) We write

$$
\begin{array}{c}
\big(\int := \mathrm{Disc} \circ \mathrm{Shp}\big) \\
\text{"shape"} \\
\bot \\
\big(\flat := \mathrm{Disc} \circ \mathrm{Pnts}\big) \\
\text{"discrete"} \\
\bot \\
\big(\sharp := \mathrm{Chtc} \circ \mathrm{Pnts}\big) \\
\text{"continuous"}
\end{array}
\quad : \quad \mathbf{H} \to \mathbf{H}
\tag{9.2}
$$

for the induced adjoint triple (Def. 8.1.40) of modalities (1.18) (*cohesive modalities*).

The following direct consequence may serve to illustrate how these axioms are put to work:

Proposition 9.1.2 (Composite cohesive modalities)**.** *The cohesive modalities (Def. 9.1.1) satisfy:*

$$\textstyle\int \circ \flat \;\simeq\; \flat \qquad \textit{and} \qquad \flat \circ \sharp \;\simeq\; \flat\,. \tag{9.3}$$

Proof. That Disc and Chtc in (9.1) are fully faithful means, equivalently (Prop. 8.1.44), that the co-unit morphisms (8.73)

$$\mathrm{Shp} \circ \mathrm{Disc} \xrightarrow{\;\simeq\;} \mathrm{id}\;, \qquad \mathrm{Pnts} \circ \mathrm{Chtc} \xrightarrow{\;\simeq\;} \mathrm{id} \tag{9.4}$$

are natural equivalences. Hence the image under $\mathrm{Disc} \circ (-) \circ \mathrm{Pnts}$ of the first of these is a natural equivalence of the form

$$\textstyle\int \circ \flat \;=\; \mathrm{Disc} \circ \mathrm{Shp} \circ \mathrm{Disc} \circ \mathrm{Pnts} \xrightarrow{\;\simeq\;} \mathrm{Disc} \circ \mathrm{Pnts} \;=\; \flat\,. \tag{9.5}$$

while the image of the second is of the form

$$\flat \circ \sharp \;=\; \mathrm{Disc} \circ \mathrm{Pnts} \circ \mathrm{Chtc} \circ \mathrm{Pnts} \xrightarrow{\;\simeq\;} \mathrm{Disc} \circ \mathrm{Pnts} \;=\; \flat\,. \tag{9.6}$$

□

Lemma 9.1.3 (Only the empty object has empty shape)**.** *Let* $\mathbf{H}$ *be a cohesive ∞-topos (Def. 9.1.1). Then* $X \in \mathbf{H}$ *is empty, i.e., equivalent to the initial object* $\varnothing$ (8.86), *precisely if its shape* (9.2) *is empty:*

$$X \;\simeq\; \varnothing \qquad \Leftrightarrow \qquad \textstyle\int X \;\simeq\; \varnothing\,. \tag{9.7}$$

Proof. In one direction, assume that $X \simeq \varnothing$. Noticing that $\varnothing$ is the initial colimit and that colimits are preserved by $\int$, this being a left adjoint (Prop. 8.1.42), it follows that $\int(\varnothing) \simeq \varnothing$.

In the other direction, assume that the shape of X is empty. Then the shape unit (8.72) is a morphism of the form

$$X \xrightarrow{\eta^{\int}_X} \int X \simeq \varnothing \tag{9.8}$$

and thus $X \simeq \varnothing$ follows as in (8.87), by universality of colimits (Ex. 8.1.49). □

9.1.1.1 Cohesive ∞-group actions

The condition that Shp preserves finite products implies the following properties.

Proposition 9.1.4 (Shape preserves groups, actions, and their homotopy quotients)**.** *Let* $\mathbf{H}$ *be a cohesive* ∞*-topos (Def. 9.1.1),* $G \in \mathrm{Grp}(\mathbf{H})$ (8.151) *and and* $(X, \rho) \in G\mathrm{Actions}(\mathbf{H})$ *(Prop. 8.2.6).*

(i) *Then the shape* $\int X$ (9.2) *of* X *is equipped with an induced* $\int G$*-action, such that the shape of the homotopy quotient* (8.161) *is the homotopy quotient of the shapes. The analogous statement holds for* $\flat$ (9.2)*:*

$$\int (X /\!\!/ G) \simeq (\int X) /\!\!/ (\int G) \quad \text{and} \quad \flat (X /\!\!/ G) \simeq (\flat X) /\!\!/ (\flat G) \,. \tag{9.9}$$

(ii) *In particular, both* $\int$ *and* $\flat$ *preserve group objects and their deloopings (Prop. 8.2.1):*

$$\int \mathbf{B} G \simeq \mathbf{B} \int G \quad \text{and} \quad \flat \mathbf{B} G \simeq \mathbf{B} \flat G \,. \tag{9.10}$$

Proof. The homotopy quotient of X by G is, equivalently, a colimit over a simplicial diagram of finite Cartesian products of copies of X and G (8.161). Hence the statement follows for every ∞-functor that commutes with simplicial colimits and with finite products. But, since $\int$ is a left adjoint, it commutes with all colimits (Prop. 8.1.42) and also with finite products, by assumption on Shp and since Disc is a right adjoint. Similarly, $\flat$ is both left and right adjoint, and hence preserves all colimits and all limits (again Prop. 8.1.42). That preservation of homotopy quotients implies preservation of ∞-groups follows by the delooping theorem (Prop. 8.2.1). □

Lemma 9.1.5 (Cohesive shape preserves some homotopy fiber products)**.** *In a cohesive* ∞*-topos* $\mathbf{H}$ *(Def. 9.1.1), the shape functor* Shp (9.1) *preserves homotopy fiber products over cohesively discrete objects. That is, for* $B \in \mathbf{B} \xhookrightarrow{\mathrm{Disc}} \mathbf{H}$ *and* $X, Y \in \mathbf{H}_{/B}$ *we have a natural equivalence*

$$\mathrm{Shp}(X \underset{B}{\times} Y) \simeq \mathrm{Shp}(X) \underset{B}{\times} \mathrm{Shp}(Y) \,. \tag{9.11}$$

Proof. This is proven in [Sc13, Thm. 3.8.19] under the assumption that $\mathbf{H}$ admits an ∞-cohesive site of definition. This assumption was shown to be unnecessary in [BP22, Lemma 3.10]. □

Lemma 9.1.6 (Shape of $\eta^{\int}$-induced action)**.** *Let* $\mathbf{H}$ *be a cohesive* ∞*-topos (Def. 9.1.1),* $G \in \mathrm{Grp}(\mathbf{H})$ *(Prop. 8.2.1) and* $(X, \rho) \in G\mathrm{Actions}(\mathbf{H})$ *(Prop. 8.2.6).*

(i) *The left-induced action (Prop. 8.2.12)*

$$\big(\widetilde{X},\widetilde{\rho}\big) \;:=\; \mathbf{B}\big(\eta^{\int}_G\big)_!\,(X,\rho) \;\in\; \big(\textstyle\int G\big)\mathrm{Actions}(\mathbf{H}) \tag{9.12}$$

along the shape unit morphism (8.72) $G \xrightarrow{\eta^{\int}_G} \int G$ *acts on an object whose shape* (9.2) *is that of* X*:*

$$\textstyle\int \widetilde{X} \;\simeq\; \int X\,, \tag{9.13}$$

whence

$$\big(\textstyle\int X\;,\;\int \rho\big) \;\in\; \big(\textstyle\int G\big)\mathrm{Actions}(\mathbf{H})\,. \tag{9.14}$$

(ii) *Similarly, the restricted-induced action (Prop. 8.2.12)*

$$\big(\widetilde{X},\widetilde{\rho}\big) \;:=\; \mathbf{B}\big(\textstyle\int\epsilon^{\flat}\big)^{*}\circ\mathbf{B}\big(\eta^{\int}_G\big)_!\,(X,\rho) \;\in\; (\flat G)\mathrm{Actions}(\mathbf{H}) \tag{9.15}$$

along the pair of group homomorphisms (using Prop. 9.1.4)

$$G \xrightarrow{\;\eta^{\int}_G\;} \textstyle\int G \xleftarrow{\;\int\epsilon^{\flat}_G\;} \flat G \tag{9.16}$$

acts on an object whose shape (9.2) *is that of* X*:*

$$\textstyle\int \widetilde{X} \;\simeq\; \int X\,. \tag{9.17}$$

Proof. By Prop. 8.2.6 and Prop. 8.2.12, the object $\widetilde{X}$ sits in a diagram of Cartesian squares (Notation 8.1.37) as shown on the left in the following (the full square in case **(i)**, the pasting decomposition for case **(ii)**):

$$\begin{array}{ccccc}
\widetilde{X} & \longrightarrow & \widetilde{X}/\!\!/\flat G & \longrightarrow & X/\!\!/G \\
& & & & \big\downarrow{\scriptstyle \rho} \\
\big\downarrow & \text{(pb)} & \big\downarrow & \text{(pb)} & \mathbf{B}G \\
& & & & \big\downarrow{\scriptstyle \mathbf{B}\eta^{\int}_G} \\
* & \longrightarrow & \mathbf{B}\flat G & \xrightarrow{\mathbf{B}\int\epsilon^{\flat}_G} & \mathbf{B}\textstyle\int G
\end{array}
\quad\overset{\textstyle\int}{\longmapsto}\quad
\begin{array}{ccccc}
\int\widetilde{X} & \longrightarrow & (\int\widetilde{X})/\!\!/(\flat G) & \longrightarrow & (\int X)/\!\!/(\int G) \\
\big\downarrow & \text{(pb)} & \big\downarrow & \text{(pb)} & \big\downarrow{\scriptstyle \int\rho} \\
* & \longrightarrow & \mathbf{B}\flat G & \xrightarrow{\mathbf{B}\int\epsilon^{\flat}_G} & \mathbf{B}\textstyle\int G
\end{array} \tag{9.18}$$

But, since the objects in the bottom row $\mathbf{B}\int G \simeq \int \mathbf{B}G$ and $\mathbf{B}\flat G \simeq \flat\,\mathbf{B}G$ (equivalences by Prop. 9.1.4) are both cohesively discrete, Lemma 9.1.5 says that the image of these squares under shape is still Cartesian. This is shown on the right in (9.18), where we have identified the shape of the various objects by using Prop. 9.1.4 and idempotency of the modality (Prop. 8.1.45). With this, the pasting law (Prop. 8.1.39) implies that the outer right square in (9.18) is itself Cartesian, hence that $\int\widetilde{X}$ is the homotopy fiber of $\int\rho$. This implies the claim, by Prop. 8.2.6. □

Proposition 9.1.7 (Automorphisms along shape-unit)**.** *Let* $\mathbf{H}$ *be a cohesive* ∞*-topos (Def. 9.1.1),* $G \in \mathrm{Grp}(\mathbf{H})$ *(Prop. 8.2.1) and* $(X,\rho) \in G\mathrm{Actions}(\mathbf{H})$ *(Prop. 8.2.6). There is a canonical homomorphism*

$$\mathrm{Aut}(X) \xrightarrow{\;\mathrm{Aut}\left(\eta^{\int}_X\right)\;} \mathrm{Aut}(\textstyle\int X) \tag{9.19}$$

from the automorphism group (Def. 8.2.13) of X *to that of the shape* (9.2) *of* X*, which is such that the shape unit* $\eta^{\int}_X$ (8.72) *is equivariant (Def. 8.2.10) with respect to the canonical automorphism action* (8.174) *on* X *and the restriction (Prop. 8.2.12) along* *this morphism* (9.1.7) *of the canonical automorphism action on* $\int X$*:*

$$(X, \rho_{\mathrm{Aut}(X)}) \xrightarrow{\eta^{\int}_X} \mathrm{Aut}(\eta^{\int}_X)^* \big(\textstyle\int X, \rho_{\mathrm{Aut}(\int X)}\big) \quad \in \mathrm{Aut}(X)\mathrm{Actions}(\mathbf{H})\,. \tag{9.20}$$

Proof. Take the morphism (9.1.7) to be the composite

$$\mathrm{Aut}(X) \xrightarrow{\eta^{\int}_{\mathrm{Aut}(X)}} \textstyle\int(\mathrm{Aut}(X)) \xrightarrow{\Omega \vdash \int\rho_{\mathrm{Aut}}} \mathrm{Aut}(\int X)\,, \qquad \mathrm{Aut}(X) \xrightarrow{\mathrm{Aut}(\eta^{\int}_X)} \mathrm{Aut}(\int X) \tag{9.21}$$

where **(a)** the left morphism is the shape unit (8.72), using Prop. 9.1.4, while **(b)** the right morphism is that which exhibits, via Prop. 8.2.14, the $\int\mathrm{Aut}(X)$-action $\int\rho_{\mathrm{Aut}}$ (9.14) on $\int X$ from Lemma 9.1.6. Then consider the following diagram of homotopy fiber sequences:

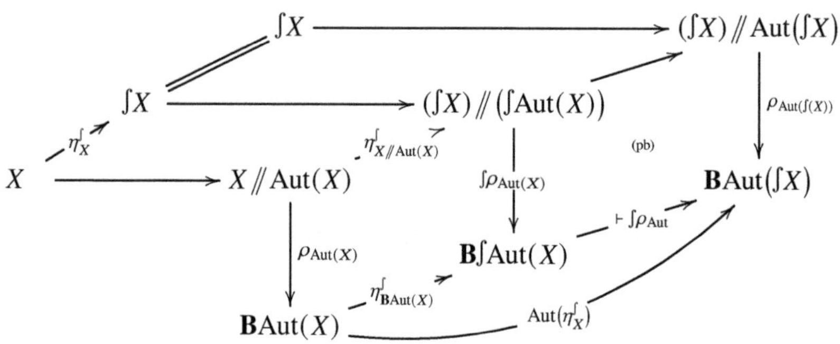

(9.22)

Here **(i)** the fiber sequence in the middle is that from the right of (9.18), **(ii)** the right part is the defining pullback from Prop. 8.2.14, while **(iii)** the left part exists by the naturality of $\eta^{\int}$. By the commutativity of the total front square, it factors through the coresponding pullback square, thus implying the claim. □

9.1.1.2 Concrete cohesive objects

Definition 9.1.8 (Concrete objects)**.** Let $\mathbf{H}$ be a cohesive ∞-topos (Def. 9.1.1).

(i) We say that an object $X \in \mathbf{H}$ is *concrete* if

(a) it is 0-truncated, $X \in \mathbf{H}_0 \hookrightarrow \mathbf{H}$ (Def. 8.1.73),

(b) its unit $\eta^{\sharp}_X$ (8.72) of the $\sharp$-modality (9.2) is (-1)-truncated (Def. 8.1.74), hence a monomorphism. By the -1-image factorization (8.137),

$$\underbrace{X \xrightarrow[\text{(-1)-conn.}]{} \sharp_{-1}X \xhookrightarrow[\text{(-1)-trunc.}]{} \sharp X}_{\text{image factorization}}\,, \qquad X \xrightarrow[\text{unit of } \sharp\text{-modality}]{\eta^{\sharp}_X} \sharp X \tag{9.23}$$

this means equivalently that X is equivalent to its -1-image under the $\sharp$-unit (8.72):

$$X \in \mathbf{H}_0 \quad \vdash \quad \begin{cases} X \text{ is concrete } \Leftrightarrow X \overset{\eta^\sharp_X}{\hookrightarrow} \sharp X \\ \qquad\qquad\quad\;\; \Leftrightarrow \sharp_{-1} X \simeq X\,. \end{cases} \tag{9.24}$$

(ii) We write

$$\mathbf{H}_{0,\sharp_{-1}} \hookrightarrow \mathbf{H}_0 \hookrightarrow \mathbf{H} \tag{9.25}$$

for the full subcategory of the 0-truncated objects on those which are concrete.

(iii) For any object, we call the (-1)-image (9.23) of the sharp unit on its 0-truncation (8.131) its *concretification*:

$$X \in \mathbf{H} \quad \vdash \quad \sharp_{-1}(\tau_0 X) \in \mathbf{H}_{0,\sharp_{-1}}\,. \tag{9.26}$$

Proposition 9.1.9 (Reflection of concrete objects). *A morphism $f : X \to Y$ out of any object X into a concrete object Y* (Def. 8.1.84) *factors uniquely through the concretification* (9.26) *of X, in that we have a natural bijection of hom-sets of this form:*

$$Y \text{ concrete} \;\vdash\; \mathbf{H}(\sharp_{-1}\tau_0 X, Y) \xrightarrow[\sim]{(X \twoheadrightarrow \sharp_{-1}\tau_0 X)^*} \mathbf{H}(\tau_0 X, Y) \xrightarrow[\sim]{(\eta^{\tau_0}_X)^*} \mathbf{H}(X, Y)\,. \tag{9.27}$$

Proof. The second isomorphism in (9.27) is the reflectivity (8.132) of 0-truncated objects. We need to discuss the first isomorphism. For ease of notation and without restriction of generality, we assume now that X is already 0-truncated.

Then by naturality of the $\sharp$-unit, the following outer square commutes, which implies the unique dashed arrow and hence the claim by the functoriality of the image factorization (Rem. 8.1.84):

$$\begin{array}{ccc} X & \xrightarrow{f} & Y \\ \downarrow & & \| \\ \sharp_{-1}X & \overset{\exists !}{\dashrightarrow} & \sharp_{-1}Y \\ \downarrow & & \downarrow \\ \sharp X & \xrightarrow{\sharp f} & \sharp Y \end{array} \qquad \eta^\sharp_X \text{ (left composite)},\; \eta^\sharp_Y \text{ (right composite)} \tag{9.28}$$

□

9.1.1.3 Cohesive charts

Definition 9.1.10 (Charts). Let $\mathbf{H}$ be a cohesive ∞-topos (Def. 9.1.1). We say that an *∞-category of cohesive charts* for $\mathbf{H}$ is an ∞-site Charts for $\mathbf{H}$ (Prop. 8.1.57)

$$\mathbf{H} \underset{\hookrightarrow}{\overset{L}{\underset{\bot}{\longleftarrow}}} \text{PreSheaves}_\infty(\text{Charts})$$

all of whose objects (under the ∞-Yoneda embedding y, Prop. 8.1.53) have contractible shape (9.2):

$$\begin{array}{ccccc} \text{Charts} & \overset{y}{\hookrightarrow} & \mathbf{H} & \xrightarrow{\text{Shp}} & \text{Groupoids}_\infty \\ U & \longmapsto & U & \longmapsto & \text{Shp}(U) \simeq * \end{array} \quad \Leftrightarrow \quad \begin{array}{ccccc} \text{Charts} & \overset{y}{\hookrightarrow} & \mathbf{H} & \xrightarrow{\int} & \mathbf{H} \\ U & \longmapsto & U & \longmapsto & \int(U) \simeq *\,. \end{array} \tag{9.29}$$

Lemma 9.1.11 (Charts are cohesively connected)**.** *Let* $\mathbf{H}$ *be a cohesive* ∞*-topos (Def. 9.1.1) with a site of* Chrt *(Def. 9.1.10). Then, for* $U \in$ Chrt *and* $\{X_i \in \mathbf{H}\}_{i \in I}$ *an indexed set of objects of* $\mathbf{H}$*, we have that every morphism from* U *into the coproduct of the* X_i *factors through one of the* X_i*:*

$$U \xrightarrow{f} \coprod_{i \in I} X_i \quad \Leftrightarrow \quad \underset{i_0 \in I}{\exists} \quad U \dashrightarrow X_{i_0} \xrightarrow{q_{X_{i_0}}} \coprod_{i \in I} X_i \,, \quad \text{(composite } U \xrightarrow{f} \textstyle\coprod_{i\in I} X_i) \tag{9.30}$$

Proof. Consider the pullbacks $U_i \xrightarrow{q_{U_i}} U$ along f of the canonical inclusions of the X_i into their coproduct, given by these Cartesian squares (Notation 8.1.37):

$$\begin{array}{ccc} U_i & \longrightarrow & X_i \\ {\scriptstyle q_{U_i}}\downarrow & \text{(pb)} & \downarrow{\scriptstyle q_{X_i}} \\ U & \xrightarrow[f]{} & \coprod_{i \in I} X_i \end{array} \tag{9.31}$$

By Prop. 8.1.48, this is such that

$$U \simeq \coprod_{i \in I} U_i \,. \tag{9.32}$$

The image of (9.32) under shape (9.2) is

$$* \simeq \int U \simeq \coprod_{i \in I} \int U_i \qquad \in \mathrm{Grpd}_\infty \overset{\mathrm{Disc}}{\hookrightarrow} \mathbf{H} \,, \tag{9.33}$$

where on the left we used the defining property (9.29) of charts, and on the right we used that the shape operation, being a left adjoint, preserves coproducts (Prop. 8.1.42). But, since $* \in \mathrm{Grpd}_\infty$ is connected, this implies that there is $i_0 \in I$ with

$$\int U_i \simeq \begin{cases} \varnothing & | \quad i \neq i_0 \\ * & | \quad i = i_0 \,. \end{cases} \tag{9.34}$$

From this, Lemma 9.1.3 implies that $U_i \simeq \varnothing$ for $i \neq i_0$ and, with (9.32), this implies

$$U_{i_0} \underset{\simeq}{\xrightarrow{q_{U_{i_0}}}} U \,. \tag{9.35}$$

Using this in (9.31) gives the desired factorization. □

Lemma 9.1.12 (Quotient by cohesively discrete ∞-group)**.** *Let* $\mathbf{H}$ *be a cohesive* ∞*-topos (Def. 9.1.1) which admits a site of* Chrt *(Def. 9.1.10). Then, for*

$$G \in \mathrm{Grp}(\mathrm{Grpd}_\infty) \overset{\mathrm{Disc}}{\hookrightarrow} \mathrm{Grp}(\mathbf{H}) \tag{9.36}$$

a cohesively discrete ∞*-group* (8.151) *and* $U \in$ Chrt*, we have an equivalence*

$$\mathbf{H}(U, * /\!\!/ G) \simeq * /\!\!/ G \quad \in \mathrm{Grpd}_\infty \,. \tag{9.37}$$

Proof. Since Disc is both a left and a right adjoint, it preserves (Prop. 8.1.42) the homotopy quotient that corresponds to the effective epimorphism $* \twoheadrightarrow * /\!\!/ G$ (Prop. 8.1.88) so that

$$* /\!\!/ G \in \mathrm{Grpd}_\infty \overset{\mathrm{Disc}}{\hookrightarrow} \mathbf{H} \tag{9.38}$$

is a cohesively discrete object. With this, we have the following sequence of natural equivalences:

$$
\begin{aligned}
\mathbf{H}\big(U,\, *\!/\!\!/G\big) &\simeq \mathbf{H}\big(U,\, \mathrm{Disc}(*\!/\!\!/G)\big) \\
&\simeq \mathrm{Grpd}_\infty\big(\mathrm{Shp}(U),\, *\!/\!\!/G\big) \\
&\simeq \mathrm{Grpd}_\infty\big(*,\, *\!/\!\!/G\big) \\
&\simeq *\!/\!\!/G\,,
\end{aligned}
\tag{9.39}
$$

where the second step is the hom-equivalence (8.71) of the Shp $\dashv$ Disc-adjunction and the third step is the condition that the chart U has contractible shape. □

Lemma 9.1.13 (Homming Charts into quotients by discrete groups)**.** *Let* $\mathbf{H}$ *be a cohesive* ∞*-topos (Def. 9.1.1) which admits* Chrt *(Def. 9.1.10). Then, for* $X \in \mathbf{H}$ *an object equipped with an* ∞*-action (Prop. 8.160) by a geometrically discrete* ∞*-group* G (9.36)*, the homotopy quotient* $X /\!\!/ G$ (8.161) *is given as an* ∞*-sheaf on* Chrt*, by assigning to* $U \in$ Chrt *the homotopy quotient of the* ∞*-groupoid of* U*-shapes plots of* X*:*

$$
X /\!\!/ G \;:\; U \longmapsto \mathbf{H}(U, X) /\!\!/ G\,. \tag{9.40}
$$

Proof. Consider the image of the homotopy fiber sequence that characterizes the given ∞-action (Prop. 8.2.6) under homming the chart U into it:

$$
\begin{array}{ccc}
X \xrightarrow{\;\mathrm{fib}(p)\;} X/\!\!/G & & \\
\quad\quad\quad\quad\downarrow{\scriptstyle p} & \xmapsto{\;\mathbf{H}(U,-)\;} & \\
\quad\quad\quad\quad *\!/\!\!/G & &
\end{array}
$$

$$
\begin{array}{l}
\mathbf{H}(U,X) \xrightarrow{\;\mathrm{fib}(\mathbf{H}(U,p))\;} \mathbf{H}(U,X)/\!\!/G \;\simeq\; \mathbf{H}\big(U, X/\!\!/G\big) \\
\quad\quad\quad\quad\quad\quad\quad\quad\quad\quad\quad \downarrow{\scriptstyle \mathbf{H}(U,p)} \\
\quad\quad\quad\quad\quad\quad\quad\quad\quad\quad\quad *\!/\!\!/G \;\simeq\; \mathbf{H}(U, *\!/\!\!/G)\,.
\end{array}
\tag{9.41}
$$

Since the hom-functor $\mathbf{H}(U,-)$ preserves limits, the result is again a homotopy fiber sequence, as shown on the right of (9.41). Moreover, by the assumption that G is geometrically discrete and that U is geometrically contractible, we have the equivalence (9.41) shown on the bottom right. This means that the fiber sequence on the right of (9.41) exhibits $\mathbf{H}(U, X/\!\!/G)$ as the homotopy quotient $\mathbf{H}(U,X)/\!\!/G$ of an ∞-action by G on $\mathbf{H}(U,X)$. □

Lemma 9.1.14 (Fixed locus in 0-truncated objects for discrete groups)**.** *Let* $\mathbf{H}$ *be a cohesive* ∞*-topos (Def. 9.1.1) with a site of* Chrt *(Def. 9.1.10). Let* $G \in \mathrm{Grp}(\mathbf{H})$ *(Prop. 8.2.1) be discrete* $G \simeq \flat G$ *and 0-truncated,* $G \simeq \tau_0 G$*, and let* $(X,\rho) \in G\mathrm{Actions}(\mathbf{H})$ *(Prop. 8.2.6) with* $X \simeq \tau_0 X$ *also 0-truncated. Then the* G*-fixed locus* $X^G \in \mathbf{H}$ *(Def. 8.2.24) is itself 0-truncated and such that, for* $U \in$ Chrt*, we have a natural equivalence*

$$
\mathbf{H}\big(U, X^G\big) \;\simeq\; \mathbf{H}(U,X)^G \;:=\; \Big\{\phi \in \mathbf{H}(U,X) \;\Big|\; \underset{g \in G}{\forall}\; g\cdot\phi = \phi\Big\} \tag{9.42}
$$

between **(a)** *the hom-set from U to X^G and* **(b)** *the naive set of fixed points in the hom-set from U to X, with respect to the restriction (Prop. 8.2.12) along $K \hookrightarrow G$ of the induced G-action* (9.41) *on the latter.*

Proof. We claim that we have the following sequence of natural equivalences:

$$\begin{aligned}
\mathbf{H}(U, X^G) &= \mathbf{H}\big(U, \mathbf{B}(G \to *)_*((X,\rho))\big) \\
&\simeq \mathbf{H}_{/\mathbf{B}G}\big(\mathbf{B}(G \to *)_*(U), X /\!\!/ G\big) \\
&\simeq \mathbf{H}_{/\mathbf{B}G}\big((*/\!\!/G) \times U\,,\, X /\!\!/ G\big) \\
&\simeq \mathbf{H}\big((*/\!\!/G) \times U\,,\, X /\!\!/ G\big) \underset{\mathbf{H}((*/\!\!/G)\times U\,,\,*/\!\!/G)}{\times} \{\mathrm{pr}_1\} \\
&\simeq \mathrm{Grpd}\Big(*/\!\!/G\,,\, \mathbf{H}\big(U\,,\, X /\!\!/ G\big)\Big) \underset{\mathrm{Grpd}\left(*/\!\!/G\,,\,\mathbf{H}(U\,,\,*/\!\!/G)\right)}{\times} \{\widetilde{\mathrm{pr}_1}\} \\
&\simeq \mathrm{Grpd}\Big(*/\!\!/G\,,\, \mathbf{H}\big(U\,,\, X\big) /\!\!/ G\Big) \underset{\mathrm{Grpd}(*/\!\!/G\,,\,*/\!\!/G)}{\times} \{\mathrm{id}\} \\
&\simeq \mathbf{H}(U, X)^G.
\end{aligned} \tag{9.43}$$

Here the first three lines are the definition of fixed loci (8.192) and the hom-equivalences (8.71) of the resulting adjunction (8.117). The fourth line is the characterization (8.110) of hom-∞-groupoids in slices (Prop. 8.1.64), the fifth line uses the tensoring (8.88) of $\mathbf{H}$ over Grpd_∞ (Prop. 8.1.50), and the sixth line follows by Prop. 9.1.13.

To see the last step in (9.43), use the explicit presentation of the groupoid $\mathbf{H}(U, X)/\!\!/ G$ as an action groupoid, by Example 8.1.31. This way the projection map in the fiber product in the sixth line in (9.43) is presented by a Kan fibration, whence this homotopy fiber product may be computed equivalently as a 1-categorical fiber product of sets of objects and of sets of morphisms, separately. Moreover, since $\{\mathrm{id}\}$ has no non-trivial morphisms and since the projection functor itself is faithful, there are in fact no non-trivial morphisms in this fiber product, which is hence just the set whose elements are precisely those functors of action groupoids which are equal to the identity on labels in G:

$$\begin{aligned}
\mathrm{Grpd}\Big(*/\!\!/G\,,\, \mathbf{H}(U,X)/\!\!/G\Big) \underset{\mathrm{Grpd}(*/\!\!/G\,,\,*/\!\!/G)}{\times} \{\mathrm{id}\} &\simeq \left\{ \begin{array}{ccc} */\!\!/G & \longrightarrow & \mathbf{H}(U,X)/\!\!/G \\ * & \longmapsto & \phi \\ \big\downarrow{\scriptstyle g \in G} & & \big\downarrow{\scriptstyle g} \\ * & \longmapsto & g \cdot \phi \end{array} \right\} \\
&\simeq \mathbf{H}(U, X)^G.
\end{aligned} \tag{9.44}$$

□

Lemma 9.1.15 (n-Truncated morphisms via n-truncated homotopy fibers)**.** *Let* $\mathbf{H}$ *be an ∞-topos which is cohesive (Def. 9.1.1). Let G be a finite group in* $\mathbf{H}$ (9.154)*. Then, for every $n \in \{-2, -1, 0, 1, \cdots\}$ and for any morphism in* $\mathbf{H}$ *to its delooping groupoid (Example 8.1.30) $X \xrightarrow{p} */\!\!/G$, the following are equivalent*

(i) *p is an n-truncated morphism (Def. 8.1.74);*

(ii) *the homotopy fiber of p (over the canonical point of $*/\!\!/G$) is an n-truncated object (Def. 8.1.73).*

Proof. Let $U \in \mathrm{Chrt}$ and consider homing it into the homotopy fiber sequence in question:

$$
\begin{array}{ccc}
X & \longrightarrow & \mathcal{X} \\
\big\downarrow & \text{(pb)} & \big\downarrow{\scriptstyle p} \\
* & \longrightarrow & * /\!\!/ G
\end{array}
\quad\Rightarrow\quad
\begin{array}{ccc}
\mathbf{H}(U, X) & \longrightarrow & \mathbf{H}(U, \mathcal{X}) \\
\big\downarrow & \text{(pb)} & \big\downarrow{\scriptstyle \mathbf{H}(U,p)} \\
* & \longrightarrow & \mathbf{H}(U, * /\!\!/ G) \simeq * /\!\!/ G \,.
\end{array}
\tag{9.45}
$$

Since the hom-functor $\mathbf{H}(U, -)$ preserves limits, the square on the right is again a homotopy pullback. Since U is a chart and G is discrete, we have the equivalence (9.37) shown on the bottom right. Since $* /\!\!/ G$ has an essentially unique point, the square on the right exhibits the essentially unique homotopy fiber of the morphism $\mathbf{H}(U, p)$. Since the charts U are generators of $\mathbf{H}$ (objects of an ∞-site of definition), the morphism p is n-truncated (Def. 8.1.74) precisely if for each chart U the homotopy fiber of $\mathbf{H}(U, p)$ is n-truncated. But the square on the right shows that this homotopy fiber is $\mathbf{H}(U, X)$, and hence this means, equivalently, that X is an n-truncated object (according to Def. 8.1.73). □

9.1.1.4 Examples of cohesive ∞-toposes

We indicate some examples of cohesive ∞-toposes (Def. 9.1.1), following [Sc13]. For more details of the constructions see spring [SS26c].

Example 9.1.16 (Discrete cohesion). The base ∞-topos Grpd_∞ is trivially a cohesive ∞-topos (Def. 9.1.1) with all operations being identities:

$$
\mathrm{Grpd}_\infty
\;
\begin{array}{c}
\xrightarrow{\;\;\mathrm{id}\;\;} \\[-2pt] \bot \\[-2pt]
\xleftarrow{\;\;\mathrm{id}\;\;} \\[-2pt] \bot \\[-2pt]
\xrightarrow{\;\;\mathrm{id}\;\;} \\[-2pt] \bot \\[-2pt]
\xleftarrow{\;\;\mathrm{id}\;\;}
\end{array}
\;
\mathrm{Grpd}_\infty \,.
\tag{9.46}
$$

For emphasis, we also call this the ∞-topos of *geometrically discrete ∞-groupoids*.

Definition 9.1.17 (Site for homotopical cohesion). A small ∞-site (8.101) is an *∞-site for homotopical cohesion* if

(i) its Grothendieck topology is trivial and

(ii) the underlying ∞-category has finite products, i.e., has a terminal object and binary Cartesian products.

Example 9.1.18 (Homotopical cohesion). The ∞-topos of ∞-sheaves (Def. 8.1.58) over an ∞-site C for homotopical cohesion (Def. 9.1.17) is cohesive (Def. 9.1.1):

$$
\mathbf{H} := \mathrm{Shv}_\infty(C)
\;
\begin{array}{c}
\xrightarrow{\;\;\underrightarrow{\lim}\;\;} \\[-2pt] \bot \\[-2pt]
\xleftarrow{\;\;\mathrm{const}\;\;} \\[-2pt] \bot \\[-2pt]
\xrightarrow{\;\;\underleftarrow{\lim}\;\;} \\[-2pt] \bot \\[-2pt]
\xleftarrow{\;\;\mathrm{Chtc}\;\;}
\end{array}
\;
\mathrm{Grpd}_\infty \,.
\tag{9.47}
$$

(i) The operation $\mathrm{Pnts} \simeq \underleftarrow{\lim}$ forms the limit of ∞-presheaves regarded as ∞-functors on C^{op} (by Prop. 8.1.52); but since C is assumed to have a terminal object, this is equivalently just the evaluation on that object:

$$
\mathrm{Pnts}(X) \;\simeq\; X(*) \;\simeq\; \mathbf{H}(*, X) \,,
\tag{9.48}
$$

where on the right we used the ∞-Yoneda lemma (Prop. 8.1.54). This makes manifest how $\mathrm{Pnts}(X)$ is the "underlying ∞-groupoid of points of X".

(ii) The operation $\mathrm{Shp} \simeq \underrightarrow{\lim}$ is the colimit of ∞-presheaves regarded as ∞-functors (by Prop. 8.1.52). Since the colimit of any representable functor is the point (Lemma 8.1.56)

$$C \xhookrightarrow{\;y\;} \mathrm{Shv}_\infty(C) \xrightarrow{\;\mathrm{Shp}\;} \mathrm{Grpd}_\infty\,, \qquad \underbrace{C \longrightarrow \mathrm{Grpd}_\infty}_{\mathrm{const}_*} \tag{9.49}$$

this means that C serves itself as a category of Chrt in this case (Def. 9.1.10).

Example 9.1.19 (Smooth cohesion)**.** The ∞-sheaf ∞-topos (Def. 8.1.58) over the site of SmthMfd (Def. 8.1.9, see [FStS12, App.]), which we call the ∞-topos of *smooth ∞-groupoids*

$$\mathrm{SmthGrpd}_\infty \;:=\; \mathrm{Shv}_\infty(\mathrm{SmthMfd})\,, \tag{9.50}$$

is cohesive (Def. 9.1.1): The adjoint quadruple (9.1) arises as in Example 9.1.18, which here happens to descend from ∞-presheaves to ∞-sheaves.

In this case, we have:

(i) A category of Chrt (Def. 9.1.10) is given (Prop. 8.1.10) by CrtSpc (Def. 8.1.5)

$$\mathrm{CrtSpc} \xhookrightarrow{\;y\;} \mathrm{Shv}_\infty(\mathrm{CrtSpc}) \xrightarrow{\;\simeq\;} \mathrm{SmthGrpd}_\infty \qquad \textstyle\int(y(\mathbb{R}^n)) \;\simeq\; *\,. \tag{9.51}$$

(ii) The concrete 0-truncated objects (Def. 9.1.8) are equivalently the *diffeological spaces* (Def. 8.1.6), including the *D-topological spaces*[1] (Def. 4.1.2) as well as smooth and possibly infinite-dimensional Fréchet manifolds (Prop. 8.1.11) as further full subcategories (8.12):

$$\begin{array}{ccccc} \mathrm{TopSpc} \xrightarrow{\mathrm{Cdfflg}} \mathrm{DTopSpc} & \searrow & & & \\ & & \mathrm{DiffSp} & \xhookrightarrow{\substack{\text{concrete}\\ \text{0-truncated}\\ \text{objects}}} & \mathrm{SmthGrpd}_\infty\,. \\ \mathrm{FréMfd} & \nearrow & & & \end{array} \tag{9.52}$$

(iii) The concrete 1-truncated objects (Def. 9.1.8) form the (2, 1)-category of diffeological groupoids with Morita/Hilsum-Skandalis morphisms (Remark 8.1.90) between them, which includes, by (9.52), the (2, 1)-categories of D-topological groupoids and of (possibly infinite-dimensional Fréchet-)Lie groupoids:

$$\begin{array}{ccccc} \mathrm{TopGrpd} \xrightarrow{\mathrm{Cdfflg}} \mathrm{DTopGrpd} & \searrow & & & \\ & & \mathrm{DiffGrpd} & \xhookrightarrow{\substack{\text{concrete}\\ \text{1-truncated}\\ \text{objects}}} & \mathrm{SmthGrpd}_\infty\,. \\ \mathrm{FréLieGrpd} & \nearrow & & & \end{array} \tag{9.53}$$

(iv) (**Smooth Oka principle** [SS26c, Thm. 3.3.53]) The cohesive shape (9.2) is given equivalently [BEBP19][Pav20][Bunk20, §3] by the *smooth ∞-path ∞-*

[1]These are the Δ-*generated spaces* of [Sm][Dug03]; see Remark 8.1.3.

groupoid:

$$\begin{array}{rcccl} \int X & \simeq & \underset{\longrightarrow}{\lim}\, \mathbf{Map}(\Delta^{\bullet}_{\mathrm{smth}}, X) & \in & \mathrm{SmthGrpd}_{\infty}\,, \\ \text{hence}\ \ \mathrm{Shp}(X) & \simeq & \underset{\longrightarrow}{\lim}\, X(\Delta^{\bullet}_{\mathrm{smth}}) & \in & \mathrm{Grpd}_{\infty}\,, \end{array} \tag{9.54}$$

where $\Delta^{\bullet}_{\mathrm{smth}}$ is the simplicial smooth manifold of extended simplices (Def. 8.1.35) and $\mathbf{Map}(-,-)$ denotes the internal hom (8.82) in $\mathrm{SmthGrpd}_{\infty}$.

(v) The cohesive shape (9.2) of **(a)** any topological space and **(b)** any finite-dimensional smooth manifold regarded, respectively, as smooth ∞-groupoids via (9.52) is equivalently (by (9.54) with Prop. 8.1.36 and by [Sc13, 4.3.29], respectively) its standard topological homotopy type $\mathrm{Shp}_{\mathrm{Top}}$ (8.1.29):

$$\textbf{(a)}\quad \mathrm{TopSpc} \xrightarrow{\ \mathrm{Cdfflg}\ } \mathrm{DiffSp} \hookrightarrow \mathrm{SmthGrpd}_{\infty} \xrightarrow{\ \mathrm{Shp}\ } \mathrm{Grpd}_{\infty}\,, \qquad \Downarrow\simeq \qquad \text{(lower arrow: } \mathrm{Shp}_{\mathrm{Top}}) \tag{9.55}$$

$$\textbf{(b)}\quad \mathrm{SmthMfd} \hookrightarrow \mathrm{DiffSp} \hookrightarrow \mathrm{SmthGrpd}_{\infty} \xrightarrow{\ \mathrm{Shp}\ } \mathrm{Grpd}_{\infty}\,, \qquad \Downarrow\simeq \qquad \text{(lower arrow: } \mathrm{Shp}_{\mathrm{Top}} \circ \mathrm{Dtplg}) \tag{9.56}$$

(vi) The cohesive shape (9.2) of a topological groupoid, when regarded, via its coreflection (8.12), as a D-topological groupoid and hence as a smooth ∞-groupoid (9.53) is equivalent (by (9.55), and since $\int$ is left adjoint and hence preserves homotopy colimits, Prop. 8.1.42) its simplicial-topological shape (Def. 8.1.33):

$$\mathrm{TopGrpd} \xrightarrow{\ \mathrm{Cdfflg}\ } \mathrm{DiffGrpd} \hookrightarrow \mathrm{SmthGrpd}_{\infty} \xrightarrow{\ \mathrm{Shp}\ } \mathrm{Grpd}_{\infty}\,. \qquad \Downarrow\simeq \qquad \text{(lower arrow: } \mathrm{Shp}_{\mathrm{sTop}}) \tag{9.57}$$

Example 9.1.20 (Spectral cohesion)**.** Let $\mathbf{H}$ be a cohesive ∞-topos (Def. 9.1.1). Then its tangent ∞-topos $T\mathbf{H} = \mathrm{SpectralBundles}(\mathbf{H})$ (Example 8.1.67) is cohesive [Sc13, 4.1.9] over the base tangent ∞-topos (8.190):

$$T\mathbf{H} \begin{array}{c} \xrightarrow{\ \ T\mathrm{Shp}\ \ } \\ \perp \\ \xleftarrow{\ \ T\mathrm{Disc}\ \ } \\ \perp \\ \xrightarrow{\ \ T\mathrm{Pnts}\ \ } \\ \perp \\ \xleftarrow{\ \ T\mathrm{Chtc}\ \ } \end{array} T\mathrm{Grpd}_{\infty}\,. \tag{9.58}$$

Remark 9.1.21 (Differential cohomology in cohesive ∞-toposes)**.** The intrinsic cohomology theory (1.21) of a cohesive ∞-topos (Def. 9.1.1) is *differential cohomology* [Sc13].

(i) In the case when $\mathbf{H} := \mathrm{SmthGrpd}_{\infty}$ (Example 9.1.19), this is a non-abelian differential cohomology theory generalizing the theory of Cartan-Ehresmann connections on smooth fiber bundles to ∞-connections on smooth ∞-bundles [SSS12][FStS12][NSS12a].

(ii) In the case when $\mathbf{H} := T\mathrm{SmthGrpd}_\infty$ is the cohesive tangent ∞-topos (Example 9.1.20) to that of smooth ∞-groupoids (Example 9.1.19), the intrinsic cohomology furthermore subsumes abelian Hopkins-Singer differential cohomology theories and variants [BNV13], as well as the twisted versions of these (Remark 8.2.23), such as twisted differential KU-theory [GS19a] and twisted differential KO-theory [GS19b].

9.1.2 Differential Geometry

We present a formulation of differential geometry internal to ∞-toposes which we call *elastic* [Sc13][Sc18], adjoining to the plain *shape* operation $\int$ of §9.1.1 a *de Rham shape* operation $\Im$, in generalization of [Si96][ST97].

Definition 9.1.22 (Elastic ∞-topos).

(i) An *elastic ∞-topos* over $\mathbf{B} = \mathrm{Grpd}_\infty$ is an ∞-topos $\mathbf{H}$ (Def. 8.1.46) whose base geometric morphism (Prop. 8.1.59), to be denoted $\mathrm{Pnts}\colon \mathbf{H} \to \mathrm{Grpd}_\infty$, is equipped with a factorization as follows, having adjoints (Def. 8.1.40) as shown:

$$
\begin{array}{lcccccc}
\text{“reduced”} & & & \xleftarrow{\quad \mathrm{Rdcd} \quad}\!\!\hookrightarrow & & & \\
 & & & \bot & & & \\
\text{“infnt shape”} & \mathrm{Shp}: & & \xrightarrow{\ \mathrm{Shp}_{\mathrm{inf}}\ } & & \times\!\!\xrightarrow{\ \mathrm{Shp}_{\Re}\ } & \\
 & & & \bot & & \bot & \\
\text{“infnt discrete”} & & \mathbf{H} & \xleftarrow{\ \mathrm{Disc}_{\mathrm{inf}}\ }\!\!\hookrightarrow & \mathbf{H}_{\Re} & \xleftarrow{\ \mathrm{Disc}_{\Re}\ }\!\!\hookrightarrow & \mathbf{B}\colon \mathrm{Disc} \\
 & & & \bot & & \bot & \\
\text{“infnt points”} & \mathrm{Pnts}: & & \xrightarrow{\ \mathrm{Pnts}_{\mathrm{inf}}\ } & \text{- - -} & \xrightarrow{\ \mathrm{Pnts}_{\Re}\ } & \\
 & & & & \bot & & \\
 & & & \longleftarrow & \mathrm{Chtc} & \hookrightarrow & \\
 & & \substack{\text{elastic}\\ \infty\text{-topos}} & & \substack{\text{reduced}\\ \text{sub-topos}} & & \substack{\text{discrete}\\ \text{sub-topos}}
\end{array}
\tag{9.59}
$$

(ii) Hence an elastic ∞-topos $\mathbf{H}$ is, in particular, a cohesive ∞-topos over $\mathbf{B}$, according to Def. 9.1.1, and so is its sub-∞-topos $\mathbf{H}_{\Re}$ of reduced objects.

(iii) We write

$$
\begin{array}{c}
\big(\ \Re \ := \ \mathrm{Rdcd} \circ \mathrm{Shp}_{\mathrm{inf}}\ \big) \\
\text{“reduced”} \\
\bot \\
\big(\ \Im \ := \ \mathrm{Disc}_{\mathrm{inf}} \circ \mathrm{Shp}_{\mathrm{inf}}\ \big) \\
\text{“étale”} \\
\bot \\
\big(\ \mathcal{L} \ := \ \mathrm{Disc}_{\mathrm{inf}} \circ \mathrm{Pnts}_{\mathrm{inf}}\ \big) \\
\text{“locally constant”}
\end{array}
\quad : \mathbf{H} \to \mathbf{H}
\tag{9.60}
$$

for the further induced modalities (1.18) (*elastic modalities*), accompanying the cohesive modalities of (9.2).

9.1.2.1 Examples of elastic ∞-toposes

We indicate some examples of elastic ∞-toposes (Def. 9.1.22), following [Sc13][Sc18].

Remark 9.1.23. We use the term “halo” for *infinitesimal neighborhoods*, hence for collections of (generalized) points which are “infinitesimally close” to a given space. This terminology is common in the school of “nonstandard analysis” (cf. [DD95, p.

6,11]) which is *not* the formalism we are using, but of course the idea of infinitesimal neighborhoods is independent of the formalization. Elsewhere we have been referring to the following *haloed Cartesian spaces* (Def. 9.1.24) as *infinitesimally thickened* Cartesian spaces (cf. [Sc13][KS17][GS26]) which expresses the same idea but is linguistically less elegant. The term "halo" is not only short but also nicely evocative of the evanescent nature of infinitesimal neighborhoods.

Definition 9.1.24 (Haloed Cartesian spaces)**.** Let $k \in \mathbb{N}$.

(i) We write

$$\begin{array}{ccc} k\mathrm{HldCrtSp} & \xhookrightarrow{\;C^\infty(-)\;} & \mathrm{CAlg}_{\mathbb{R}}^{\mathrm{op}} \\ \mathbb{R}^n \times \mathbb{D}_W & \longmapsto & C^\infty(\mathbb{R}^n) \otimes_{\mathbb{R}} (\mathbb{R} \oplus W) \end{array} \tag{9.61}$$

for the full subcategory of that of commutative $\mathbb{R}$-algebras on those which are tensor products of **(a)** the algebra of real-valued smooth functions on a Cartesian space $\mathbb{R}^n$, with **(b)** a finite-dimensional real algebra with a maximal ideal W that is nilpotent of order $k+1$, in that $W^{k+1} = 0$.

(ii) We write

$$\begin{array}{ccccc} \infty\mathrm{HldCrtSp} & := & \bigcup\limits_{k \in \mathbb{N}} k\mathrm{HldCrtSp} & \xhookrightarrow{\;C^\infty(-)\;} & \mathrm{CAlg}_{\mathbb{R}}^{\mathrm{op}} \\ \mathbb{R} \times \mathbb{D}_W & & \longmapsto & & C^\infty(\mathbb{R}^n) \otimes_{\mathbb{R}} W \end{array} \tag{9.62}$$

for the analogous full subcategory where each W is (finite dimensional and) nilpotent of some finite order.

(iii) We regard these categories as equipped with the coverage (Grothendieck pre-topology) whose covers are the families of morphisms of the form

$$\Big\{ \mathbb{R}^n \times \mathbb{D} \xrightarrow{\;f_i \times \mathrm{id}\;} \mathbb{R}^n \times \mathbb{D} \Big\}_{i \in I}$$

$$\text{such that } \Big\{ \mathbb{R}^n \xrightarrow{\;f_i\;} \mathbb{R}^n \Big\}_{i \in I} \text{ is a cover in CrtSpc (Def. 8.1.5).} \tag{9.63}$$

Lemma 9.1.25 (Coreflections of haloed Cartesian spaces)**.** *Regarding the category* kHldCrtSp *from Def. 9.1.24:*

(i) *For $k = 0$, this is equivalently the category of plain Cartesian spaces of Def. 8.1.5:*

$$0\mathrm{HldCrtSp} \simeq \mathrm{CrtSpc}\,. \tag{9.64}$$

(ii) *For any $k \in \mathbb{N}$, the evident full inclusion of* kHldCrtSp *into* $(k+1)$HldCrtSp *is co-reflective*

$$\infty\mathrm{HldCrtSp} \underset{\mathrm{Shp}_{\mathrm{inf},\infty}}{\overset{\mathrm{Rdcd}_\infty}{\underset{\longrightarrow}{\overset{\hookleftarrow}{\bot}}}} \cdots \underset{\mathrm{Shp}_{\mathrm{inf},2}}{\overset{\mathrm{Rdcd}_2}{\underset{\longrightarrow}{\overset{\hookleftarrow}{\bot}}}} 2\mathrm{HldCrtSp} \underset{\mathrm{Shp}_{\mathrm{inf},1}}{\overset{\mathrm{Rdcd}_1}{\underset{\longrightarrow}{\overset{\hookleftarrow}{\bot}}}} 1\mathrm{HldCrtSp} \underset{\mathrm{Shp}_{\mathrm{inf}}}{\overset{\mathrm{Rdcd}}{\underset{\longrightarrow}{\overset{\hookleftarrow}{\bot}}}} \mathrm{CrtSpc} \tag{9.65}$$

with

$$C^\infty\big(\mathrm{Shp}_{\mathrm{inf},k}(\mathbb{R}^n \times \mathbb{D}_W)\big) \simeq C^\infty(\mathbb{R}^n) \otimes_{\mathbb{R}} (\mathbb{R} \oplus W)/W^{k+1}. \tag{9.66}$$

Proof. Statement **(i)** follows as a special case of the general fact, sometimes known as *Milnor's exercise* (since the key idea is hinted at in [MSt74, Prob. 1-C]), that passage to their real algebras of smooth functions embeds smooth manifolds fully faithfully into the opposite or real algebras (a general proof is in [KMS93, §35.10], see also [Gr05]; for general perspective see [Nes03, §6]):

$$\mathrm{SmthMfd} \xhookrightarrow{\;C^\infty(-)\;} \mathrm{CAlg}_{\mathbb{R}}^{\mathrm{op}} \,. \tag{9.67}$$

Statement **(ii)** follows readily from the definition, using the fact that algebra homomorphisms preserve the order of nilpotency. □

Example 9.1.26 (Haloed smooth ∞-groupoids). For $k \in \mathbb{N} \sqcup \{\infty\}$, the ∞-sheaf ∞-topos (Def. 8.1.58) over the site of k-haloed Cartesian spaces (Def. 9.1.24)

$$k\mathrm{HldSmthGrpd}_\infty \;:=\; \mathrm{Shv}_\infty(k\mathrm{HldCrtSp}) \tag{9.68}$$

is elastic (Def. 9.1.22), with $(\mathrm{Rdcd} \dashv \mathrm{Shp}_{\mathrm{inf}})$ in (9.59) given by the Kan extension of the co-reflections of sites from Lemma 9.1.25:

$$\begin{array}{ccccc}
 & \xleftarrow{\;\;\mathrm{Rdcd}\;\;} & & & \\
 & \bot & & & \\
 & \xrightarrow{\;\;\mathrm{Shp}_{\mathrm{inf}}\;\;} & & \xrightarrow{\;\;\mathrm{Shp}_{\mathfrak{R}}\;\;} & \\
 & \bot & & \bot & \\
k\mathrm{HldSmthGrpd}_\infty & \xleftarrow{\;\;\mathrm{Disc}_{\mathrm{inf}}\;\;} & \mathrm{SmthGrpd}_\infty & \xleftarrow{\;\;\mathrm{Disc}_{\mathfrak{R}}\;\;} & \mathrm{Grpd}_\infty \,. \\
 & \bot & & \bot & \\
 & \xrightarrow{\;\;\mathrm{Pnts}_{\mathrm{inf}}\;\;} & & \xrightarrow{\;\;\mathrm{Pnts}_{\mathfrak{R}}\;\;} & \\
 & & \xleftarrow{\;\;\mathrm{Chtc}\;\;} & &
\end{array} \tag{9.69}$$

(i) Here for $k = 1$ we will, for short, abbreviate

$$\mathrm{HldSmthGrpd}_\infty \;:=\; 1\mathrm{HldSmthGrpd}_\infty \,. \tag{9.70}$$

(ii) For the case $k = \infty$, the underlying 1-topos is equivalently Dubuc's "Cahiers topos" [Du79a] [Ko86][KS17] (cf. [GS26, §4.1] for translation between our and the traditional definition).

(iii) For any k, we have:

(a) The full sub-∞-topos of reduced objects (9.59) is (by Lemma 9.1.25) that of smooth ∞-groupoids from Example 9.1.19

$$k\mathrm{HldSmthGrpd}_\infty \xleftarrow{\;\;\mathrm{Disc}_{\mathrm{inf}}\;\;} \mathrm{SmthGrpd}_\infty \,. \tag{9.71}$$

(b) The 0-truncated concrete objects (Def. 9.1.8) are still equivalently the *diffeological spaces* (Def. 8.1.6) as was the case in (9.52)

$$\begin{array}{ccc}
\mathrm{DTopSpc} & \searrow & & \\
 & & \mathrm{DiffSp} \xhookrightarrow{\;\text{0-truncated concrete objects}\;} k\mathrm{HldSmthGrpd}_\infty \\
\mathrm{FréMfd} & \nearrow & &
\end{array} \tag{9.72}$$

and, more generally, the 1-truncated concrete objects are still the *diffeological groupoids*, as was the case in (9.53):

$$\begin{array}{ccc}
\mathrm{DTopGrpd} & \searrow & & \\
 & & \mathrm{DiffGrpd} \xhookrightarrow{\;\text{1-truncated concrete objects}\;} k\mathrm{HldSmthGrpd}_\infty \,. \\
\mathrm{FréLieGrpd} & \nearrow & &
\end{array} \tag{9.73}$$

(c) A category of charts (Def. 9.1.10) for $\mathrm{HldSmthGrpd}_\infty$ is given by $k\mathrm{HldCrtSp}$ (Def. 9.1.24) itself.

9.1.2.2 Étale geometry

Definition 9.1.27 (Étale-over-X modality)**.** Let $\mathbf{H}$ be an elastic ∞-topos (Def. 9.1.22) and $X \in \mathbf{H}$ an object. We say that the *étale-over-X* modality on the slice ∞-topos over X (Def. 8.1.62) is the ∞-functor

$$
\begin{array}{ccc}
\mathbf{H}_{/X} & \xrightarrow{\;\Im_X\;} & \mathbf{H}_{/X} \\
Y & & Y \times_{\Im X} \Im Y \\
{\scriptstyle f}\big\downarrow & \longmapsto & \big\downarrow{\scriptstyle (\eta^{\Im}_X)^*(\Im f)} \\
X & & X
\end{array}
\qquad
\begin{array}{ccc}
X \times_{\Im X} \Im Y & \longrightarrow & \Im X \\
\big\downarrow & \text{(pb)} & \big\downarrow{\scriptstyle \Im f} \\
Y & \xrightarrow{\;\eta^{\Im}_Y\;} & \Im Y
\end{array}
\quad (X \dashrightarrow X \times_{\Im X} \Im Y,\; X \xrightarrow{\eta^{\Im}_X} \Im X,\; X \xrightarrow{f} Y)
\tag{9.74}
$$

which sends any morphism f into X to the pullback of its image under the plain étale modality $\Im$ (9.60) along its unit morphism (8.72), hence to the left vertical morphism in the Cartesian square shown on the right.

Definition 9.1.28 (Local diffeomorphism)**.** Let $\mathbf{H}$ be an elastic ∞-topos (Def. 9.1.22). We say that a morphism $Y \xrightarrow{f} X$ in $\mathbf{H}$ is a *local diffeomorphism* if it is étale-over-X (Def. 9.1.27)

$$
\Im_X(f) \simeq X \,, \tag{9.75}
$$

hence (see Prop. 9.1.34 for this implication) if the naturality square of the unit (8.72) of the $\Im$-modality (9.60) is a Cartesian square:

$$
\begin{array}{c}
Y \\
{\scriptstyle f}\big\downarrow{\scriptstyle \text{ét}} \\
X
\end{array}
\quad \Leftrightarrow \quad
\begin{array}{ccc}
Y & \xrightarrow{\;\eta^{\Im}_Y\;} & \Im Y \\
{\scriptstyle f}\big\downarrow & \text{(pb)} & \big\downarrow{\scriptstyle \Im f} \\
X & \xrightarrow[\;\eta^{\Im}_X\;]{} & \Im X \,.
\end{array}
\tag{9.76}
$$

Lemma 9.1.29 (Closure of class of local diffeomorphisms)**.** *Let $\mathbf{H}$ be an elastic ∞-topos (Def. 9.1.22). The class of local diffeomorphisms in $\mathbf{H}$ (Def. 9.1.28)*

(i) *satisfies left-cancellation: given a pair of composable morphisms f, g where g is a local diffeomorphism, then f is so precisely if the composite $g \circ f$ is:*

$$
\begin{array}{ccccc}
Z & & \xrightarrow{\;f\;} & & Y \\
& {\scriptstyle g\circ f}\searrow & & \swarrow{\scriptstyle g}\;{\scriptstyle \text{ét}} & \\
& & X & &
\end{array}
\;\Rightarrow\; \Big(f \text{ is local diffeo} \;\Leftrightarrow\; g \circ f \text{ is local diffeo} \Big). \tag{9.77}
$$

(ii) *is closed under pullbacks: if in a Cartesian square the right vertical morphism is a local diffeomorphism, then so is the left morphism*

$$
\begin{array}{ccc}
Y' \times_X Y & \longrightarrow & Y \\
{\scriptstyle g^* f}\big\downarrow & \text{(pb)} & {\scriptstyle \text{ét}}\big\downarrow{\scriptstyle f} \\
Y' & \xrightarrow[\;g\;]{} & X
\end{array}
\quad \Rightarrow \quad g^* f \text{ is a local diffeo.} \tag{9.78}
$$

Proof. This is a routine argument:

(i) For two composable morphisms, consider the pasting of their $\eta^{\Im}$-naturality squares

$$\begin{array}{ccc} Z & \xrightarrow{\eta^{\Im}_Z} & \Im Z \\ {\scriptstyle f}\downarrow & \text{(pb)} & \downarrow{\scriptstyle \Im f} \\ Y & \xrightarrow{\eta^{\Im}_Y} & \Im Y \\ {\scriptstyle g}\downarrow & \text{(pb)} & \downarrow{\scriptstyle \Im g} \\ X & \xrightarrow[\eta^{\Im}_X]{} & \Im X\,. \end{array} \tag{9.79}$$

By the functoriality of $\Im$, the total rectangle is the $\eta^{\Im}$-naturality square of $g \circ f$. But, by the pasting law (Prop. 8.1.39) and the assumption that the bottom square is Cartesian, the total rectangle is Cartesian precisely if so is the top square.

(ii) For two morphisms with the same codomain, consider the pasting of their pullback square with the $\eta^{\Im}$-naturality square of one of them, as shown on the left here:

$$\begin{array}{ccccc} Y' \times_X Y & \xrightarrow{f^* g} & Y & \xrightarrow{\eta^{\Im}} & \Im Y \\ {\scriptstyle g^* f}\downarrow & \text{(pb)} & \downarrow{\scriptstyle f} & \text{(pb)} & \downarrow{\scriptstyle \Im f} \\ Y' & \xrightarrow[g]{} & X & \xrightarrow[\eta^{\Im}_X]{} & \Im X \end{array}$$

$$\simeq \quad \begin{array}{ccccc} Y \times_X Y' & \xrightarrow{\eta^{\Im}_{(Y \times_X Y')}} & \Im(Y \times_X Y') & \xrightarrow{\Im(f^* f)} & \Im Y' \\ {\scriptstyle g^* f}\downarrow & & \downarrow{\scriptstyle \Im(g^* f)} & \text{(pb)} & \downarrow{\scriptstyle \Im g} \\ Y' & \xrightarrow[\eta^{\Im}_{Y'}]{} & \Im Y' & \xrightarrow[\Im(f')]{} & \Im X\,. \end{array} \tag{9.80}$$

By the naturality of $\eta^{\Im}$, this pasting diagram on the left is equivalent to that shown on the right. Moreover, if f is a local diffeomorphisms, it follows that three of the squares are pullbacks (the rightmost one by using that $\Im$ is right adjoint and thus preserves pullbacks, Prop. 8.1.42), as shown. With that, the pasting law (Prop. 8.1.39) implies, first, that the total rectangle on the left is a pullback, hence also that on the left, and then that the remaining square on the right is a pullback. This means that $g^* f$ is a local diffeomorphism. □

Definition 9.1.30 (Local neighborhood)**.** Let $\mathbf{H}$ be an elastic ∞-topos (Def. 9.1.22). For $Y \xrightarrow{f} X$ a morphism in $\mathbf{H}$, we say that the corresponding *local neighborhood* of Y in X is the purely étale aspect of f, hence is the object $N_f X \in \mathbf{H}_{/X}$ given by $\Im_{/X}(f) \simeq (\eta^{\Im}_X)^*(\Im f)$, hence given by the following homotopy pullback square:

$$\begin{array}{ccc} N_f X & \longrightarrow & \Im X \\ {\scriptstyle \Im_{/X}(f)}\downarrow & \text{(pb)} & \downarrow{\scriptstyle \Im f} \\ Y & \xrightarrow[\eta^{\Im}_X]{} & \Im Y\,. \end{array} \tag{9.81}$$

Definition 9.1.31 (Tangent bundle)**.** Let $\mathbf{H}$ be an elastic ∞-topos (Def. 9.1.22). Then for $X \in \mathbf{H}$ any object, we say that its *infinitesimal tangent bundle* is

$$TX \;:=\; X \underset{\Im X}{\times} X \;\in\; \mathbf{H}_{/X}\,, \tag{9.82}$$

hence the left morphism in this Cartesian square:

$$\begin{array}{ccc} TX & \longrightarrow & X \\ {\scriptstyle (\eta^{\Im}_X)^*(\eta^{\Im}_X)_!(\mathrm{id}_X)}\big\downarrow & {\scriptstyle (\mathrm{pb})} & \big\downarrow {\scriptstyle \eta^{\Im}_X} \\ X & \underset{\eta^{\Im}_X}{\longrightarrow} & \Im X\,. \end{array} \tag{9.83}$$

Example 9.1.32 (Local neighborhood of a point)**.** Let $\mathbf{H}$ be an elastic ∞-topos (Def. 9.1.22). For $X \in \mathbf{H}$ any object and $* \xrightarrow{x} X$ any point, the homotopy fiber of the tangent bundle (Def. 9.1.31) over x is equivalent to the local neighborhood of x (Def. 9.1.30):

$$T_x X \;\simeq\; N_x X\,. \tag{9.84}$$

This follows immediately from the definitions by the pasting law (Prop. 8.1.39):

$$\begin{array}{ccccc} N_x X \simeq T_x X & \longrightarrow & TX & \longrightarrow & X \\ \big\downarrow & {\scriptstyle (\mathrm{pb})} & \big\downarrow & {\scriptstyle (\mathrm{pb})} & \big\downarrow {\scriptstyle \eta^{\Im}_X} \\ * & \underset{x}{\longrightarrow} & X & \underset{\eta^{\Im}_X}{\longrightarrow} & \Im X\,. \end{array} \tag{9.85}$$

Proposition 9.1.33 (Pullback along local diffeomorphisms preserves tangent bundles)**.** *In an elastic ∞-topos (Def. 9.1.22), pullback along a local diffeomorphism $Y \underset{\text{ét}}{\xrightarrow{f}} X$ (Def. 9.1.28) preserves tangent bundles (Def. 9.1.31) in that*

$$f^*(TX) \;\simeq\; TY \qquad \textit{via:} \qquad \begin{array}{ccc} TY & \xrightarrow{Tf} & TX \\ \big\downarrow & {\scriptstyle (\mathrm{pb})} & \big\downarrow \\ Y & \underset{f}{\xrightarrow{\text{ét}}} & X\,. \end{array} \tag{9.86}$$

Proof. Consider the pasting of the defining Cartesian squares, shown on the left here:

$$\begin{array}{ccccc} f^*TX & \longrightarrow & TX & \longrightarrow & X \\ \big\downarrow & {\scriptstyle (\mathrm{pb})} & \big\downarrow & {\scriptstyle (\mathrm{pb})} & \big\downarrow {\scriptstyle \eta^{\Im}_X} \\ Y & \underset{f}{\xrightarrow{\text{ét}}} & X & \underset{\eta^{\Im}_X}{\longrightarrow} & \Im X \end{array} \;\simeq\; \begin{array}{ccccc} TY & \longrightarrow & Y & \xrightarrow{f} & X \\ \big\downarrow & {\scriptstyle (\mathrm{pb})} & \big\downarrow {\scriptstyle \eta^{\Im}_X} & {\scriptstyle (\mathrm{pb})} & \big\downarrow {\scriptstyle \eta^{\Im}_X} \\ Y & \underset{\eta^{\Im}_X}{\longrightarrow} & \Im Y & \underset{\Im f}{\longrightarrow} & \Im X\,. \end{array} \tag{9.87}$$

By the pasting law (Prop. 8.1.39), the total rectangle on the left is itself Cartesian. Moreover, the bottom composite morphism on the left is equivalent to the bottom composite morphism on the right, by the naturality of $\eta^{\Im}_X$. Therefore, using again the pasting law (Prop. 8.1.39), the total rectangle on the left is equivalent to the pasting of the two consecutive Cartesian squares shown on the right. These identify, in the top row, the middle object Y by (9.76) and thus the left object TY by (9.83). □

9.1.2.3 Étale toposes

Definition 9.1.34 (Étale ∞-topos)**.** Let $\mathbf{H}$ be an elastic ∞-topos (Def. 9.1.22) and $X \in \mathbf{H}$. Then we say that the *étale ∞-topos* of X, to be denoted $\mathbf{\acute{E}t}_X$, is the full sub-∞-category (Def. 8.1.1) of the slice ∞-topos over X (Prop. 8.1.62) on those morphisms that are local diffeomorphisms (Def. 9.1.28):

$$\mathbf{\acute{E}t}_X \;:=\; (\mathbf{H}_{/X})_{\Im_X} \hookrightarrow \mathbf{H}_{/X}\,. \tag{9.88}$$

Proposition 9.1.35 (Reflections of étale toposes)**.** *Let $\mathbf{H}$ be an elastic ∞-topos (Def. 9.1.22) and $X \in \mathbf{H}$ an object. Then the étale topos $\mathbf{\acute{E}t}_X$ from Def. 9.1.34:*

(i) *is indeed an ∞-topos (Def. 8.1.46);*

(ii) *its defining full inclusion* (9.88) *has both a left- and a right-adjoint (Def. 8.1.40):*

$$\mathbf{\acute{E}t}_X \underset{\underset{\text{LcllCnstnt}_X}{\longleftarrow}}{\overset{\overset{\text{Etl}_X}{\longleftarrow}}{\underset{\bot}{\overset{\bot}{\hookrightarrow}}}} \mathbf{H}_{/X} \qquad (i_X) \tag{9.89}$$

(iii) *whose induced adjoint modality* (1.18)

$$\begin{array}{c}\big(\,\Im_X \;:=\; i_X \circ \text{Étl}_X\,\big) \\ \text{“étale over } X\text{”} \\ \bot \\ \big(\,\mathcal{L}_X \;:=\; i_X \circ \text{LcllCnstnt}_X\,\big) \\ \text{“locally constant over } X\text{”}\end{array} \;:\; \mathbf{H}_{/X} \to \mathbf{H}_{/X} \tag{9.90}$$

is on the left that of Def. 9.1.27:

$$\text{Étl}_X : \left(\begin{array}{c} Y \\ \big\downarrow{\scriptstyle p} \\ X \end{array}\right) \mapsto \left(\begin{array}{c} (\eta^{\Im}_X)^*(\Im Y) \\ \big\downarrow{\scriptstyle (\eta^{\Im}_X)^*(\Im p)} \\ X \end{array}\right) \quad \text{i.e.:} \quad \begin{array}{ccc} (\eta^{\Im}_X)^*(\Im Y) & \xrightarrow{(\Im p)^*(\eta^{\Im}_X)} & \Im Y \\ \big\downarrow{\scriptstyle (\eta^{\Im}_X)^*(\Im p)} & \text{(pb)} & \big\downarrow{\scriptstyle \Im p} \\ X & \xrightarrow[\eta^{\Im}_X]{} & \Im X\,. \end{array} \tag{9.91}$$

Proof. First to see that (9.91) is well-defined as a functor to $\mathbf{\acute{E}t}_X$ (this proceeds as in [CHM85, 3.3][CJKP97, 3][CRi20, 7.3]): We need to check that $(\eta^{\Im}_X)^*(\Im p)$ is a local diffeomorphism (Def. 9.1.28). For this, it is sufficient to have equivalences

$$\Im\big((\eta^{\Im}_X)^*(\Im p)\big) \;\simeq\; \Im p\,, \tag{9.92}$$

and

$$(\Im p)^*(\eta^{\Im}_X) \;\simeq\; \eta^{\Im}_X \tag{9.93}$$

because then the Cartesian square on the right of (9.91) exhibits this property.

But (9.92) follows by applying $\Im$ to the square on the right of (9.91), by idempotency (Prop. 8.1.45), and since equivalences are preserved by pullback (Example 8.1.38). With this, (9.93) follows from the naturality of the $\Im$-unit, by the universal factorization shown dashed in the following diagram:

$$\begin{tikzcd} (\eta^{\Im}_X)^*(\Im Y) \arrow[dr, dashed, "\simeq"] \arrow[drr, bend left, "\eta^{\Im}_{(\eta^{\Im}_X)^*(\Im Y)}"] \arrow[ddr, bend right, "(\eta^{\Im}_X)^*(\Im p)"'] & & \\ & (\eta^{\Im}_X)^*(\Im Y) \arrow[r, "(\Im p)^*(\eta^{\Im}_X)"] \arrow[d, "(\eta^{\Im}_X)^*(\Im p)"] \arrow[dr, phantom, "\text{(pb)}"] & \Im Y \arrow[d, "\Im p"] \\ & X \arrow[r, "\eta^{\Im}_X"'] & \Im X\,. \end{tikzcd} \tag{9.94}$$

Notice that, similarly, there is a natural transformation

$$\begin{tikzcd} Y \arrow[rr, "\eta^{\mathrm{Etl}_X}_Y"] \arrow[dr, "p"'] & & \mathrm{Etl}_X(Y) \arrow[dl, "\text{ét}"] \\ & X & \end{tikzcd} \tag{9.95}$$

induced as the universal factorization shown dashed in the following diagram:

$$\begin{tikzcd} Y \arrow[dr, dashed, "\simeq"] \arrow[drr, bend left, "\eta^{\Im}_Y"] \arrow[ddr, bend right, "p"'] & & \\ & (\eta^{\Im}_X)^*(\Im Y) \arrow[r, "(\Im p)^*(\eta^{\Im}_X)"] \arrow[d, "(\eta^{\Im}_X)^*(\Im p)"] \arrow[dr, phantom, "\text{(pb)}"] & \Im Y \arrow[d, "\Im p"] \\ & X \arrow[r, "\eta^{\Im}_X"'] & \Im X \end{tikzcd} \tag{9.96}$$

and notice that this in an $\Im$-equivalence:

$$\Im\big(\eta^{\mathrm{Etl}_X}_{Y_1}\big) \quad \text{is an equivalence}\,. \tag{9.97}$$

Condition (9.97) follows by applying $\Im$ to the whole left part of the diagram on the right of (9.98), using idempotency (Prop. 8.1.45) and that equivalences are preserved by pullback (Example 8.1.38).

Second, to see that (9.91) defines a left adjoint to the inclusion: We need to check the corresponding hom-equivalence (8.71), shown on the left here:

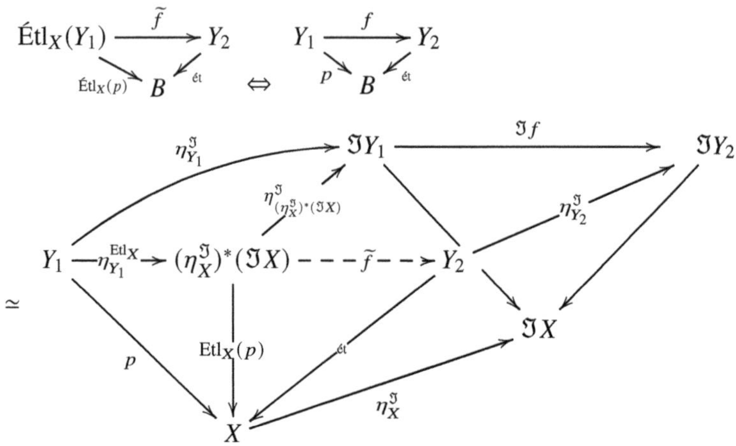

(9.98)

On the bottom of (9.98) we show an induced factorization: The square sub-diagram on the right of (9.98) is Cartesian by the assumption that we are homming into a local diffeomorphism, while the square in the middle is Cartesian by (9.94). Thus, given f, the morphism $\widetilde{f}$ is induced by the universal property of the right Cartesian square. Conversely, given $\widetilde{f}$, precomposition with the $\eta^{\mathrm{Etl}X}_{Y_1}$ (9.96) gives a morphism f. To see that this correspondence is an equivalence, we just need to observe that $\Im(\widetilde{f}) \simeq \Im f$. This follows by (9.97).

Thus we have established the existence of the left adjoint $\mathrm{Étl}_X$. With this, to see the right adjoint $\mathrm{LclCnst}_X$ as well as the fact that $\mathbf{Ét}$ is an ∞-topos, it is now sufficient to show that $\mathbf{Ét}_X \xhookrightarrow{i_X} \mathbf{H}_{/X}$ preserves colimits: Because, by the reflection $\mathrm{Étl}_X$ this implies, first, that $\mathbf{Ét}_X$ is a presentable ∞-category, in fact an ∞-topos (by Prop. 8.1.57, since it is then an accessibly embedded reflective subcategory of the slice $\mathbf{H}_{/X}$, which is an ∞-topos by Prop. 8.1.62); and thus, second, the existence of the right adjoint by the adjoint ∞-functor theorem (Prop. 8.1.43).

So to see that i_X preserves colimits, consider any small $\mathcal{I} \in \mathrm{Cat}_\infty$ and a diagram

$$Y_\bullet : \mathcal{I} \longrightarrow \mathbf{Ét}_X \xhookrightarrow{i_X} \mathbf{H}_{/X} . \tag{9.99}$$

Since i_X is fully faithful by construction, it is sufficient to show that the colimit of this diagram formed in $\mathbf{H}_{/X}$ is itself in the image of i_X. This colimit, in turn, is computed in $\mathbf{H}$ (by Example 8.1.68) with its morphism q to X universally induced, and this we need to show to be a local diffeomorphism (Def. 9.1.28). Hence we need to show that the following square on the left is Cartesian:

$$
\begin{array}{ccc}
\varinjlim Y_\bullet & \xrightarrow{\eta^{\Im}_{\varinjlim Y_\bullet}} & \Im(\varinjlim Y_\bullet) \\
{\scriptstyle q}\big\downarrow & \scriptstyle(\mathrm{pb}) & \big\downarrow{\scriptstyle \Im q} \\
X & \xrightarrow[\eta^{\Im}_X]{} & \Im X
\end{array}
\;\Leftrightarrow\;
\begin{array}{ccc}
\varinjlim Y_\bullet & \xrightarrow{(\eta^{\Im}_{Y_\bullet})} & \varinjlim(\Im Y_\bullet) \\
{\scriptstyle q}\big\downarrow & \scriptstyle(\mathrm{pb}) & \big\downarrow{\scriptstyle \Im q} \\
X & \xrightarrow[\eta^{\Im}_X]{} & \Im X
\end{array}
\tag{9.100}
$$

$$
\Leftrightarrow\;\underset{i\in I}{\forall}\;
\begin{array}{ccc}
Y_i & \xrightarrow{\eta^{\Im}_{Y_i}} & \Im Y_i \\
{\scriptstyle q_i}\big\downarrow & \scriptstyle(\mathrm{pb}) & \big\downarrow{\scriptstyle \Im q_i} \\
X & \xrightarrow[\eta^{\Im}_X]{} & \Im X\,.
\end{array}
$$

But, since $\Im$ is a left adjoint and hence preserves colimits (Prop. 8.1.42), this is equivalent to the square in the middle being Cartesian. Finally, by universality of colimits (8.80) in the ∞-topos $\mathbf{H}$, this is equivalent to all the squares on the right being Cartesian. This is the case, by the assumption (9.99). □

Remark 9.1.36 (Local and global ∞-section functors). Let $\mathbf{H}$ be an elastic ∞-topos (Def. 9.1.22) and $X \in \mathbf{H}$. Then we may think of the étale ∞-topos $\mathbf{Ét}_X$ (Def. 9.1.34, Prop. 9.1.35) as the internal construction of the ∞-topos of ∞-*sheaves over* X. Under this interpretation:

(i) the ∞-functor LcllCnst (9.89) has the interpretation of sending any ∞-bundle $E \twoheadrightarrow X$ (Notation 8.1.61) to its ∞-*sheaf of local sections* $\underline{E} := \mathrm{LcllCnst}_X(E)$;

(ii) the direct image of the base geometric morphism (8.103) has the interpretation of sending any ∞-sheaf to its ∞-groupoid of global sections:

$$
\underset{\text{over }X}{\overset{\infty\text{-bundles}}{\mathbf{H}_{/X}}}
\;
\underset{\underset{\text{form }\infty\text{-sheaf of local sections}}{\underline{(-)}:=\mathrm{LcllCnstnt}_X}}{\overset{i_X}{\underset{\longrightarrow}{\overset{\longleftarrow}{\bot}}}}
\;
\underset{\text{on }X}{\overset{\infty\text{-sheaves}}{\mathbf{Ét}_X}}
\;
\underset{\underset{\text{form }\infty\text{-groupoid of global sections}}{\Gamma_X}}{\overset{\Delta_X}{\underset{\longrightarrow}{\overset{\longleftarrow}{\bot}}}}
\;
\mathrm{Grpd}_\infty\,.
\qquad \underset{\Gamma_X}{\mathbf{H}_{/X}\longrightarrow \mathrm{Grpd}_\infty}
\tag{9.101}
$$

Notice that the global sections of the ∞-sheaf of local sections of an ∞-bundle E is the global sections of that ∞-bundle (as in Remark 8.2.21):

$$
\Gamma_X(\underline{E}) \;\simeq\; \Gamma_X(E) \tag{9.102}
$$

(by the essential uniqueness of the base geometric morphism (Prop. 8.1.59) and the fact that the base geometric morphism on ∞-bundles forms global sections, Remark 8.2.22).

9.1.2.4 Étale groupoids

Definition 9.1.37 (Étale groupoid). Let $\mathbf{H}$ be an elastic ∞-topos (Def. 9.1.22).

(i) We say that $X_\bullet \in \mathrm{Grpd}(\mathbf{H})$ (Def. 8.1.86) is an *étale groupoid* if all its face maps are local diffeomorphisms (Def. 9.1.28):

$$X_\bullet \text{ is étale groupoid} \quad \Leftrightarrow \quad \underset{\substack{n \in \mathbb{N} \\ 0 \le i \le n}}{\forall} \; X_{n+1} \xrightarrow[\text{ét}]{d_i} X_n \,. \tag{9.103}$$

(ii) We write

$$\text{ÉtGrpd}(\mathbf{H}) \hookrightarrow \text{Grpd}(\mathbf{H}) \quad \in \text{Cat}_\infty \tag{9.104}$$

for the full sub-∞-category of that of all groupoids (8.144) on those that are étale groupoids.

As a variant of Prop. 8.1.88, we have:

Proposition 9.1.38 (Étale groupoids are equivalent to stacks with étale atlases). *Let* **H** *be an elastic ∞-topos (Def. 9.1.22) and* $X_\bullet \in \text{Grpd}(\mathbf{H})$ *(Def. 8.1.86). Then the following conditions are equivalent:*

(i) *The groupoid* $X_\bullet$ *is an étale groupoid (Def. 9.1.37).*

(ii) *The associated atlas* $X_0 \xrightarrow{a} \mathcal{X}$ *(via Prop. 8.1.88) is a local diffeomorphism (Def. 9.1.28).*

$$\begin{array}{ccccl}
\vdots & & \vdots & & \\
X \times_{\mathcal{X}} X & \simeq & X_1 & & \text{“étale groupoid”} \\
{\scriptstyle \text{ét}}\downarrow \uparrow {\scriptstyle \text{ét}} \downarrow {\scriptstyle \text{ét}} & & {\scriptstyle s}\downarrow \uparrow {\scriptstyle e} \downarrow {\scriptstyle t} & & \\
X_0 & = & X_0 & & \\
{\scriptstyle a}\downarrow{\scriptstyle \text{ét}} & & \downarrow & & \text{“étale atlas”} \\
\mathcal{X} & \simeq & \varinjlim X_\bullet & & \text{“étale stack”}
\end{array} \tag{9.105}$$

Proof. By definition of local diffeomorphisms, we need to demonstrate the logical equivalence shown on the left:

$$\underset{n_1 \xrightarrow{\phi} n_2}{\forall} \begin{array}{ccc}
X_{n_1} & \xrightarrow{\eta^{\Im}_{X_{n_1}}} & \Im X_{n_1} \\
{\scriptstyle X_\phi}\downarrow & \text{(pb)} & \downarrow{\scriptstyle \Im X_\phi} \\
X_{n_2} & \xrightarrow[\eta^{\Im}_{X_{n_2}}]{} & \Im X_{n_2}
\end{array}
\Leftrightarrow
\begin{array}{ccc}
X_0 & \xrightarrow{\eta^{\Im}_{X_0}} & \Im X_0 \\
{\scriptstyle a}\downarrow & \text{(pb)} & \downarrow{\scriptstyle \Im a} \\
\varinjlim X_\bullet & \xrightarrow[\eta^{\Im}_{\varinjlim X_\bullet}]{} & \Im \varinjlim X_\bullet
\end{array}$$

$$\Leftrightarrow
\begin{array}{ccc}
X_0 & \xrightarrow{\eta^{\Im}_{X_0}} & \Im X_0 \\
{\scriptstyle a}\downarrow & \text{(pb)} & \downarrow{\scriptstyle \Im a} \\
\varinjlim X_\bullet & \xrightarrow[\varinjlim \eta^{\Im}_{X_\bullet}]{} & \varinjlim \Im X_\bullet \,.
\end{array} \tag{9.106}$$

But since $\Im$ preserves all limits and colimits (being a left and a right adjoint, Prop. 8.1.42), we have **(a)** also the logical equivalence shown on the right of (9.106); and **(b)** that $\Im X_\bullet$ is itself a groupoid with atlas $\Im a$ and that $X_\bullet \xrightarrow{\eta^{\Im}_{X_\bullet}} \Im X_\bullet$ is a morphism in Grpd(**H**) (8.144). By **(a)**, it is now sufficient to prove the composite logical equivalence in (9.106). By **(b)**, this follows with Prop. 8.1.91. □

Proposition 9.1.39 (Tangent stacks). *Let* $\mathbf{H}$ *be an elastic* ∞*-topos (Def. 9.1.22) and* $X_\bullet \in \text{ÉtaleGroupoids}(\mathbf{H})$ *(Def. 9.1.37) with étale atlas* $X \xrightarrow{\text{ét}} \mathcal{X}$ *(via Prop. 9.1.38). Then:*

(i) *the system of tangent bundles* $TX_\bullet$ *(Def. 9.1.31) is itself an étale groupoid (Def. 9.1.37), the tangent groupoid;*

(ii) *its atlas (under Prop. 9.1.38) is the differential* $TX_0 \xrightarrow{Ta} T\mathcal{X}$ *of the given atlas, hence the tangent stack is:*

$$T\mathcal{X} \simeq \varinjlim TX_\bullet \,. \tag{9.107}$$

Proof. **(i)** That $TX_\bullet$ is itself a groupoid (Def. 8.1.86) follows because both the tangent bundle construction $T(-)$ (9.83) as well as the groupoid Segal conditions (8.143) are pullback constructions, hence limits, which commute over each other. To see that $TX_\bullet$ is an étale groupoid, consider the following diagram:

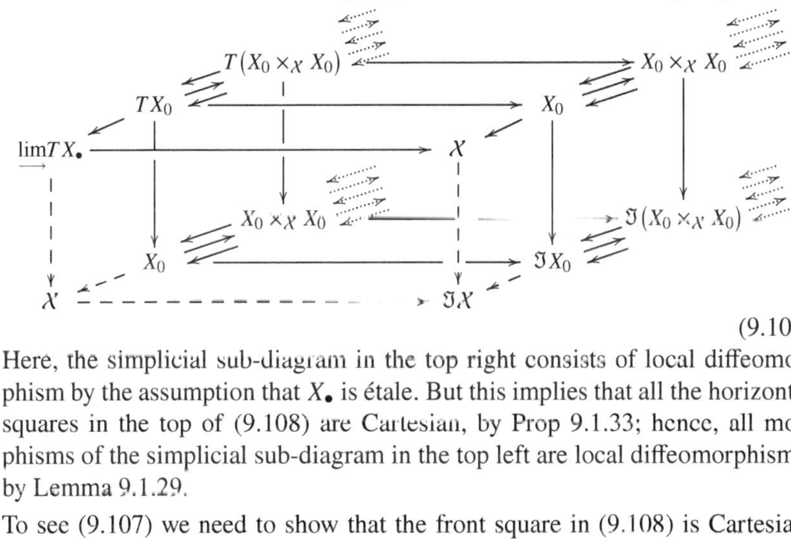

(9.108)

Here, the simplicial sub-diagram in the top right consists of local diffeomorphism by the assumption that $X_\bullet$ is étale. But this implies that all the horizontal squares in the top of (9.108) are Cartesian, by Prop 9.1.33; hence, all morphisms of the simplicial sub-diagram in the top left are local diffeomorphisms, by Lemma 9.1.29.

(ii) To see (9.107) we need to show that the front square in (9.108) is Cartesian. Observe:

(a) All horizontal squares in (9.108) are Cartesian: the top ones by the above argument for **(i)**, the bottom ones by the assumption that $X_\bullet$ is étale.

(b) All solid vertical squares in (9.108) are also Cartesian, by definition (9.83) of tangent bundles.

(c) The object $\mathcal{X}$ in the bottom front left of (9.108) is not just the colimit of the simplicial sub-diagram in the bottom left, but in fact of the full left sub-diagram (because of the colimit of the top left sub-diagram in the front top left). Similarly, the object $\Im\mathcal{X}$ is in fact the colimit over the full right sub-diagram in (9.108) (using that $\Im$ preserves colimits, being a left adjoint, Prop. 8.1.42).

Now **(a)** and **(b)** verify the assumption of Prop. 8.1.48 applied to the diagram (9.108), regarded as a natural transformation from its left part to its right part;

and with **(c)**, the conclusion of Prop. 8.1.48 says that the front square in (9.108) is Cartesian. □

Lemma 9.1.40 (Degreewise local diffeomorphisms of étale groupoids)**.** *Let* **H** *be an elastic ∞-topos (Def. 9.1.22) and* $X_\bullet, Y_\bullet \in$ *ÉtaleGroupoids*(**H**) *(Def. 9.1.37). If a morphism* $X_\bullet \xrightarrow{f_\bullet} Y_\bullet$ *is such that for all* $n \in \mathbb{N}$*, the component* $X_n \xrightarrow{f_n} Y_n$ *is a local diffeomorphism (Def. 9.1.28), then induced morphism on stacks* $\mathcal{X} \xrightarrow{\varinjlim f_\bullet} \mathcal{Y}$ *is also a local diffeomorphism (Def. 9.1.38).*

Proof. Consider the following diagram:

(9.109)

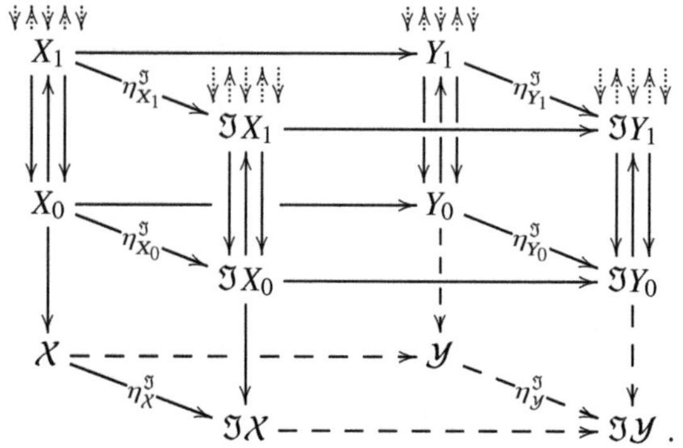

Observe that:

(a) all solid $\eta^{\Im}$-naturality squares in this diagram are Cartesian, by the assumption that the rear part of the diagram is a degreewise local diffeomorphism of étale groupoids.

(b) $\mathcal{Y}$ is not just the colimit of the partial diagram $Y_\bullet$ in the rear right, but in fact is also the colimit of the full non-dashed rear part of the diagram (using that $\mathcal{X}$ is the colimit of the rear left part). Similarly, $\Im\mathcal{Y}$ is the colimit of the non-dashed front part of the diagram (using that $\Im$ preserves limits and colimits, being a left and a right adjoint, Prop. 8.1.42).

Hence if we regard the diagram as a natural transformation from its rear to its front part, then Prop. 8.1.48 applies and says that also the bottom dashed square is Cartesian, and hence that $\mathcal{X} \to \mathcal{Y}$ is a local diffeomorphism. □

Definition 9.1.41 (Étalification of groupoids)**.** Let **H** be an elastic ∞-topos (Def. 9.1.22) and $X_\bullet \in \mathrm{Grpd}(\mathbf{H})$ (Def. 8.1.86). Notice that, by Prop. 8.1.88 for all $n \in \mathbb{N}$ we have for all $0 \le i \le n$ that all face maps $X_{n+1} \xrightarrow{d_i} X_n$ are in fact equivalent to each other, being related by an automorphism of X_{n+1} given by permutation of fiber product factors (8.145)

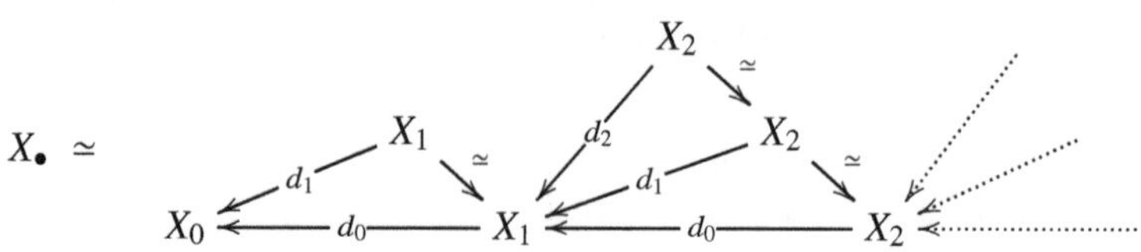

(9.110)

(and similarly for the degeneracy maps). Therefore, we may regard $X_\bullet$ as a diagram in the slice $\mathbf{H}_{X_0}$. and apply $\mathcal{L}_{X_0}$ (9.90) to this diagram (9.110) to obtain

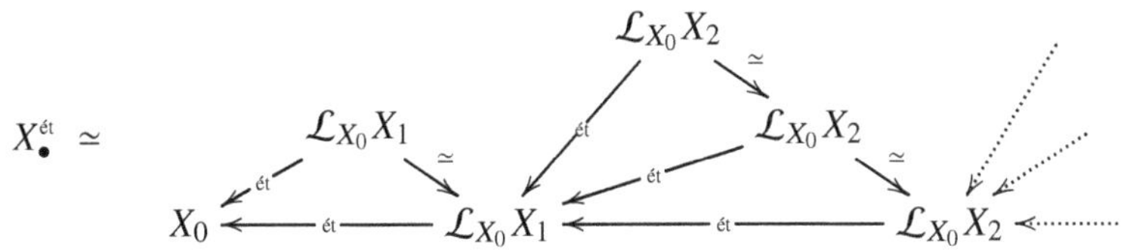

(9.111)

Observe that:

(a) the simplicial diagram (9.111) is again a groupoid, since the right adjoint functor $\mathcal{L}_{X_0}$ preserves the characterizing fiber products (8.143) (by Prop. 8.1.42);

(b) this groupoid is étale (Def. 9.1.37), since the morphisms of the form $\mathcal{L}_{X_0}X_n \longrightarrow X_0$ in (9.111) are local diffeomorphisms by construction, whence all other morphisms $\mathcal{L}_{X_0}X_{n_1} \longrightarrow \mathcal{L}_{X_0}X_{n_2}$ are local diffeomorphisms by the left-cancellation property (9.77).

Hence, we say that:

(i) The simplicial diagram (9.111) is the *étalification* of the groupoid $X_\bullet$.

$$X_\bullet^{\text{ét}} \in \text{ÉtGrpd}(\mathbf{H}) \,. \tag{9.112}$$

(ii) If the corresponding atlas of $X_\bullet$ (via Prop. 8.1.88) is denoted $X_0 \twoheadrightarrow \mathcal{X}$, then we write

$$X_0 \overset{\text{ét}}{\twoheadrightarrow} \mathcal{X}^{\text{ét}} \tag{9.113}$$

for the corresponding étale atlas (via Prop. 9.1.38) of the étalified groupoid (9.112).

9.1.3 Super Geometry

We present a formulation of super-geometry internal to ∞-toposes which we call *solid* [Sc13][Sc18].

9.1.3.1 Super-geometry

Definition 9.1.42 (Solid ∞-topos).

(i) An ∞-topos $\mathbf{H}$ (Def. 8.1.46) over $\mathbf{B} = \text{Grpd}_\infty$ is a *solid ∞-topos* if its base geometric morphism (Prop. 8.1.40), to be called $\text{Pnts} : \mathbf{H} \to \mathbf{B}$, is equipped with a factorization as follows, with adjoints (Def. 8.1.40) as shown:

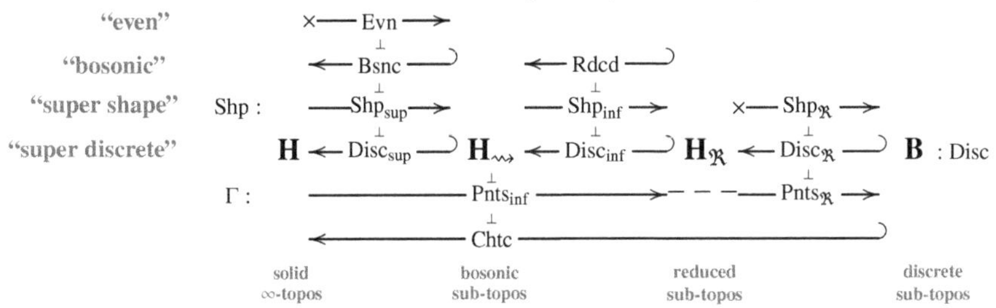

(9.114)

(ii) In particular, a solid ∞-topos is also an elastic ∞-topos (Def. 9.1.22), as is its sub-∞-topos $\mathbf{H}_{\rightsquigarrow}$ of bosonic objects.

(iii) We write

$$
\begin{array}{c}
\big(\underset{\text{“even”}}{\rightrightarrows} := \mathrm{Bsn} \circ \mathrm{Evn} \big) \\
\perp \\
\big(\underset{\text{“bosonic”}}{\rightsquigarrow} := \mathrm{Bsn} \circ \mathrm{Shp}_{\mathrm{sup}} \big) \\
\perp \\
\big(\underset{\text{“rheonomic”}}{\mathrm{Rh}} := \mathrm{Disc}_{\mathrm{sup}} \circ \mathrm{Shp}_{\mathrm{sup}} \big)
\end{array}
\quad : \quad \mathbf{H} \to \mathbf{H} \tag{9.115}
$$

for the further induced modalities (1.18) (*solid modalities*) accompanying the elastic modalities (9.60) and the cohesive modalities (9.2).

9.1.3.2 Examples of solid ∞-toposes

We indicate an example of a solid ∞-topos (Def. 9.1.42). In generalization of Def. 9.1.24, we have the following:

Definition 9.1.43 (∞-Haloed super Cartesian spaces).

(i) Write

$$
\begin{array}{ccc}
\mathrm{SprHldCrtSp} & \xhookrightarrow{\;C^\infty(-)\;} & \mathrm{sCAlg}_{\mathbb{R}}^{\mathrm{op}} \\
\mathbb{R}^{n|q} \times \mathbb{D}_W & \longmapsto & C^\infty(\mathbb{R}^n) \otimes_{\mathbb{R}} \wedge^\bullet_{\mathbb{R}}(\mathbb{R}^q) \otimes_{\mathbb{R}} (\mathbb{R} \oplus \mathrm{W})
\end{array}
\tag{9.116}
$$

for (as in [KS97][KS00]) the full subcategory of the opposite of super-commutative super-algebras over the real numbers on those which are tensor products of

(a) algebras $C^\infty(\mathbb{R}^n)$ of smooth functions on a Cartesian space $\mathbb{R}^n$, for $d \in \mathbb{N}$;
(b) Grassmann algebras $\wedge^\bullet_{\mathbb{R}}\mathbb{R}^q$ on $q \in \mathbb{N}$ generators in odd degree;
(c) finite-dimensional $\mathbb{R} \oplus W in$CAlg with a single nilpotent maximal ideal W.

(ii) We regard this as a site via the the coverage (i.e., a Grothendieck pre-topology) whose covers are of the form

$$
\Big\{ \underbrace{\mathbb{R}^n \times \mathbb{R}^{0|q}}_{\mathbb{R}^{n|q}} \times \mathbb{D} \xrightarrow{\;f_i \times \mathrm{id} \times \mathrm{id}\;} \mathbb{R}^n \times \mathbb{R}^{0|q} \times \mathbb{D} \Big\}_{i \in I}
\tag{9.117}
$$

such that $\big\{ \mathbb{R}^n \xrightarrow{\;f_i\;} \mathbb{R}^n \big\}_{i \in I}$ is a cover in CrtSpc (Def. 8.1.5).

Lemma 9.1.44 (Reflections of super-commutative algebras into commutative algebras). *The canonical inclusion of* ∞HldCrtSp *(Def. 9.1.24) into* SprHldCrtSp *(Def. 9.1.43) has a left and a right adjoint (Def. 8.1.40)*

$$
\mathrm{SprHldCrtSp}
\;\;
\begin{array}{c}
\xrightarrow{\;\mathrm{Evn}\;} \\
\perp \\
\xhookleftarrow{\;\mathrm{Bsnc}\;} \\
\perp \\
\xrightarrow{\;\mathrm{Shp}_{\mathrm{sup}}\;}
\end{array}
\;\;
\infty\mathrm{HldCrtSp}
\tag{9.118}
$$

where:

(i) *The left adjoint* Evn *in* (9.118) *is given in terms of super-algebras of smooth functions* (9.116) *by passage to the sub-algebra of even-graded elements:*

$$\begin{aligned} C^\infty\Big(\mathrm{Evn}(\mathbb{R}^{n|q}\times\mathbb{D})\Big) &\simeq C^\infty(\mathbb{R}^{n|q}\times\mathbb{D})_{\mathrm{even}} \\ &\simeq C^\infty(\mathbb{R}^{n}\times\mathbb{D})\otimes_{\mathbb{R}} C^\infty(\mathbb{R}^{0|q})_{\mathrm{even}}\,. \end{aligned} \tag{9.119}$$

(ii) *The right adjoint* $\mathrm{Shp}_{\mathrm{sup}}$ *in* (9.118) *is given in terms of super-algebras of smooth functions* (9.116) *by passage to the quotient algebra by the ideal of odd-graded elements:*

$$\begin{aligned} C^\infty\Big(\mathrm{Shp}_{\mathrm{sup}}(\mathbb{R}^{n|q}\times\mathbb{D})\Big) &\simeq C^\infty(\mathbb{R}^{n|q}\times\mathbb{D})/C^\infty(\mathbb{R}^{n|q}\times\mathbb{D})_{\mathrm{odd}} \\ &\simeq C^\infty(\mathbb{R}^{n}\times\mathbb{D})\otimes_{\mathbb{R}} \underbrace{C^\infty(\mathbb{R}^{0|q})/C^\infty(\mathbb{R}^{0|q})_{\mathrm{odd}}}_{\simeq\mathbb{R}} \\ &\simeq C^\infty(\mathbb{R}^{n}\times\mathbb{D}) \end{aligned} \tag{9.120}$$

and hence directly by

$$\mathrm{Shp}_{\mathrm{sup}}(\mathbb{R}^{n|q}\times\mathbb{D}) \simeq \mathbb{R}^{n}\times\mathbb{D}\,. \tag{9.121}$$

Proof. By regarding the situation under the defining embedding as being in $\mathrm{sCAlg}_{\mathbb{R}}$ (Def. 9.1.43), it is equivalent to the statement that the canonical inclusion of commutative algebras into super-commutative super-algebras has a right and a left adjoint given by passage to the even sub-algebra and to the quotient by the odd ideal, respectively:

$$\mathrm{sCAlg}_{\mathbb{R}} \begin{array}{c} \xrightarrow{\;A\,\mapsto\, A/A_{\mathrm{odd}}\;} \\ \bot \\ \hookleftarrow \\ \bot \\ \xrightarrow{\;A\,\mapsto\, A_{\mathrm{even}}\;} \end{array} \mathrm{CAlg}_{\mathbb{R}}\,. \tag{9.122}$$

This follows readily by inspection from the fact that homomorphisms of super algebras preserve super-degree, by definition. One place where this adjoint triple has been made explicit before is [CR12, Example 3.18]. □

Example 9.1.45 (Super-haloed smooth ∞-groupoids). The ∞-category of ∞ sheaves (Def. 8.1.58)

$$\mathrm{SprHldSmthGrpd}_\infty := \mathrm{Shv}_\infty(\mathrm{SprHldCrtSp}) \tag{9.123}$$

over the site from Def. 9.1.43 is a solid ∞-topos (Def. 9.1.42).

(i) Its bosonic (9.115) sub-∞-topos is that of ∞HldSmthGrpd (Example 9.1.26), and its reduced (9.59) sub-∞-topos that of $\mathrm{SmthGrpd}_\infty$ (Example 9.1.19):

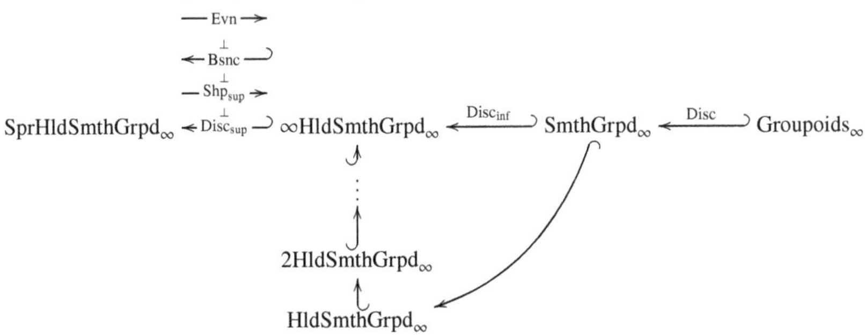

(9.124)

where the adjoint triple $(\mathrm{Evn} \dashv \mathrm{Bsnc} \dashv \mathrm{Shp}_{\mathrm{sup}})$ arises by left Kan extension from that of Lemma 9.1.44.

(ii) The full inclusion of SmthMfd, inherited from (9.52), extends to a full inclusion of super-manifolds (as in [CCF11, 4.6][HKST11, 2]):

$$\mathrm{SmthMfd} \xhookrightarrow{\mathrm{Disc_{sup}}} \mathrm{SuperManifolds} \hookrightarrow \mathrm{SprHldSmthGrpd}_\infty . \tag{9.125}$$

(iii) Accordingly, super-Lie groups (e.g. [Ya93][CCF11, 7]) embed faithfully into all group objects (Prop. 8.2.1):

$$\underset{\text{Lie groups}}{\mathrm{Grp}(\mathrm{SmthMfd})} \xhookrightarrow{\mathrm{Disc_{sup}}} \underset{\text{super Lie groups}}{\mathrm{Grp}(\mathrm{SuperManifolds})} \hookrightarrow \mathrm{Grp}(\mathrm{SprHldSmthGrpd}_\infty). \tag{9.126}$$

(iv) In particular, for $d \in \mathbb{N}$ and $\mathbf{N} \in \mathrm{Spin}(d,1)\mathrm{Representations}_{\mathbb{R}}$, the corresponding *supersymmetry* groups, i.e., the *super-Poincaré group* and its underlying translational *super-Minkowski group* (e.g. [Fr99, §3]) are group objects

$$\underset{\substack{\text{super-Minkowski}\\ \text{super Lie group}}}{\mathbb{R}^{d,1|\mathbf{N}}} \hookrightarrow \underset{\substack{\text{super-Poincaré}\\ \text{super Lie group}}}{\mathrm{Iso}(\mathbb{R}^{d,1|\mathbf{N}})} \twoheadrightarrow \mathrm{Spin}(d,1) \in \mathrm{Grp}(\mathrm{SprHldSmthGrpd}_\infty). \tag{9.127}$$

Remark 9.1.46 (Superspace cohomology theory in solid ∞-toposes)**.** The intrinsic cohomology (1.21) in the solid ∞-topos of $\mathrm{SprHldSmthGrpd}_\infty$ (Example 9.1.45)

(i) includes the super-rational cohomology of super-Minkowski spacetimes (9.127) that governs the fundamental (κ-symmetric) super p-brane sigma-models of string/M-theory [FSS15a][FSS17][FSS18], review in [FSS19].

(ii) Its enhancement to *twisted* super-rational cohomology of super-Minkowski spacetimes (9.127), which happens (by Remark 8.2.23) in the intrinsic cohomology of the tangent ∞-topos $T(\mathrm{SprHldSmthGrpd}_\infty)$ (Example 8.1.67), encodes the double dimensional reduction from fundamental M-branes to D-branes [BSS18].

(iii) Its enhancement to *proper equivariant* super-rational cohomology of super-Minkowski spacetimes (9.127), which happens (by Remark 11.1.4 and Theorem 11.1.9) in the intrinsic cohomology of the singular-solid ∞-topos $\mathrm{SnglrSprHldSmthGrpd}_\infty$ (Example 9.2.2), encodes also the black (solitonic) super p-branes [HSS18].

Lemma 9.1.47 (In super-geometric groupoids étale implies bosonic)**.** *In the solid ∞-topos of* $\mathrm{SprHldSmthGrpd}_\infty$ *(Ex. 9.1.43) we have a natural equivalence*

$$\rightsquigarrow \circ \Im \;\simeq\; \Im \tag{9.128}$$

saying that $\Im$-modal objects (9.60) *are bosonic* (9.115).

Proof. Observe that on $\mathrm{SprHldCrtSp} \xhookrightarrow{y} \mathrm{SprHldSmthGrpd}_\infty$ (Def. 9.1.43), we have a natural equivalence

$$\Re \circ \rightrightarrows \;\simeq\; \Re \tag{9.129}$$

saying that the reduction (9.60) of the even aspect (9.115) of the space is equivalently the reduced aspect.

To see this, consider $\mathbb{R}^{n|q} \times \mathbb{D}_W \in \mathrm{SprHldCrtSp}$ and use, by Example 9.1.45 with Lemma 9.1.44, the operation $\Re \circ \rightrightarrows$ is given in terms of the defining super-algebras of functions (9.1.43) by passage to the reduced algebra of the even subalgebra

$$
\begin{aligned}
C^\infty\Big(\Re \circ \rightrightarrows (\mathbb{R}^{n|q} \times \mathbb{D}_W)\Big) &\simeq \Big(\big(C^\infty(\mathbb{R}^n) \otimes_{\mathbb{R}} (\wedge^\bullet_{\mathbb{R}} \mathbb{R}^q) \otimes_{\mathbb{R}} (\mathbb{R} \oplus W)\big)_{\mathrm{even}}\Big)_{\mathrm{red}} \\
&\simeq \Big(C^\infty(\mathbb{R}^n) \otimes_{\mathbb{R}} \underbrace{(\wedge^\bullet_{\mathbb{R}} \mathbb{R}^q)_{\mathrm{even}}}_{\simeq \mathbb{R} \oplus \wedge^2 \mathbb{R}^q \oplus \wedge^4 \mathbb{R}^q \oplus \cdots} \otimes_{\mathbb{R}} (\mathbb{R} \oplus W)\Big)_{\mathrm{red}} \\
&\simeq \Big(C^\infty(\mathbb{R}^n) \otimes_{\mathbb{R}} \big(\mathbb{R} \oplus (W \oplus \wedge^2 \mathbb{R}^q \oplus \wedge^4 \mathbb{R}^q \oplus \cdots)\big)\Big)_{\mathrm{red}} \\
&\simeq C^\infty(\mathbb{R}^n) \otimes_{\mathbb{R}} \underbrace{\big(\mathbb{R} \oplus (W \oplus \wedge^2 \mathbb{R}^q \oplus \wedge^4 \mathbb{R}^q \oplus \cdots)\big)_{\mathrm{red}}}_{\simeq \mathbb{R}} \\
&\simeq C^\infty(\mathbb{R}^n) .
\end{aligned}
\tag{9.130}
$$

Here in the last step we used that every non-unit element in the Grassmann algebra is nilpotent. But, by (9.120) and (9.66), we also have

$$
\begin{aligned}
C^\infty\big(\Re(\mathbb{R}^{n|q} \times \mathbb{D}_W)\big) &\simeq C^\infty\big(\mathrm{Shp}_{\mathrm{inf}} \circ \mathrm{Shp}_{\mathrm{sup}}(\mathbb{R}^{n|q} \times \mathbb{D}_W)\big) \\
&\simeq C^\infty\big(\mathrm{Shp}_{\mathrm{inf}}(\mathbb{R}^n \times \mathbb{D}_W)\big) \\
&\simeq C^\infty(\mathbb{R}^n) ,
\end{aligned}
\tag{9.131}
$$

where in the first step we used the elastic structure (9.114) $\Re := \mathrm{Bsnc} \circ \mathrm{Rdcd} \circ \mathrm{Shp}_{\mathrm{inf}} \circ \mathrm{Shp}_{\mathrm{sup}}$ leaving the two full embeddings on the left notationally implicit. Since all these equivalences are natural, this implies (9.129). With this, we have the following sequence of natural equivalences for general $X \in \mathbf{H} := \mathrm{SprHldSmthGrpd}_\infty$:

$$
\begin{aligned}
\mathbf{H}\big(\mathbb{R}^{n|q} \times \mathbb{D} , \rightsquigarrow \circ \Im(X)\big) &\simeq \mathbf{H}\Big(\Re \circ \rightrightarrows (\mathbb{R}^{n|q} \times \mathbb{D}) , X\Big) \\
&\simeq \mathbf{H}\Big(\Re(\mathbb{R}^{n|q} \times \mathbb{D}) , X\Big) \\
&\simeq \mathbf{H}\big(\mathbb{R}^{n|q} \times \mathbb{D} , \Im X\big) ,
\end{aligned}
\tag{9.132}
$$

where the first and the last steps are the defining hom-equivalences (8.71) while the middle step is (9.129). Thus the statement (9.128) follows by the ∞-Yoneda lemma (Prop. 8.1.54). □

9.2 Singularities

Given a cohesive ∞-topos $\mathbf{H}_\cup$ as in §9.1.1, we construct here a new ∞-topos $\mathbf{H}$ (Def. 9.2.3), to be called *singular-cohesive*, with the following properties:

(i) **H** contains (9.151) for each finite group G, an object $\underset{\curlyvee}{G} \in \mathbf{H}$, to be thought of as the generic G-orbi-singularity (*Figure D*).

(ii) **H** carries (Prop. 9.2.5) an adjoint triple of modalities (1.18) to be read as follows

$$\begin{array}{c} \vee \\ \text{“singular”} \\ \perp \\ \cup \\ \text{“smooth”} \\ \perp \\ \curlyvee \\ \text{“orbi-singular”} \end{array} \quad : \quad \mathbf{H} \to \mathbf{H}\,, \tag{9.133}$$

with $\mathbf{H}_{\cup}$ being the full sub-∞-category of smooth objects in **H**,

(iii) such that (Prop. 9.2.18):

$\vee(\underset{\curlyvee}{G}) \simeq *$ “The purely singular aspect of an orbi-singularity is the quotient of a point, hence a point.”

$\cup(\underset{\curlyvee}{G}) \simeq * /\!\!/ G$ “The purely smooth aspect of an orbi-singularity is the homotopy quotient of a point.”

$\curlyvee(\underset{\curlyvee}{G}) \simeq \underset{\curlyvee}{G}$ “An orbi-singularity is purely orbi-singular.”

Essentially this list of conditions might completely characterize **H** as in Def. 9.2.3. Here we leave a fully axiomatic characterization of singular cohesion as an open problem and are content with making the following definitions:

9.2.1 Singular cohesive geometry

Definition 9.2.1 (The 2-site of singularities)**.**

(i) We write

$$\mathrm{Snglrt} := \mathrm{Grpd}_{\leq 1,\mathrm{cn},\mathrm{fin}} \hookrightarrow \mathrm{Grpd}_{\infty} \tag{9.134}$$

for the full sub-∞-category of ∞-groupoids on the connected 1-truncated objects whose π_1 is finite.

(ii) A skeleton of this (2, 1)-category has, of course, as objects the delooping groupoids (Example 8.1.30) $* /\!\!/ G$ that are presented by a single object and a *finite* group G of automorphisms of that object.

(iii) When regarded as objects of Snglrt in (9.134), we will denote these by “$\curlyvee$” attached to the symbol for the group:

$$\begin{array}{ccc} \underset{\curlyvee}{G} & \in & \mathrm{Snglrt} \\ \downarrow & & \downarrow \\ * /\!\!/ G & \in & \mathrm{Grpd}_{\infty} \end{array} \tag{9.135}$$

(iv) The hom-∞-groupoids between these singularities are, equivalently, the action groupoids (Example 8.1.31) whose objects are group homomorphisms and whose morphisms are conjugation actions on these:

$$\begin{aligned} \mathrm{Snglrt}(\underset{\curlyvee}{G_1}, \underset{\curlyvee}{G_2}) &:= \mathrm{Grpd}_{\infty}(* /\!\!/ G_1, * /\!\!/ G_2) \\ &\simeq \mathrm{Grp}(G_1, G_2) /\!\!/_{\mathrm{conj}} G_2\,. \end{aligned} \tag{9.136}$$

(v) We regard Snglrt as an ∞-site with trivial Grothendieck topology, so that ∞-sheaves on Snglrt are ∞-presheaves (8.92).

Remark 9.2.2 (The global orbit category)**.** The category Snglrt in Def. 9.2.1 is sometimes known in the literature as the "global orbit category" (though at other times this term is used for its wide but non-full subcategory of the faithful morphisms). It has elsewhere been denoted: "Orb case ①" (in [HG07, 4.1]), "Glob" (in [Re14, 2.2]), "Orb" (in [Kö16, 2.1][Ju20, 3.2]), and (up to equivalence) "$\mathbf{O}_{\mathrm{gl}}$" (in [Schw17][Kö16, 2.2]). The terminology in Def. 9.2.1 is meant to be more suggestive of the role this category plays in the theory, from the perspective of cohesive homotopy theory. In fact, the (global) orbit category is often taken to contain not just all finite groups but all compact Lie groups, with the hom-spaces then being the geometric realization of the topological mapping groupoids. We restrict to discrete groups (hence finite if compact) for reasons explained in Remark 9.2.20. This restriction is also amplified in [DHLPS19].

Definition 9.2.3 (Singular-cohesive ∞-topos)**.** Consider a cohesive ∞-topos (Def. 9.1.1), now to be denoted with "$\scriptstyle\cup$"-subscripts

$$
\mathbf{H}_{\cup}
\;
\begin{array}{c}
\xleftarrow{\;\;\mathrm{Shp}\;\;} \\
\bot \\
\xleftarrow{\;\;\mathrm{Dsc}\;\;} \\
\bot \\
\xrightarrow{\;\;\mathrm{Pnts}\;\;} \\
\bot \\
\xleftarrow{\;\;\mathrm{Chtc}\;\;}
\end{array}
\;
\mathbf{B}_{\cup} \;:=\; \mathrm{Grpd}_{\infty}
\tag{9.137}
$$

and assumed to have a site of Chrt (Def. 9.1.10). The corresponding *singular-cohesive* ∞-topos is that of $\mathbf{H}_{\cup}$-valued ∞-sheaves (8.92) over the site of Snglrt (Def. 9.2.1):

$$
\begin{array}{ccc}
\mathbf{H} := \mathrm{Shv}_{\infty}(\mathrm{Snglrt},\, \mathbf{H}_{\cup}) & \underset{\mathrm{Smth}}{\overset{\mathrm{NnOrbSnglr}}{\underset{\bot}{\rightleftarrows}}} & \mathbf{H}_{\cup} \\
\mathrm{Disc}\uparrow \dashv \downarrow\mathrm{Pnts} & & \mathrm{Disc}\uparrow \dashv \downarrow\mathrm{Pnts} \\
\mathbf{B} := \mathrm{Shv}_{\infty}(\mathrm{Snglrt},\, \mathbf{B}_{\cup}) & \underset{\mathrm{Smth}}{\overset{\mathrm{NnOrbSnglr}}{\underset{\bot}{\rightleftarrows}}} & \mathbf{B}_{\cup}\,,
\end{array}
\tag{9.138}
$$

where horizontally we are showing the base geometric morphisms (Prop. 8.1.59) of sheaves over the site Snglrt, while vertically we are showing the base geometric morphism (9.1) of $\mathbf{H}_{\cup}$ over $\mathbf{B}_{\cup}$ extended objectwise over Snglrt, by functoriality.

Lemma 9.2.4 (Singularities are 2-site for homotopical cohesion)**.** *The 2-site* Snglrt *(Def. 9.2.1) is an ∞-site for homotopical cohesion, in the sense of Def. 9.1.17.*

Proof. It is immediately checked that

(i) the terminal object is given by the trivial group:

$$
* \;\simeq\; \underset{\curlyvee}{1}
\tag{9.139}
$$

(ii) Cartesian product is direct product of groups:

$$
\underset{\curlyvee}{G_1} \times \underset{\curlyvee}{G_2} \;\simeq\; \underset{\curlyvee}{G_1 \times G_2}\,.
\tag{9.140}
$$

□

Proposition 9.2.5 (Singular cohesion)**.** *A singular-cohesive ∞-topos (Def. 9.2.3)*

$$
\begin{array}{ccccc}
 & & \mathbf{H} & & \\
 & \overset{\mathrm{Smth}}{\swarrow} & & \overset{\mathrm{Pnts}}{\searrow} & \\
\mathbf{H}_{\rotatebox[origin=c]{70}{$\subset$}} & & & & \mathbf{B} \\
 & \underset{\mathrm{Pnts}}{\searrow} & & \underset{\mathrm{Smth}}{\swarrow} & \\
 & & \mathbf{B}_{\rotatebox[origin=c]{70}{$\subset$}} & &
\end{array}
\tag{9.141}
$$

is itself cohesive (Def. 9.1.1) in two ways:

(i) *over the singular-base ∞-topos* $\mathbf{B}$ *by the cohesion of* $\mathbf{H}_{\rotatebox[origin=c]{70}{$\subset$}} \to \mathbf{B}_{\rotatebox[origin=c]{70}{$\subset$}}$ (9.1) *applied object-wise over all* Snglrt

$$
\begin{array}{lccc}
\text{“shape”} & & \xleftarrow{\quad}\!\!\times\;\mathrm{Shp}\;\longrightarrow & \\
 & & \bot & \\
\text{“discrete”} & & \longleftarrow\;\mathrm{Disc}\;\hookrightarrow & \\
 & & \bot & \\
\text{“points”} & \mathbf{H} & \longrightarrow\;\mathrm{Pnts}\;\longrightarrow & \mathbf{B}\,; \\
 & & \bot & \\
\text{“chaotic”} & & \longleftarrow\;\mathrm{Chtc}\;\hookrightarrow &
\end{array}
\tag{9.142}
$$

(ii) *over the non-singular cohesive base ∞-topos* $\mathbf{H}_{\rotatebox[origin=c]{70}{$\subset$}}$ *(Def. 9.1.1) in that the global section geometric morphism* $\mathbf{H} \xrightarrow{\mathrm{Smth}} \mathbf{H}_{\rotatebox[origin=c]{70}{$\subset$}}$ *of* (9.138) *is part of a cohesive adjoint quadruple, to be denoted*

$$
\begin{array}{lccc}
\text{“singular”} & & \times\!\!\longleftarrow\;\mathrm{Snglr}\;\longrightarrow & \\
 & & \bot & \\
\text{“not orbi-singular”} & & \longleftarrow\;\mathrm{NnOrbSnglr}\;\hookrightarrow & \\
 & & \bot & \\
\text{“smooth”} & \mathbf{H} & \longrightarrow\;\mathrm{Smth}\;\longrightarrow & \mathbf{H}_{\rotatebox[origin=c]{70}{$\subset$}}\,. \\
 & & \bot & \\
\text{“orbi-singular”} & & \longleftarrow\;\mathrm{OrbSnglr}\;\hookrightarrow &
\end{array}
\tag{9.143}
$$

Proof. The first statement is immediate. The second statement follows via Lemma 9.2.4 by Example 9.1.18. □

Notation 9.2.6 (Singular-elastic/solid ∞-topos)**.** Let $\mathbf{H}$ be a singular-cohesive ∞-topos (Def. 9.2.3) with an underlying smooth cohesive ∞-topos $\mathbf{H}_{\rotatebox[origin=c]{70}{$\subset$}} \hookrightarrow \mathbf{H}$. Then

(i) if $\mathbf{H}_{\rotatebox[origin=c]{70}{$\subset$}}$ is in fact an elastic ∞-topos (Def. 9.1.22), we say that $\mathbf{H}$ is a *singular-elastic ∞-topos*;

(ii) if $\mathbf{H}_{\rotatebox[origin=c]{70}{$\subset$}}$ is in fact a solid ∞-topos (Def. 9.1.42), we say that $\mathbf{H}$ is a *singular-solid ∞-topos*.

Definition 9.2.7 (Singular-cohesive modalities)**.** Given a singular cohesive ∞-topos (Def. 9.2.3), with its singular cohesion from Prop. 9.2.5, we write

$$
\begin{array}{c}
\big(\rotatebox[origin=c]{70}{$<$} \;:=\; \mathrm{NnOrbSnglr} \circ \mathrm{Snglr}\big) \\
\text{“singular”} \\
\bot \\
\big(\rotatebox[origin=c]{70}{$\subset$} \;:=\; \mathrm{NnOrbSnglr} \circ \mathrm{Smth}\big) \\
\text{“smooth”} \\
\bot \\
\big(\rotatebox[origin=c]{70}{$\prec$} \;:=\; \mathrm{OrbSnglr} \circ \mathrm{Smth}\big) \\
\text{“orbi-singular”}
\end{array}
\quad : \quad \mathbf{H} \to \mathbf{H}
\tag{9.144}
$$

for the adjoint triple of modalities $\mathbf{H} \longrightarrow \mathbf{H}$ induced (1.18) via (9.143); accompanying the cohesive modalities (9.2) induced via (9.142).

The above terminology reflects the difference (see Figure D) between a plain singularity $\rotatebox[origin=c]{70}{$<$}$ (singular but not orbi-singular) as opposed to its enhancement to an actual orbifold singularity $\curlyvee$. We record the following elementary but important consequence:

Proposition 9.2.8 (Smooth orbi-singular is smooth). *The singularity modalities (Def. 9.2.7) satisfy:*

$$\vee \circ \cup \simeq \cup \quad \textit{and} \quad \cup \circ \curlyvee \simeq \cup\,. \tag{9.145}$$

Proof. As in Prop. 9.1.2. □

Lemma 9.2.9 (Objectwise application of singularity modalities). *The singular-modalities in* (9.143) *are computed objectwise over* Chrt, *as in Example 9.1.18, followed by ∞-sheafification* L_{Chrt} (8.100)*:*

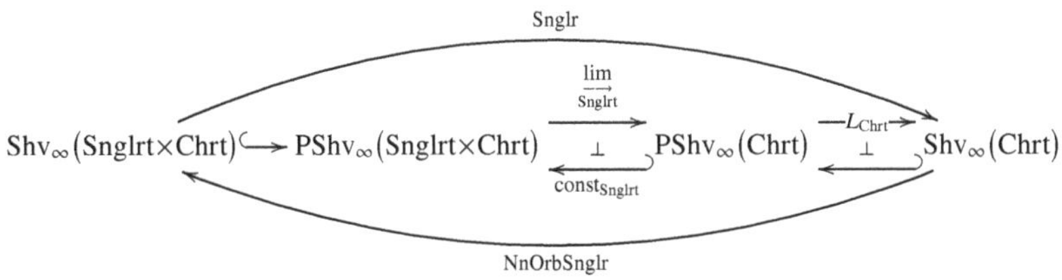

(9.146)

Proof. By essential uniqueness of adjoints (8.71). □

9.2.2 Examples of singular-cohesive ∞-toposes

First of all there is the *geometrically discrete* base case:

Example 9.2.10 (Singular ∞-groupoids). For $\mathbf{H}_{\cup} := \mathrm{Grpd}_\infty$ the base ∞-topos of plain ∞-groupoids (8.48), the singular-cohesive ∞-topos from Def. 9.2.3

$$\mathrm{SingularGroupoids}_\infty := \mathrm{Shv}_\infty(\mathrm{Snglrt},\, \mathrm{Grpd}_\infty) \tag{9.147}$$

is that of traditional unstable global homotopy theory [Schw18, §1s], as discussed in this form in [Re14, §4.1] (here with evaluation on all finite groups instead of all compact Lie groups).

What we will be mainly concerned with here are the following *smooth geometric* versions of singular cohesion (Ex. 9.2.11), which combine the smooth structure of cohesive, elastic, and solid ∞-toposes with orbi-singular geometry:

Example 9.2.11 (Singular-smooth ∞-groupoids).

(i) We call the singular-cohesive ∞-topos (Def. 9.2.3) over those of smooth ∞-groupoids (Example 9.1.19) the ∞-topos of *singular-smooth ∞-groupoids*:

$$\begin{aligned} \mathrm{SnglrSmthGrpd}_\infty &:= \mathrm{Shv}_\infty(\mathrm{Snglrt},\, \mathrm{SmthGrpd}_\infty) \\ &\simeq \mathrm{Shv}_\infty(\mathrm{CrtSpc} \times \mathrm{Snglrt})\,. \end{aligned} \tag{9.148}$$

(ii) We call the singular-elastic ∞-topos (Def. 9.2.6) over $\mathrm{HldSmthGrpd}_\infty$ (Example 9.1.26)

$$\begin{aligned} \mathrm{SnglrHldSmthGrpd}_\infty &:= \mathrm{Shv}_\infty(\mathrm{Snglrt},\, \mathrm{HldSmthGrpd}_\infty) \\ &\simeq \mathrm{Shv}_\infty(\mathrm{HldCrtSp} \times \mathrm{Snglrt})\,. \end{aligned} \tag{9.149}$$

(iii) We call the singular-solid ∞-topos (Def. 9.2.6) over $\mathrm{SprHldSmthGrpd}_\infty$ (Example 9.1.45)

$$\begin{aligned}\mathrm{SnglrSprHldSmthGrpd}_\infty &:= \mathrm{Shv}_\infty\big(\mathrm{Snglrt},\ \mathrm{SprHldSmthGrpd}_\infty\big)\\ &\simeq \mathrm{Shv}_\infty\big(\mathrm{SprHldCrtSp} \times \mathrm{Snglrt}\big)\,.\end{aligned} \tag{9.150}$$

For the second lines of (9.148), (9.148), and (9.150), see Lemma 9.2.16.

Remark 9.2.12 (On the nature of these singular-smooth ∞-topoi)**.** It is worth commenting on the nature of the examples in Ex. 9.2.11 (and generally of the specific nature of our Def. 9.2.3):

(i) The objects $\mathbb{R}^n \times \prec^{G} \in \mathrm{CrtSpc} \times \mathrm{Snglrt}$ of the site appearing in (9.148) may be understood as very simple elementary orbifolds, namely those of the form $\mathbb{R}^n /\!\!/ G$ for *trivial* G-action. In the perspective of toposes as categories of spaces that admit *probes* by objects of a site (cf. §6), the factor $\prec^{G}$ serves to detect orbisingular (G-fixed) loci in generalized singular-cohesive spaces (as in Ex. 9.2.10), and then the factor $\mathbb{R}^n$ serves to detect the smooth structure *of that locus*.

(ii) The product structure of these simple elementary probes ensures the commutativity of the cohesive with the singular modalities as in (9.138) (further expanded on in §9.2.4), which is the driving force behind the synthetic axiomatics of orbifold geometry developed in §10 and which is exactly what is needed for the construction of orbi/equivariant nonabelian differential cohomology as explained in §7.

(iii) On the other hand, in other contexts, where the interest is neither in synthetic axiomatization of orbifold differential geometry nor in their nonabelian differential cohomology, the construction in (9.148) suggests an evident variant where the site $\mathrm{CrtSpc} \times \mathrm{Snglrt}$ of simple elementary orbifolds is replaced by a suitable site of *all* orbifolds. This variant ∞-topos is considered in [CCL26].

9.2.3 Basic properties of singular cohesion

Definition 9.2.13 (Orbi-singularities)**.** Let $\mathbf{H}$ be singular-cohesive ∞-topos (Def. 9.2.3).

(i) We regard the objects $\prec^{G} \in \mathrm{Snglrt}$ (9.135) as objects of $\mathbf{H}$ under the ∞-Yoneda-embedding (Prop. 8.1.53) and the inclusion (9.138) of discrete objects:

$$\prec^{G} \in \mathrm{Snglrt} \overset{y}{\hookrightarrow} \mathrm{Shv}_\infty\big(\mathrm{Snglrt}, \mathbf{B}_{\cup}\big) \overset{\mathrm{Disc}}{\hookrightarrow} \mathrm{Shv}_\infty\big(\mathrm{Snglrt}, \mathbf{H}_{\cup}\big) = \mathbf{H}\,. \tag{9.151}$$

(ii) More generally, for

$$G \in \mathrm{Grp}(\mathrm{Grpd}_\infty) \xrightarrow{\mathrm{Grp(Disc)}} \mathrm{Grp}(\mathbf{H}_{\cup}) \tag{9.152}$$

any discrete ∞-group (9.142), we also write

$$\prec^{G} := \curlyvee(\mathbf{B}G) \in \mathbf{H} \tag{9.153}$$

for the orbi-singularization (9.143) of its delooping (8.151).

Lemma 9.2.17 shows that the two notations in Def. 9.2.13 are consistent with each other.

Remark 9.2.14 (Finite groups in singular cohesion). Given a singular-cohesive ∞-topos (Def. 9.2.3), the images of a finite group G under the following sequence of inclusions are naturally all denoted by the same symbol:

$$\begin{array}{ccccccccc} \mathrm{Grp}^{\mathrm{fin}} & \hookrightarrow & \mathrm{Grp}(\mathrm{Set}) & \hookrightarrow & \mathrm{Grp}(\mathrm{Grpd}_\infty) & \xhookrightarrow{\mathrm{Grp}(\mathrm{Disc})} & \mathrm{Grp}(\mathbf{H}_{\cup}) & \xhookrightarrow{\mathrm{Grp}(\mathrm{NnOrbSnglr})} & \mathrm{Grp}(\mathbf{H}) \\ G & \longmapsto & G & \longmapsto & G & \longmapsto & G & \longmapsto & G\,. \end{array} \tag{9.154}$$

With this understood, we also have identifications as follows (where now the ambient ∞-categories are implicit from the context):

$$\ast /\!\!/ G \;\simeq\; \mathrm{Disc}(\ast /\!\!/ G) \quad \text{and} \quad \prec^{G} \;\simeq\; \mathrm{Disc}(\prec^{G}) \tag{9.155}$$

where on the right we are recalling the definition (9.151).

Similarly:

Remark 9.2.15 (Smooth charts in singular cohesion). Consider a singular-cohesive ∞-topos (Def. 9.2.3) with an ∞-site Chrt of charts (Def. 9.1.10). Then images of the charts $U \in \mathrm{Chrt}$ under the ∞-Yoneda embedding (Prop. 8.1.53), and further under NnOrbSnglr (9.138), are naturally denoted by the same symbol:

$$\begin{array}{ccccc} S & \xrightarrow{\;y\;} & \mathbf{H}_{\cup} & \xrightarrow{\mathrm{NnOrbSnglr}} & \mathbf{H} \\ U & \longmapsto & U & \longmapsto & U\,. \end{array} \tag{9.156}$$

Lemma 9.2.16 (∞-Yoneda on product site). *Consider a singular-cohesive ∞-topos* $\mathbf{H}$ *(Def. 9.2.3) with an ∞-site* Chrt *of cohesive charts (Def. 9.1.10) for* $\mathbf{H}_{\cup}$.

(i) *Then a site (Def. 8.1.58) for the full singular-cohesive* $\mathbf{H}$ *is the Cartesian product site*

$$\mathrm{SingularCharts} \;:=\; \mathrm{Chrt} \times \mathrm{Snglrt} \tag{9.157}$$

in that

$$\mathbf{H} \;\simeq\; \mathrm{Shv}_\infty(\mathrm{Chrt} \times \mathrm{Snglrt})\,. \tag{9.158}$$

(ii) *Under the ∞-Yoneda embedding (Prop. 8.1.53), objects in the product site map to the Cartesian product of their prolonged Yoneda embeddings (in the sense of Remark 9.2.14 and Remark 9.2.15):*

$$\begin{array}{ccc} \mathrm{Chrt} \times \mathrm{Snglrt} & \xrightarrow{\;\;y\;\;} & \mathbf{H} \\ (U, \prec^{G}) & \longmapsto & U \times \prec^{G}\,, \end{array} \tag{9.159}$$

where on the right we are using the abbreviated notation from (9.151) *and* (9.156).

Proof. On the one hand, we have a natural equivalence

$$\mathbf{H}\Big(y(U_1, \prec^{G_1})\,,\; y(U_2, \prec^{G_2})\Big) \;\simeq\; \mathrm{Chrt}(U_1, U_2) \times \mathrm{Snglrt}(\,\ast /\!\!/ G_1\,,\; \ast /\!\!/ G_2) \tag{9.160}$$

by fully-faithfulness of the ∞-Yoneda embedding (Prop. 8.1.53) and by the definition of product sites. On the other hand, we have a sequence of natural equivalences

$$
\begin{aligned}
&\mathbf{H}\Big(y(U_1, \orbisingular^{G_1}) \,,\; U_2 \times \orbisingular^{G_2}\Big) \\
&\simeq \mathbf{H}\Big(y(U_1, \orbisingular^{G_1}),\, \mathrm{NnOrbSnglr}(U_2) \times \mathrm{Disc}(\orbisingular^{G_2})\Big) \\
&\simeq \mathbf{H}\Big(y(U_1, \orbisingular^{G_1}),\, \mathrm{NnOrbSnglr}(U_2)\Big) \times \mathbf{H}\Big(y(U_1, \orbisingular^{G_1}),\, \mathrm{Disc}(\orbisingular^{G_2})\Big) \\
&\simeq \mathbf{H}_{\cup}\Big(\mathrm{Snglr}\big(y(U_1, \orbisingular^{G_1})\big),\, U_2\Big) \times \mathbf{B}\Big(\textstyle\int\big(y(U_1, \orbisingular^{G_1})\big),\, \orbisingular^{G_2}\Big) \\
&\simeq \mathbf{H}_{\cup}(U_1, U_2) \times \mathbf{B}(\orbisingular^{G_1},\, \orbisingular^{G_2}) \\
&\simeq \mathrm{Chrt}(U_1, U_2) \times \mathrm{Snglrt}(\orbisingular^{G_1},\, \orbisingular^{G_2}).
\end{aligned}
\tag{9.161}
$$

Here the first step is by definition, the second step is the universal property of the Cartesian product, and the third step is the hom-equivalence (8.71) of the adjunctions $\mathrm{Snglr} \dashv \mathrm{NnOrbSnglr}$ and $\int \dashv \mathrm{Disc}$, respectively. In the fourth step, we use (9.29) and (9.163), respectively. The last step is the fully-faithfulness of the ∞-Yoneda embedding (Prop. 8.1.53). Since both (9.160) and (9.161) are natural in $\big(U', (*/\!\!/G)_{\curlyvee}\big)$, and since their right-hand sides coincide, it follows by the ∞-Yoneda embedding (Prop. 8.1.53) that also the representatives of the left-hand sides coincide: $y\big(U_2, \orbisingular^{G_2}\big) \simeq U_2 \times \orbisingular^{G_2}$. □

Lemma 9.2.17 (Images and pre-images of orbi-singularities)**.** *Let* $\mathbf{H}$ *be a singular-cohesive* ∞*-topos (Def. 9.2.3). Then the images and pre-images of the generic singularities* $\orbisingular^{G}$ (9.151) *under the functors* (9.143) *exhibiting the singular cohesion are as follows (see Figure D):*

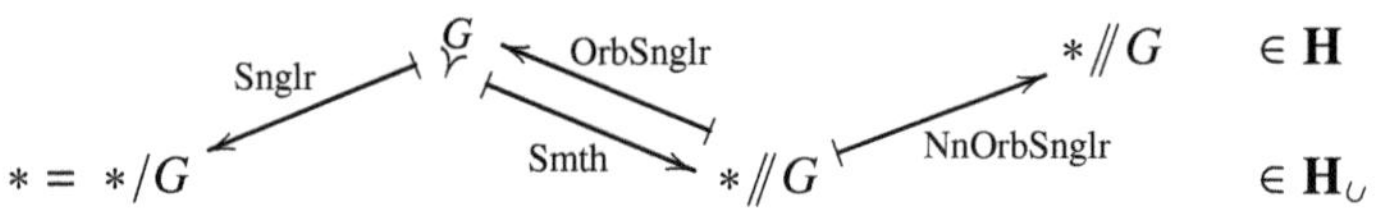

(9.162)

Proof. By the singular cohesion established in the proof of Prop. 9.2.5 we have that:

(i) the functor $\mathrm{Snglr} \simeq \varinjlim$ is the colimit functor (Prop. 8.1.52),

(ii) the functor $\mathrm{Smth} \simeq \mathrm{Snglrt}(\orbisingular^{1}, -)$ is the hom-functor (8.2) out of the terminal object (9.139).

Using this, we deduce the claim:

(i) Since colimits of representable ∞-functors are equivalent to the point (Lemma 8.1.56) we have

$$
\mathrm{Snglr}(\orbisingular^{G}) \;\simeq\; * \;\simeq\; */G\,. \tag{9.163}
$$

(ii) Observing that (9.136) reduces to $\mathrm{Snglrt}(\orbisingular^{1}, \orbisingular^{G}) \simeq */\!\!/G$ we have

$$
\mathrm{Smth}(\orbisingular^{G}) \;\simeq\; */\!\!/G\,. \tag{9.164}
$$

(iii) With this and by the various adjunctions we have, for $U \in \mathbf{H}_{\cup}$ any geometrically contractible generator (9.29) and $K \in \mathrm{Grp}^{\mathrm{fin}}$ any finite group, the following sequence of natural equivalences:

$$
\begin{aligned}
\mathbf{H}\Big(U \times \overset{K}{\rotatebox[origin=c]{70}{$\prec$}}, \mathrm{OrbSnglr}(\ast /\!\!/ G)\Big) &\simeq \mathbf{H}_{\rotatebox[origin=c]{70}{$\subset$}}\Big(\underbrace{\mathrm{Smth}(U \times \overset{K}{\rotatebox[origin=c]{70}{$\prec$}})}_{\simeq\, U \times \mathrm{Smth}(\overset{K}{\rotatebox[origin=c]{70}{$\prec$}})}, \ast /\!\!/ G\Big) \\
&\simeq \mathbf{H}_{\rotatebox[origin=c]{70}{$\subset$}}\big(U \times (\ast /\!\!/ K), \underbrace{\ast /\!\!/ G}_{\simeq \mathrm{Disc}(\ast /\!\!/ G)}\big) \\
&\simeq \mathrm{Grpd}_{\infty}\big(\underbrace{\mathrm{Shp}(U)}_{\simeq \ast} \times (\ast /\!\!/ K), \ast /\!\!/ G\big) \\
&\simeq \mathrm{Snglrt}\big(\overset{K}{\rotatebox[origin=c]{70}{$\prec$}}, \overset{G}{\rotatebox[origin=c]{70}{$\prec$}}\big) \\
&\simeq \mathbf{B}\Big(\underbrace{\mathrm{Shp}(U)}_{\simeq \ast} \times \overset{K}{\rotatebox[origin=c]{70}{$\prec$}}, \overset{G}{\rotatebox[origin=c]{70}{$\prec$}}\Big) \\
&\simeq \mathbf{B}\Big(\mathrm{Shp}(U \times \overset{K}{\rotatebox[origin=c]{70}{$\prec$}}), \overset{G}{\rotatebox[origin=c]{70}{$\prec$}}\Big) \\
&\simeq \mathbf{H}\Big(U \times \overset{K}{\rotatebox[origin=c]{70}{$\prec$}}, \mathrm{Disc}\big(\overset{G}{\rotatebox[origin=c]{70}{$\prec$}}\big)\Big) \\
&\simeq \mathbf{H}\big(U \times \overset{K}{\rotatebox[origin=c]{70}{$\prec$}}, \overset{G}{\rotatebox[origin=c]{70}{$\prec$}}\big),
\end{aligned}
\tag{9.165}
$$

where in several steps we recognized geometric discreteness, by (9.155) in Remark 9.2.14.

But, by Lemma 9.2.16, this chain of natural equivalences in total is a natural equivalence of the form

$$
\mathbf{H}\Big(y\big(U, \overset{K}{\rotatebox[origin=c]{70}{$\prec$}}\big), \mathrm{OrbSnglr}\big(\ast /\!\!/ G\big)\Big) \simeq \mathbf{H}\Big(y\big(U, \overset{K}{\rotatebox[origin=c]{70}{$\prec$}}\big), \overset{G}{\rotatebox[origin=c]{70}{$\prec$}}\Big). \tag{9.166}
$$

From this, the ∞-Yoneda embedding (Prop. 8.1.53) implies that $\mathrm{OrbSnglr}\big(\ast /\!\!/ G\big) \simeq \overset{G}{\rotatebox[origin=c]{70}{$\prec$}}$. □

It is useful to re-express this in terms of the modalities:

Proposition 9.2.18 (Orbi-singularities are orbi-singular)**.** *Let* $\mathbf{H}$ *be a singular-cohesive* ∞*-topos (Def. 9.2.3) and consider a finite group* $G \in \mathrm{Grp}^{\mathrm{fin}}$ (9.154)*. Then the images of the generic orbi-singularity* $\overset{G}{\rotatebox[origin=c]{70}{$\prec$}} \in \mathbf{H}$ (9.151) *under the modalities* (9.144) *are (see Figure D):*

$$
\rotatebox[origin=c]{70}{$<$}\big(\overset{G}{\rotatebox[origin=c]{70}{$\prec$}}\big) \simeq \ast, \qquad \rotatebox[origin=c]{70}{$\subset$}\big(\overset{G}{\rotatebox[origin=c]{70}{$\prec$}}\big) \simeq \ast /\!\!/ G, \qquad \rotatebox[origin=c]{70}{$\prec$}\big(\overset{G}{\rotatebox[origin=c]{70}{$\prec$}}\big) \simeq \overset{G}{\rotatebox[origin=c]{70}{$\prec$}}. \tag{9.167}
$$

Proof. This follows directly with Lemma 9.2.17 and the definition (9.144). For example:

$$
\rotatebox[origin=c]{70}{$\prec$}\big(\overset{G}{\rotatebox[origin=c]{70}{$\prec$}}\big) \simeq \underbrace{\mathrm{OrbSnglr} \circ \overbrace{\mathrm{Smth}\big(\overset{G}{\rotatebox[origin=c]{70}{$\prec$}}\big)}^{\simeq \ast /\!\!/ G}}_{\simeq \overset{G}{\rotatebox[origin=c]{70}{$\prec$}}}. \tag{9.168}
$$

□

In the same vein, we also have the following immediate but important property:

Proposition 9.2.19 (Orbi-singularities are geometrically discrete). *Let* $\mathbf{H}$ *be a singular-cohesive* ∞*-topos (Def. 9.2.3) and consider a finite group* $G \in \mathrm{Grp}^{\mathrm{fin}}$ (9.154).

(i) *Then the basic orbi-singularity* $\overset{G}{\curlyvee} \in \mathbf{H}$ (9.151) *is geometrically discrete* (9.2) *and thus also pure shape:*

$$\flat \overset{G}{\curlyvee} \simeq \overset{G}{\curlyvee}, \qquad \int \overset{G}{\curlyvee} \simeq \overset{G}{\curlyvee}. \tag{9.169}$$

(ii) *The same is true for* $\mathrm{Smth}(* /\!\!/ G)_{\curlyvee} \simeq * /\!\!/ G$*:*

$$\flat(* /\!\!/ G) \simeq * /\!\!/ G, \qquad \int(* /\!\!/ G) \simeq * /\!\!/ G. \tag{9.170}$$

Proof. Both statements follow immediately from the definitions and the fact that G is finite and hence geometrically discrete (9.154). □

Remark 9.2.20 (Need for discrete/finite groups in Snglrt). It is to make Lemma 9.2.17 and hence Prop. 9.2.18 true that Def. 9.2.1 requires the global orbit category Snglrt to consist of finite groups, instead of more general compact Lie groups (Remark 9.2.2): If Snglrt were to contain non-discrete compact Lie groups G, then the same argument as in Lemma 9.2.17 would give in (9.167) the following more general formula:

$$\cup \overset{G}{\curlyvee} \simeq * /\!\!/ \flat G \tag{9.171}$$

(where on the right we think of the Lie group G as being cohesive via (9.52)). Since the condition $G \simeq \flat G$ characterizes discrete groups, this would break Prop. 10.1.2, in that then the shape of the orbi-singularization of a topological groupoid would take non-traditional values on non-discrete groups in the global orbit category.

The following lemma further illustrates the nature of orbi-singular cohesion:

Lemma 9.2.21 (Smooth 0-truncated objects are orbi-singular). *Let* $\mathbf{H}$ *be a singular-cohesive* ∞*-topos (Def. 9.2.3). Then if* $X \in \mathbf{H}_{\cup,0}$ *is smooth* (9.144) *and 0-truncated (Def. 8.1.73), it is also orbi-singular* (9.144)*:*

$$\tau_0(X) \simeq X \;\; \text{and} \;\; \cup(X) \simeq X \qquad \Rightarrow \qquad \curlyvee(X) \simeq X. \tag{9.172}$$

Proof. Since X is smooth, there exists $X_{\cup} \in \mathbf{H}_{\cup}$ such that $X \simeq \mathrm{Smth}(X_{\cup})$. Observe that X being 0-truncated implies that $X_{\cup}$ is 0-truncated, (by using in Def. 8.1.73 the hom-equivalence (8.71) of the right adjoint Smth).

Now let $\mathcal{S}$ be any site (8.100) for $\mathbf{H}_{\cup}$. Then, for $U \in \mathcal{S} \hookrightarrow \mathbf{H}_{\cup}$ and $G \in \mathrm{Grp}^{\mathrm{fin}}$, we have the following sequence of natural equivalences, using the various adjoint functors, their idempotency, and respect for products:

$$
\begin{aligned}
(\Upsilon X)\big(\mathrm{Smth}(U) \times \underset{\Upsilon}{G}\big) &\simeq (\Upsilon \mathrm{Smth}(X_{\cup}))\big(\mathrm{Smth}(U) \times \underset{\Upsilon}{G}\big) \\
&\simeq \mathbf{H}\big(\mathrm{Smth}(U) \times \underset{\Upsilon}{G}, \Upsilon \mathrm{Smth}(X_{\cup})\big) \\
&\simeq \mathbf{H}\big(\mathrm{Smth}(U) \times \cup(\underset{\Upsilon}{G}), \mathrm{Smth}(X_{\cup})\big) \\
&\simeq \mathbf{H}\big(\mathrm{Smth}(U \times (* /\!\!/ G)), \mathrm{Smth}(X_{\cup})\big) \\
&\simeq \mathbf{H}_{\cup}\big(U \times (* /\!\!/ G), X_{\cup}\big) \\
&\simeq \mathrm{Grpd}_{\infty}\Big(* /\!\!/ G, \mathbf{H}_{\cup}\big(U, X_{\cup}\big)\Big) \\
&\simeq \mathrm{Grpd}_{\infty}\big(*, \mathbf{H}_{\cup}(U, X_{\cup})\big) \\
&\simeq \mathbf{H}_{\cup}\big(U, X_{\cup}\big) \\
&\simeq \mathbf{H}\big(\mathrm{Smth}(U), \mathrm{Smth}(X_{\cup})\big) \\
&\simeq \mathbf{H}\big(\mathrm{Smth}(U) \times \underset{\Upsilon}{G}, \mathrm{Smth}(X_{\cup})\big) \\
&\simeq \mathbf{H}\big(\mathrm{Smth}(U) \times \underset{\Upsilon}{G}, X\big) \\
&\simeq X\big(\mathrm{Smth}(U) \times \underset{\Upsilon}{G}\big).
\end{aligned}
\tag{9.173}
$$

Here the first and the last step use the ∞-Yoneda embedding (Prop. 8.1.53), while the middle step uses the fact that $X_{\cup}$ is 0-truncated, hence that $\mathbf{H}_{\cup}(U, X_{\cup})$ is 0-truncated (i.e. a set), to find that there is in fact no dependency on G. Hence the claim follows by the ∞-Yoneda embedding (Prop. 8.1.53), in view of Lemma 9.2.16. □

Remark 9.2.22 (Degenerate case of orbi-singular)**.** The natural language statement of Lemma 9.2.21 shows that the modality Υ "orbi-singular" (9.143) really means: "All singularities *that are present* are orbi-singularities", which becomes a trivially satisfied condition when there are no singularities, such as for smooth and 0-truncated objects.

9.2.4 Interplay between geometric and singular cohesion

Lemma 9.2.23 (Smooth commutes with shape)**.** *In a singular-cohesive ∞-topos (Def. 9.2.3), the smooth-modality* (9.144) *commutes with all three cohesive modalities* (9.2) *(as per Prop. 9.2.5):*

$$
\cup \circ \int \;\simeq\; \int \circ \cup\,, \qquad \cup \circ \flat \;\simeq\; \flat \circ \cup\,, \qquad \cup \circ \sharp \;\simeq\; \sharp \circ \cup\,. \tag{9.174}
$$

Proof. Under the defining identification $\mathbf{H} \simeq \mathrm{Shv}_{\infty}(\mathrm{Snglrt}, \mathbf{H}_{\cup})$, let $\mathcal{X} \in \mathbf{H}$ be any object regarded as a $\mathbf{H}_{\cup}$-valued ∞-presheaf on Snglrt:

$$
\mathcal{X} \;:\; \underset{\Upsilon}{K} \;\longmapsto\; \mathcal{X}(\underset{\Upsilon}{K}) \;\in\; \mathbf{H}_{\cup}\,. \tag{9.175}
$$

Observe then (by Example 9.1.18 via Lemma 9.2.4) that $\cup$ turns such a presheaf into the constant presheaf on its value at the terminal object $\underset{\Upsilon}{1}$:

$$
(\cup \mathcal{X}) \;:\; \underset{\Upsilon}{K} \;\longmapsto\; \mathcal{X}(\underset{\Upsilon}{1})\,. \tag{9.176}
$$

On the other hand, the geometric modalities operate objectwise over Snglrt (Remark 9.2.9):

$$
(\textstyle\int \mathcal{X}) \;:\; \underset{\Upsilon}{K} \;\longmapsto\; \textstyle\int(\mathcal{X}(\underset{\Upsilon}{K}))\,. \tag{9.177}
$$

With this, we have the following sequence of natural equivalences for $\mathcal{X} \in \mathbf{H}$ and $\prec^{K} \in \mathrm{Snglrt}$:

$$\begin{aligned}(\cup \int \mathcal{X})(\prec^{K}) &\simeq (\int \mathcal{X})(\prec^{1}) \\ &\simeq \int(\mathcal{X}(\prec^{1})) \\ &\simeq \int((\cup \mathcal{X})(\prec^{K})) \\ &\simeq (\int \cup \mathcal{X})(\prec^{K}) .\end{aligned} \tag{9.178}$$

Hence the claim follows by the ∞-Yoneda embedding (Prop. 8.1.53). The argument for $\flat$ and $\sharp$ is analogous. □

Remark 9.2.24 (Dichotomy between naive and proper orbifold cohomology via singular-cohesion)**.** In contrast to Lemma 9.2.23, the orbi-singular modality $\prec$ (9.144) does *not* commute with the cohesive shape modality $\int$ (9.2), in general. This phenomenon is the very source of the *proper equivariant* structure seen in singular-cohesive ∞-toposes, reflected in the following dichotomy between geometric- and homotopy fixed points of an orbi-space and in the distinction between proper- and Borel-equivariant cohomology:

	$\prec \circ \int$	$\int \circ \prec$	
Def. 9.2.25 (i)	Homotopy fixed-points	Geometric fixed-points	Def. 9.2.25 (ii)
Def. 11.1.1	Borel-equivariant cohomology	Proper equivariant cohomology	Def. 11.1.2
Def. 11.2.3	Tangentially twisted cohomology	Tangentially twisted proper orbifold cohomology	Def. 11.2.5

Definition 9.2.25 (Geometric- and homotopy-fixed points)**.** Let $\mathbf{H}$ be a singular-cohesive ∞-topos (Def. 9.2.3), $G \in \mathrm{Grp}(\mathbf{H})$ (Prop. 8.2.1) being discrete $G \simeq \flat G$ and 0-truncated $G \simeq \tau_0 G$, and $(X, \rho) \in G\mathrm{Actions}(\mathbf{H})$ (Prop. 8.2.6) with smooth $X \simeq \cup X$, hence

$$X \in \mathbf{H}_{\cup} \xhookrightarrow{\mathrm{NnOrbSinglr}} \mathbf{H} . \tag{9.179}$$

For any subgroup $K \subset G$, the ∞-groupoid of $\prec^{K}$-points in the slice (Prop. 8.1.62) over $\prec^{G}$ (9.153)...

(i) ...of the orbi-singularization (9.143) of the shape (9.142) of $X/\!\!/G$ is the *homotopy fixed point space* of X

$$\mathrm{HmtpFxdPntSpc}^{K}(X) := \mathbf{H}_{/\prec^{G}}\left(\prec^{K},\ \prec\int(X/\!\!/G)\right). \tag{9.180}$$

(ii) ...of the shape (9.142) of the orbi-singularization (9.143) of $X/\!\!/G$ is the *geometric fixed point space* of X

$$\mathrm{GmtrcFxdPntSpc}^{K}(X) := \mathbf{H}_{/\prec^{G}}\left(\prec^{K},\ \int\prec(X/\!\!/G)\right). \tag{9.181}$$

On the right we are using Prop. 9.2.18 and Prop. 9.2.19 to see that both expressions indeed live in the slice over $\prec^{G}$.

Proposition 9.2.26 (Homotopy-fixed point spaces are fixed loci in shapes). *The homotopy-fixed point spaces* (9.180) *of the G-space X in Def. 9.2.25 are, equivalently, the fixed-loci (Def. 8.2.24) of the shape* $\mathrm{Shp}(X) \in \mathrm{Grpd}_\infty$ (9.1) *of X:*

$$\mathrm{HmtpFxdPntSpcs}^K(X) \simeq \big(\mathrm{Shp}(X)\big)^K \quad \in \mathrm{Grpd}_\infty \tag{9.182}$$

with respect to the induced $G \simeq \int G$-action (using Prop. 9.1.4, discreteness of G, and cohesion in the form of Prop. 9.1.2).

Proof. We claim a sequence of natural equivalences as follows:

$$\begin{aligned}
&\mathrm{HmtpFxdPntSpc}^K(X)\\
&\equiv \mathbf{H}_{/\overset{G}{\Yright}}\Big(\overset{K}{\Yright},\ \Yright\textstyle\int(X /\!\!/ G)\Big)\\
&\simeq \mathbf{H}_{/\overset{G}{\Yright}}\Big(\overset{K}{\Yright},\ \Yright(\textstyle\int X) /\!\!/ G\Big)\\
&\simeq \mathbf{H}_{/\mathrm{OrbSnglr}(*/\!\!/G)}\Big(\mathrm{OrbSinglr}(*/\!\!/K),\ \mathrm{OrbSnglr}\big((\textstyle\int X)/\!\!/G\big)\Big)\\
&\simeq (\mathbf{H}_{\cup})_{/*/\!\!/G}\big(*/\!\!/K,\ (\textstyle\int X)/\!\!/G\big)\\
&\simeq (\mathrm{Grpd}_\infty)_{/*/\!\!/G}\big(*/\!\!/K,\ \mathrm{Shp}(X)/\!\!/G\big)\\
&\simeq (\mathrm{Grpd}_\infty)_{/*/\!\!/K}\big(*/\!\!/K,\ \mathrm{Shp}(X)/\!\!/K\big)\\
&\simeq (\mathrm{Shp}(X))^K .
\end{aligned} \tag{9.183}$$

Here, the first step is the definition (9.180), and the second step uses Prop. 9.1.4, discreteness of G, and cohesion in the form of Prop. 9.1.2. In the third step, we observe with $\overset{K}{\Yright} \simeq \Yright(*/\!\!/K)$ (Lemma 9.2.17) and $\Yright := \mathrm{OrbSinglr} \circ \mathrm{Smth}$ (9.144) that all objects and morphisms are in the image of OrbSnglr, and in the fourth step we use that this functor is fully faithful, by Prop. 9.2.5. In the fifth step, we similarly observe that all objects and morphisms are, in fact, furthermore, in the image of Disc (by assumption on G and by definition of $\int := \mathrm{Disc} \circ \mathrm{Shp}$ (9.2)), which is fully faithful by the axioms of cohesion (9.1). The sixth step observes the universal factorization through the pullback

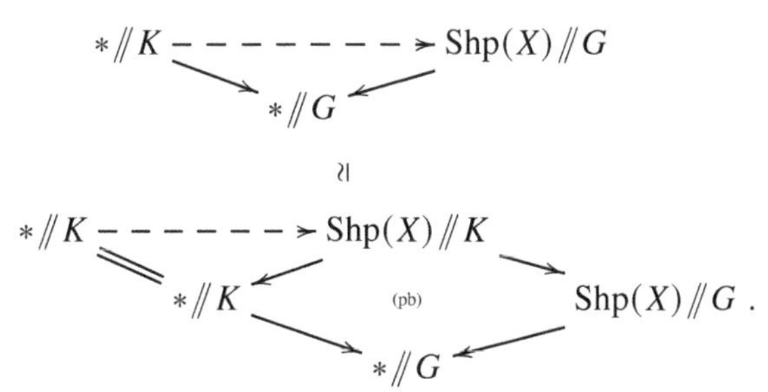

(9.184)

The pullback, in turn, is the homotopy quotient of the restricted action, as shown, by Prop. 8.2.12. With this, the last step follows by Example 8.2.26. In summary, the composite of the sequence of equivalences (9.183) gives the statement (9.182). □

Example 9.2.27 (Geometric fixed points generally differ from homotopy fixed points)**.** As in Example 9.2.11, let $\mathbf{H} := \mathrm{SnglrSmthGrpd}_\infty$. For $n \in \mathbb{N}$, $n \geq 1$, consider the Cartesian space $\mathbb{R}^n \in \mathrm{SmthMfd} \hookrightarrow \mathbf{H}$, via (9.52), and regard it as equipped with the additive translation action of $\mathbb{Z}^n$ induced from the left action of the additive group $(\mathbb{R}^n, +)$ on itself, under the canonical inclusion $(\mathbb{Z}^n, +) \hookrightarrow (\mathbb{R}^n, +)$:

$$(\mathbb{R}^n, \rho_\ell) \in \mathbb{Z}^n\mathrm{Actions}(\mathbf{H}) . \tag{9.185}$$

So the quotient of this action $\mathbb{R}^n /\!\!/ \mathbb{Z}^n \simeq \mathbb{R}^n/\mathbb{Z}^n \simeq \mathbb{T}^n \in \mathrm{SmthMfd} \hookrightarrow \mathbf{H}$ is the standard n-torus. We then have for the two notions of fixed-point spaces from Def. 9.2.25:

(i) The *Homotopy-fixed point space* (9.180) of the action (9.185) is equivalent to the point (by Prop. 9.2.26 and (9.51)):

$$\mathrm{HmtpFxdPntSpc}^{\mathbb{Z}^n}(\mathbb{R}^n) \simeq \big(\underbrace{\textstyle\int \mathbb{R}^n}_{\simeq *} \big)^{\mathbb{Z}^n} \simeq * \tag{9.186}$$

(ii) The *geometric fixed point space* (9.181) of the action (9.185) is empty

$$\mathrm{GmtrcFxdPntSpc}^{\mathbb{Z}^n}(\mathbb{R}^n) \simeq \big(\mathbb{R}^n\big)^{\mathbb{Z}^n} \simeq \varnothing . \tag{9.187}$$

This follows by Lemma 10.1.7, using that no element of the set underlying $\mathbb{R}^n$ is fixed by the action of $\mathbb{Z}^n$.

Part V

Geometric Orbifold Cohomology

10

Orbifold Geometry

Within an ambient context of singular-cohesive homotopy theory (§9), we now formulate the two geometric aspects of orbifolds:

- §10.1 – as cohesive spaces with orbi-singularities,
- §10.2 – as cohesive spaces locally equivalent to a given model space.

In the end, we combine both aspects to form the *proper ∞-categories of orbifolds*: this is Def. 10.2.46.

10.1 Higher Orbispaces

We observe (Prop. 10.1.2) that the shape of the orbi-singularization of a topological groupoid, regarded in singular-smooth homotopy theory (Ex. 9.2.11), is the corresponding *orbispace* in global equivariant homotopy theory.

Remark 10.1.1 (Orbispaces in topology and in global equivariant homotopy theory).

(i) Orbispaces in topology. The term *orbispace* was originally introduced [Hae90] to mean the topological version of orbifolds, i.e., Satake's original concept [Sa56] but disregarding any differentiable structure. From the perspective of étale groupoids/stacks, this means to consider topological groupoids/stacks instead of Lie groupoids/differentiable stacks. So this usage of the term "orbispace" serves to complete the following table:

Smooth manifold	Topological manifold	
Orbifold	Orbispace	*(geometric sense)*
Lie groupoid	Topological groupoid	
Differentiable stack	Topological stack	

In this sense, orbispaces have been discussed, e.g., in [Hae84][Hae91, §5] [Ch01][He01].

(ii) Orbispaces in global equivariant homotopy theory. In [HG07] it was suggested to change perspective and to instead regard these topological groupoids $\mathcal{X}_{\text{top}}$ via the systems of homotopy types of all their geometric fixed-point spaces,

DOI: 10.1201/9781003675969-10

by the following formula [HG07, 4.2] (beware the differing conventions, as per Remark 9.2.2):

$$G \longmapsto \underset{\substack{\text{topological} \\ \text{mapping groupoid}}}{\overset{\text{homotopy type of (fat) geometric realization of}}{\big\|\mathbf{Map}\big(\mathbf{B}G\,,\,\mathcal{X}_{\mathrm{top}}\big)\big\|}} \qquad \underset{(\textit{equivariant homotopical sense})}{\text{orbispace}} \tag{10.1}$$

This is a global-equivariant version of how topological G-spaces are incarnated in G-equivariant homotopy theory via Elmendorf's theorem (recalled as Prop. 8.1.27), and has served to motivate the development of global equivariant homotopy theory [Schw18].

In the course of this development, homotopy theorists adopted the term "orbispace" to refer not to the topological groupoid $\mathcal{X}_{\mathrm{top}}$ (as [Hae90] originally did) but rather to the global equivariant homotopy type that is represented via (10.1). Usage of the term *orbispace* in this sense of global homotopy theory is, after [HG07], in [Re14][Kö16][Schw17][Lu19, 3][Ju20]. In [Ju20, 3.15] formula 10.1 is used (following suggestions in [Schw17, Introd.][Schw18, pp. ix-x]) to define (abelian, non-geometric) cohomology of orbifolds with coefficients in global equivariant spectra.

Our Prop. 10.1.2 shows that these two different meanings of the term "orbispace" in the literature are disentangled as well as unified by the notion of singular cohesion (Def. 9.2.3) in that orbispaces in the sense **(ii)** are the shape $\int$ (9.1) of the orbi-singularization $\prec$ (9.144) of the topological groupoids in **(i)**:

$$\begin{array}{ccc} \mathrm{TopGrpd} & \xrightarrow{\ \int \circ \prec\ } & \mathrm{Orbispaces} \\ \mathcal{X}_{\mathrm{top}} & \longmapsto & \Big(\prec^{G} \;\mapsto\; \big\|\mathbf{Map}(\mathbf{B}G, \mathcal{X}_{\mathrm{top}})\big\| \Big). \end{array} \tag{10.2}$$

Hence Prop. 10.1.2 means that, before passing to their pure shape, we may think of the orbi-singularizations of objects in singular-cohesive ∞-toposes as *cohesive orbispaces*, lifting the concept of plain orbispaces in the sense **(ii)** from plain homotopy theory to geometric (differential, étale) homotopy theory, hence back to sense **(i)** and beyond.

The crucial fact underlying the phenomenon (10.2), both in Prop. 10.1.2 and in Lemma 10.1.7, is that the probe of an orbi-singular object $\prec\mathcal{X}_{\cup}$ by a generic orbi-singularity $\prec^{K}$ (9.135) is, by adjunction (9.144), equivalently the probe of the underlying smooth object by the smooth aspect of $\prec^{K}$, hence is, by (9.167) in Prop. 9.2.18, the geometric G-fixed locus in $\mathcal{X}_{\cup}$:

$$\prec^{G} \longrightarrow \prec\mathcal{X}_{\cup} \quad \overset{(9.144)}{\Leftrightarrow} \quad \cup\prec^{G} \longrightarrow \mathcal{X}_{\cup} \quad \overset{(9.167)}{\Leftrightarrow} \quad * /\!\!/ G \longrightarrow \mathcal{X}_{\cup}\,. \tag{10.3}$$

Equivalently, since $\prec^{G} \simeq \prec(* /\!\!/ G)$ (Lemma 9.2.17) the composite correspondence (10.3) is fully-faithfulness of $\prec$.

10.1.1 Ex.: Topological groupoids as cohesive orbispaces

Proposition 10.1.2 (Shape of orbi-singularized topological groupoid is orbispace). *Let* $\mathbf{H} := \mathrm{SnglrSmthGrpd}_\infty$ *(Ex. 9.2.11), and let*

$$\begin{array}{ccccc} \mathrm{TopGrpd} & \xrightarrow{\mathrm{Cdfflg}} & \mathrm{SmthGrpd}_\infty & \xrightarrow{\mathrm{NnOrbSnglr}} & \mathbf{H} \\ \mathcal{X}_{\mathrm{top}} & \longmapsto & & & \mathcal{X}_{\cup} \end{array} \tag{10.4}$$

be a topological groupoid, regarded via the embeddings (9.53) *and* (9.143). *If* $\mathcal{X}_{\cup}$ *is such that both its spaces of objects and of morphisms are retracts of cell complexes (for instance: both are CW-complexes* (8.5)*) then the shape* (9.142) *of its orbi-singularization* (9.144) *is, as an* ∞*-presheaf* (9.138) *of* ∞*-groupoids on* Snglrt (9.2.1) *(i.e., on the global orbit category, Remark 9.2.2)*

$$\int \curlyvee \mathcal{X}_{\cup} \;\in\; \mathrm{Sh}_\infty(\mathrm{Snglrt}) \xhookrightarrow{\mathrm{Disc}} \mathbf{H} \tag{10.5}$$

given by the assignment (10.3)

$$\int \curlyvee \mathcal{X}_{\cup} \;:\; \curlyvee^{G} \;\longmapsto\; \big\| \mathbf{Map}(\mathbf{B}G, \mathcal{X}_{\mathrm{top}}) \big\| \,, \tag{10.6}$$

where on the right we have the fat geometric realization of the topological functor groupoid [Se74] (see [HG07, 2.3]), with $\mathbf{B}G \;\simeq\; * /\!\!/ G$ *(Ex. 8.1.30) regarded as a finite topological groupoid.*

Proof. Recall from (9.51) in Ex. 9.1.19 that Chrt := CrtSpc (Def. 8.1.5) is a site of cohesive charts (Def. 9.1.10) for $\mathrm{SmthGrpd}_\infty$. We claim that for $\mathbb{R}^n \in \mathrm{CrtSpc}$ and $\curlyvee^{G} \in \mathrm{Snglrt}$ (Def. 9.2.1), hence $\mathbb{R}^n \times \curlyvee^{G} \in \mathrm{CrtSpc} \times \mathrm{Snglrt}$ (Lemma 9.2.16), we have thc following sequence of natural equivalences:

$$\begin{aligned} \mathbf{H}(\mathbb{R}^n \times \curlyvee^{G},\, \curlyvee \mathcal{X}_{\cup}) &= \mathbf{H}(\mathbb{R}^n \times \curlyvee^{G},\, \mathrm{OrbSnglr}(\mathcal{X}_{\cup})) \\ &\simeq \mathbf{H}_{\cup}\big(\underbrace{\mathrm{Smth}(\mathbb{R}^n \times \curlyvee^{G})}_{\simeq \mathbb{R}^n \times \mathbf{B}G} ,\, \mathcal{X}_{\cup} \big) \\ &\simeq \mathbf{H}_{\cup}\Big(\mathbb{R}^n ,\, \mathbf{Map}(\mathbf{B}G ,\, \mathcal{X}_{\cup}) \Big) \\ &\simeq \mathbf{H}_{\cup}\Big(\mathbb{R}^n ,\, \mathrm{Cdfflg}\mathbf{Map}(\mathbf{B}G ,\, \mathcal{X}_{\mathrm{top}}) \Big). \end{aligned} \tag{10.7}$$

Here the first step is (10.12), the second is the hom-equivalence (8.71) of the adjunction Smth ⊣ OrbSnglr (9.143) and using under the brace that Smth preserves products (by Prop. 8.1.42), that $\mathbb{R}^n$ is already smooth, and that $\mathrm{Smth}(\curlyvee^{G}) \simeq (* /\!\!/ G)$ by (9.162). The third step is Lemma 8.1.16.

Since also the composite of all these natural equivalences is thus natural, the ∞-Yoneda lemma (Prop. 8.1.54) implies that

$$\curlyvee \mathcal{X}_{\cup} \;:\; \curlyvee^{K} \;\longmapsto\; \mathrm{Cdfflg}\,\mathrm{Map}(\mathbf{B}G ,\, \mathcal{X}_{\cup})\,. \tag{10.8}$$

Now, since $\int$ acts objectwise over $\curlyvee^{K}$ (9.142), we find from this that

$$\begin{aligned} \int \curlyvee \mathcal{X}_{\cup} \;:\; \curlyvee^{K} \;\longmapsto\;& \int \mathrm{Cdfflg}\,\mathbf{Map}(\mathbf{B}G ,\, \mathcal{X}_{\mathrm{top}}) \\ \simeq\;& \mathrm{Shp}_{\mathrm{sTop}}\Big(\mathbf{Map}(\mathbf{B}G ,\, \mathcal{X}_{\mathrm{top}}) \Big) \\ \simeq\;& \big\| \mathrm{Map}(\mathbf{B}G ,\, \mathcal{X}_{\mathrm{top}}) \big\| \,. \end{aligned}$$

Here the first step is (9.57) and the last step follows by Prop. 8.1.34. □

10.1.2 Cohesive G-orbispaces

We now discuss in more detail the analogue of Prop. 10.1.2 in **(a)** the special case of global quotient stacks $\mathcal{X}_{\cup} \simeq X /\!\!/ G$ by a discrete group G, but **(b)** in the full generality of X being any 0-truncated cohesive space (not necessarily a topological space, but for instance a smooth manifold or diffeological space (9.52) or even a non-concrete object).

Remark 10.1.3 (Good orbifolds and good cohesive orbispaces)**.** The traditional orbifolds that arise as global quotients $\mathcal{X}_{\cup} \simeq X /\!\!/ G$ of a smooth manifold X by the action of a *discrete* group G are called *good orbifolds* (e.g. [Ka08, 6]). Therefore, the cohesive G-orbispaces discussed now (Def. 10.1.4) could be called (after forgetting their slicing over $\underset{\curlyvee}{G}$) the *good cohesive orbispaces.*

Definition 10.1.4 (Cohesive G-orbispace)**.** Let $\mathbf{H}$ be a singular-cohesive ∞-topos (Def. 9.2.3) and $G \in \mathrm{Grp}(\mathbf{H})$ (Prop. 8.2.1) discrete $G \simeq \flat G$. We say that a *cohesive G-orbispace* is an object

$$\begin{array}{c} \mathcal{X} \\ \big\downarrow{\scriptstyle p} \\ \underset{\curlyvee}{G} \end{array} \in \mathbf{H}_{/\underset{\curlyvee}{G}} \tag{10.9}$$

in the slice over the G-orbi-singularity (9.153) that is:

$$\begin{array}{llll} \textbf{(a)} & \text{orbi-singular:} & \prec(p) \simeq p & \text{(Def. 9.2.7)}\,, \\ \textbf{(b)} & \text{0-truncated:} & (\tau_0)_{/\underset{\curlyvee}{G}}(p) \simeq p & \text{(Def. 8.1.73)}\,. \end{array} \tag{10.10}$$

Definition 10.1.5 (Universal covering space of a G-orbi-singular space)**.** Given a Cohesive G-orbispace $\mathcal{X} \in \mathbf{H}_{/\underset{\curlyvee}{G}}$ (Def. 10.1.4), we say that its *universal covering space* $X \in \mathbf{H}$ the homotopy fiber of the defining morphism to $\underset{\curlyvee}{G}$ over its essentially unique point:

$$\begin{array}{ccc} X & \xrightarrow{\mathrm{fib}(p)} & \mathcal{X} \\ & & \big\downarrow{\scriptstyle p} \\ & & \underset{\curlyvee}{G}\,. \end{array} \tag{10.11}$$

Proposition 10.1.6 (Properties of universal covering spaces)**.** *Let* $\mathbf{H}$ *be a singular-cohesive ∞-topos (Def. 9.2.3). Given a G-orbi-singular space* $\mathcal{X} \in \mathbf{H}_{/\underset{\curlyvee}{G}}$ *(Def. 10.1.4), its universal covering space X (Def. 10.1.5)*

(i) *is:*

$$\begin{array}{llll} \textbf{(a)} & \textit{0-truncated:} & \tau_0(X) \simeq X & \textit{(Def. 8.1.73)}\,, \\ \textbf{(b)} & \textit{smooth:} & \cup(X) \simeq X & \textit{(Def. 9.1.1)}\,, \end{array}$$

(ii) *and is equipped with a G-action (Prop. 8.2.6) such that $\mathcal{X}$ is the orbi-singularization* (9.144) *of the corresponding homotopy quotient:*

$$\mathcal{X} \simeq \prec\big(X /\!\!/ G\big)\,. \tag{10.12}$$

Proof. **(i)** That X is **(a)** 0-truncated follows from the condition that p is 0-truncated and using Lemma 9.1.15. To see that X is **(b)** smooth, observe that by the defining assumption (10.10) that p is orbi-singular, it is the image under OrbSnglr (9.143) of a morphism $p_{\cup}$ in $\mathbf{H}_{\cup}$:

$$X \xrightarrow{\mathrm{fib}(p)} \mathcal{X} \quad \downarrow p \quad \mathcal{G} \quad \simeq \quad \mathrm{OrbSnglr}\left(\begin{array}{ccc} X_{\cup} & \xrightarrow{\mathrm{fib}(p_{\cup})} & \mathcal{X}_{\cup} \\ & & \downarrow p_{\cup} \\ & & * /\!\!/ G \end{array} \right). \tag{10.13}$$

We claim that in fact $X \simeq \mathrm{NnOrbSinglr}(X_{\cup})$, whence $X \simeq \cup(X)$: First, since OrbSnglr is a right adjoint it preserves homotopy fibers (Prop. 8.1.42), $\mathrm{fib}(p) \simeq \mathrm{OrbSnglr}\big(\mathrm{fib}(p_{\cup})\big)$; hence, we have $X \simeq \mathrm{OrbSnglr}(X_{\cup})$. It follows, in particular, that $X_{\cup}$ is 0-truncated, since $X \simeq \mathrm{OrbSnglr}(X_{\cup})$ is 0-truncated by part **(a)**, and using that OrbSnglr is fully faithful. From this it follows that $\mathrm{OrbSnglr}(X_{\cup}) \simeq \mathrm{NnOrbSinglr}(X_{\cup})$, by Lemma 9.2.21. Together this gives the claim **(b)**.

With this, part **(ii)** now follows by comparison with (8.160). □

10.1.3 Shape of Cohesive G-orbispaces.

We derive the following formula (10.15) in Prop. 11.1.6 which generalizes the embedding of G-spaces into global equivariant homotopy theory, discussed in [Re14, p. 7][Lu19, 3.2.17], from topological G-spaces to general cohesive G-spaces. In §11.1 this serves to prove that the intrinsic cohomology of good cohesive orbispaces subsumes proper equivariant cohomology (Theorem 11.1.9).

Lemma 10.1.7 (Shape of Cohesive G-orbispaces). *Let* $\mathbf{H}$ *be a singular-cohesive* ∞*-topos (Def. 9.2.3).* (8.48), $G \in \mathrm{Grp}(\mathbf{H})$ *(Prop. 8.2.1) be a 0-truncated* $G \simeq \tau_0 G$ *and discrete* $G \simeq \flat G$ *and let* $X \in \mathbf{H}$ *be smooth* $X \simeq \cup X$ *and 0-truncated* $X \simeq \tau_0 X$ *and equipped with a* G*-action* $(X, \rho) \in G\mathrm{Actions}(\mathbf{H})$ *(Prop. 8.2.6).*

(i) *Then the orbi-singularization* (9.143) *of the corresponding homotopy quotient* (8.160)

$$\mathcal{X} \;:=\; \curlyvee(X /\!\!/ G) \;\in\; \mathbf{H} \;:=\; \mathrm{Shv}_\infty\big(\mathrm{Snglrt}, \mathbf{H}_{\cup}\big)\,, \tag{10.14}$$

when regarded as an $\mathbf{H}_{\cup}$*-valued* ∞*-presheaf on* Snglrt (9.138), *assigns to a singularity* $\underset{\curlyvee}{K}$ (9.151) *the disjoint union of fixed loci* $X^{\phi(K)}$ *(Def. 8.2.24) of the smooth covering space* X *(Def. 10.1.5) for all group homomorphisms* $\phi : K \to G$ *homotopy-quotiented* (8.161) *by the residual* G*-action (Prop. 10.1.6):*

$$\mathcal{X} \;:\; \underset{\curlyvee}{K} \;\longmapsto\; \left(\coprod_{\phi \in \mathrm{Grp}(K,G)} X^{\phi(K)} \right) /\!\!/ G\,. \tag{10.15}$$

(ii) *Moreover, its shape* (9.142)

$$\mathrm{Shp}\big(\curlyvee(X /\!\!/ G)\big) \;\in\; \mathrm{SingularGroupoids} \;:=\; \mathrm{Shv}_\infty\big(\mathrm{Snglrt}\big) \tag{10.16}$$

assigns to a singularity $\underset{\curlyvee}{K}$ (9.151) *the cohesive shape* (9.1) *of these disjoint unions of fixed loci (Def. 8.2.24) of the smooth covering space* X *(Def. 10.1.5) homotopy-quotiented by its* G*-action (Prop. 10.1.6):*

$$\mathrm{Shp}(\mathcal{X}) \;:\; \underset{\curlyvee}{K} \;\longmapsto\; \mathrm{Shp}\left(\coprod_{\phi \in \mathrm{Grp}(K,G)} X^{\phi(K)} \right) /\!\!/ G\,. \tag{10.17}$$

Proof. We claim that for $U \in \mathrm{Chrt}$ (Def. 9.1.10) and $\rotatebox[origin=c]{70}{$\prec$}^{K} \in \mathrm{Snglrt}$ (Def. 9.2.1), hence $U \times \rotatebox[origin=c]{70}{$\prec$}^{K} \in \mathrm{Chrt} \times \mathrm{Snglrt}$ (Lemma 9.2.16), we have the following sequence of natural equivalences:

$$
\begin{aligned}
\mathbf{H}\big(U \times \rotatebox[origin=c]{70}{$\prec$}^{K}, \mathcal{X}\big) &= \mathbf{H}\big(U \times \rotatebox[origin=c]{70}{$\prec$}^{K}, \mathrm{OrbSnglr}(X /\!\!/ G)\big) \\
&\simeq \mathbf{H}_{\cup}\big(\underbrace{\mathrm{Smth}\big(U \times \rotatebox[origin=c]{70}{$\prec$}^{K}\big)}_{\simeq U \times (\ast /\!\!/ K)}, X /\!\!/ G\big) \\
&\simeq \mathrm{Grpd}_{\infty}\big((\ast /\!\!/ K), \mathbf{H}_{\cup}(U, X /\!\!/ G)\big) \\
&\simeq \mathrm{Grpd}_{1}\big((\ast /\!\!/ K), \mathbf{H}_{\cup}(U, X) /\!\!/ G\big) \\
&\simeq \Big(\coprod_{\phi \in \mathrm{Grp}(K,G)} \mathbf{H}_{\cup}(U, X)^{\phi(K)}\Big) /\!\!/ G \\
&\simeq \Big(\coprod_{\phi \in \mathrm{Grp}(K,G)} \mathbf{H}_{\cup}\big(U, X^{\phi(K)}\big)\Big) /\!\!/ G \\
&\simeq \Big(\mathbf{H}_{\cup}\Big(U, \coprod_{\phi \in \mathrm{Grp}(K,G)} X^{\phi(K)}\Big)\Big) /\!\!/ G \\
&\simeq \mathbf{H}_{\cup}\Big(U, \Big(\coprod_{\phi \in \mathrm{Grp}(K,G)} X^{\phi(K)}\Big) /\!\!/ G\Big).
\end{aligned}
\tag{10.18}
$$

Here the first step is (10.12), the second is the adjunction $\mathrm{Smth} \dashv \mathrm{OrbSnglr}$ (9.143) and using under the brace that Smth preserves products (by Prop. 8.1.42), that U is already smooth by assumption, and that $\mathrm{Smth}\big(\rotatebox[origin=c]{70}{$\prec$}^{K}\big) \simeq (\ast /\!\!/ K)$ by (9.162). The third step is the tensoring of $\mathbf{H}$ over ∞-groupoids (Prop. 8.1.50) (using the geometric discreteness $(\ast /\!\!/ K) \simeq \mathrm{Disc}(\ast /\!\!/ K)$ by Remark 9.2.14) The fourth step uses the geometric contractibility of U and the discreteness of G to identify $\mathbf{H}_{\cup}(U, X /\!\!/ G) \simeq \mathbf{H}_{\cup}(U, X) /\!\!/ G$ (Lemma 9.1.13). The fifth is the general observation of Ex. 8.1.32 about hom-groupoids between quotient groupoids of sets. The sixth step uses Prop. 9.1.14 to find that the fixed points in the set of maps are the maps into the fixed-point locus. After this key step, we just re-organize term: The seventh step uses the connectedness of U (Lemma 9.1.11) to find that a coproduct of homs out of U is a hom into the coproduct. Finally, the eighth step uses again Lemma 9.1.13.

(i) The composite equivalence (10.7) implies the first claim (10.15) by the ∞-Yoneda embedding (Prop. 8.1.53), using Lemma 9.2.16.

(ii) From this, the second claim (10.17) follows, using that Shp acts objectwise over Snglrt (9.142), and preserves homotopy quotients by discrete groups (Prop. 9.1.4).

□

Remark 10.1.8 (Relevance of 0-truncated orbi-singular spaces).

(i) The crucial assumption that makes the proof of Lemma 10.1.7 work is **(a)** that G is discrete and **(b)** that X is 0-truncated. This is what yields 1-groupoidal homs in the middle step of (10.7) and thus the form of the expression in the next step, as on the right-hand side of (8.58).

(ii) Without the assumption of $\mathcal{X}$ being 0-truncated over $* /\!\!/ G$, the proof of Lemma 10.1.7 would proceed verbatim up to that middle step, but then would break as the nontrivial morphisms present in $\mathcal{X}$ would then mix with those of the action by G.

(iii) Lemma 10.1.7 shows that this subtlety is closely related to the cohesive nature of the problem: We either have a space which is 0-truncated but carries cohesive (i.e. geometric) structure, or we turn it into its cohesive shape which is un-truncated but geometrically discrete.

Singular quotient of Cohesive G-orbispaces.

Proposition 10.1.9 (Singular quotient of G-orbi-singular space). *Let* $\mathbf{H}$ *be a singular-cohesive* ∞*-topos (Def. 9.2.3),* $G \in \mathrm{Grp}(\mathbf{H})$ *being discrete* $G \simeq \flat G$ *and 0-truncated* $G \simeq \tau_0 G$*. Let* $\mathcal{X}$ *be a* G*-orbi-singular space (Def. 10.1.4) with universal covering space* $X \in \mathbf{H}_{\cup,0} \hookrightarrow \mathbf{H}$ *equipped with its induced* G*-action (Def. 10.1.5, Prop. 10.1.6) and such that all the isotropy groups of the actions are finite. Then the singularization* (9.143) *of* $\mathcal{X}$ *is the plain* G*-quotient of* X

$$\mathrm{Snglr}(\mathcal{X}) \;\simeq\; X/G \quad \in \mathbf{H}_{\cup,0} \hookrightarrow \mathbf{H}_{\cup} \tag{10.19}$$

(i.e., the quotient of the G*-action formed in the 1-topos* $\mathbf{H}_{\cup,0}$ *of 0-truncated objects).*

Proof. For $U \in \mathrm{Chrt}$, write

$$\mathbf{H}_{\cup}(U, X) /\!\!/ G \;\simeq\; \coprod_{c} \big(* /\!\!/ H_c \big) \quad \in \mathrm{Grpd}_1 \tag{10.20}$$

for the essentially unique decomposition of the groupoid on the left into its connected components

$$c \in \pi_0\Big(\mathbf{H}_{\cup}(U, X) /\!\!/ G \Big) \;\simeq\; \mathbf{H}_{\cup}(U, X)/G\,, \tag{10.21}$$

each of which is equivalent to the delooping groupoid (Ex. 8.1.30) of its fundamental group

$$H_c \;:=\; \pi_1\Big(\mathbf{H}_{\cup}(U, X) /\!\!/ G\,,\, c \Big) \;\in \mathrm{Grp}\,. \tag{10.22}$$

Now, by Lemma 10.1.7 and re-instantiating the last few manipulations in (10.7), we have that over each $U \in \mathrm{Chrt}$ the incarnation of the G-orbi-singular space $\mathcal{X}$ as an ∞-presheaf on Snglrt is given by:

$$\begin{aligned} \mathcal{X}(U) \;:\; \prec^{K} \;\longmapsto\;\; & \mathrm{Grpd}_1\Big(* /\!\!/ K\,,\, \mathbf{H}_{\cup}(U, X) /\!\!/ G \Big) \\ \simeq\; & \mathrm{Grpd}_1\Big(* /\!\!/ K\,,\, \coprod_{c} \big(* /\!\!/ H_c \big) \Big) \\ \simeq\; & \coprod_{c} \mathrm{Grpd}_1\Big(* /\!\!/ K\,,\, * /\!\!/ H_c \Big) \\ \simeq\; & \coprod_{c} \mathrm{Snglrt}\Big(\prec^{K}\,,\, \prec^{H_c} \Big). \end{aligned} \tag{10.23}$$

Here the first step is (10.20), the second step uses that the delooping groupoids $* /\!\!/ H_c$ of the isotropy groups H_c are connected and the last step observes the definition of Snglrt (Def. 9.2.1). By the ∞-Yoneda embedding (Prop. 8.1.53) over the site of Snglrt

(9.134) this means that

$$\mathcal{X}(U) \simeq \coprod_{c} \prec^{H_c} \;\in\; \mathrm{Shv}_\infty(\mathrm{Snglrt}) \,. \tag{10.24}$$

With this, we find that $\mathrm{Snglr}(\mathcal{X}) \in \mathrm{PreSheaves}_\infty(\mathrm{Chrt})$ is given by

$$\begin{aligned} \mathrm{Snglr}(\mathcal{X}) \;:\; U \longmapsto\;& \mathrm{Snglr}(\mathcal{X}(U)) \\ \simeq\;& \mathrm{Snglr}\Big(\coprod_c \prec^{H_c}\Big) \\ \simeq\;& \coprod_c \mathrm{Snglr}\Big(\prec^{H_c}\Big) \\ \simeq\;& \coprod_c * \\ \simeq\;& \pi_0\Big(\mathbf{H}_{\smile}(U, X) /\!\!/ G\Big) \\ \simeq\;& \mathbf{H}_{\smile}(U, X)/G \,. \end{aligned} \tag{10.25}$$

Here the first line is the object-wise application of Snglr (Remark 9.2.9), while the next line is (10.24). From there we use that Snglr, being a left adjoint, preserves coproducts (Prop. 8.1.42) and then that it takes the elementary singularities to points, by Lemma 9.2.17, here using the assumption that the isotropy groups H_c are finite. Finally, we identify (10.21). But this resulting assignment is just that of $X/G \in \mathrm{PreSheaves}(\mathrm{Chrt})$:

$$X/G \;:\; U \longmapsto \mathbf{H}(U, X)/G \tag{10.26}$$

and hence the claim follows. □

Remark 10.1.10. The assumption in Prop. 10.1.9 that the isotropy groups are finite is necessary for the statement to hold:[1] For instance, the singularization (9.143) of $\mathcal{X} := * /\!\!/ \mathbb{Z} = \mathbf{B}\mathbb{Z}$ is instead the object itself:

$$\begin{aligned} \prec(* /\!\!/ \mathbb{Z}) &\simeq \underset{\underset{\prec^K}{\longrightarrow}}{\lim} \mathrm{Grpd}\big(* /\!\!/ K,\; * /\!\!/ \mathbb{Z}\big) \\ &\simeq \underset{\underset{\prec^K}{\longrightarrow}}{\lim} (* /\!\!/ \mathbb{Z}) \\ &\simeq * /\!\!/ \mathbb{Z} \,. \end{aligned}$$

Here in the second line, we used that the only group homomorphism from the finite group K (by Def. 9.2.1) to $\mathbb{Z}$ is the trivial one (and in the third line that the colimit of a constant contravariant functor on a category with a terminal object is the constant value itself). But this also shows that if one wants the singularization of orbispaces like $* /\!\!/ \mathbb{Z}$ to still be the point then one would need to accordingly enlarge the category Snglrt (Def. 9.2.1) to include such "infinitely singular" objects.

10.1.4 Examples of Cohesive G-orbispaces

We make explicit two classes of examples of cohesive G-orbispaces (Def. 10.1.4): Fréchet-smooth orbispaces and topological orbispaces.

[1] We thank David Jaz Myers for amplifying this point.

Example 10.1.11 (Fréchet smooth G-orbispaces)**.** Consider

$$X \in \text{FréchetManifolds} \hookrightarrow \text{SmthGrpd}_\infty \tag{10.27}$$

a (possibly infinite-dimensional Fréchet-)smooth manifold regarded as a 0-truncated concrete smooth ∞-groupoid (9.52). Given a $G \in \text{Grp}(\mathbf{H})$ (9.154) being discrete $G \simeq \flat G$, a smooth action ρ of G on X is equivalently a homotopy fiber sequence in SmthGrpd_∞ of this form (Prop. 8.2.6):

$$\begin{array}{ccc} X & \xrightarrow{\text{fib}(\rho)} & X /\!\!/ G \\ & & \downarrow{\scriptstyle \rho} \\ & & * /\!\!/ G \end{array} . \tag{10.28}$$

Here the homotopy quotient (8.160)

$$X /\!\!/ G \in \text{LieGroupoids} \hookrightarrow \text{SmthGrpd}_\infty \tag{10.29}$$

is the corresponding (possibly infinite-dimensional Fréchet-)Lie groupoid, regarded as a smooth ∞-groupoid via the embedding (9.53). Its orbi-singularization (9.143) is a G-orbi-singular space, in the sense of Def. 10.1.4, in the ∞-topos $\text{SnglrSmthGrpd}_\infty$ (9.148):

$$\begin{array}{c} \mathcal{X} \\ \downarrow \\ \underset{\prec}{G} \end{array} := \text{OrbSnglr}\begin{pmatrix} X /\!\!/ G \\ \downarrow \\ * /\!\!/ G \end{pmatrix} . \tag{10.30}$$

This orbi-singular smooth groupoid (10.30) is what we suggest is the proper incarnation of the quotient orbifold that is presented by the smooth manifold X with its G-action. Notice that (see *Figure G*):

(i) its *purely smooth aspect* is the Lie groupoid

$$\cup(\mathcal{X}) \simeq X /\!\!/ G \in \text{LieGroupoids} \hookrightarrow \text{SnglrSmthGrpd}_\infty , \tag{10.31}$$

(by Prop. 10.1.6) which is the incarnation of this orbifold, according to [MP97][PS10]

(ii) its *purely singular aspect* is the diffeological space

$$\vee(\mathcal{X}) \simeq X/G \in \text{DiffeologicalSpaces} \hookrightarrow \text{SingularSmoothGroupoid}_\infty \tag{10.32}$$

(by Prop. 10.1.9) which is the incarnation of this orbifold, according to [IKZ10].

However, it is only the full orbi-singular object $\mathcal{X}$ which is structured enough to have proper (Bredon-)equivariant cohomology. This is the content of Theorem 11.1.9.

Example 10.1.12 (Topological G-orbispaces)**.** For G a finite group, let $G \circlearrowright X_{\text{top}}$ be a topological G-space (Def 8.1.18) with Borel construction

$$\begin{array}{ccc} X_{\text{top}} & \longrightarrow & X \underset{G}{\times} EG \\ & & \downarrow \\ & & BG \end{array} \tag{10.33}$$

Via its continuous diffeology (8.12), this is equivalently a 0-truncated (and concrete) object in $\mathbf{H}_\cup := \text{SmthGrpd}_\infty$ (Ex. 9.1.19)

$$X := \text{Cdfflg}(X_{\text{top}}) \in \mathbf{H}_{\cup,0} \tag{10.34}$$

equipped with a smooth G-action (Prop. 8.2.6)

$$\begin{array}{ccc} X & \longrightarrow & X /\!\!/ G \\ & & \downarrow \\ & & * /\!\!/ G\,. \end{array} \tag{10.35}$$

The orbi-singularization (9.143) of the corresponding homotopy quotient is a G-orbi-singular space (Def. 10.1.4)

$$\begin{array}{c} \mathcal{X} \\ \downarrow \\ \overset{G}{\prec} \end{array} \;:=\; \mathrm{OrbSnglr}\left(\begin{array}{c} \mathrm{Cdfflg}(X_{\mathrm{top}}) /\!\!/ G \\ \downarrow \\ * /\!\!/ G \end{array} \right). \tag{10.36}$$

Proposition 10.1.13 (Shape of good orbifolds)**.** *Consider a finite-dimensional smooth G-orbifold, as in Ex. 10.1.11 (a good orbifold, Remark 10.1.3)*

$$\mathcal{X} \;:=\; \mathrm{OrbSnglr}\big(X /\!\!/ G\big). \tag{10.37}$$

Then its cohesive shape (9.143) $\mathrm{Shp}(\mathcal{X}) \;\in\; \mathrm{Shv}_\infty(\mathrm{Snglrt})$ *is, over any singularity* $\overset{K}{\prec}$ (9.135), *the topological shape* (8.1.29) *of the G-Borel construction on the disjoint union of all K-fixed subspaces $X_{\mathrm{top}}^{\phi(K)} \subset X_{\mathrm{top}}$* (8.39) *in the underlying* (8.12) *D-topological G-space (Def. 8.1.18):*

$$\mathrm{Shp}(\mathcal{X}) \;:\; \overset{K}{\prec} \;\longmapsto\; \mathrm{Shp}_{\mathrm{Top}}\Bigg(\bigg(\coprod_{\phi \in \mathrm{Grp}(K,G)} \big(\mathrm{Dtplg}(X)\big)^{\phi(K)} \bigg) \underset{G}{\times} EG \Bigg). \tag{10.38}$$

Proof. With Lemma 10.1.7, the task is reduced to showing that, for $\phi(K) \subset G$ any specified subgroup, we have an equivalence

$$\mathrm{Shp}\big(X^{\phi(K)}\big) \;\simeq\; \mathrm{Shp}_{\mathrm{Top}}\big((\mathrm{Dtplg}(X))^{\phi(K)}\big) \quad \in \mathrm{Grpd}_\infty \tag{10.39}$$

between the cohesive shape (9.1) of the orbi-singular homotopy quotient of X by G and the ordinary topological shape (8.1.29) of the D-topological space underlying X. But this is (9.56) in Ex. 9.1.19, given by [Sc13, 4.3.29]. □

Proposition 10.1.14 (Shape of topological G-orbi spaces)**.** *Consider the topological G-orbi-singular space, as in Ex. 10.1.12,*

$$\mathcal{X} \;:=\; \mathrm{OrbSnglr}\big(\mathrm{Cdfflg}(X_{\mathrm{top}}) /\!\!/ G\big). \tag{10.40}$$

Then its cohesive shape (9.143) $\mathrm{Shp}(\mathcal{X}) \;\in\; \mathrm{Shv}_\infty(\mathrm{Snglrt})$ *is, over any singularity* $\overset{K}{\prec}$ (9.135), *the topological space* (8.1.29) *of the G-Borel construction on the disjoint union of all K-fixed subspaces $X_{\mathrm{top}}^{\phi(K)} \subset X_{\mathrm{top}}$* (8.39)*:*

$$\mathrm{Shp}(\mathcal{X}) \;:\; \overset{K}{\prec} \;\longmapsto\; \mathrm{Shp}_{\mathrm{Top}}\Bigg(\bigg(\coprod_{\phi \in \mathrm{Grp}(K,G)} X_{\mathrm{top}}^{\phi(K)} \bigg) \underset{G}{\times} EG \Bigg). \tag{10.41}$$

Proof. With Lemma 10.1.7, the task is reduced to showing that, for $\phi(K) \subset G$ any specified subgroup, we have an equivalence

$$\mathrm{Shp}\big(\mathrm{Cdfflg}(X_{\mathrm{top}})^{\phi(K)}\big) \;\simeq\; \mathrm{Shp}_{\mathrm{Top}}\big(X_{\mathrm{top}}^{\phi(K)}\big) \quad \in \mathrm{Grpd}_\infty \tag{10.42}$$

between the cohesive shape (9.1) of the orbi-singular homotopy quotient by G of the continuous-diffeological space and the ordinary topological shape (8.1.29) But this is item (9.55) in Ex. 9.1.19, given by combining the result (9.54) of [BEBP19][Bunk20, §3] with Prop. 8.1.36 from [CW14]. □

10.2 Higher Orbifolds

We introduce a general theory of orbi-singular spaces, whose underlying smooth cohesive groupoid is locally diffeomorphic to a fixed local model space V. Since, for $V = \mathbb{R}^n \in \mathrm{HldSmthGrpd}_\infty$, these are ordinary n-folds (i.e., ordinary n-dimensional manifolds for any n, see Ex. 10.2.5), or, more generally, étale ∞-groupoids with atlases by n-folds (Ex. 10.2.6), including ordinary orbifolds, we generally speak of *V-folds*, with a hat tip to [Sa56]. Externally these are V-étale ∞-stacks (Remark 10.2.3) but their theory internal to the ambient elastic ∞-topos (such as the construction of their frame bundles in Prop. 10.2.14) is elegant and finitary and lends itself to full formalization in homotopy type theory [Ch24] (see p. 10). The *proper* incarnation (see Remark 10.2.48) of these V-folds as orbifolds is via their orbi-singularization (Def. 10.2.46, Remark 10.2.48).

Remark 10.2.1 (The implicit order of jet geometry). Where we speak internally of infinitesimal concepts such as *frame bundles* and their structure groups in the following, their external nature depends on the choice of interpreting elastic ∞-topos $\mathbf{H}$ (Def. 9.1.22):

For orbifolds regarded in $\mathbf{H} = \mathrm{HldSmthGrpd}_\infty$ (9.70) these come out as ordinary frame bundle and ordinary general linear groups (Ex. 10.2.13), while in $\mathbf{H} \equiv k\mathrm{HldSmthGrpd}_\infty$ (9.68) they come out as *k-jet frame bundles* and *k-jet groups* (cf. Ex. 10.2.17). Similarly, the notion of *integrability* of G-structures depends on k (cf. Ex. 10.2.32).

This highlights that our synthetic formalism in the form presented here does not internally fix the order of infinitesimality that is expressed by the elastic modalities. On the one hand this is a boon: It means that every theorem established abstractly with the elastic modalities translates externally into an infinite sequence of theorems, applicable to each order of infinitesimality. On the other hand it is desirable to be able to speak internally of first-order infinitesimality, for instance in order to witness already internally the crucial difference between torsion-free and flat G-structures (cf. Ex. 10.2.32).

Such an internal axiomatization of first-order haloes may be possible via the interplay with the solid modalities, since the odd-graded infinitesimals of standard supergeometry have a special "1st order nature" to them (in that their super-commutativity is equivalently the statement that first-order super-infinitesimals are closed under summation). But here we leave this question open.

10.2.1 V-folds and V-étale groupoids

Definition 10.2.2 (V-folds)**.** Let $\mathbf{H}$ be an elastic ∞-topos $\mathbf{H}$ (Def. 9.1.22).

(i) Given $V \in \mathrm{Grp}(\mathbf{H})$ (Prop. 8.2.1), we say that an object $X \in \mathbf{H}$ is a *V-fold* if there exists a correspondence between V and X

$$V \xleftarrow{\;\text{ét}\;} U \xrightarrow{\;\text{ét}\;} X \qquad (10.43)$$

such that

(a) both morphisms are local diffeomorphisms (Def. 9.1.28) and

(b) the right one is, in addition, an effective epimorphism (Def. 8.1.80), then called a *V-atlas of* X (8.148).

(ii) We write

$$V\mathrm{Folds}(\mathbf{H}) \subset \mathbf{H} \qquad (10.44)$$

for the full sub-∞-category of V-folds in $\mathbf{H}$ and we write

$$V\mathrm{Folds}(\mathbf{H})^{\text{ét}} \subset \mathbf{H} \qquad (10.45)$$

for its wide subcategory on those morphisms which are local diffeomorphisms (Def. 9.1.28).

Remark 10.2.3 (V-folds and V-étale groupoids)**.** By Prop. 9.1.38, a V-fold (Def. 10.2.2) is a stack (8.148) whose choice of V-atlas (10.43) realizes it as an étale groupoid (Def. 9.1.37) with space of objects locally diffeomorphic over V:

$$\begin{array}{ccccc} & & \vdots & & \vdots \\ & & U \times_X U & \simeq & U_1 \\ & & {\scriptstyle \mathrm{pr}_1}\downarrow \uparrow{\scriptstyle \Delta} \downarrow{\scriptstyle \mathrm{pr}_2} & & {\scriptstyle s}\downarrow \uparrow{\scriptstyle e} \downarrow{\scriptstyle t} \\ V & \xleftarrow{\;\text{ét}\;} & U & = & U_0 \\ & & {\scriptstyle a}\downarrow{\scriptstyle \text{ét}} & & \downarrow \\ & & X & \simeq & \underset{\longrightarrow}{\lim}\, U_\bullet \end{array} \qquad (10.46)$$

"V-étale groupoid"

"V-atlas"

"V-fold"

Example 10.2.4 (V is a V-fold)**.** Let $\mathbf{H}$ be an elastic ∞-topos $\mathbf{H}$ (Def. 9.1.22) and $V \in \mathrm{Grp}(\mathbf{H})$ (Prop. 8.2.1). Then the underlying object $V \in \mathbf{H}$ itself is a V-fold (Def. 10.2.2): A V-atlas (10.43) is given by the identity morphisms

$$V \xleftarrow[\text{ét}]{\;\mathrm{id}\;} V \xrightarrow[\text{ét}]{\;\mathrm{id}\;} V\,. \qquad (10.47)$$

Example 10.2.5 (Smooth manifolds are $\mathbb{R}^n$-folds)**.** For $k \in \mathbb{N}$ with $k \geq 1$, let $\mathbf{H} = k\mathrm{HldSmthGrpd}_\infty$ (Ex. 9.1.26). Then, for every $n \in \mathbb{N}$, the object

$$V := \mathbb{R}^n \in \mathrm{CrtSpc} \hookrightarrow k\mathrm{HldSmthGrpd}_\infty \qquad (10.48)$$

canonically carries the structure of a group object $(\mathbb{R}^n, +) \in \mathrm{Grp}(\mathbf{H})$, via addition in $\mathbb{R}^n$ regarded as a vector space. Now every smooth manifold

$$X \in \mathrm{SmthMfd} \hookrightarrow k\mathrm{HldSmthGrpd}_\infty \tag{10.49}$$

of dimension n is a V-fold, hence an $\mathbb{R}^n$-fold in the sense of Def. 10.2.2: For any choice of atlas in the traditional sense of manifold theory, namely an open cover

$$\left\{U_j \xrightarrow{\phi_i} X\right\}_{j \in J} \tag{10.50}$$

by local diffeomorphisms ϕ_i from open subsets of Cartesian space

$$U_j \overset{\iota_j}{\hookrightarrow} \mathbb{R}^n , \tag{10.51}$$

a V-atlas (10.43) is obtained by setting:

$$\mathbb{R}^n \xleftarrow[\text{ét}]{(\iota_j)_{j\in J}} \coprod_{j \in J} U_j \xrightarrow[\text{ét}]{(\phi_j)_{j\in J}} X . \tag{10.52}$$

Example 10.2.6 (Differentiable étale stacks are $\mathbb{R}^n$-folds [Sc13, Prop. 4.5.56]). Let $\mathbf{H} = \mathrm{HldSmthGrpd}_\infty$ (Ex. 9.1.26) and take $V = (\mathbb{R}^n, +)$ as in Ex. 10.2.5. Then a diffeological groupoid $X \in \mathbf{H}$ (9.72) is a V-fold (Def. 10.2.2) for $V = \mathbb{R}^n$ (10.48) if it is an n-dimensional *differentiable étale stack* in that:

(i) it admits an atlas (effective epimorphism) $X_0 \twoheadrightarrow X$ from a smooth n-manifold X_0 (via (9.52) and (9.71))

(ii) its source and target morphisms with respect to this atlas are local diffeomorphisms.

Generally, a smooth ∞-groupoid presented by a Kan simplicial smooth manifold is an $\mathbb{R}^n$-fold in the sense of Def. 10.2.2 if it presents an *étale ∞-groupoid* in that all its simplicial face maps are local diffeomorphisms.

Examples 10.2.7 (Super-manifolds are $\mathbb{R}^{n|q}$-folds). Let $\mathbf{H} = \mathrm{SprHldSmthGrpd}_\infty$ (Ex. 9.1.26). Then, for every $n, q \in \mathbb{N}$, the super-Cartesian space (Def. 9.1.43)

$$V := \mathbb{R}^{n|q} \in \mathrm{SprHldCrtSp} \hookrightarrow \mathrm{SprHldSmthGrpd}_\infty \tag{10.53}$$

carries the structure of a group object, whose bosonic aspect (9.114) is (10.48). The corresponding V-folds (Def. 10.2.2) are the $(n|q)$-dimensional supermanifolds (9.125).

Example 10.2.8 (General super étale ∞-stacks). Let $\mathbf{H} = \mathrm{SprHldSmthGrpd}_\infty$ (Ex. 9.1.45). Then for any $V \in \mathrm{Grp}(\mathbf{H})$ the corresponding V-étale ∞-stacks (Remark 10.2.3) realize a flavor of *super étale ∞-stacks*, locally modeled on V. Lemma 9.1.47 implies that, generally, the bosonic part $\overset{\rightsquigarrow}{X}$ of a super étale ∞-stack is a bosonic étale ∞-stack locally modeled on the bosonic part $\overset{\rightsquigarrow}{V}$ of V:

$$\begin{array}{ccccc} & V\mathrm{Folds}(\mathbf{H}) & \xrightarrow{\rightsquigarrow} & \overset{\rightsquigarrow}{V}\mathrm{Folds}(\mathbf{H}) & \\ \substack{\text{supergeometric} \\ \text{étale } \infty\text{-stack}} & X & \longmapsto & \overset{\rightsquigarrow}{X} & \substack{\text{underlying bosonic} \\ \text{étale } \infty\text{-stack}} \end{array} \tag{10.54}$$

10.2.2 Quotients of V-folds

Proposition 10.2.9 (Orbifolding of a V-fold is a V-fold)**.** *Let* $\mathbf{H}$ *be an elastic* ∞*-topos* $\mathbf{H}$ *(Def. 9.1.22),* $V, G \in \mathrm{Grp}(\mathbf{H})$ *(Prop. 8.2.1) with* $G \simeq \flat G$ *discrete, and* $(X, \rho) \in G\mathrm{Actions}(\mathbf{H})$ *(Prop. 8.2.6). Then if* X *is a* V*-fold (Def. 10.2.2) so is its homotopy quotient* $X /\!\!/ G$ (8.161)*. Specifically, if* $U \xrightarrow{\text{ét}} X$ *is a* V*-atlas for* X (10.43)*, then a* V*-atlas for* $V /\!\!/ G$ *is given by composition with the homotopy fiber inclusion map* $\mathrm{fib}(\rho)$ (8.160)*:*

$$
\begin{array}{ccccc}
 & & U & & \\
 & \overset{\text{ét}}{\swarrow} & & \overset{\text{ét}}{\searrow} & \\
V & & & & X \xrightarrow{\mathrm{fib}(\rho)} X /\!\!/ G\,.
\end{array}
\tag{10.55}
$$

Proof. We need to show that the composite morphism on the right of (10.55) is **(a)** an effective epimorphism and **(b)** a local diffeomorphism. Since both of these classes of morphisms are closed under composition (Lemma 8.1.82 and Lemma 9.1.29), it is sufficient to show that $\mathrm{fib}(\rho)$ itself has these two properties.

For **(a)** observe that, by definition of homotopy fibers (8.160), we have a Cartesian square

$$
\begin{array}{ccc}
X & \xrightarrow{\mathrm{fib}(\rho)} & X /\!\!/ G \\
\downarrow & \text{(pb)} & \downarrow{\scriptstyle \rho} \\
* & \twoheadrightarrow & \mathbf{B}G
\end{array}
\tag{10.56}
$$

Here, the bottom morphism is an effective epimorphism (Ex. 8.2.2). Since these are preserved by homotopy pullback, also $\mathrm{fib}(\rho)$ is an effective epimorphism.

For **(b)** consider the image of this square (10.56) under $\Im$. Since $\Im$ is both a right and a left adjoint it preserves Cartesian squares and homotopy quotients (by Prop. 8.1.42), while it preserves discrete objects by elasticity (9.59) and idempotency (Prop. 8.1.44, Prop. 8.1.45). Therefore, the diagram

$$
\begin{array}{ccc}
\Im X & \xrightarrow{\Im \mathrm{fib}(\rho) \simeq \mathrm{fib}(\Im \rho)} & (\Im X) /\!\!/ G \\
\downarrow & \text{(pb)} & \downarrow{\scriptstyle \Im \rho} \\
* & \twoheadrightarrow & \mathbf{B}G
\end{array}
\tag{10.57}
$$

is Cartesian. Consider finally the pasting composite of this second square (10.57) with the naturality square of $\eta^{\Im}$ on $\mathrm{fib}(\rho)$:

$$
\begin{array}{ccc}
X & \xrightarrow{\mathrm{fib}(\rho)} & X /\!\!/ G \\
{\scriptstyle \eta^{\Im}_X}\downarrow & & \downarrow{\scriptstyle \eta^{\Im}_{X/\!\!/G}} \\
\Im X & \longrightarrow & (\Im X) /\!\!/ G \\
\downarrow & \text{(pb)} & \downarrow{\scriptstyle \Im \rho} \\
* & \longrightarrow & \mathbf{B}G
\end{array}
\quad (\text{right composite } = \rho)
\tag{10.58}
$$

Here the composite morphism on the right is equivalent to ρ, as shown, by the naturality of $\eta^{\Im}$ and using that the object $\mathbf{B}G$, being discrete, is $\Im$-modal. Therefore, the total outer rectangle of (10.58) is Cartesian by (10.56). Moreover, the bottom square of (10.58) is Cartesian by (10.57). Therefore the pasting law (Prop. 8.1.39) implies that the top square of (10.58) is Cartesian. But this means (9.76) that $\mathrm{fib}(\rho)$ is a local diffeomorphism. □

Proposition 10.2.10 (Induced G-action on the tangent bundle)**.** *Let* $\mathbf{H}$ *be an elastic* ∞*-topos* $\mathbf{H}$ *(Def. 9.1.22),* $V, G \in \mathrm{Grp}(\mathbf{H})$ *(Prop. 8.2.1), with* $G \simeq \flat G$ *discrete,* $(X, \rho) \in G\mathrm{Actions}(\mathbf{H})$ *(Prop. 8.2.6) and* $X \in V\mathrm{Folds}(X)$ *(Def. 10.2.2). Then the tangent bundle* TX *(Def. 9.1.31) carries an essentially unique* G*-action* $T\rho$ *such that:*

(i) *the defining projection* $TX \to X$ *is* G*-equivariant (Def. 8.2.10);*

(ii) *the homotopy quotient of* TX *is the tangent bundle of the orbifolded* V*-fold* $X /\!\!/ G$ *(Prop. 10.2.9):*

$$(TX) /\!\!/ G \simeq T(X /\!\!/ G) \quad \in \mathbf{H}_{/X/\!\!/G}\,. \tag{10.59}$$

Proof. Consider the following diagram:

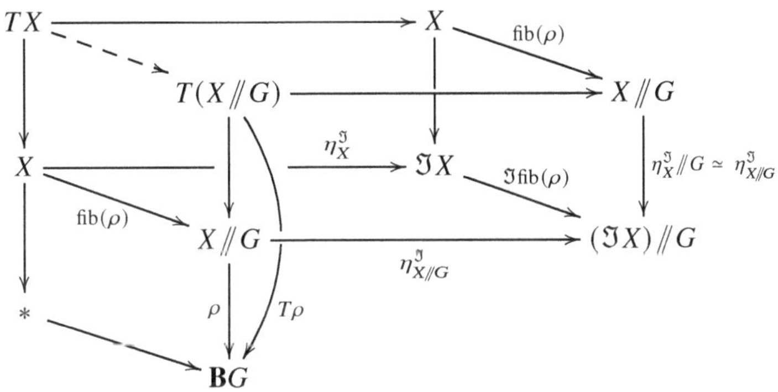

(10.60)

Here the bottom left square is that characterizing the G-action on X, by (8.160); while the bottom and right squares are both the naturality square of $\eta^{\Im}$ on the morphism $\mathrm{fib}(\rho)$ (where we use that $\Im$ commutes with taking the homotopy quotient by the discrete group G). Now observe that:

(a) The bottom and right squares are pullback squares since $\mathrm{fib}(\rho)$ is a local diffeomorphism (Def. 9.1.28) by Prop. 10.2.9.

(b) The front and back squares are pullback squares by the definition of tangent bundles (Def. 9.1.31).

In particular, the solid part of the diagram is homotopy-commutative, so that, by the universal property of the front pullback square, the dashed morphism exists, essentially uniquely, such as to make the top and the top left square homotopy-commutative. Further observe, by repeatedly applying the pasting law (Prop. 8.1.39), that:

(c) The top left square is a homotopy pullback since the back, right and front squares are pullbacks by (a) and (b).

(d) The total left rectangle is a pullback, since the top one is so, by (c), and the bottom one is so, by the action property (8.160).

Thus, again by the action property (8.160), the total left rectangle exhibits a G-action on TX whose homotopy quotient is as claimed (10.59), and its factorization into two pullback squares as shown exhibits the projection $TX \to X$ as a homomorphism of G-actions, hence as being G-equivariant (Def. 8.2.10). □

Proposition 10.2.11 (Induced G-action on local neighborhood of fixed point)**.** *Let* $\mathbf{H}$ *be an elastic* ∞*-topos* $\mathbf{H}$ *(Def. 9.1.22),* $V, G \in \mathrm{Grp}(\mathbf{H})$ *(Prop. 8.2.1), with* $G \simeq \flat G$ *discrete,* $(X, \rho) \in G\mathrm{Actions}(\mathbf{H})$ *(Prop. 8.2.6) with* $X \in V\mathrm{Folds}(X)$ *(Def. 10.2.2) and* $* \xrightarrow{x} X$ *a homotopy fixed point (Def. 8.2.24). Then the induced* G*-action* $T\rho$ *on the tangent bundle* TX*, from Prop. 10.2.10, restricts to a* G*-action* $T_x\rho$ *on the local neighborhood* T_xX *(Ex. 9.1.32) of the homotopy fixed point* x*.*

Proof. Consider the following diagram:

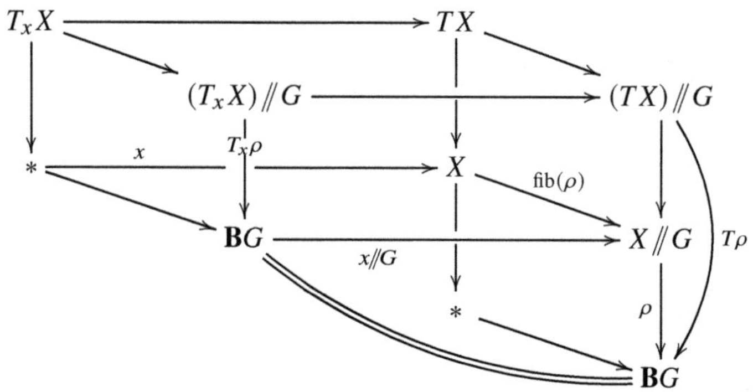

(10.61)

Here the squares on the right are from (10.60) and are thus both homtopy Cartesian. The rear square is the homotopy pullback square defining the tangent fiber, and we define the front square to be a homotopy pullback, giving us the object denoted $(T_xX) /\!\!/ G$. We need to show that this object really is the homotopy quotient of the restricted action. But the bottom horizontal square homotopy-commutes, exhibiting the homotopy fixed point by (8.191), so that, by applying the pasting law (Prop. 8.1.39) to the top vertical squares, it follows that also the top left square is Cartesian. This already identifies $(T_xX) /\!\!/ G$ as the homotopy quotient of some G-action on T_xX, by Prop. 8.2.6. To see that this is indeed the restricted action, observe that the front triangle commutes, again by (8.191), so that the total diagram exhibits the fiber inclusion $T_xX \to TX$ as being a homomorphism G-actions $T_x\rho \to T_\rho$ (by Prop. 8.2.6). □

10.2.3 Frame bundles

Definition 10.2.12 (Structure group of V-folds)**.** Let $\mathbf{H}$ be an elastic ∞-topos (Def. 9.1.22) and $V \in \mathrm{Grp}(\mathbf{H})$ (Prop. 8.2.1), to be regarded as the local model space of V-folds (Def. 10.2.2). Then:

(i) We say that the automorphism group (Def. 8.2.13) of the local neighborhood (Ex. 9.1.32) of the neutral element $* \xrightarrow{e} V$ (Ex. 8.2.3)

$$\mathrm{Aut}(T_eV) \quad \in \mathrm{Grp}(T_eV) \tag{10.62}$$

is the *structure group of* V*-folds.*

(ii) We write

$$(T_e V,\, \rho_{\mathrm{Aut}}) \;\in\; \mathrm{Aut}(T_e V)\mathrm{Actions}(\mathbf{H}) \tag{10.63}$$

for its canonical action (8.174).

Example 10.2.13 (Ordinary general linear group)**.** Let $\mathbf{H} = \mathrm{HldSmthGrpd}_\infty$ (Ex. 9.1.26) and let

$$V := (\mathbb{R}^n, +) \;\in\; \mathrm{Grp}(\mathrm{SmthMfd}) \hookrightarrow \mathrm{Grp}(\mathbf{H}) \tag{10.64}$$

via the full inclusion (9.72), with $\mathbb{R}^n$ regarded as a group under addition of tuples of real numbers. Then the structure group of $\mathbb{R}^n$-folds, according to Def. 10.2.12, is the traditional general linear group, regarded as a Lie group:

$$\mathrm{Aut}(T_0\mathbb{R}^n) \;\simeq\; \mathrm{GL}(n)\,. \tag{10.65}$$

Proposition 10.2.14 (Frame bundle)**.** *Let* $\mathbf{H}$ *be an elastic* ∞*-topos* $\mathbf{H}$ *(Def. 9.1.22),* $V \in \mathrm{Grp}(\mathbf{H})$ *(Prop. 8.2.1) and* $X \in \mathbf{H}$ *a* V*-fold (Def. 10.2.2). Then the tangent bundle of* X *(Def. 9.1.31) is a fiber bundle (Def. 8.2.18) with typical fiber the local neighborhood* T_eV *(Def. 9.1.30) of the neutral element* $* \xrightarrow{e} V$*, hence is the associated bundle of an* $\mathrm{Aut}(T_eV)$*-principal* (10.62) *bundle (Prop. 8.2.15), to be called the* frame bundle *of* X*:*

$$\begin{array}{ccc} \overset{\text{tangent bundle}}{TX} & \longrightarrow & (T_eV)/\!\!/\mathrm{Aut}(T_eV) \\ \downarrow & \text{(pb)} & \downarrow \\ X & \underset{\vdash\,\mathrm{Frm}(X)}{\longrightarrow} & \mathbf{B}\mathrm{Aut}(T_eV) \end{array} \qquad \begin{array}{ccc} \overset{\text{frame bundle}}{\mathrm{Frm}(X)} & \longrightarrow & * \\ \downarrow & \text{(pb)} & \downarrow \\ X & \underset{\vdash\,\mathrm{Frm}(X)}{\longrightarrow} & \underset{\text{structure group}}{\mathbf{B}\mathrm{Aut}(T_eV)}\,. \end{array} \tag{10.66}$$

Proof. By Prop. 9.1.33 the tangent bundles over any V-atlas (10.43) for X form two Cartesian squares as follows:

$$\begin{array}{ccccc} & & TU & & \\ & \swarrow & & \searrow & \\ V \times T_eV \simeq TV & & \text{(pb)} \;\downarrow\; \text{(pb)} & & TX \\ \downarrow & & U & & \downarrow \\ & \underset{\text{ét}}{\swarrow} & & \underset{\text{ét}}{\searrow} & \\ V & & & & X \end{array} \tag{10.67}$$

Moreover, by Prop. 10.2.20 the tangent bundle of V is trivial, as shown on the left. Since Cartesian products are preserved by homotopy pullback, the left square implies that also $TU \simeq U \times T_eV$ is trivial. But with this the existence of the right square is the defining characterization for TX to be a T_eV-fiber bundle. □

Remark 10.2.15 (Frame bundles are well-defined)**.** The frame bundle (Def. 10.2.14) of a V-fold (Def. 10.2.2) is independent, up to a contractible space of equivalences, of the choice of V-atlas (10.43) in the construction (10.67): This follows as a special case of the essential independence of classifying maps of fiber bundles from the choice of trivializing cover, as in Prop. 8.2.19, using that not only the class of effective epimorphisms but also that of local diffeomorphisms is closed under pullback and composition (Lemma 9.1.29).

Proposition 10.2.16 (V-fold is $\mathrm{Aut}(T_e V)$-quotient of its frame bundle)**.** *Let* $\mathbf{H}$ *be an elastic* ∞*-topos* $\mathbf{H}$ *(Def. 9.1.22),* $V \in \mathrm{Grp}(\mathbf{H})$ *(Prop. 8.2.1), and* $X \in V\mathrm{Folds}(\mathbf{H})$ *(Def. 10.2.2). Then* X *is equivalent to the homotopy quotient* (8.161) *of its own frame bundle (Prop. 10.2.14) by* $\mathrm{Aut}(T_e V)$*:*

$$X \simeq \mathrm{Frm}(X) /\!\!/ \mathrm{Aut}(T_e V) . \tag{10.68}$$

Proof. This is immediate from the equivalence between principal bundles and homotopy quotient projections (Remark 8.2.16) applied to the frame bundle (10.66). □

Example 10.2.17 (Frame bundles on smooth manifolds)**.** Let $\mathbf{H} = \mathrm{HldSmthGrpd}_\infty$ (Ex. 9.1.26) and $X \in \mathrm{SmthMfd} \hookrightarrow \mathbf{H}$ a smooth manifold (9.72) regarded as an $\mathbb{R}^n$-fold according to Ex. 10.2.5.

(i) Then its frame bundle, according to Prop. 10.2.14, is the $\mathrm{GL}(n)$-principal bundle on X which is the frame bundle in the traditional sense of differential geometry.

(ii) For the same manifold but regarded in $\mathbf{H} = k\mathrm{HldSmthGrpd}_\infty$ with $k \geq 1$, we instead get the corresponding jet version of the frame bundle (see e.g. [KMS93, 12.12]).

10.2.4 Framed V-folds

Definition 10.2.18 (Framing)**.** Let $\mathbf{H}$ be an elastic ∞-topos $\mathbf{H}$ (Def. 9.1.22). A *framing* of an object $X \in \mathbf{H}$ is a trivialization of its tangent bundle Def. 9.1.31, hence an equivalence

$$TX \simeq X \times T_x X \quad \in \mathbf{H}_{/X} \tag{10.69}$$

for $* \xrightarrow{x} X$ any point.

Remark 10.2.19 (Framing on a V-fold)**.** If X is a V-fold (Def. 10.2.2) then a *framing* on V in the sense of Def. 10.2.18 is equivalent, by Prop. 10.2.14, to a trivialization of the frame bundle, hence to a trivialization of its classifying map (10.66):

$$\begin{array}{ccc} TX & \xrightarrow[\simeq]{\mathrm{fr}} & X \times T_e V \\ & \searrow \quad \swarrow & \\ & X & \end{array} \quad \Leftrightarrow \quad \begin{array}{ccc} & * & \\ X \nearrow & \Uparrow^{\simeq}\ \vdash \mathrm{fr} & \searrow \mathbf{B}\mathrm{Aut}(T_e V) \\ & \xrightarrow[\vdash\, \mathrm{Frm}(X)]{} & \end{array} . \tag{10.70}$$

Proposition 10.2.20 (Groups carry canonical framings by left-translation)**.** *In an elastic* ∞*-topos* $\mathbf{H}$ *(Def. 9.1.22) every group object* $V \in \mathrm{Grp}(\mathbf{H})$ *(Prop. 8.2.1) carries a canonical framing (Def. 10.2.18), which we call the framing by left translation:*

$$TV \xrightarrow[\simeq]{\mathrm{fr}_\ell} V \times T_e V \quad \in \mathbf{H}_{/V} . \tag{10.71}$$

Proof. Since $\Im$ preserves group structure (as in Prop. 9.1.4), the defining homotopy fiber product of the tangent bundle of V (9.83) sits in a Mayer-Vietoris sequence

(Prop. 8.2.5) as shown in the first square of the following:

$$
\begin{array}{ccccc}
TV & \longrightarrow & \longrightarrow & \longrightarrow & * \\
\big\downarrow & & \text{(pb)} & & \big\downarrow{\scriptstyle \vdash e} \\
V \times V & \xrightarrow{(\eta^{\Im}_V, \eta^{\Im}_V) = \eta^{\Im}_{V \times V}} & \Im V \times \Im V & \xrightarrow{(-)\cdot(-)^{-1}} & \Im V
\end{array}
$$

$$
\simeq \quad
\begin{array}{ccccc}
TV & \longrightarrow & T_e V & \longrightarrow & * \\
\big\downarrow & \text{(pb)} & \big\downarrow & \text{(pb)} & \big\downarrow{\scriptstyle \vdash e} \\
V \times V & \xrightarrow{(-)\cdot(-)^{-1}} & V & \xrightarrow{\eta^{\Im}_V} & \Im V
\end{array}
\tag{10.72}
$$

Using that $\Im$ preserves products (by Prop. 8.1.42) and using the naturality of its unit transformation $\eta^{\Im}$ (8.72), this Cartesian square on top is equivalent to the total rectangle shown at the bottom. By the pasting law (Prop. 8.1.39), this is the pasting of two Cartesian squares, the right one of which exhibits the local neighborhood $T_e V$ (Def. 9.1.30) as shown. To see what the Cartesian property of the left square on the right says, consider pasting to it the top square appearing in the diagram (8.157) which exhibits the group division $(-) \cdot (-)^{-1}$ in Ex. 8.2.4:

$$
\begin{array}{ccc}
TV & \longrightarrow & T_e V \\
\big\downarrow & \text{(pb)} & \big\downarrow \\
V \times V & \xrightarrow{(-)\cdot(-)^{-1}} & V \\
{\scriptstyle \mathrm{pr}_1}\big\downarrow & \text{(pb)} & \big\downarrow \\
V & \longrightarrow & *
\end{array}
\tag{10.73}
$$

Since both squares are Cartesian, the pasting law (Prop. 8.1.39) says that the total rectangle is Cartesian. This is the equivalence (10.71). □

Proposition 10.2.21 (Canonical framing on group is equivariant under group automorphisms). *Consider an elastic ∞-topos* $\mathbf{H}$ *(Def. 9.1.22),* $V, G \in \mathrm{Grp}(\mathbf{H})$ *(Prop. 8.2.1). with 0-truncated* $V \simeq \tau_0 V$ *and* $(V, \rho_G) \in G\mathrm{Actions}(\mathbf{H})$ *(Prop. 8.2.6) acting by group-automorphisms (Prop. 8.2.29) hence by restriction* $\rho_G = \mathbf{B}i^* \rho_{\mathrm{Aut_{Grp}}}$ *(Prop. 8.2.12) along a group homomorphism* $G \xrightarrow{i} \mathrm{Aut}_e(V)$*, to the group-automorphism group* $\mathrm{Aut_{Grp}}(V)$ *(Def. 8.2.28). Then the canonical framing* fr_ℓ *on V from Prop. 10.2.20 is G-equivariant (Def. 8.2.10), in that it lifts to a morphism of G-actions (Prop. 8.2.6) of the form*

$$
(TV, T\rho) \xrightarrow{\mathrm{fr}_\ell} (V, \rho) \times (T_e V, T_e \rho) \quad \in G\mathrm{Actions}(\mathbf{H}), \tag{10.74}
$$

where $T\rho$ is the induced action on TV from Prop. 10.2.10, and $T_e\rho$ is the induced action on $T_e V$ from Prop. 10.2.11 (which exists since group-automorphisms of V are in particular pointed automorphisms of V (Def. 8.2.27).

Proof. Consider the following diagram:

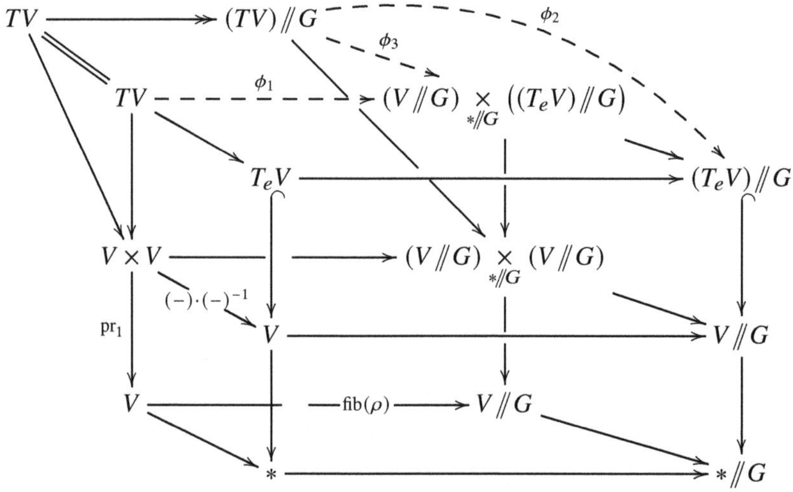

(10.75)

Here

- the bottom square is the Cartesian square (8.160) which exhibits the action on V,
- the middle horizontal square is the Cartesian square which exhibits the equivariance under group-automorphisms of the group division operation (Prop. 8.2.30),
- the total left rectangle is the Cartesian square from (10.73) which exhibits the canonical framing,
- the total front face is the pasting of
  - on the bottom: the Cartesian square (8.160) which exhibits the action on V,
  - on the top: the Cartesian square which is the pasting of the top and the top-right squares in (10.61) equibiting the action on $T_e V$

  and hence is itself Cartesian,
- the bottom and the total right squares are the defining Cartesian squares of the fiber products, and hence, by the pasting law, also their pasting to the total right square is Cartesian,
- the total vertical rear square, with the dashed morphism ϕ_1 on top, is the one thus induced from the universal property of the fiber product, and is itself Cartesian, by the pasting law (Prop. 8.1.39), (using, by the above items, that the left, right and front squares are Cartesian and that the diagram of squares commutes),
- the slanted square in the rear is the pasting of the Cartesian square on the left of (10.60), that exhibits the induced G-action on TV, with the diagonal square on $\mathrm{fib}(\rho)$.

Now observe that inside this big diagram (10.75) we find the following solid homotopy-commutative sub-diagram

$$\begin{array}{ccc} TV & \longrightarrow & (T_e V) \mathbin{/\!\!/} G \\ \big\downarrow & \overset{\phi_2}{\nearrow} & \big\downarrow \\ (TV) \mathbin{/\!\!/} G & \longrightarrow & V \mathbin{/\!\!/} G\,. \end{array} \tag{10.76}$$

Here the left morphism is an effective epimorphism (by Lemma 8.2.7), and the right morphism is (-1)-truncated by the assumption that V is 0-truncated (Lemma xyz). Therefore, the connected/truncated factorization system (Prop. 8.1.83) implies an essentially unique lift ϕ_2, as shown. This, in turn, implies the morphism ϕ_3 in (10.75), again by the universal property of the homotopy fiber product.

Now, since both the slanted as well as the vertical total rear squares are Cartesian, the diagram (10.75) shows that the contravariant base change (Prop. 8.1.65) of ϕ_3 along $\mathrm{fib}(\rho)$ is an equivalence. But since $\mathrm{fib}(\rho)$ is an effective epimorphism (Lemma 8.1.71) , base change along it is conservative (Prop. 8.1.71), and hence it follows that ϕ_3 itself is already an equivalence.

With that identification, the total cube in (10.75) exhibits the G-equivariance of the framing. □

Proposition 10.2.22 (Orbifolding of framed V-folds)**.** *Let* $\mathbf{H}$ *be an elastic* ∞*-topos* $\mathbf{H}$ *(Def. 9.1.22),* $V, G \in \mathrm{Grp}(\mathbf{H})$ *(Prop. 8.2.1) with* $G \simeq \flat G$ *discrete, and* $(X, \rho_X), (T_e V, \rho_{T_e V}) \in G\mathrm{Actions}(\mathbf{H})$ *(Prop. 8.2.6) for* X *a* V*-fold (Def. 10.2.2) equipped with a framing* fr *(Def. 10.2.18). Then the following are equivalent:*

(i) *The framing is* G*-equivariant (Def. 8.2.10) with respect to the induced action on* TX *(from Prop. 10.2.10) and the product action* $\rho_X \times \rho_{T_e V}$ *on* $X \times T_e V$*, hence lifts to a morphism*

$$(TX, \rho_{TX}) \xrightarrow[\simeq]{\mathrm{fr}} (X, \rho_X) \times (T_e V, \rho_{T_e V}) \quad \in G\mathrm{Actions}(\mathbf{H}) \tag{10.77}$$

(ii) *The classifying map (10.66) of the frame bundle (Def. 10.2.14) of the orbifolded* V*-fold* $X \mathbin{/\!\!/} G$ *(Prop. 10.2.9) factors through* $\mathbf{B}G$ *as*

$$X \mathbin{/\!\!/} G \xrightarrow{\;\rho_X\;} \mathbf{B}G \xrightarrow{\;\vdash \rho_{T_e V}\;} \mathbf{B}\mathrm{Aut}(T_e V) \qquad \text{(curved arrow } \vdash \mathrm{Frm}(X \mathbin{/\!\!/} G) \text{ from } X \mathbin{/\!\!/} G \text{ to } \mathbf{B}\mathrm{Aut}(T_e V), \ \Downarrow) \tag{10.78}$$

Proof. Consider the following diagram:

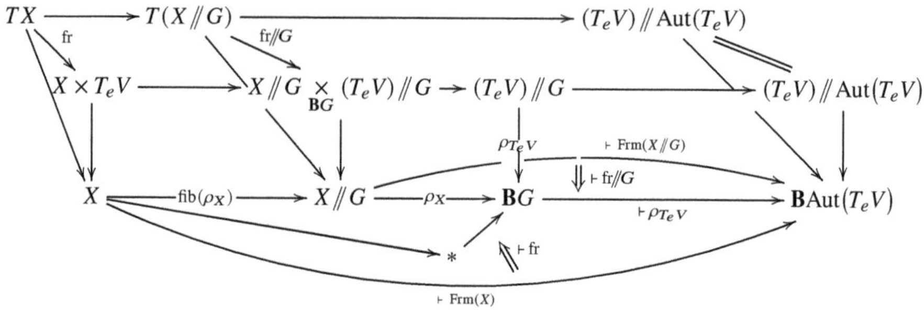

(10.79)

Note that here:

(a) The total outer part of the diagram exhibits the given framing fr via its classifying homotopy $\vdash$ fr, according to Remark 10.2.19.

(b) The front squares in the middle and on the right are the pullback squares that defines the diagonal G-action and the classification of the $\rho_{T_e V}$-action respectively. Hence also their pasting composite is a pullback, by the pasting law (Prop. 8.1.39).

(i) First to see that G-equivariance of fr implies the factorization (10.78): By the characterization of G-actions (8.160) G-equivariance of fr means, equivalently, that fr is the morphism on homotopy fibers over $\mathbf{B}G$ induced from an equivalence fr$/\!\!/G$ on homotopy quotients. But, by **(b)** and Prop. 10.2.10, such an equivalence is classified by a homotopy of the form (10.78).

(ii) Now to see that, conversely, the existence of a homotopy "$\vdash$ fr$/\!\!/G$" of the form (10.78) implies the existence of a G-equivariant framing fr (quotation marks now since we yet have to show that the two are related in this way). For this, we have to show that the morphism on homotopy fibers induced by fr$/\!\!/G$ is a framing fr. But, by the nature of the G-action on TX from Prop. 10.2.10, the nature of the diagonal G-action exhibited by the middle front square, and using the pasting law (Prop. 8.1.39), this means to show that the left front and rear squares are homotopy pullbacks. For the front left square this follows by the factorization of $\rho_X \circ \mathrm{fib}(\rho_X)$ through the point, using **(a)**, **(b)**, and the pasting law (Prop. 8.1.39). For the rear left square, this follows by Prop. 9.1.33, since fib(ρ) is a local diffeomorphism by Prop. 10.2.9. □

10.2.5 G-Structures

Definition 10.2.23 (G-Structure coefficients)**.** Let $\mathbf{H}$ be an elastic ∞-topos (Def. 9.1.22) and $V \in \mathrm{Grp}(\mathbf{H})$ (Prop. 8.2.1). Then a *coefficient for G-structure*

$$(G, \phi) \;\in\; \mathrm{Grp}(\mathbf{H})_{/\mathrm{Aut}(T_e V)} \tag{10.80}$$

is a group G equipped with a homomorphism of groups $G \xrightarrow{\phi} \mathrm{Aut}(T_e V)$ to the structure group (Def. 10.2.12) of V-folds. Under delooping (8.151), this is equivalently a morphism in $\mathbf{H}$ of the form $\mathbf{B}G \xrightarrow{\mathbf{B}\phi} \mathbf{B}\mathrm{Aut}(T_e V)$.

Definition 10.2.24 (G-structures on V-folds)**.** Let $\mathbf{H}$ be an elastic ∞-topos (Def. 9.1.22), $V \in \mathrm{Grp}(\mathbf{H})$ (Prop. 8.2.1), $(G, \phi) \in \mathrm{Grp}(\mathbf{H})_{/\mathrm{Aut}(T_e V)}$ (Def. 10.2.23), and $X \in V\mathrm{Folds}(\mathbf{H})$ (Def. 10.2.2).

(i) We say that

- *a (G, ϕ)-structure on X* (often just *G-structure* if ϕ is understood),
- or *(G, ϕ)-structure on its frame bundle* (Def. 10.2.14),
- or *reduction of the structure group* (10.2.12) *along ϕ*

is a lift (τ, g) of the frame bundle classifying map (10.66) through $\mathbf{B}\phi$:

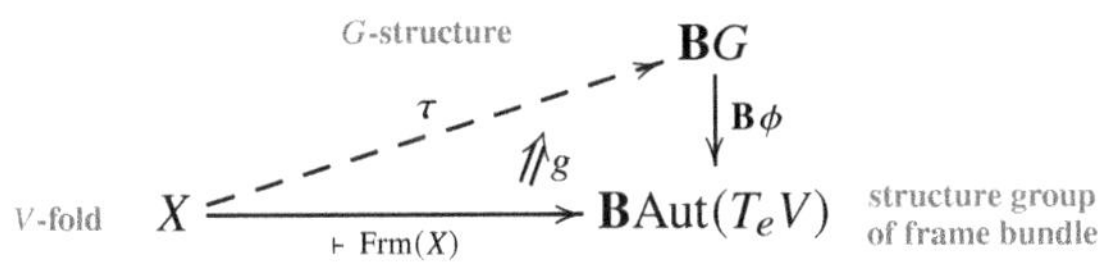

(10.81)

(ii) We say that the *G-frame bundle* $G\mathrm{Frm}(X)$ of a V-fold X equipped with such a (G, ϕ)-structure is the G-principal bundle which is classified (via Prop. 8.2.15): by τ, hence the object in the following diagram:

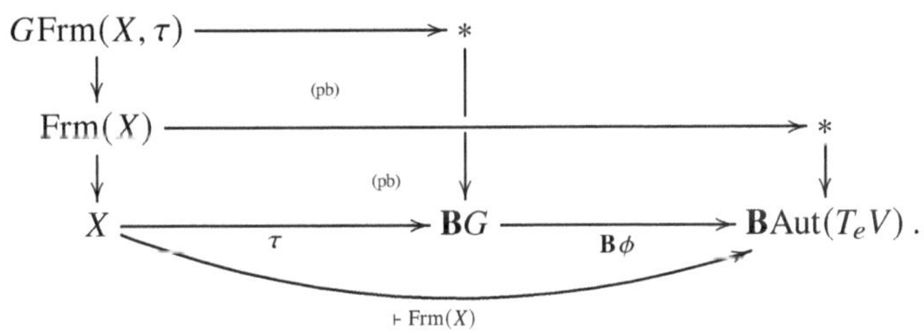

(10.82)

(iii) We write

$$(G, \phi)\mathrm{Structures}_X(\mathbf{H}) := \mathbf{H}_{/\mathbf{B}\mathrm{Aut}(T_e V)}\big(\vdash \mathrm{Frm}(X)\,,\, \mathbf{B}\phi\big) \in \mathrm{Grpd}_\infty \qquad (10.83)$$

for the ∞-groupoid of (G, ϕ)-structures on the V fold X.

In direct generalization of Prop. 10.2.16 we have:

Proposition 10.2.25 (G-structured V-fold is G-quotient of its G-frame bundle)**.** *Let $\mathbf{H}$ be an elastic ∞-topos (Def. 9.1.22), $V \in \mathrm{Grp}(\mathbf{H})$ (Prop. 8.2.1), $(G, \phi) \in \mathrm{Grp}(\mathbf{H})_{/\mathrm{Aut}(T_e V)}$ (Def. 10.2.23), $X \in V\mathrm{Folds}(\mathbf{H})$ (Def. 10.2.2), and $(\tau, g) \in (G, \phi)\mathrm{Structures}_X(\mathbf{H})$ (Def. 10.2.24). Then:*

(i) *X is equivalently the homotopy quotient* (8.161) *of its G-frame bundle* (10.82) *by G:*

$$X \simeq G\mathrm{Frm}(X, \tau) /\!\!/ G\,. \qquad (10.84)$$

(ii) *the classifying map of the G-frame bundle on X exhibits the action of G on $G\mathrm{Frm}(X, \tau)$ according to* (8.160).

Proof. This is immediate from the equivalence between principal bundles and homotopy quotient projections (Remark 8.2.16) applied to the G-frame bundle (10.82):

$$
\begin{array}{c}
G\mathrm{Frm}(X,\tau) \\
\Big\downarrow{\scriptstyle \mathrm{fib}(\rho)\,\simeq\,\mathrm{fib}(\tau)} \\
G\mathrm{Frm}(X,\tau)/\!\!/G \;\simeq\; X \xrightarrow{\;\tau\;} \mathbf{B}G\,. \\
\underbrace{\qquad\qquad\qquad\qquad\qquad}_{\rho_G}
\end{array}
\tag{10.85}
$$

□

Example 10.2.26 (G-structure induced from framing)**.** Let $\mathbf{H}$ be an elastic ∞-topos (Def. 9.1.22), $V \in \mathrm{Grp}(\mathbf{H})$ (Prop. 8.2.1), and $X \in V\mathrm{Folds}(\mathbf{H})$ (Def. 10.2.2). Then a framing on X (Def. 10.2.18) induces a (G,ϕ)-structure (Def. 10.2.24) for any $(G,\phi) \in \mathrm{Grp}(\mathbf{H})_{/\mathrm{Aut}(T_eV)}$, given by the pasting

$$
\begin{array}{ccccc}
 & & * & \longrightarrow & \mathbf{B}G \\
 & \nearrow & \Uparrow{\scriptstyle \vdash \mathrm{fr}} & \searrow \;\Rightarrow & \Big\downarrow{\scriptstyle \mathbf{B}\phi} \\
X & & \xrightarrow{\quad\vdash \mathrm{Frm}(X)\quad} & & \mathbf{B}\mathrm{Aut}(T_eV)
\end{array}
\tag{10.86}
$$

of the homotopy $\vdash$ fr (10.70) which classifies the framing (Remark 10.2.19) with the homotopy that exhibits the group homomorphism ϕ as a morphism of pointed objects (Prop. 8.2.1).

Example 10.2.27 (Canonical G-structure)**.** Let $\mathbf{H}$ be an elastic ∞-topos $\mathbf{H}$ (Def. 9.1.22), and $V \in \mathrm{Grp}(\mathbf{H})$ (Prop. 8.2.1). Then V itself, regarded as a V-fold by Ex. 10.2.4, carries a (G,ϕ)-structure (Def. 10.2.24) for any $(G,\phi) \in \mathrm{Grp}(\mathbf{H})_{/\mathrm{Aut}(T_eV)}$, induced via Ex. 10.2.26 from its canonical framing fr_ℓ (10.71) via left-translation (Prop. 10.2.20). We call this the *canonical* (G,ϕ)*-structure* on V:

$$
\begin{array}{ccc}
 & & \mathbf{B}G \\
 & \overset{\tau_V}{\nearrow} \;\; \Uparrow{\scriptstyle g_V} & \Big\downarrow{\scriptstyle \mathbf{B}\phi} \\
V & \longrightarrow & \mathbf{B}\mathrm{Aut}(T_eV)
\end{array}
\quad := \quad
\begin{array}{ccccc}
 & & * & \longrightarrow & \mathbf{B}G \\
 & \nearrow & \Uparrow{\scriptstyle \vdash \mathrm{fr}_\ell} & \searrow \;\Rightarrow & \Big\downarrow{\scriptstyle \mathbf{B}\phi} \\
V & & \xrightarrow{\quad\vdash \mathrm{Frm}(V)\quad} & & \mathbf{B}\mathrm{Aut}(T_eV)
\end{array}
\tag{10.87}
$$

10.2.6 Local isometries

Lemma 10.2.28 (G-structures pull back along local diffeomorphisms)**.** *Let* $\mathbf{H}$ *be an elastic* ∞*-topos (Def. 9.1.22),* $V \in \mathrm{Grp}(\mathbf{H})$ *(Prop. 8.2.1), and* $(G,\phi) \in \mathrm{Grp}(\mathbf{H})_{/\mathrm{Aut}(T_eV)}$ *(Prop. 8.2.1, Def. 8.2.13, Ex. 9.1.32). Then pre-composition constitutes a contravariant* ∞*-functor ("pullback of* (G,ϕ)*-structures")*

$$
\begin{array}{ccccc}
\big(V\mathrm{Folds}(\mathbf{H})^{\mathrm{\acute{e}t}}\big)^{\mathrm{op}} & \longrightarrow & \mathrm{Grpd}_\infty & & \\
X_1 & \longmapsto & (G,\phi)\mathrm{Structures}_{X_1}(\mathbf{H}) & \ni & \tau\circ f \\
{\scriptstyle \mathrm{\acute{e}t}}\Big\downarrow{\scriptstyle f} & & \Big\uparrow{\scriptstyle f^*} & & \Big\uparrow \\
X_2 & \longmapsto & (G,\phi)\mathrm{Structures}_{X_2}(\mathbf{H}) & \ni & \tau
\end{array}
\tag{10.88}
$$

from the ∞*-category* (10.45) *of* V*-folds and local diffeomorphisms, which assigns to any* V*-fold its* ∞*-groupoid* (10.83) *of* (G,ϕ)*-structures (Def. 10.2.24).*

Proof. We need to show that for (τ, g) a (G, ϕ)-structure on X_2, the composite

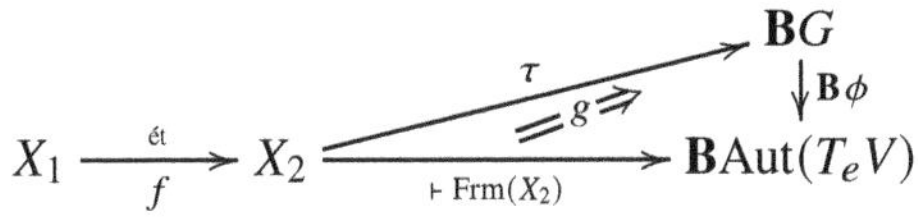

(10.89)

is a (G, ϕ)-structure on X_1. For this we need to exhibit a natural equivalence

$$\big(\vdash \mathrm{Frm}(X_2)\big) \circ f_1 \quad \simeq \quad \vdash \mathrm{Frm}(X_1) \tag{10.90}$$

so that

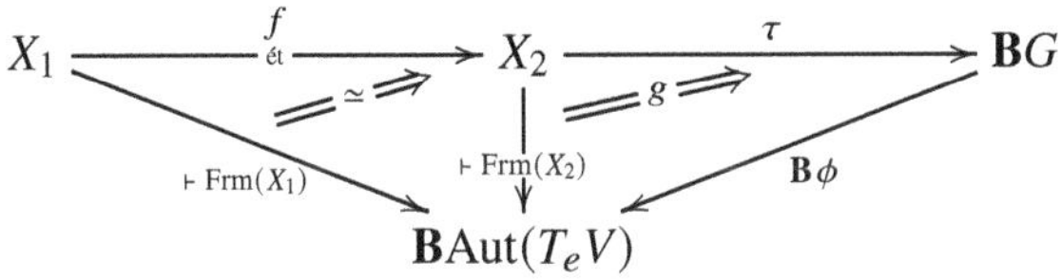

(10.91)

But this exists by Prop. 9.1.33. □

Definition 10.2.29 (Local isometries between G-structured V-folds)**.** Let $\mathbf{H}$ be an elastic ∞-topos (Def. 9.1.22), $V \in \mathrm{Grp}(\mathbf{H})$ (Prop. 8.2.1), and $(G, \phi) \in \mathrm{Grp}(\mathbf{H})_{/\mathrm{Aut}(T_e V)}$ (Prop. 8.2.1, Def. 8.2.13, Ex. 9.1.32).

(i) For $X_1, X_2 \in V\mathrm{Folds}$ (Def. 10.2.2) and $(\tau_i, g_i) \in (G, \phi)\mathrm{Structures}_{X_i}(\mathbf{H})$ (10.83), we say a *local isometry*, to be denoted

$$\big(X_1, (\tau_1, g_1)\big) \xrightarrow[\text{met}]{(f, \sigma)} \big(X_2, (\tau_2, g_2)\big) \tag{10.92}$$

is a pair

$$X_1 \xrightarrow[\text{ét}]{f} X_2\,, \qquad f^*(\tau_2, g_2) \xrightarrow[\simeq]{\sigma} (\tau_1, g_1)\,, \tag{10.93}$$

consisting of a local diffeomorphism (Def. 9.1.28) and an equivalence of (G, ϕ)-structures (10.83) between that on its domain V-fold and the pullback (10.88) of the (G, ϕ)-structure on its codomain V-fold.

(ii) Equivalently, by (10.88), a local isometry (10.93) is a morphism between (G, ϕ)-structured V-folds regarded as objects in the iterated slice ∞-topos (Ex. 8.1.63)

(a) over $\mathbf{B}\mathrm{Aut}(T_e V)$ via their classifying maps of their frame bundles (10.66)

(b) over $(\mathbf{B}G, \mathbf{B}\phi)$ via their (G, ϕ)-structure (10.81)

of this form:

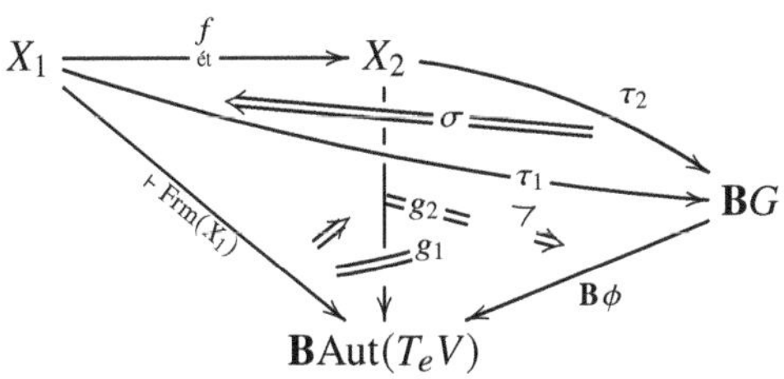

(10.94)

$$\in \Big(\mathbf{H}_{/\mathbf{B}\mathrm{Aut}(T_e V)}\Big)_{/(\mathbf{B}G, \mathbf{B}\phi)}\Big(\big(X_1, (\tau_1, g_1)\big)\,,\ \big(X_2, (\tau_2, g_2)\big)\Big).$$

(iii) Hence we write

$$(G,\phi)\mathrm{StrctrdVFolds}(\mathbf{H}) \longrightarrow \big(\mathbf{H}_{/\mathbf{B}\mathrm{Aut}(T_e V)}\big)_{/\mathbf{B}G} \in \mathrm{Cat}_\infty \qquad (10.95)$$

for the sub-∞-category of this iterated slice on 1-morphisms of the form (10.94).

10.2.7 Integrability of G-structures

Definition 10.2.30 (Integrable G-structure)**.** Let $\mathbf{H}$ be an elastic ∞-topos (Def. 9.1.22), $V \in \mathrm{Grp}(\mathbf{H})$ (Prop. 8.2.1), $(G,\phi) \in \mathrm{Grp}(\mathbf{H})_{/\mathrm{Aut}(T_e V)}$ (Def. 10.2.23).

(i) Given $(X,(\tau_X, g_X)) \in (G,\phi)\mathrm{StrctrdVFolds}(\mathbf{H})$ (Def. 10.2.29), we say that (τ, g) is an *integrable* (G,ϕ)-structure on the V-fold X if there exists a correspondence of local isometries (10.93) between V equipped with its canonical (G,ϕ)-structure (τ_V, g_V) (Def. 10.2.27) and $(X,(\tau_X, g_X))$:

$$\big(V,(\tau_V,g_V)\big) \xleftarrow{\ \mathrm{met}\ } \big(U,(\tau_U,g_U)\big) \xrightarrow{\ \mathrm{met}\ } \big(X,(\tau_X,g_X)\big) \qquad (10.96)$$

such that the right leg is, in addition, an effective epimorphism (Def. 8.1.80), then called a $(V,(\tau_V, g_V))$-*atlas of* $(X,(\tau_X, g_X))$ (8.148). (Underlying this, forgetting the (G,ϕ)-structures, is a V-atlas (10.43).)

(ii) We write

$$\mathrm{Intgrbl}(G,\phi)\mathrm{StrctrdVFolds}(\mathbf{H}) \hookrightarrow (G,\phi)\mathrm{StrctrdVFolds}(\mathbf{H}) \in \mathrm{Cat}_\infty \quad (10.97)$$

for the full sub-∞-category of that of (G,ϕ)-structured V-folds (10.95) on those that are integrable.

Definition 10.2.31 (Locally integrable G-structure)**.** Let $\mathbf{H}$ be an elastic ∞-topos (Def. 9.1.22), $V \in \mathrm{Grp}(\mathbf{H})$ (Prop. 8.2.1), $(G,\phi) \in \mathrm{Grp}(\mathbf{H})_{/\mathrm{Aut}(T_e V)}$ (Def. 10.2.23), $X \in V\mathrm{Folds}(\mathbf{H})$ (Def. 10.2.2), and $(\tau, g) \in (G,\phi)\mathrm{Structures}_X(\mathbf{H})$ (Def. 10.2.24). We say that (τ, g) is a *locally integrable* (G,ϕ)-structure if, for each point $* \xrightarrow{x} X$, there is a local diffeomorphism ϕ_x of the local neighborhood (Def. 9.1.30) of $* \xrightarrow{e} V$ onto a local neighborhood of x such that the restriction of (τ, g) along ϕ is equivalent to the canonical (G,ϕ)-structure (Def. 10.2.27) on $T_e V$:

$$\underset{* \xrightarrow{x} X}{\forall} \quad \underset{\substack{T_e V \xrightarrow[\text{ét}]{\phi_x} X \\ e \nwarrow \ \nearrow x \\ *}}{\exists} \quad : \quad \phi_x^*(\tau, g) \simeq (\tau_{T_e V}, g_{T_e V}). \qquad (10.98)$$

Another way to say this: We have a correspondence of local isometries as in (10.96), but with the right leg required to be an effective epimorphism only under $\flat$.

Example 10.2.32 (G-Structures on smooth manifolds and orbifolds)**.**

(i) Let $\mathbf{H} = \mathrm{HldSmthGrpd}_\infty$ (Ex. 9.1.26), $G \in \mathrm{LieGroups} \hookrightarrow \mathrm{Grp}(\mathbf{H})$ (see (9.72)), and $X \in \mathrm{SmthMfd} \hookrightarrow \mathbf{H}$ regarded as an $\mathbb{R}^n$-fold according to Ex. 10.2.5. In this case, the structure group of X (Def. 10.2.12) is the ordinary general linear group $\mathrm{GL}_{\mathbb{R}}(n)$ (Ex. 10.2.13). Therefore, a G-structure on X in the sense of Def. 10.2.24 is (by Ex. 10.2.17) a G-structure in the traditional sense of differential geometry [St64, VII][Kob72][Mol77]; and it is integrable according to Def. 10.2.30 if it is "flat" in the traditional sense of [Gu65] and locally integrable according to Def. 10.2.31 precisely if it is "uniformly 1-flat" in the traditional sense of [Gu65], namely, if it is torsion-free (review in [Lot01]). Examples are shown in Table 10.1 on p. 265.

(ii) For $k > 1$ and $\mathbf{H} = k\mathrm{HldSmthGrpd}_\infty$ (Def. 9.1.26), the local integrability condition of Def. 10.2.31 is of the form of the "uniformly k-flatness"-condition of [Gu65]. But beware that, according to Def. 10.2.24, but in contrast to [Gu65], in this case the G-structure itself is not on the plain frame bundle but on the order-k jet frame bundle (by Ex. 10.2.17).

Table 10.1 Examples of G-structures (cf. Ex. 10.2.32)

$G \xrightarrow{\phi} \mathrm{GL}_{\mathbb{R}}(n)$	**(G,ϕ)-structure**	**Locally integrable**	**Integrable**	see
$\mathrm{Sp}_{\mathbb{R}}(n) \hookrightarrow \mathrm{GL}_{\mathbb{R}}(n)$	almost symplectic	symplectic	symplectic	[St64, VII.2]
$\mathrm{GL}_{\mathbb{C}}(n/2) \hookrightarrow \mathrm{GL}_{\mathbb{R}}(n)$	almost complex	complex	complex	
$\mathrm{O}(n) \hookrightarrow \mathrm{GL}_{\mathbb{R}}(n)$	Riemannian	torsion-free Riemannian	flat Riemannian	
$\mathrm{O}(n-1,1) \hookrightarrow \mathrm{GL}_{\mathbb{R}}(n)$	Lorentzian	torsion-free Lorentzian	flat Lorentzian	[LPZ13]
$\mathrm{O}(n) \rtimes \mathbb{R} \hookrightarrow \mathrm{GL}_{\mathbb{R}}(n)$	$\mathrm{CO}(n)$ structure	conformal	flat conformal	[AG98]
$\mathrm{CR}(\frac{n}{2}-1) \hookrightarrow \mathrm{GL}_{\mathbb{R}}(n)$	$\mathrm{CR}(n)$-structure	Cauchy-Riemann	flat Cauchy-Riemann	[DT06]
$\mathrm{GL}_{\mathbb{H}}(n/4) \hookrightarrow \mathrm{GL}_{\mathbb{R}}(n)$	$\mathrm{GL}_{\mathbb{H}}(\frac{n}{4})$-structure	hypercomplex	flat hypercomplex	[Jo95]
$\mathrm{U}(n/2) \hookrightarrow \mathrm{GL}_{\mathbb{R}}(n)$	hermitian almost complex	Kähler	Kähler	[Mor07, 11.1]
$\mathrm{SU}(n/2) \hookrightarrow \mathrm{GL}_{\mathbb{R}}(n)$	$\mathrm{SU}(n)$ structure	Calabi-Yau	Calabi-Yau	[Pri15, 1.3]
$\mathrm{Sp}(\frac{n}{4})\cdot\mathrm{Sp}(1) \hookrightarrow \mathrm{GL}_{\mathbb{R}}(n)$	almost unimodular quaternionic	quaternionic Kähler	flat quaternionic Kähler	[AM93a] [AM93b]
$\mathrm{Sp}(n/4) \hookrightarrow \mathrm{GL}_{\mathbb{R}}(n)$	almost Hyperkähler	Hyperähler	flat Hyperkähler	
$G_2 \hookrightarrow \mathrm{GL}_{\mathbb{R}}(7)$	G_2-structure	torsion-free G_2-structure	flat/interable G_2-structure	[Br05]
$\mathrm{Spin}(7) \hookrightarrow \mathrm{GL}_{\mathbb{R}}(8)$	$\mathrm{Spin}(7)$-structure	torsion-free $\mathrm{Spin}(7)$-structure	flat $\mathrm{Spin}(7)$-structure	[Br87] [Jo01]

10.2.8 Haefliger groupoids

Definition 10.2.33 (Haefliger groupoid)**.** Let $\mathbf{H}$ be an elastic ∞-topos (Def. 9.1.22) and $V \in \mathrm{Grp}(\mathbf{H})$ (Prop. 8.2.1).

(i) With no further structure,

(a) The *V-Haefliger groupoid* is the étale groupoid (Def. 9.1.37)

$$\mathrm{Haef}_\bullet(V) \in \mathrm{\acute{E}taleGroupoids}(\mathbf{H}), \tag{10.99}$$

which is the étalification (Def. 9.1.41) of the Atiyah groupoid (Def. 8.2.17) of the frame bundle (Def. 10.2.12) of V regarded as a V-fold (Ex. 10.2.4):

$$\mathrm{Haef}_\bullet(V) \;:=\; \mathrm{At}^{\text{ét}}_\bullet\big(\mathrm{Frm}(V)\big). \tag{10.100}$$

(b) The *V-Haefliger stack* of V is the corresponding V-fold (according to Remark 10.2.3):

$$\mathcal{H}\mathrm{aef}(V) \;:=\; \mathcal{A}\mathrm{t}^{\text{ét}}\big(\mathrm{Frm}(V)\big) \;\in\; V\mathrm{Folds}\,. \tag{10.101}$$

(ii) Given, in addition, $(G, \phi) \in \mathrm{Grp}(\mathbf{H})_{/\mathrm{Aut}(T_e V)}$ (Def. 10.2.23), with $G\mathrm{Frm}(V) \to V$ denoting the G-frame bundle (10.82) corresponding to the canonical (G, ϕ)-structure on V (Def. 10.2.27), we say

(a) the $\big(V, (G, \phi)\big)$*-Haefliger groupoid* is the étale groupoid (Def. 9.1.37)

$$\mathrm{Haef}_\bullet\big(V, (G, \phi)\big) \;\in\; \text{ÉtaleGroupoids}(\mathbf{H}) \tag{10.102}$$

which is the étalification (Def. 9.1.41) of the Atiyah groupoid (Def. 8.2.17) of the G-frame bundle (10.82):

$$\mathrm{Haef}_\bullet\big(V, (G, \phi)\big) \;:=\; \mathrm{At}^{\text{ét}}_\bullet\big(G\mathrm{Frm}(V)\big)\,. \tag{10.103}$$

(b) The $\big(V, (G, \phi)\big)$*-Haefliger stack* of V is the corresponding V-fold (according to Remark 10.2.3):

$$\mathcal{H}\mathrm{aef}\big(V, (G, \phi)\big) \;:=\; \mathcal{A}\mathrm{t}^{\text{ét}}\big(G\mathrm{Frm}(V)\big) \;\in\; V\mathrm{Folds}\,. \tag{10.104}$$

Proposition 10.2.34 (Haefliger stack represents V-fold structure)**.** *Let* $\mathbf{H}$ *be an elastic ∞-topos (Def. 9.1.22) $V \in \mathrm{Grp}(\mathbf{H})$ (Prop. 8.2.1) and $X \in \mathbf{H}$. Then the following are equivalent:*

(i) *X is a V-fold (Def. 10.2.2);*

(ii) *X admits a local diffeomorphism to the V-Haefliger stack (Def. 10.2.33).*

Proof. First consider the implication **(i)** $\Rightarrow$ **(ii)**: Assuming X is a V-fold, consider a V-atlas (10.43) $V \xleftarrow{\text{ét}} U \xrightarrow{\text{ét}} X$. By Prop. 9.1.33 (and as in the proof of Prop. 10.2.14), the pullbacks of the frame bundles of V and of X along this V-atlas to U coincide there, which means that we have a homotopy-commutative square of their classifying maps (10.66) as shown on the bottom left of the following diagram:

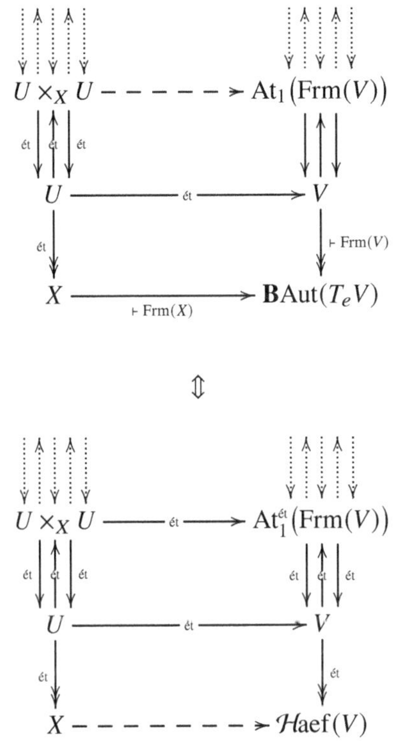

(10.105)

By passing to nerves (Ex. 8.1.87) of the vertical morphisms, this induces a morphism of groupoids as shown on the top left. But $U_\bullet$ is an étale groupoid (by Prop. 9.1.38), and $U \longrightarrow V$ is a local diffeomorphism by definition of V-atlases, so that the top left part of the left diagram in (10.105) is in the étale slice over V (Def. 9.1.34). Therefore, the adjunction (9.89) of Prop. 9.1.35 implies that the top part of the diagram on the left of (10.105) factors through the étalification (Def. 9.1.41) as shown in the top part on the right. With this we get the dashed morphism on the right by passing to colimits over the vertical simplicial diagrams (as in Prop. 9.1.38).

It only remains to see that the dashed morphism on the right is itself a local diffeomorphism. For this observe that all the horizontal morphisms are local diffeomorphisms, using the assumptions and then left-cancellability (Lemma 9.1.29). Therefore, the statement follows Lemma 9.1.40.

For the converse implication **(ii)** ⇒ **(i)**: Given a local diffeomorphism as shown dashed on the right of (10.105), we need to produce a V-atlas for X. So now define the bottom square on the right of (10.105) to be the pullback of the étale atlas of the Haefliger stack along the given morphism. This does make the top left span of the square a V-atlas by the fact that the classes of local diffeomorphisms and of effective epimorphisms are both closed under pullback (by Lemma 8.1.82 and Lemma 9.1.29). □

Proposition 10.2.35 (G-Structured Haefliger stack represents integrable G-structure). *Let* $\mathbf{H}$ *be an elastic* ∞*-topos (Def. 9.1.22),* $V \in \mathrm{Grp}(\mathbf{H})$ *(Prop. 8.2.1),* $(G, \phi) \in \mathrm{Grp}(\mathbf{H})_{/\mathrm{Aut}(T_e V)}$ *(Def. 10.2.23). The* $\big(V, (G, \phi)\big)$*-Haefliger groupoid (Def. 10.2.33), carries a canonical integrable* (G, ϕ)*-structure (Def. 10.2.30)*

$$(\tau_{\mathcal{H}}, g_{\mathcal{H}}) \in (G, \phi)\mathrm{Structures}_{\mathcal{H}\mathrm{aef}(V)}(\mathbf{H}) \tag{10.106}$$

such that the operation of pullback of (10.88) *along local diffeomorphism (Lemma 10.2.28) constitutes a natural bijection*

$$\begin{array}{ccc} \pi_0\, \mathrm{Intgrbl}(G, \phi)\mathrm{Strc}V\mathrm{Folds}(\mathbf{H}) & \simeq & \pi_0\, \acute{\mathrm{E}}\mathrm{t}_{\mathcal{H}\mathrm{aef}(V,(G,\phi))} \\ \big(X, (\tau, g)\big) & \longmapsto & \Big(X \xrightarrow{\vdash(\tau, g)} \mathcal{H}\mathrm{aef}\big(V, (G, \phi)\big)\Big) \end{array} \tag{10.107}$$

between the sets of equivalence classes of:

(i) *integrably* (G, ϕ)*-structured* V*-folds (Def. 10.2.30),*

(ii) *local diffeomorphisms into the* $\big(V, (G, \phi)\big)$*-Haefliger stack, hence objects in its étale topos (Def. 9.1.34).*

Proof. We proceed as in the proof of Prop. 10.2.34, but lifting the diagram there from $\mathbf{H}$ to the iterated slice $\big(\mathbf{H}_{/\mathbf{B}\mathrm{Aut}(T_e V)}\big)_{/\mathbf{B}G}$ (10.94).

(i) First, consider an integrably G-structured V-fold $\big(X, (\tau, g)\big)$. We describe the construction of a local diffeomorphism into the Haefliger stack from this: Pick any $\big(V, (\tau_V, g_V)\big)$-atlas $\big(V, (\tau_V, g_V)\big) \xleftarrow{\mathrm{met}} \big(U, (\tau_U, g_U)\big) \xrightarrow{\mathrm{met}} \big(X, (\tau_X, g_X)\big)$ (10.96). By Def. 10.2.24, this is equivalently a choice of equivalence between the pullbacks to U of the G-structures on V and on X. Regarded in the iterated slice (10.94), this equivalently means that we have a square in $\big(\mathbf{H}_{/\mathbf{B}\mathrm{Aut}(T_e V)}\big)_{/\mathbf{B}G}$

(10.94), as shown in the following:

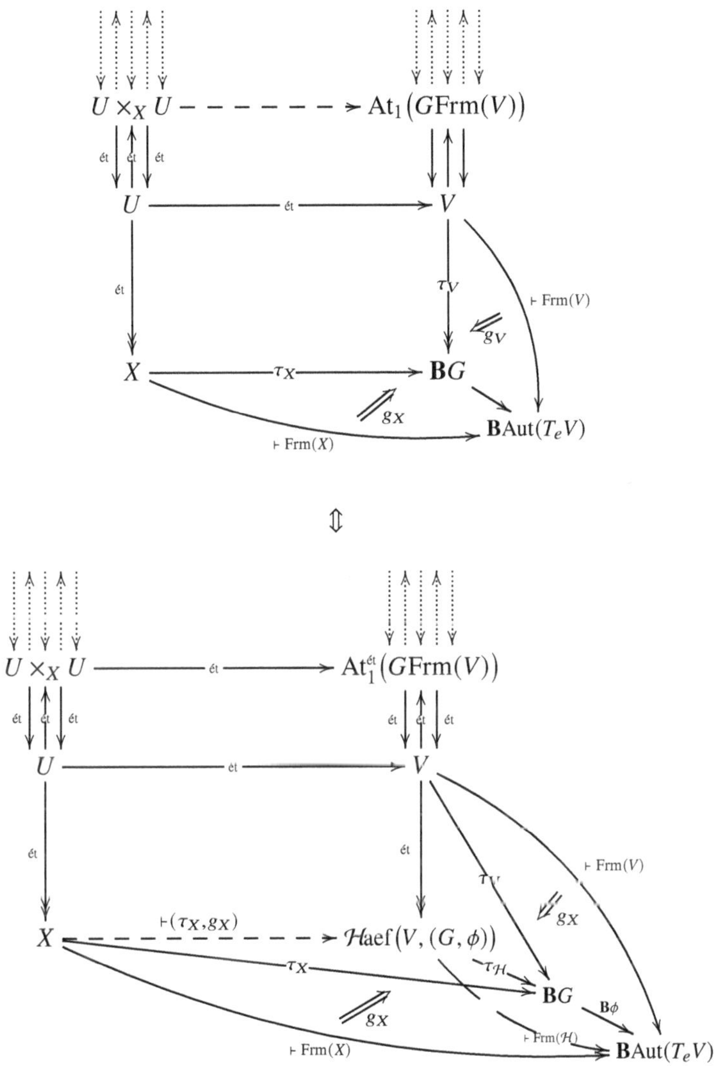

(10.108)

Now we proceed as follows:

- **(a)** Observing (with Prop. 8.1.69) that fiber products in the iterated slice are actually given by the plain fiber products in **H** equipped with canonical morphisms to the slicing objects, we find that passing to nerves (Ex. 8.1.87) of the vertical morphisms on the left of (10.108) yields a morphism from the étale groupoid induced by the given V-cover of X to the Atiyah groupoid of $G\mathrm{Frm}(X)$ (Def. 8.2.17) – just as in (10.105), but now equipped with coherent maps to $\mathbf{B}\phi$.
- **(b)** Therefore, we obtain the factorization through the $(V, (G, \phi))$-Haefliger groupoid (the étalification of the Atiyah groupoid of the G-frame bundle shown

on the top right of (10.108)) just as in (10.105), but now, in addition, coherently equipped with maps to $\mathbf{B}\phi$.

- **(c)** After this étalification we may identify these maps: Since those on V remain unchanged by étalification over V, these still give the canonical (G,ϕ)-structure (τ_V, g_V), as shown on the far right of (10.108). But since now the vertical simplicial morphisms are all local diffeomorphisms, pullback along which preserves (G,ϕ)-structure (by Lemma 10.2.28) and in particular preserves tangent- and frame bundles (by Prop. 9.1.33), it follows that all stages of the $\big(V,(G,\phi)\big)$-Haefliger groupoid in the top right are now equipped with the classifying map of their frame bundles.
- **(d)** Since colimits in the slice are given by colimits in the underlying topos (by Ex. 8.1.68), the colimit over the simplicial sub-diagram on the far right of (10.108) still yields the $\big(V,(\tau,g)\big)$-Haefliger stack (10.104), as shown, now equipped with canonical maps to $\mathbf{B}\phi$.
- **(e)** We claim that the induced map from the Haefliger stack to $\mathbf{B}\mathrm{Aut}(T_eV)$, denoted $\vdash \mathrm{Frm}(\mathcal{H})$ in (10.108), is indeed the classifying map of the frame bundle of the Haefliger stack:

$$\vdash \mathrm{Frm}(\mathcal{H}) \;\simeq\; \vdash \mathrm{Frm}\Big(\mathcal{H}\mathrm{aef}\big(V,(G,\phi)\big)\Big). \tag{10.109}$$

  This follows because:

  - by **(c)** above, the component maps of the colimiting map classify the frame bundles of the stages of the simplicial nerve;
  - therefore, the colimiting map classifies the colimit of the frame bundles of the simplicial nerve, by Prop. 8.1.72;
  - but the colimit of the tangent bundles of the étale cover is the tangent bundle of the corresponding étale stack, by Prop. 9.1.39.
- **(f)** In particular, this implies that the induced homotopy which fills the bottom right part of (10.108):

$$\vdash \mathrm{Frm}(\mathcal{H}) \xRightarrow{\;g_{\mathcal{H}}\;} \mathbf{B}\phi \circ \tau_{\mathcal{H}}, \tag{10.110}$$

  canonically given by the colimit construction in the iterated slice, constitutes a (G,ϕ)-structure on the $\big(V,(G,\phi)\big)$-Haefliger stack.
- **(g)** In conclusion, the dashed morphism on the right of (10.108) exists and is a local diffeomorphism, as in the proof of Prop. 10.2.34; but, by construction in the iterated slice, it is now exhibited as a local isometry to the Haefliger stack equipped with the induced (G,ϕ)-structure (10.110).

(ii) The converse construction is now immediate: Given a local diffeomorphism of the form shown dashed on the right of (10.108), pulling back the étale atlas of the Haefliger stack along it yields a V-atlas for X (just as in the proof of this converse step in Prop. 10.2.34), and pulling (via Lemma 10.2.28) the (G,ϕ)-structure (10.110) around the resulting Cartesian square makes this a $\big(V,(G,\phi)\big)$-atlas that exhibits X as equipped with an integrable (G,ϕ)-structure. This construction is clearly injective on equivalence classes by ∞-functoriality of the pullback construction (10.88) of (G,ϕ)-structures, and it is surjective on equivalence classes by item (i) above. Hence, it is a bijection on equivalence classes, as claimed. □

10.2.9 Tangential structures

Closely akin to G-structures (Def. 10.2.24) are *tangential structures* (Def. 10.2.36) where not the structure group itself is lifted, but only its shape:

Definition 10.2.36 (Tangential structure). Let $\mathbf{H}$ be an elastic ∞-topos (Def. 9.1.22), $V \in \mathrm{Grp}(\mathbf{H})$ (Prop. 8.2.1), $(G, \phi) \in \mathrm{Grp}(\mathbf{H})_{/\int \mathrm{Aut}(T_e V)}$ (Def. 10.2.23), and $X \in V\mathrm{Folds}(\mathbf{H})$ (Def. 10.2.2).

(i) We say that a *tangential* (G, τ)*-structure* on X is a lift (τ, g) through $\mathbf{B}\phi$ of the composite of the frame bundle classifying map (10.66) with the shape-unit (8.72):

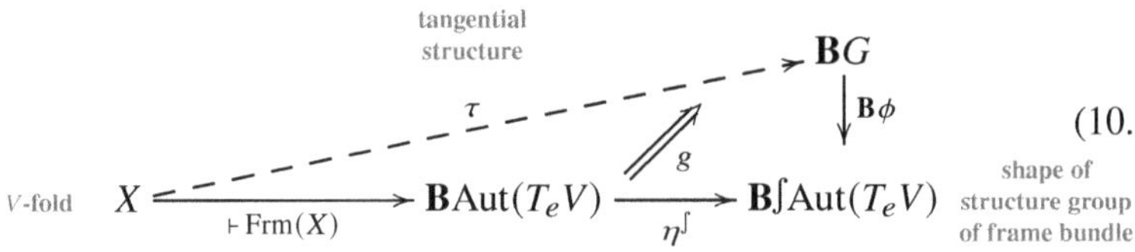

(10.111)

(ii) We write

$$\mathrm{Tangential}(G, \phi)\mathrm{Structures}_X(\mathbf{H}) \;:=\; \mathbf{H}_{/\mathbf{B}\int \mathrm{Aut}(T_e V)}\big(\eta^{\int} \circ \vdash \mathrm{Frm}(X)\,,\, \mathbf{B}\phi\big) \tag{10.112}$$

for the ∞*-groupoid of* (G, ϕ)*-tangential structures* on the V-fold X.

Example 10.2.37 (Tangential structures on smooth manifolds).
Let $\mathbf{H} = \mathrm{HldSmthGrpd}_\infty$ (Ex. 9.1.26) $G \in \mathrm{LieGroups} \hookrightarrow \mathrm{Grp}(\mathbf{H})$ (see (9.72)) and $X \in \mathrm{SmthMfd} \hookrightarrow \mathbf{H}$ regarded as an $\mathbb{R}^n$-fold according to Ex. 10.2.5. In this case, the structure group of X (Def. 10.2.12) is the ordinary general linear group $\mathrm{GL}_{\mathbb{R}}(n)$ (Ex. 10.2.13). Hence here tangential structure in the general sense of Def. 10.2.36 is tangential structure in the traditional sense of differential topology (popularized under this name in [GMTW06, 5], originally introduced as "(B, f)-structure" [La63][St68, II], review in [Ko96, 1.4]).

Example 10.2.38 (Cohesive refinement of tangential structure). Every (G, ϕ)-structure (Def. 10.2.24) induces tangential $(\int G, \int \phi)$-structure (Def. 10.2.36) by composition with the naturality square of $\eta^{\int}$ on $\mathbf{B}\phi$:

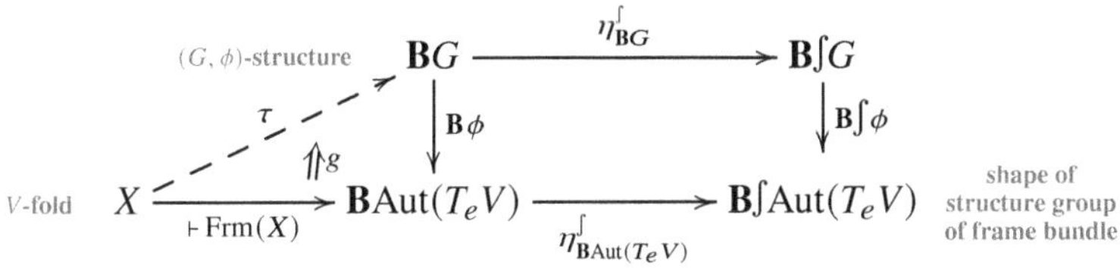

(10.113)

Conversely, realizing a tangent structure as obtained from a G-structure this way means to find a geometric (differential) refinement.

Example 10.2.39 (Orientation structure)**.** Let $\mathbf{H} = \mathrm{HldSmthGrpd}_\infty$ (Ex. 9.1.26) and $X \in \mathbf{H}$ an $\mathbb{R}^n$-fold (Def. 10.2.2), hence an ordinary manifold (Ex. 10.2.5) or, more generally, an ordinary étale Lie groupoid (Ex. 10.2.6). With the general linear and the (special) orthogonal group regarded as smooth groups via (9.72)

$$\mathrm{SO}(n) \xrightarrow{i_{\mathrm{SO}}} \mathrm{O}(n) \xrightarrow{i_{\mathrm{O}}} \mathrm{GL}(n) \quad \in \quad \mathrm{Grp}(\mathrm{SmthMfd}) \longrightarrow \mathrm{Grp}(\mathbf{H}) \qquad (10.114)$$

we have:

(i) an $\mathrm{O}(n)$-structure (Def. 10.2.24) on X is equivalently a Riemannian structure (Ex. 10.2.32);

(ii) but a tangential $\int\mathrm{O}(n)$-structure (Def. 10.111) is equivalently *no structure*, since $\int\mathrm{O}(n) \underset{\simeq}{\xrightarrow{\int i_{\mathrm{O}}}} \int\mathrm{GL}(n)$ is an equivalence of underlying shapes (since $\mathrm{O}(n)$ is the maximal compact subgroup of $\mathrm{GL}(n)$),

(iii) while a tangential $\int\mathrm{SO}(n)$-structure (Def. 10.111) is an *orientation* of X.

(iv) A differential refinement, in the sense of Ex. 10.2.38, of such an orientation structure is an oriented Riemannian structure (via its induced volume form).

Example 10.2.40 (Higher Spin structure [SSS09][SSS12])**.** Let $\mathbf{H} = \mathrm{HldSmthGrpd}_\infty$ (Ex. 9.1.26) and $X \in \mathbf{H}$ an $\mathbb{R}^n$-fold (Def. 10.2.2), hence an ordinary manifold (Ex. 10.2.5) or, more generally, an ordinary étale Lie groupoid (Ex. 10.2.6). The sequence of groups (10.114) in Ex. 10.2.39 is, under shape, the beginning of the *Whitehead tower* of $\int O(n) \simeq \int\mathrm{GL}(n)$. The tangential structures (Def. 10.2.36, Ex. 10.2.37) corresponding to the stages in this tower are the *Spin structure* and its higher analogues:

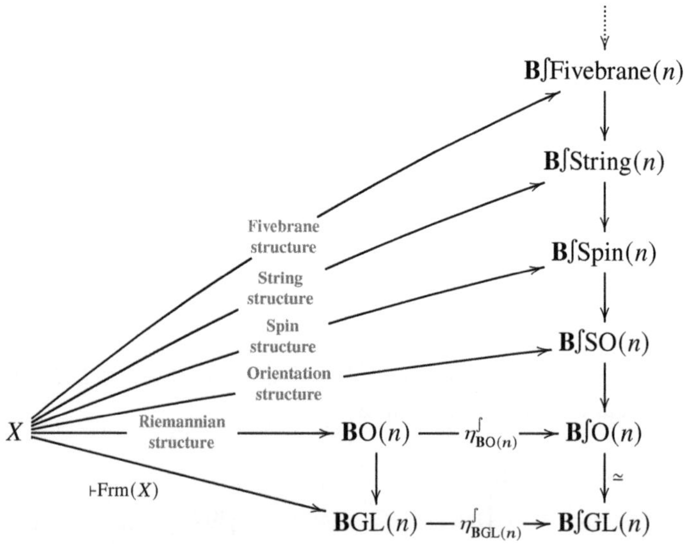

(10.115)

10.2.10 Flat V-folds

Definition 10.2.41 (Flat V-folds)**.** Let $\mathbf{H}$ be an elastic ∞-topos (Def. 9.1.22), $V \in \mathrm{Grp}(\mathbf{H})$ (Prop. 8.2.1), and $X \in V\mathrm{Folds}(\mathbf{H})$ (Def. 10.2.2). We say that X is *flat* if the classifying map (10.66) of its frame bundle (Prop. 10.2.14) factors through the $\flat$-counit (8.73), hence if it carries (G, ϕ)-structure (Def. 10.2.24) for $(G, \phi) = (\flat \mathrm{Aut}(T_e V), \epsilon^{\flat}_{\mathrm{Aut}(T_e V)})$:

$$
\begin{array}{ccc}
 & & \flat \mathbf{B}\mathrm{Aut}(T_e V) \\
 & \overset{\tau}{\nearrow} \;\; \Uparrow & \big\downarrow {\scriptstyle \epsilon^{\flat}_{\mathbf{B}\mathrm{Aut}(T_e V)}} \\
X & \xrightarrow[\vdash \mathrm{Frm}(X)]{} & \mathbf{B}\mathrm{Aut}(T_e V)
\end{array}
\tag{10.116}
$$

By the universal property of $\epsilon^{\flat}$ and since $\flat$ commutes with $\mathbf{B}$, this means equivalently that X carries G-structure for any discrete group $G \simeq \flat\flat G$.

Proposition 10.2.42 (Flat frame bundles are V-folds)**.** *Let* $\mathbf{H}$ *be an elastic* ∞*-topos (Def. 9.1.22),* $V \in \mathrm{Grp}(\mathbf{H})$ *(Prop. 8.2.1), and* $X \in V\mathrm{Folds}(\mathbf{H})$ *(Def. 10.2.2). If* X *is flat (Def. 10.2.41), then*

(i) *its flat frame bundle* $(\flat \mathrm{Aut}(T_e V))\mathrm{Frm}(X)$ (10.82) *is itself a* V*-fold (Def. 10.2.2),*

(ii) *the bundle morphism is a local diffeomorphism (Def. 9.1.28):*

$$
(\flat \mathrm{Aut}(T_e V))\mathrm{Frm}(X) \xrightarrow{\text{ét}} X \,. \tag{10.117}
$$

Proof. First consider **(ii)**: We need to show that the left square in the following pasting diagram is Cartesian:

$$
\begin{array}{ccccc}
(\flat \mathrm{Aut}(T_e V))\mathrm{Frm}(X) & \xrightarrow{\eta^{\Im}_{(\flat \mathrm{Aut}(T_e V))\mathrm{Frm}(X)}} & \Im\big((\flat \mathrm{Aut}(T_e V))\mathrm{Frm}(X)\big) & \longrightarrow & \Im * \\
{\scriptstyle p}\big\downarrow & & {\scriptstyle \Im p}\big\downarrow & {\scriptstyle (\mathrm{pb})} & \big\downarrow \\
X & \xrightarrow[\eta^{\Im}_X]{} & \Im X & \xrightarrow[\Im \tau]{} & \Im \flat \mathrm{Aut}(T_e V)
\end{array}
\tag{10.118}
$$

Here the right square is Cartesian, by definition (10.82), and since $\Im$, being a right adjoint, preserves Cartesian squares (by Prop. 8.1.42). Hence, by the pasting law (Prop. 8.1.39), it is sufficient to show that the total rectangle is Cartesian. But, by the naturality of $\eta^{\Im}$, the total rectangle is equivalent to that of the following pasting diagram:

$$
\begin{array}{ccccc}
(\flat \mathrm{Aut}(T_e V))\mathrm{Frm}(X) & \longrightarrow & * & \xrightarrow{\eta^{\Im}_*} & \Im * \\
\big\downarrow & {\scriptstyle (\mathrm{pb})} & \big\downarrow & {\scriptstyle (\mathrm{pb})} & \big\downarrow \\
X & \xrightarrow[\tau]{} & \flat \mathbf{B}\mathrm{Aut}(T_e V) & \xrightarrow[\eta^{\Im}_{\flat \mathbf{B}\mathrm{Aut}(T_e V)}]{} & \Im \flat \mathbf{B}\mathrm{Aut}(T_e V)
\end{array}
\tag{10.119}
$$

Here the left square is Cartesian by the definition (10.82), while the right square is Cartesian since its two horizontal morphisms are equivalences by elasticity. Hence, the total rectangle is Cartesian by the pasting law (Prop. 8.1.39).

Regarding **(i)**: We need to exhibit a V-atlas (10.43) for the flat frame bundle. So let $V \xleftarrow{\text{ét}} U \xrightarrow{\text{ét}} X$ be a V-atlas for X, and consider the following pullback diagram:

$$
\begin{array}{ccc}
U \underset{X}{\times} (\flat\mathrm{Aut}(T_e V))\mathrm{Frm}(X) & \overset{\text{ét}}{\twoheadrightarrow} & (\flat\mathrm{Aut}(T_e V))\mathrm{Frm}(X) \\
{\scriptstyle \text{ét}}\big\downarrow & \text{(pb)} & \big\downarrow{\scriptstyle \text{ét}} \\
U & \underset{\text{ét}}{\twoheadrightarrow} & X \\
{\scriptstyle \text{ét}}\big\downarrow & & \\
V & &
\end{array}
\qquad (10.120)
$$

Observe that all four morphisms in the square are effective epimorphisms (Def. 8.1.80) and local diffeomorphisms (Def. 9.1.28): The bottom one by definition, the right one by **(ii)**, and hence the other two since both classes of morphisms are closed under pullback (Lemma 8.1.82 and Lemma 9.1.29). Finally, since the class of local diffeomorphisms is also closed under composition (Lemma 9.1.29), the total vertical morphism is a local diffeomorphism, and hence the total outer diagram is a V-atlas of the flat frame bundle. □

Proposition 10.2.43 ($\flat G$-frame bundles are V-folds)**.** *Let* $\mathbf{H}$ *be an elastic* ∞*-topos (Def. 9.1.22),* $V \in \mathrm{Grp}(\mathbf{H})$ *(Prop. 8.2.1),* $X \in V\mathrm{Folds}(\mathbf{H})$ *(Def. 10.2.2),* $(G, \phi) \in \mathrm{Grp}(\mathbf{H})_{/\mathrm{Aut}(T_e V)}$ *(Prop. 8.2.1, Def. 8.2.13, Ex. 9.1.32) with* $G \simeq \flat G$ *discrete and* $(\tau, g) \in (G, \phi)\mathrm{Structures}_X(\mathbf{H})$*. Then the corresponding* G*-frame bundle* (10.82) *is itself a* V*-fold:*

$$
G \simeq \flat G \quad \Leftrightarrow \quad G\mathrm{Frm}(X) \in V\mathrm{Folds}(\mathbf{H})\,. \qquad (10.121)
$$

Proof. The proof proceeds verbatim as that for Prop. 10.2.42, just with the structure group restricted along $\flat G \to \flat\mathrm{Aut}(T_e V)$. □

In summary, we have found the general abstract version of the local model spaces of orbifolds:

Proposition 10.2.44 (Local orbifold model spaces)**.** *Let* $\mathbf{H}$ *be an elastic* ∞*-topos (Def. 9.1.22),* $G, V \in \mathrm{Grp}(\mathbf{H})$ *(Prop. 8.2.1), with* $G \simeq \flat G$ *discrete, and* $(V, \rho) \in G\mathrm{Actions}(\mathbf{H})$ *(Prop. 8.2.6) a restriction (Prop. 8.2.12) of the action* (V, ρ_{Aut}) *by group-automorphisms (Prop. 8.2.29). Then the homotopy quotient* (8.161)

$$
V /\!\!/ G \in \mathbf{H} \qquad (10.122)
$$

of V *regarded with its canonical framing (Prop. 10.2.20)*

(i) *is a flat* V*-fold (Def. 10.2.41);*

(ii) *with* G*-structure (Def. 10.2.24)*

(iii) *whose* G*-frame bundle* (10.82) *is* G*-equivariantly (Def. 8.2.10) equivalent to* V *itself:*

$$
G\mathrm{Frm}(V /\!\!/ G) \simeq V\,. \qquad (10.123)
$$

Proof. First observe that $V /\!\!/ G$ is a V-fold, by Prop. 10.2.9 applied to Ex. 10.2.4. That this is flat **(i)** is implied by **(ii)**, since G is assumed to be discrete. For **(ii)** and **(iii)** observe that the canonical framing on V is G-equivariant, by Prop. 10.2.21, so that Prop. 10.2.22 implies G-structure on $V /\!\!/ G$ classified by the action morphism

ρ itself. But this means that its homotopy fiber, hence the corresponding G-frame bundle (Def. 10.82) is V itself, by (8.160) (and in accord with Prop. 10.2.43). □

Example 10.2.45 (Ordinary orbifold singularities)**.** Let $\mathbf{H} := \mathrm{HldSmthGrpd}_\infty$ (Ex. 9.1.26) and $V := (\mathbb{R}^n, +)$ as in Ex. 10.2.5. Then a group automorphism of V is a *linear* isomorphism, hence $\mathrm{Aut}_{\mathrm{Grp}}(\mathbb{R}^n, +) \simeq \mathrm{GL}(n)$. Therefore, in this case the assumptions of Prop. 10.2.44 hold precisely for V, a *linear representation* of the discrete group G, and thus we recover the traditional local orbifold models $V /\!\!/ G$ from [Sa56] (in their incarnation as étale groupoids).

10.2.11 Orbi-V-folds

Finally, we may now easily promote V-folds to orbifolds proper and hence promote the ∞-category of étale stacks to an ∞-category of higher proper orbifolds:

Definition 10.2.46 (Orbi-V-folds)**.** Let $\mathbf{H}$ be a singular-elastic ∞-topos (Def. 9.2.6) and $V \in \mathrm{Grp}(\mathbf{H}_{\cup})$. We say that an *orbi-$V$-fold* is an object $\mathcal{X} \in \mathbf{H}$ which is the orbi-singularization (Def. 9.2.7) of a V-fold (Def. 10.2.2, hence of an étale ∞-stack modeled on V, cf. Exp. 10.2.6).

(i) We write $V\mathrm{Orbfld}(\mathbf{H}) \subset \mathbf{H}$ for the full sub-∞-category on orbi-V-folds:

$$\mathcal{X} \in V\mathrm{Orbfld}(\mathbf{H}) \qquad \Leftrightarrow \qquad \cup\mathcal{X} \in V\mathrm{Folds}(\mathbf{H}) \,. \tag{10.124}$$

This means, equivalently, that the orbi-V-folds in $\mathbf{H}$ are the orbi-singularizations (9.143) of the V-folds in $\mathbf{H}_{\cup}$:

$$\begin{array}{ccc} V\mathrm{Folds}(\mathbf{H}_{\cup}) & \underset{\text{OrbSnglr}}{\overset{\text{Smth}}{\rightleftarrows}}\;(\simeq) & V\mathrm{Orbfld}(\mathbf{H}) \\ \\ \mathrm{Smth}(\mathcal{X}) & \longleftarrow\!\shortmid & \mathcal{X} \\ \| & & \| \\ X & \longmapsto & \mathrm{OrbSnglr}(X) \end{array} \tag{10.125}$$

(ii) Similarly, given, in addition, $(G, \phi) \in \mathrm{Grp}(\mathbf{H})_{/\mathrm{Aut}(T_e V)}$ (Def. 10.2.23), we write $(G, \phi)\mathrm{Strctrd}V\mathrm{Orbfld}(\mathbf{H}) \subset \mathbf{H}$ for the full sub-∞-category on (G, ϕ)-structured orbi-V-folds (Def. 10.2.29):

$$\begin{array}{ccc} (G, \phi)\mathrm{Strctrd}V\mathrm{Folds}(\mathbf{H}_{\cup}) & \underset{\text{OrbSnglr}}{\overset{\text{Smth}}{\rightleftarrows}}\;(\simeq) & (G, \phi)\mathrm{Strctrd}V\mathrm{Orbfld}(\mathbf{H}) \\ \\ (\mathrm{Smth}(\mathcal{X}), (\tau, g)) & \longleftarrow\!\shortmid & (\mathcal{X}, (\tau, g)) \\ \| & & \| \\ (X, (\tau, g)) & \longmapsto & (\mathrm{OrbSnglr}(X), (\tau, g)) \end{array} \tag{10.126}$$

Remark 10.2.47 (Coefficients for orbifold cohomology)**.** The point of Def. 10.2.46 is that, by regarding a V-fold in the elastic ∞-topos $\mathbf{H}_{\cup}$ equivalently as an orbi-V-fold in the larger singular-elastic ∞-topos $\mathbf{H}$, a larger class of coefficients for intrinsic cohomology theories (1.21) becomes available, notably coefficients of the form $\int \curlyvee(A /\!\!/ G)$ (see Lemma 10.1.7). This is what gives rise, in §11, to proper orbifold cohomology (Def. 11.2.5) in contrast to the coarser cohomology of underlying étale groupoids (Def. 11.2.1).

Remark 10.2.48 (The proper ∞-category of higher orbifolds)**.** While (10.125) is an equivalence of abstract ∞-categories,

(i) it is not an equivalence of sub-∞-categories of the ambient singular-elastic ∞-topos $\mathbf{H}$:

$$
\begin{array}{ccccc}
\overset{\substack{\infty\text{-category of}\\ \text{of étale groupoids}}}{V\mathrm{Folds}(\mathbf{H}_{\cup})} & \not\simeq & \overset{\substack{\text{proper}\\ \infty\text{-category}\\ \text{of orbifolds}}}{V\mathrm{Orbfld}(\mathbf{H})} & \in & (\mathrm{Cat}_{\infty})_{/\mathbf{H}}\,. \\
& {\scriptstyle \mathrm{Smth}}\searrow & & \swarrow{\scriptstyle \mathrm{OrbSnglr}} & \\
& & \mathbf{H} & &
\end{array}
\qquad (10.127)
$$

(ii) To bring out this distinction, also in view of Remark 10.2.47, we call $V\mathrm{Orbfld}(\mathbf{H})$ (Def. 10.2.46) the *proper ∞-category of orbifolds*, in contrast to the ∞-category $V\mathrm{Folds}(\mathbf{H}_{\cup})$ (10.44) of étale ∞-groupoids.

(iii) It is a happy coincidence that *proper* is also the technical adjective chosen in [DHLPS19] for equivariant homotopy theories presented by ∞-presheaves over categories of orbits with compact – hence finite if discrete – isotropy groups: In this terminology the singular-cohesive ∞-topos $\mathbf{H}$ is, according to Def. 9.2.3, indeed a *proper* global equivariant homotopy theory.

Example 10.2.49 (Subcategories of smooth and of flat orbifolds)**.** Let $\mathbf{H}$ be an elastic ∞-topos (Def. 9.1.22), $V \in \mathrm{Grp}(\mathbf{H})$ (Prop. 8.2.1) and $(G, \phi) \in \mathrm{Grp}(\mathbf{H})_{/\mathrm{Aut}(T_e V)}$ (Prop. 8.2.1, Def. 8.2.13, Ex. 9.1.32). We have fully faithful inclusions into the ∞-category of (G, ϕ)-structured orbi-V-folds (Def. 10.2.46)

$$
\begin{array}{ccccc}
& & (G,\phi)\mathrm{Strctrd}V\mathrm{Orbfld}(\mathbf{H}) & & \\
& \overset{\text{smooth orbifolds}}{\underset{i_{\cup}}{\nearrow}} & & \overset{\text{flat orbifolds}}{\underset{i_{\flat}}{\nwarrow}} & \\
(G,\phi)\mathrm{Strctrd}V\mathrm{Folds}(\mathbf{H}_0) & & & & (\flat G, \phi \circ \epsilon^{\flat})\mathrm{Strctrd}V\mathrm{Orbfld}(\mathbf{H})
\end{array}
\qquad (10.128)
$$

of

(i) smooth (G, ϕ)-structured V-folds, via Lemma 9.2.21;

(ii) flat $(\flat G, \phi \circ \epsilon^{\flat})$-structured V-folds (Def. 10.2.41).

11

Orbifold Cohomology

With an internal higher topos-theoretic characterization of orbifolds in hand (from §10), we immediately obtain an induced notion of (differential, geometric, étale) *orbifold cohomology*, namely as the intrinsic cohomology (1.21) of the ambient singular-cohesive ∞-topos. Here we discuss how this new intrinsic notion of orbifold cohomology
– subsumes proper equivariant cohomology theory (§11.1)
– and unifies it with tangentially twisted cohomology (§11.2).

The main result here is a general construction of orbifold étale cohomology which we show to naturally unify
(i) tangentially twisted cohomology of smooth but curved spaces with
(ii) RO-graded proper equivariant cohomology of flat but singular spaces.

As fundamental examples, we present a new model of twisted orbifold K-theory (following [SS26c, Ex. 4.5.4]) as well as tangentially twisted orbifold Cohomotopy.

11.1 Equivariant Cohomology

11.1.1 Proper equivariant cohomology

Definition 11.1.1 (Borel equivariant cohomology)**.** Let $\mathbf{H}_{\cup}$ be a cohesive ∞-topos (Def. 9.1.1) $G \in \mathrm{Grp}(\mathbf{H}_{\cup})$ (Prop. 8.2.1) and $(X, \tau), (A, \rho) \in G\mathrm{Actions}(\mathbf{H}_{\cup})$ (Prop. 8.2.6). Then the *Borel equivariant cohomology* of X with coefficients in A is the intrinsic cohomology (1.21) in the slice $\mathbf{H}_{/\mathbf{B}G}$ (Prop. 8.1.62) of the homotopy quotient (8.161) of X with coefficients in the shape (9.2) of the homotopy quotient of A:

$$
\begin{aligned}
\overset{\text{Borel equivariant cohomology}}{H_{\mathrm{Borel}}(X, A)} \;&:=\; \pi_0\, \mathbf{H}_{/\mathbf{B}G}\big((X /\!\!/ G),\, (A /\!\!/ G)\big) \\
&=\; \pi_0 \left\{ \begin{array}{ccc} (X /\!\!/ G) & \overset{\overset{\text{cocycle}}{c}}{\dashrightarrow} & (A /\!\!/ G) \\ & \Leftarrow & \\ {\scriptstyle \tau} \searrow & & \swarrow {\scriptstyle \rho} \\ & \mathbf{B}G & \end{array} \right\}.
\end{aligned}
\tag{11.1}
$$

Definition 11.1.2 (Proper equivariant cohomology)**.** Let $\mathbf{H}$ be a singular-cohesive ∞-topos (Def. 9.2.3), $G \in \mathrm{Grp}(\mathbf{H}_\flat)$ (Prop. 8.2.1) a discrete ∞-group, and

DOI: 10.1201/9781003675969-11

$(X, \tau), (A, \rho) \in G\mathrm{Actions}(\mathbf{H})$ (Prop. 8.2.6). Then we say that the *proper equivariant cohomology* of X with coefficients in A is the intrinsic cohomology (1.21) in the slice $\mathbf{H}_{/\curlyvee \mathbf{B}G}$ (Prop. 8.1.62) of the orbi-singularization (9.144) of the homotopy quotient (8.161) of X with coefficients in the shape (9.2) of the orbi-singularization of the homotopy quotient of A:

$$\begin{aligned} \underset{\text{proper equivariant cohomology}}{H_G(X, A)} \;&:=\; \pi_0\, \mathbf{H}_{/\curlyvee \mathbf{B}G}\Big(\curlyvee(X/\!\!/G)\,,\, \int \curlyvee(A/\!\!/G)\Big) \\ &=\; \left\{ \begin{array}{ccc} \curlyvee(X/\!\!/G) & \overset{\text{cocycle}}{\underset{c}{\dashrightarrow}} & \int \curlyvee(A/\!\!/G) \\ & \Longleftarrow & \\ {\scriptstyle \curlyvee(\tau)} \searrow & & \swarrow {\scriptstyle (\eta^{\int}_{\curlyvee \mathbf{B}G})^{-1} \circ \int \curlyvee(\rho)} \\ & \curlyvee \mathbf{B}G & \end{array} \right\}_{/\sim} . \end{aligned} \tag{11.2}$$

11.1.2 Recovering traditional G-equivariant cohomology

We discuss how, in the case of a finite group G, traditional G-equivariant cohomology (see §8.1.8) is a special case of proper equivariant cohomology (Def. 11.1.2). We take the key observation from [Re14] (Prop. 11.1.6).

Definition 11.1.3 (G-equivariant cohesive ∞-topos [SS26c, Def. 3.3.64]). Let $\mathbf{H}_{\smallsmile}$ be a cohesive ∞-topos (Def. 9.1.1) and $G \in \mathrm{Grp}^{\mathrm{fin}}$ a finite group (9.154). We write

$$G\mathbf{H}_{\smallsmile} \;:=\; \mathrm{Shv}_\infty\big(G\mathrm{Orb},\, \mathbf{H}_{\smallsmile}\big) \;=\; \mathrm{Func}_\infty\big(G\mathrm{Orb}^{\mathrm{op}},\, \mathbf{H}_{\smallsmile}\big) \tag{11.3}$$

for the ∞-topos of $\mathbf{H}_{\smallsmile}$-valued ∞-sheaves on the G-orbit category (Def. 8.1.25), to be called the corresponding *G-equivariant cohesive ∞-topos.*

Remark 11.1.4 (Proper equivariant cohomology theory in singular ∞-toposes)**.** In the case $\mathbf{H}_{\smallsmile} \simeq \mathrm{Grpd}_\infty$ (8.48), Def. 11.1.3 reduces to the ∞-category $G\mathrm{Grpd}_\infty$ (Def. 8.1.21) of traditional G-equivariant homotopy theory (recalled in §8.1.8). The intrinsic cohomology (1.21) of the ∞-topos $G\mathrm{Grpd}_\infty$ – or of its tangent ∞-topos $T\big(G\mathrm{Grpd}_\infty\big)$ (Ex. 8.1.67) in the twisted abelian case (Remark 8.2.23) – is *proper equivariant cohomology* (following terminology in [DHLPS19]), including G-Bredon cohomology [Bre67a][Bre67b] (review in [Blu17, §1.4][tD79, §7]), G-equivariant K-theory [Se68][AS69] (which is proper equivariant by [ASe04, A3.2][FHT11, A.5][DL98]), G-equivariant Cohomotopy theory [Se71][tD79, §8][SS20][BSS19], etc.

Hence, by Remark 9.1.21, to the extent that the objects of the cohesive ∞-topos $\mathbf{H}_{\smallsmile}$ in Def. 11.1.3 are ∞-groupoids equipped with further geometric or differential-geometric structure, the intrinsic cohomology theory (1.21) in $G\mathbf{H}_{\smallsmile}$ (11.3) is an enhancement of plain G-equivariant cohomology to a flavor of *proper G-equivariant differential cohomology* theory (by Remark 9.1.21).

Proposition 11.1.5 (Cohesive Elmendorf theorem)**.** *Consider a cohesive ∞-topos $\mathbf{H}_{\smallsmile}$ (Def. 9.1.1) with an ∞-site* Chrt *of charts (Def. 9.1.10). Then for $G \in \mathrm{Grp}^{\mathrm{fin}}$ a finite group, we have an equivalence of ∞-categories*

$$G\mathbf{H}_{\smallsmile} \;\simeq\; \mathrm{Shv}_\infty\big(\mathrm{Chrt}, G\mathrm{Grpd}_\infty\big)\,, \tag{11.4}$$

where $G\mathrm{Grpd}_\infty$ *is the* ∞*-category of D-topological* G*-spaces (Def. 8.1.21).*

Proof. Consider the following sequence of ∞-functors:

$$\begin{aligned} G\mathbf{H}_{\cup} &:= \mathrm{Shv}_\infty\big(G\mathrm{Orb}, \mathbf{H}_{\cup}\big) \\ &= \mathrm{Shv}_\infty\big(G\mathrm{Orb}, \mathrm{Shv}_\infty(\mathrm{Chrt})\big) \\ &\xrightarrow{\simeq} \mathrm{Shv}_\infty\big(G\mathrm{Orb} \times \mathrm{Chrt}\big) \\ &\xrightarrow{\simeq} \mathrm{Shv}_\infty\big(\mathrm{Chrt}, \mathrm{Shv}_\infty(G\mathrm{Orb})\big) \\ &\xrightarrow{\simeq} \mathrm{Shv}_\infty\big(\mathrm{Chrt}, G\mathrm{Grpd}\big). \end{aligned}$$

That the first and second of these ∞-functors are equivalences follows by the product/hom-adjunction for ∞-functors. With that, the last equivalence follows, objectwise, by Elmendorf's theorem (Prop. 8.1.27). □

Proposition 11.1.6 (G-equivariant homotopy theory embeds into G-singular cohesion)**.** *Let* $\mathbf{H}$ *be a singular-cohesive* ∞*-topos (Def. 9.2.3) over* Grpd_∞ (8.48), *and let* $G \in \mathrm{Grp}^{\mathrm{fin}}$ *be a finite group* (9.154).

(i) *Then there is a full sub-*∞*-category inclusion*

$$G\mathbf{H}_{\cup} \xhookrightarrow[\simeq]{\Delta_G} \mathbf{H}_{/\overset{G}{\curlyvee}} \tag{11.5}$$

of the G*-equivariant non-singular cohesive* ∞*-topos (Def. 11.1.3) into the slice of* $\mathbf{H}$ *(Prop. 8.1.62) over the generic* G*-orbi singularity* (9.151).

(ii) *This is such that, when pre-composed with the cohesive Elmendorf equivalence (Prop. 11.1.5), a cohesive sheaf (on* Chrt*) of* GGrpd (8.43) *presented* (8.46) *by D-topological* G*-spaces* X_U *(Def. 8.1.18) is sent to the presheaf on* Snglrt *that is given as follows:*

$$\begin{array}{c} \mathrm{Shv}_\infty\big(\mathrm{Chrt},\ G\mathrm{Grpd}_\infty\big) \simeq G\mathbf{H}_{\cup} \xrightarrow{\ \underset{\curlyvee}{\textstyle\sum}\Delta_G\ } \mathrm{Shv}_\infty\big(\mathrm{Chrt} \times \mathrm{Snglrt}\big) \\ \Big(U \mapsto \mathrm{Shp}_{G\mathrm{Top}}(X_U)\Big) \longmapsto \left(\big(U,\ \overset{K}{\curlyvee}\big) \mapsto \mathrm{Shp}_{\mathrm{Top}}\left(\left(\underset{\substack{\phi \in \\ \mathrm{Grp}(K,G)}}{\bigsqcup} X_U^{\cdot\phi(K)}\right) \underset{G}{\times} EG\right)\right) \end{array} \tag{11.6}$$

where on the right we have the topological shape (8.1.29) *of the Borel construction by the residual* G*-action on the fixed point subspaces* $X_U^{\phi(K)} \subset X_U$ (8.39).

Proof. For $\mathbf{H}_{\cup} \simeq \mathrm{Grpd}_\infty$ this is [Re14, Prop. 3.5.1]; our expression $\mathrm{Shp}_{\mathrm{Top}}\big(X_U^{\phi(K)} \underset{G}{\times} EG\big)$ is, up to convention of notation, the expression for $B\mathrm{Fun}(H, G \curvearrowright X_U)$ that is spelled out in [Re14, p. 7][Lu19, 3.2.17] (using that our G is discrete). The generalization here follows immediately by applying this equivalence objectwise in the ∞-site Chrt. □

The following is our key class of examples:

Example 11.1.7 (Cohesive shape of G-orbi-singular space is G-homotopy type)**.** In the cohesive ∞-topos $\mathbf{H}_{\cup} := \mathrm{SmthGrpd}_\infty$ (Ex. 9.1.19) consider a 0-truncated object

$X \in \mathbf{H}_{\cup,0}$ equipped with a G-action (Def. 8.160) of a discrete group G, and with corresponding Cohesive G-orbispace (Prop. 10.1.6)

$$\mathcal{X} := \mathrm{OrbSnglr}(X /\!\!/ G)$$

in $\mathbf{H} := \mathrm{SnglrSmthGrpd}_\infty$ (Ex. 9.2.11), which is either of:

(i) a smooth G-orbifold (Ex. 10.1.11):

$$X \in \mathrm{SmoothManifolds} \hookrightarrow \mathrm{DiffeologicalSpaces} \hookrightarrow \mathbf{H}_{\cup}$$

(ii) a topological G-orbi space (Ex. 10.1.12):

$$X \in \mathrm{TopSpc} \xrightarrow{\mathrm{Cdfflg}} \mathrm{DTopSpc} \hookrightarrow \mathbf{H}_{\cup}$$

Then the cohesive shape (9.143) of the G-orbi-singular space $\mathcal{X} \in \mathbf{H}$ is equivalent, under the identification of Prop. 11.1.6, to the G-topological shape (8.46) of the underlying topological G-space of X:

(i) By Prop. 10.1.13, comparing (10.38) with (11.6) we have:

$$\begin{array}{ccc} G\mathrm{SmoothManifolds} & \xrightarrow[\mathrm{OrbSnglr}((-)/\!\!/G)]{\text{form Fréchet-smooth orbifold}} & \mathrm{SnglrSmthGrpd}_{\infty/\mathcal{G}} \\ {\scriptstyle \text{form } G\text{-topological shape}}\Big\downarrow{\scriptstyle \mathrm{Shp}_{G\mathrm{Top}}(\mathrm{Dtplg}(-))} & & {\scriptstyle \mathrm{Shp}}\Big\downarrow{\scriptstyle \text{form cohesive shape}} \\ G\mathrm{Grpd}_\infty & \underset{\text{include } G\text{-equivariant homotopy theory}}{\overset{\Delta_G}{\hookrightarrow}} & \mathrm{SingularGroupoids}_{\infty/\mathcal{G}} \end{array} \tag{11.7}$$

(ii) By Prop. 10.1.14, comparing (10.41) with (11.6), we have:

$$\begin{array}{ccc} G\mathrm{TopSpc} & \xrightarrow[\mathrm{OrbSnglr}(\mathrm{Cdfflg}(-)/\!\!/G)]{\text{form topological } G\text{-orbi space}} & \mathrm{SnglrSmthGrpd}_{\infty/\mathcal{G}} \\ {\scriptstyle \text{form } G\text{-topological shape}}\Big\downarrow{\scriptstyle \mathrm{Shp}_{G\mathrm{Top}}} & & {\scriptstyle \mathrm{Shp}}\Big\downarrow{\scriptstyle \text{form cohesive shape}} \\ G\mathrm{Grpd}_\infty & \underset{\text{include } G\text{-equivariant homotopy theory}}{\overset{\Delta_G}{\hookrightarrow}} & \mathrm{SingularGroupoids}_{\infty/\mathcal{G}} \end{array} \tag{11.8}$$

Lemma 11.1.8 (Δ_G commutes with Disc)**.** *The construction Δ_G from Prop. 11.1.6 commutes with embedding of discrete cohesive structure* (9.142)*:*

$$\begin{array}{ccccc} & & \mathrm{Shv}_\infty(\mathrm{Snglrt}, \mathrm{Grpd}_\infty)_{/\mathcal{G}} & & \\ & \overset{\Delta_G}{\nearrow} & & \overset{\mathrm{Disc}}{\searrow} & \\ G\mathrm{Grpd} & & & & \mathrm{Shv}_\infty(\mathrm{Snglrt}, \mathbf{H}_{\cup})_{/\mathcal{G}} \\ & \underset{\mathrm{Disc}}{\searrow} & & \underset{\Delta_G}{\nearrow} & \\ & & G\mathbf{H}_{\cup} & & \end{array}$$

Theorem 11.1.9 (Cohomology of good orbispaces is proper equivariant cohomology)**.** *Consider the singular-cohesive ∞-topos $\mathbf{H} := \mathrm{SnglrSmthGrpd}_\infty$ (Ex. 9.2.11), and let $G \in \mathrm{Grp}^{\mathrm{fin}}$ be a discrete group* (9.154)*. Then the intrinsic cohomology* (1.21)

(i) *of a G-orbi-singular space $\mathcal{X} \in \mathbf{H}_{/\mathcal{G}}$ (Def. 10.1.4) which is either*

(a) *a topological G-orbi-space (Ex. 10.1.12) with universal covering space (Def. 10.1.5) $X_{G\mathrm{top}} \in G\mathrm{TopSpc}$ (8.36);*

(b) *a Fréchet-smooth G-orbifold (Ex. 10.1.11) with universal covering space (Def. 10.1.5)* $X \in$ *FréchetManifolds and underlying G-topological space* $X_{G\mathrm{top}} := \mathrm{Dtplg}(X)$ (8.12)*;*

(ii) *with coefficients in a cohesively discrete G-∞-groupoid A* (8.43) *(hence the G-topological shape* (8.46) *of some topological G-space* $A_{G\mathrm{top}}$*) regarded as a geometrically discrete orbi-singular ∞-groupoid* $\mathcal{A}$ *via* (11.5)*:*

$$\begin{array}{ccccccc} G\mathrm{TopSpc} & \xrightarrow{\mathrm{Shp}_{G\mathrm{Top}}} & G\mathrm{Grpd}_\infty & \xrightarrow{\mathrm{Disc}} & G\mathbf{H}_{\cup} & \xrightarrow{\Delta_G} & \mathbf{H}_{/\mathcal{G}} \\ A_{\mathrm{top}} & \longmapsto & A & \longmapsto & & & \mathcal{A} \end{array}$$

equals the proper G-equivariant cohomology (Def. 8.1.23) of $X_{G\mathrm{top}}$ *with coefficients in A:*

$$\begin{array}{lccc} & \mathbf{H}_{/\mathcal{G}}(\mathcal{X}, \mathcal{A}) & \simeq & G\mathrm{Grpd}_\infty\big(\mathrm{Shp}_{G\mathrm{Top}}(X_{G\mathrm{top}}),\, A\big) \\ \textit{hence:} & \pi_n \mathbf{H}_{/\mathcal{G}}(\mathcal{X}, \mathcal{A}) & \simeq & H_G^{-n}(X_{G\mathrm{top}}, A) \end{array}$$

intrinsic equivariant differential cohomology in ∞-topos of singular smooth ∞-groupoids

proper G-equivariant cohomology

Proof. **(i)** By Ex. 10.1.12 the topological G-orbi space $\mathcal{X}$ is given by

$$\mathcal{X} \simeq \mathrm{OrbSnglr}\big(\mathrm{Cdfflg}(X) /\!\!/ G\big)\,.$$

With this, we compute as follows:

$$\begin{aligned} \mathbf{H}_{/\mathcal{G}}(\mathcal{X}, \mathcal{A}) &= \mathbf{H}_{/\mathcal{G}}\Big(\mathrm{OrbSnglr}\big(\mathrm{Cdfflg}(X_{\mathrm{top}}) /\!\!/ G\big), \Delta_G \mathrm{Disc}(A)\Big) \\ &\simeq \mathbf{H}_{/\mathcal{G}}\Big(\mathrm{OrbSnglr}\big(\mathrm{Cdfflg}(X_{\mathrm{top}}) /\!\!/ G\big), \mathrm{Disc}(\Delta_G A)\Big) \\ &\simeq (\mathrm{Grpd}_\infty)_{/\mathcal{G}}\Big(\mathrm{Shp}\big(\mathrm{OrbSnglr}(\mathrm{Cdfflg}(X_{\mathrm{top}}) /\!\!/ G)\big), \Delta_G A\Big) \\ &\simeq (\mathrm{Grpd}_\infty)_{/\mathcal{G}}\big(\Delta_G X, \Delta_G A\big) \\ &\simeq G\mathrm{Grpd}\big(\mathrm{Shp}_{G\mathrm{Top}}(X_{\mathrm{top}}), A\big)\,. \end{aligned} \tag{11.9}$$

Here the first step, after unwinding the definitions, is Lemma 11.1.8, the second step is the Shp ⊣ Disc-adjunction (9.142), the third step is Prop. 10.1.14, and the last step is Prop. 11.1.6.

(ii) By Ex. 10.1.11 the Fréchet-smooth G-orbifold $\mathcal{X}$ is given by

$$\mathcal{X} \simeq \mathrm{OrbSnglr}\big(X /\!\!/ G\big)\,.$$

With this, we compute just as in (11.9), only that now in the third step we use Prop. 10.1.13. □

Example 11.1.10 (Orientifold cohomology). Take the singular elastic ∞-topos $\mathbf{H} = \mathrm{SnglrJetSmthGrpd}_\infty$ (Ex. 9.2.11) and $V = (\mathbb{R}^n, +) \in \mathbf{H}_{\cup}$ (10.48). Then $\mathcal{X}_{\cup} \in V\mathrm{Flds}(\mathbf{H}_{\cup})$ (Def. 10.2.2) is an ordinary n-dimensional orbifold or, more generally, an n-dimensional étale ∞-stack (by Ex. 10.2.6) with structure group (Def. 10.2.12) the ordinary general linear group $\mathbf{Aut}(T_e V) \simeq \mathrm{GL}(n)$ (by Ex. 10.2.13). Hence, the composition of the delooping (8.151) of the ordinary determinant group homomorphism $\mathrm{GL}(n) \xrightarrow{\det} \mathbb{Z}_2$ with the classifying map $\vdash \mathrm{Frm}(\mathcal{X}_{\cup})$ (10.66) of the frame bundle of X (Def. 10.2.14) realizes $\mathcal{X}_{\cup}$ as an object in the slice ∞-topos (Prop. 8.1.62) over $\mathbf{B}\mathbb{Z}_2$. Consequently, it realizes its orbi-singularization $\mathcal{X} := \curlyvee \mathcal{X}_{\cup} \in \mathbf{H}$ (9.2.7) as an object in the slice over $\curlyvee^{\mathbb{Z}_2}$ (9.135):

$$
\begin{array}{c} \mathcal{X}_{\cup} \\ {\scriptstyle \mathbf{B}\mathrm{det} \circ \vdash \mathrm{Frm}(\mathcal{X}_{\cup})} \big\downarrow \\ \mathbf{B}\mathbb{Z}_2 \end{array} \in \big(\mathbf{H}_{\cup}\big)_{/\mathbf{B}\mathbb{Z}_2} \quad \Leftrightarrow \quad \begin{array}{c} \mathcal{X} \\ {\scriptstyle \curlyvee(\mathbf{B}\mathrm{det} \circ \vdash \mathrm{Frm}(\mathcal{X}_{\cup}))} \big\downarrow \\ \curlyvee^{\mathbb{Z}_2} \end{array} \in \big(\mathbf{H}_{\cup}\big)_{/\curlyvee^{\mathbb{Z}_2}} . \tag{11.10}
$$

This is the incarnation of the orbifold as an *orbi-orientifold* [DFM11][FSS15, 4.4][SS20]. In particular, if the covering space (Def. 10.1.5)

$$X := \mathrm{fib}\big(\mathbf{B}\mathrm{det} \circ \vdash \mathrm{Frm}(\mathcal{X}_{\cup})\big)$$

happens to be an $\mathbb{R}^n$-fold (Ex. 10.2.5), we have just a plain *orientifold* (without further orbifolding), and then the intrinsic cohomology (1.21) of $\mathcal{X}$ regarded in the slice over $\curlyvee^{\mathbb{Z}_2}$(11.10) is, by Theorem 11.1.9 the proper $\mathbb{Z}_2$-equivariant cohomology of X, such as, for instance, real K-theory [At66] (see [Mas11] for the perspective in proper equivariant cohomology) or $\mathbb{Z}_2$-equivariant cohomotopy [tD79, 8.4][SS20].

11.2 Orbifold Cohomology

We introduce general *étale cohomology* of étale ∞-stacks (Def. 11.2.1), which is sensitive to geometric G-structure and to tangential structure (Def. 11.2.3). Promoting this to the *proper* incarnation of orbifolds (Remark 10.2.48), we finally obtain *tangentially twisted proper orbifold cohomology* (Def. 11.2.5) which we prove unifies tangentially twisted topological cohomology away from orbifold singularities with proper equivariant cohomology at the singularities (Thm. 11.2.6, cf. Fig. 1.7 on p. 21).

We construct two fundamental classes of examples and relate them to each other:

- §11.2.3 We discuss tangentially-twisted proper orbifold *Cohomotopy* theories (Def. 11.2.18) and observe, as an application, that these subsume the relevant cohomology theories for the physics of fractional quantum anomalous Hall systems and of M-theory, according to “Hypothesis H” (Remark 11.2.20).
- §5.1 We discuss a new model for orbifold *topological K-theory* and show how it receives orbifold cohomology operations from orbifold Cohomotopy.

11.2.1 Cohomology of V-étale ∞-stacks

Definition 11.2.1 (Étale cohomology)**.** Let $\mathbf{H}$ be an elastic ∞-topos (Def. 9.1.22), $V \in \mathrm{Grp}(\mathbf{H})$ (Prop. 8.2.1), $(G, \phi) \in \mathrm{Grp}(\mathbf{H})_{/\mathrm{Aut}(T_e V)}$ (Def. 10.2.23), and $X \in \mathrm{Intgrbl}(G, \phi)\mathrm{StrctrdVFlds}(\mathbf{H})$ (Def. 10.2.30). The *étale cohomology* of $\big(X, (\tau, g)\big)$

is its intrinsic cohomology (1.21) when regarded (via Prop. 10.2.34)

$$\begin{array}{ccc} \mathrm{Intgrbl}(G,\phi)\mathrm{StrctrdVFlds}(\mathbf{H}) & \longrightarrow & \Big(\big(\mathbf{H}_{/\mathbf{B}\mathrm{Aut}(T_e V)}\big)_{/(\mathbf{B}G,\mathbf{B}\phi)}\Big)_{/\left(\mathcal{H}\mathrm{aef}\left(V,(G,\phi)\right),(\tau_{\mathcal{H}},g_{\mathcal{H}})\right)} \\ \big(X,(\tau,g)\big) & \longmapsto & \Big(\big(X,(\tau,g)\big) \xrightarrow[\text{met}]{\vdash(\tau,g)} \mathcal{H}\mathrm{aef}\big(V,(G,\phi)\big)\Big) \end{array}$$

in the iterated slice of (10.94) over the $\big(V,(G,\phi)\big)$-Haefliger stack (Def. 10.2.33) equipped with its canonical (G,ϕ)-structure $(\tau_{\mathcal{H}}, g_{\mathcal{H}})$ (Prop. 10.2.34), hence is G-*structure-twisted cohomology* (Remark 8.2.21):

$$\begin{aligned} &\overset{\text{étale cohomology}}{H^{(\tau,g)}\big(X,A\big)} := \\ &\Big(\big(\mathbf{H}_{/\mathbf{B}\mathrm{Aut}(T_e V)}\big)_{/(\mathbf{B}G,\mathbf{B}\phi)}\Big)_{/\left(\mathcal{H}\mathrm{aef}\left(V,(G,\phi)\right),(\tau_{\mathcal{H}},g_{\mathcal{H}})\right)}\Big(\big(X,(\tau,g)\big),\,\big(A,p\big)\Big) \\ &= \left\{ \begin{array}{ccc} X & \overset{\text{cocycle}}{\underset{c}{\dashrightarrow}} & A \\ & \Leftarrow & \\ {\scriptstyle \vdash(\tau,g)} \searrow & & \swarrow {\scriptstyle p} \\ & \mathcal{H}\mathrm{aef}(V,(G,\phi)) & \end{array} \right\}. \end{aligned} \tag{11.11}$$

Remark 11.2.2 (Étale cohomology is geometric)**.** As the notation in Def. 11.2.1 indicates, étale cohomology is a "geometric cohomology theory" in that it does depend (in general) on the G-structure g on the V-fold X (for instance, its complex- or symplectic- or Riemannian- or Lorentzian structure, by Ex. 10.2.32).

Next, we focus attention on the special case where the cohomology theories are not sensitive to the metric part g of a G-structure (τ, g), but just to its tangential structure τ.

Definition 11.2.3 (Tangentially twisted cohomology)**.** Let $\mathbf{H}$ be an elastic ∞-topos (Def. 9.1.22), $V \in \mathrm{Grp}(\mathbf{H})$ (Prop. 8.2.1), $(G,\phi) \in \mathrm{Grp}(\mathbf{H})_{/\mathrm{Aut}(T_e V)}$ (Def. 10.2.23), $(A,\rho) \in G\mathrm{Actions}(\mathbf{H})$ and $X \in (G,\phi)\mathrm{StrctrdVFlds}(\mathbf{H})$ (10.95). Then, for $A \in \mathbf{H}_{/\mathbf{B}\int G}$, the *tangentially twisted cohomology* of V with coefficients in A is (see Remark 8.2.21)

$$\begin{aligned} \overset{\substack{\text{tangentially twisted}\\\text{cohomology}}}{H^{\int\tau}\big(X,A\big)} &:= \mathbf{H}_{/\int\mathrm{Aut}(T_e V)}\big((X,\eta^{\int}\circ\tau),(A /\!\!/ G,\rho)\big) \\ &= \left\{ \begin{array}{ccc} X & \overset{\text{cocycle}}{\underset{c}{\dashrightarrow}} & (\textstyle\int A) /\!\!/ (\textstyle\int G) \\ & \Leftarrow & \\ {\scriptstyle \eta^{\int}\circ\tau} \searrow & & \swarrow {\scriptstyle \int\rho} \\ & \mathbf{B}\textstyle\int G & \end{array} \right\}. \end{aligned} \tag{11.12}$$

Remark 11.2.4 (Need for G-Structure vs. tangential structure)**.**

(i) The notion of tangentially twisted cohomology in Definition 11.2.3 makes sense more generally for V-folds equipped only with tangential structure (Def. 10.2.36) instead of full G-structure (Def. 10.2.24) (hence only with a reduction of the shape of their structure group, instead of the actual structure group (Def. 10.2.12)), and it only needs A to be equipped with a $\int G$-action.

(ii) We state the definition in the more restrictive form above just in order to bring out the following promotion of this notion to its proper orbifold version (Remark 10.2.48), in Def. 11.2.5. The process of orbi-singularization is in fact sensitive to the full G-structure, and not just to its tangential shape. More precisely, it is sensitive to the *geometric fixed point spaces* of the G-structure and not just to its homotopy fixed point spaces (as per Remark 9.2.24 Ex. 9.2.27).

11.2.2 Tangentially twisted proper orbifold cohomology

We now promote tangentially twisted cohomology of V-folds (Def. 11.2.3) to a *proper orbifold cohomology* theory in the sense of Def. 10.2.46.

Definition 11.2.5 (Tangentially twisted proper orbifold cohomology)**.** Let

- $\mathbf{H}$ be a singular-elastic ∞-topos (Def. 9.2.6).
- $V \in \mathrm{Grp}(\mathbf{H}_{\cup})$ (Prop. 8.2.1).
- $(G, \phi) \in \mathrm{Grp}(\mathbf{H}_{\cup})_{/\mathrm{Aut}(T_e V)}$ (Prop. 8.2.1, Def. 8.2.13, Ex. 9.1.32).
- $\mathcal{X}_{\cup} \in V\mathrm{Flds}(\mathbf{H}_{\cup})$ (Def. 10.2.2).
- $(\tau, g) \in (G, \phi)\mathrm{Structures}_{\mathcal{X}_{\cup}}(\mathbf{H}_{\cup})$ (Def. 10.2.24).
- $(A, \rho) \in G\mathrm{Actions}(\mathbf{H}_{\cup})$.

and set $\mathcal{A} := \Upsilon(A /\!\!/ G)$ and $\mathcal{X} := \Upsilon\mathcal{X}_{\cup}$.
The *tangentially twisted proper orbifold cohomology* of $\mathcal{X}$ with coefficients in $\int\mathcal{A}$ is (see Rem. 8.2.21)

$$
\begin{aligned}
H^{\int\Upsilon\tau}(\mathcal{X}, \mathcal{A}) \;&:=\; \pi_0\, \mathbf{H}_{/\int\Upsilon\mathbf{B}G}\Big((\mathcal{X}, \eta^{\int} \circ \Upsilon(\tau)),\ (\int\mathcal{A}, \int\Upsilon\rho) \Big) \\
&= \left\{
\begin{array}{ccc}
\mathcal{X} & \overset{\text{cocycle}\; c}{\dashrightarrow} & \int\Upsilon(A /\!\!/ G) \\
& {\scriptstyle \eta^{\int}\circ\Upsilon(\tau)} \searrow \quad \swarrow {\scriptstyle \int\Upsilon(\rho)} & \\
& \int\Upsilon\mathbf{B}G &
\end{array}
\right\}_{/\sim} .
\end{aligned}
$$

Theorem 11.2.6 (Tangentially twisted orbifold cohomology at and away from singularities)**.** *Consider the tangentially twisted orbifold cohomology of Def. 11.2.5 restricted to* **(1)** *smooth and* **(2)** *flat orbifolds, according to Ex. 10.2.49. Then* (cf. Fig. 1.7 on p. 21)*:*

(i) *The tangentially twisted orbifold cohomology of flat orbifolds for 0-truncated coefficients A is naturally equivalent to the proper equivariant cohomology (Def. 11.1.2) of the total space of their $\flat G$-frame bundle* (10.82)*:*

$$
\underset{\substack{\text{flat}\\ \text{orbifold}}}{\overset{\substack{\text{tangentially twisted}\\ \text{orbifold cohomology}}}{H^{\int\Upsilon\tau}(i_{\flat}\mathcal{X}, \mathcal{A})}} \;\simeq\; \overset{\substack{\text{proper}\\ \text{equivariant cohomology}}}{H_{\flat G}\big(\underbrace{(\flat G)\mathrm{Frm}(\mathcal{X}_{\cup})}_{\flat G\text{-frame bundle}},\, A\big)} .
$$

(ii) *The tangentially twisted orbifold cohomology of smooth (non-orbi-singular) orbifolds is equivalently the tangentially twisted cohomology (Def. 11.2.1) of* *the underlying V-folds:*

$$
\underset{\substack{\text{smooth}\\ \text{orbifold}}}{\overset{\substack{\text{tangentially twisted}\\ \text{orbifold cohomology}}}{H^{\int \rotatebox[origin=c]{70}{$\prec$} \tau}\big(i_{\cup}\mathcal{X}, \mathcal{A}\big)}} \;\simeq\; \underset{\substack{\text{0-truncated}\\ V\text{-fold}}}{\overset{\substack{\text{tangentially twisted}\\ V\text{-fold cohomology}}}{H^{\int \tau}\big(\mathcal{X}_{\cup}\,,\, A\big)}}.
$$

Proof. The case **(i)** means that the classifying map of the G-structure in question factors as follows, where we use Prop. 10.2.25 to identify the leftmost morphism ρ as exhibiting the action (8.160) of $\flat G$ on $(\flat G)\mathrm{Frm}(X)$:

$$
(\flat G)\mathrm{Frm}(X) /\!\!/ \flat G \xrightarrow{\;\rho\;} \mathbf{B}\flat G \xrightarrow{\;\epsilon^{\flat}_{\mathbf{B}G}\;} \mathbf{B}G \longrightarrow \mathbf{B}\mathrm{Aut}(T_e V)\,.
$$
$$
\text{(composites: } \vdash\tau : (\flat G)\mathrm{Frm}(X) /\!\!/ \flat G \to \mathbf{B}G, \quad \vdash\mathrm{Frm}(X) : (\flat G)\mathrm{Frm}(X) /\!\!/ \flat G \to \mathbf{B}\mathrm{Aut}(T_e V)\text{)}
$$

Now we observe:

(a) with Def. 9.2.3 that $\int$ acts objectwise over Snglrt,

(b) with Prop. 8.1.55 that the pullback of presheaves over Snglrt is computed objectwise,

(c) and with Lemma 10.1.7 that $\rotatebox[origin=c]{70}{$\prec$}(A /\!\!/ G)$ is objectwise over Snglrt a homotopy quotient by G,

so that Lemma 9.1.6 applies objectwise over Snglrt to give the pullback square shown on the right here:

$$
\begin{array}{ccccc}
 & & \int \rotatebox[origin=c]{70}{$\prec$}(A /\!\!/ \flat G) & \longrightarrow & \int \rotatebox[origin=c]{70}{$\prec$}(A /\!\!/ G) \\
 & \nearrow & \downarrow & \text{(pb)} & \downarrow \\
\rotatebox[origin=c]{70}{$\prec$}\big((\flat G)\mathrm{Frm}(X) /\!\!/ (\flat G)\big) & \xrightarrow{\;\rotatebox[origin=c]{70}{$\prec$}\rho\;} & \rotatebox[origin=c]{70}{$\prec$}\mathbf{B}\flat G & \longrightarrow & \int \rotatebox[origin=c]{70}{$\prec$}\mathbf{B}G\,.
\end{array}
$$
$$
\text{(bottom composite: } \eta^{\int} \circ \rotatebox[origin=c]{70}{$\prec$}\tau\text{; the diagonal arrow is dashed)}
$$

By the universal property of the pullback, this means that every cocycle factors naturally as shown by the dashed morphism. But by Def. 11.1.2 this dashed morphism is equivalently a cocycle in proper equivariant cohomology, as claimed.

The case **(ii)** means (using Lemma 9.2.21) that the orbi-singular space $\mathcal{X}$ is in fact smooth

$$
\mathcal{X} \;\simeq\; \cup\mathcal{X} \;\simeq\; \mathrm{NnOrbSnglr}(\mathcal{X}_{\cup})\,.
$$

Therefore, we have the following natural equivalences of spaces of dashed morphisms:

$$
\begin{array}{ccccc}
\left(\begin{array}{ccc} & & \int \rotatebox[origin=c]{70}{$\prec$}(A /\!\!/ G) \\ & \nearrow & \downarrow \\ \cup\mathcal{X} \simeq \mathcal{X} & \longrightarrow & \int \rotatebox[origin=c]{70}{$\prec$}\mathbf{B}G \end{array}\right)
& \Leftrightarrow &
\left(\begin{array}{ccc} & & \cup\int \rotatebox[origin=c]{70}{$\prec$}(A /\!\!/ G) \\ & \nearrow & \downarrow \\ \cup\mathcal{X} & \longrightarrow & \cup\int \rotatebox[origin=c]{70}{$\prec$}\mathbf{B}G \end{array}\right)
& \Leftrightarrow &
\left(\begin{array}{ccc} & & \int(A /\!\!/ G) \\ & \nearrow & \downarrow \\ \mathcal{X}_{\cup} & \longrightarrow & \int\mathbf{B}G \end{array}\right) \\
\Cap & & & & \Cap \\
\mathbf{H}_{/\int \rotatebox[origin=c]{70}{$\prec$}\mathbf{B}G}\Big(\mathcal{X}, \int \rotatebox[origin=c]{70}{$\prec$}(A /\!\!/ G)\Big) & & \simeq & & \mathbf{H}_{/\int\mathbf{B}G}\Big(\mathcal{X}_{\cup}\,,\, \int(A /\!\!/ G)\Big)\,.
\end{array}
\tag{11.13}
$$

Here the first equivalence is by the adjunction NnOrbSnglr $\dashv$ Smth and the fully faithfulness of NnOrbSnglr (9.142). The second step uses $\cup \circ \int \simeq \int \circ \cup$ (Lemma 9.2.23) and $\cup \circ \curlyvee \simeq \cup$ (Remark 9.2.8) But on the right of (11.13) we see the tangentially twisted cohomology of $\mathcal{X}_{\cup}$, as claimed. □

11.2.3 J-Twisted orbifold Cohomotopy

We discuss now the example of tangentially twisted proper orbifold cohomology (Def. 11.2.5) where the coefficients are (shapes of) spheres, specifically of *Tate V-spheres* (Def. 11.2.9), In this case, the tangential twist is the *J-homomorphism* (Def. 11.2.14) whence we speak of *J-twisted Cohomotopy theory* (Def. 11.2.18).

Definition 11.2.7 (Complement of neutral element). Let $\mathbf{H}$ be an ∞-topos (Def. 8.1.46) and $V \in \mathrm{Grp}(\mathbf{H})$ (Prop. 8.2.1). Let $(V, \rho_{\mathrm{Aut}_{\mathrm{Grp}}}) \in \mathrm{Aut}_{\mathrm{Grp}}(V)\mathrm{Actions}(\mathbf{H})$ denote the group-automorphism action on V (Prop. 8.2.29).

(i) Consider those subobjects (Def. 8.1.77) of the homotopy quotient $V /\!\!/ \mathrm{Aut}_{\mathrm{Grp}}$ (8.205) whose pullback along the morphism

$$* /\!\!/ \mathrm{Aut}_{\mathrm{Grp}}(V) \xrightarrow{\; e /\!\!/ \mathrm{Aut}_{\mathrm{Grp}}(V) \;} V /\!\!/ \mathrm{Aut}_{\mathrm{Grp}}(V)\,,$$

which exhibits the neutral element as a fixed point of the group-automorphism action (Prop. 8.2.29), is empty. These are the subobjects forming the poset in the top left of the following Cartesian square (of ∞-categories):

$$\begin{array}{ccc} \mathrm{SubObjects}_{\not e}\big(V /\!\!/ \mathrm{Aut}_{\mathrm{Grp}}(V)\big) & \longrightarrow & * \\ \big\downarrow & \text{(pb)} & \big\downarrow{\scriptstyle \varnothing} \\ \mathrm{SubObjects}\big(V /\!\!/ \mathrm{Aut}_{\mathrm{Grp}}(V)\big) & \xrightarrow[\;(e /\!\!/ \mathrm{Aut}_{\mathrm{Grp}}(V))^*\;]{} & \mathrm{SubObjects}(*) \end{array} \tag{11.14}$$

(ii) Consider next the union of these subobjects, hence the colimit over the left vertical functor in (11.14), which we denote as follows:

$$\begin{aligned} &\big(V \setminus \{e\}\big) /\!\!/ \mathrm{Aut}_{\mathrm{Grp}}(V) := \\ &\underrightarrow{\lim}\Big(\mathrm{SubObjects}_{\not e}\big(V /\!\!/ \mathrm{Aut}_{\mathrm{Grp}}(V)\big) \hookrightarrow \mathrm{SubObjects}\big(V /\!\!/ \mathrm{Aut}_{\mathrm{Grp}}(V)\big)\Big)\,. \end{aligned} \tag{11.15}$$

(iii) We call the homotopy fiber $V \setminus \{e\}$ of the canonical morphism from this object (11.15) to $\mathbf{B}\mathrm{Aut}_{\mathrm{Grp}}(V)$ the *complement of the neutral element of V*

$$\begin{array}{ccc} V \setminus \{e\} & \xrightarrow{\;\mathrm{fib}(\rho_{\mathrm{Aut}_{\mathrm{Grp}}} \setminus \{e\})\;} & \big(V \setminus \{e\}\big) /\!\!/ \mathrm{Aut}_{\mathrm{Grp}}(V) \\ & & \big\downarrow \\ & & V /\!\!/ \mathrm{Aut}_{\mathrm{Grp}}(V) \\ & & {\scriptstyle \rho_{\mathrm{Aut}_{\mathrm{Grp}}}}\big\downarrow \\ & & \mathbf{B}\mathrm{Aut}_{\mathrm{Grp}}(V) \end{array} \qquad \big(\rho_{\mathrm{Aut}_{\mathrm{Grp}}} \setminus \{e\}\big) \tag{11.16}$$

(iv) We regard the complement of the neutral element as equipped with the $\mathrm{Aut}_{\mathrm{Grp}}(V)$-action which is exhibited by the homotopy fiber sequence (11.16) (by Prop. 11.16):

$$\left(V \setminus \{e\}\,,\, \rho_{\mathrm{Aut}_{\mathrm{Grp}}} \setminus \{e\}\right)\right) \in \mathrm{Aut}_{\mathrm{Grp}}(V)\mathrm{Actions}(\mathbf{H})\,.$$

Proposition 11.2.8 (Basic properties of complement of neutral element)**.** *Let* **H** *be an* ∞*-topos (Def. 8.1.46) and* $V \in \mathrm{Grp}(\mathbf{H})$ *(Prop. 8.2.1). Then the complement* $V \setminus \{e\}$ *of the neutral element (Def. 11.2.7)*

(i) *is a subobject (Def. 8.1.77) of V*

$$V \setminus \{e\} \hookrightarrow V \tag{11.17}$$

(ii) *which is disjoint from the neutral element:*

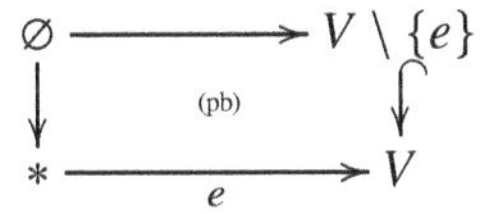

Proof. For **(i)** we use the pasting law (Prop. 8.1.39) and the homotopy fiber characterization of the group-automorphism action (8.207) to decompose (11.16) as the pasting of two Cartesian squares, as follows:

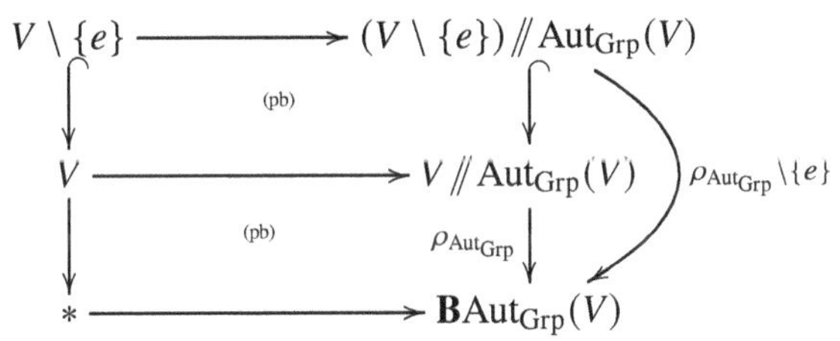

Since monomorphisms are preserved by pullback (by Prop. 8.1.83), this shows the first claim from the construction (11.15).

For **(ii)** we paste to the middle horizontal morphism in this diagram the square (8.191) which exhibits the neutral element as a fixed point of the group-automorphism action (Prop. 8.2.29), and then we pull back the right vertical morphism along the boundary of that square, as shown in the following:

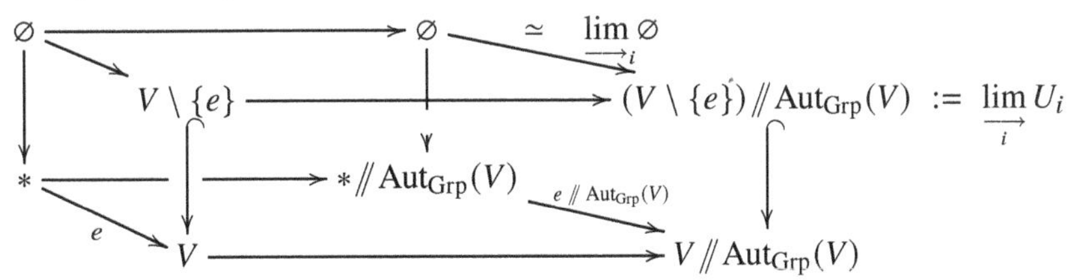

Here the right square is Cartesian since colimits in an ∞-topos are preserved by pullback (8.80) and using the definition (11.14), as indicated in the top right. Similarly, the rear square is Cartesian, since pullback preserves the initial object (this being the empty colimit, Ex. 8.1.49). With this, and since the front square is Cartesian by **(i)**, the pasting law (Prop. 8.1.39) implies that also the left square is Cartesian, which was to be shown. □

Definition 11.2.9 (Tate V-sphere)**.** Let $\mathbf{H}$ be an ∞-topos (Def. 8.1.46) and $V \in \mathrm{Grp}(\mathbf{H})$ (Prop. 8.2.1). Then we say that the *Tate V-sphere* is the homotopy cofiber

$$\mathbf{S}^V \;:=\; V/(V \setminus \{e\})$$

of the inclusion (11.17) of the complement of the neutral element into V (Def. 11.2.7), hence the object in this homotopy pushout square:

$$\begin{array}{ccc} V \setminus \{e\} & \hookrightarrow & V \\ \downarrow & \text{(po)} & \downarrow \\ * & \longrightarrow & \mathbf{S}^V . \end{array} \tag{11.18}$$

Example 11.2.10 (Tate sphere in unstable motivic homotopy theory)**.** For $\mathbf{H} := \mathrm{Shv}_\infty(\mathrm{Schemes}_{\mathrm{Nis}})$ and $V := \mathbb{A}^1$ the Tate V-sphere of Def. 11.2.9 is the Tate sphere in the traditional sense of (unstable) motivic homotopy theory, see [VRO07, 2.22].

Example 11.2.11 (Tate spheres with shape of ordinary spheres)**.** Let $\mathbf{H} = \mathrm{JetsOfSmoothGroupoids}_\infty$ (Def. 9.1.26) and $V := (\mathbb{R}^n, +)$ as in Ex. 10.2.5. Then $\mathrm{Aut}_{\mathrm{Grp}}(\mathbb{R}^n, +) = \mathrm{GL}(n)$ (as in Ex. 10.2.13) and the complement of the neutral element (Def. 11.2.7) is the ordinary complement $\mathbb{R}^n \setminus \{0\}$, whose shape is that of the ordinary $n-1$-sphere:

$$\textstyle\int (\mathbb{R}^n \setminus \{0\}) \;\simeq\; \int S^{n-1}. \tag{11.19}$$

Hence the Tate $\mathbb{R}^n$-sphere (Def. 11.2.9) is the homotopy pushout shown on the left here:

$$\begin{array}{ccc} \mathbb{R}^n \setminus \{e\} & \hookrightarrow & \mathbb{R}^n \\ \downarrow & \text{(pb)} & \downarrow \\ * & \longrightarrow & S^{(\mathbb{R}^n)} \end{array} \quad \overset{\int}{\longmapsto} \quad \begin{array}{ccc} \int S^{n-1} & \hookrightarrow & * \\ \downarrow & \text{(pb)} & \downarrow \\ * & \longrightarrow & \int \mathbf{S}^{(\mathbb{R}^n)} . \end{array}$$

Since the shape modality (9.2) is left adjoint, it preserves homotopy pushouts (Prop. 8.1.42), so that the shape of the Tate $\mathbb{R}^n$-sphere is that of the ordinary n-sphere:

$$\textstyle\int \mathbf{S}^{\mathbb{R}^n} \;\simeq\; \int S^n. \tag{11.20}$$

In contrast, the Tate $\mathbb{R}^n$-sphere itself is the "germ of a smooth sphere".

Proposition 11.2.12 (Canonical action on Tate V-sphere)**.** *Let* $\mathbf{H}$ *be an ∞-topos (Def. 8.1.46) and $V \in$* $\mathrm{Grp}(\mathbf{H})$ *(Prop. 8.2.1). The Tate V-sphere (Def. 11.2.9) inherits a canonical action (Prop. 8.2.6) of the group-automorphism group* $\mathrm{Aut}_{\mathrm{Grp}}(V)$ *(Def. 8.2.28), associated (via Prop. 8.2.14) to a group homomorphism*

$$\mathrm{Aut}_{\mathrm{Grp}}(V) \longrightarrow \mathrm{Aut}(\mathbf{S}^V) \tag{11.21}$$

whose homotopy quotient (8.161) *is given by the following homotopy pushout*

$$\begin{array}{ccc} (V \setminus \{e\}) /\!\!/ \mathrm{Aut}_{\mathrm{Grp}}(V) & \hookrightarrow & V /\!\!/ \mathrm{Aut}_{\mathrm{Grp}}(V) \\ \downarrow & \text{(po)} & \downarrow \\ * /\!\!/ \mathrm{Aut}_{\mathrm{Grp}}(V) & \longrightarrow & \mathbf{S}^V /\!\!/ \mathrm{Aut}_{\mathrm{Grp}}(V) \end{array} \tag{11.22}$$

of the defining morphisms in (11.16).

Proof. Since the forgetful ∞-functor $\mathbf{H}_{/\mathbf{B}\mathrm{Aut}_{\mathrm{Grp}}(V)} \longrightarrow \mathbf{H}$ preserves colimits (Ex.

8.1.68), the diagram (11.18) extends to a diagram over $\mathbf{B}\mathrm{Aut}_{\mathrm{Grp}}(V)$. Pulling this back along the point inclusion (8.153) and using that colimits in an ∞-topos are preserved by pullback (8.80), we find that the homotopy fiber of $\mathbf{S}^V /\!\!/ \mathrm{Aut}_{\mathrm{Grp}}(V) \to \mathbf{B}\mathrm{Aut}_{\mathrm{Grp}}(V)$ is given by the defining homotopy pushout (11.18) of the Tate V-sphere. □

Definition 11.2.13 (Linear group)**.** Let $\mathbf{H}$ be an elastic ∞-topos (Def. 9.1.22) and $V \in \mathrm{Grp}(\mathbf{H})$ (Prop. 8.2.1).

(i) We say that V is a *linear group* if it is equipped with an equivalence

$$\mathrm{Aut}(T_e V) \xrightarrow[\simeq]{\exp} \mathrm{Aut}_{\mathrm{Grp}}(V) \quad \in \mathrm{Grp}(\mathbf{H}) \tag{11.23}$$

between **(a)** the plain automorphism group of the local neighborhood of the neutral element (Def. 10.2.12) and **(b)** the group-automorphism group of V (Def. 8.2.28).

(ii) We write

$$\mathrm{LinGrp}(\mathbf{H}) \in \mathrm{Cat}_\infty$$

for the ∞-category of linear groups in $\mathbf{H}$.

Definition 11.2.14 (Tate J-homomorphism)**.** Let $\mathbf{H}$ be an elastic ∞-topos (Def. 9.1.22) and $V \in \mathrm{LinGrp}(\mathbf{H})$ (Prop. 11.2.13).

(i) The *Tate J-homomorphism* is the composite

$$\mathbf{J}_V \;:\; \mathrm{Aut}(T_e V) \xrightarrow[\simeq]{\exp} \mathrm{Aut}_{\mathrm{Grp}}(V) \longrightarrow \mathrm{Aut}(\mathbf{S}^V) \tag{11.24}$$

of **(a)** the defining equivalence (11.23) with **(b)** the homomorphism (11.21) which reflects the canonical $\mathrm{Aut}_{\mathrm{Grp}}(V)$-action on the Tate V-sphere (Def. 11.2.12).

(ii) The corresponding $\mathrm{Aut}(T_e V)$-actions on $\mathbf{S}^V$ and on $\int(\mathbf{S}^V)$, by restriction along (11.24) and (11.26) of the canonical automorphism actions (Prop. 8.2.14), we denote, respectively, by

$$(\mathbf{S}^V, \rho_{\mathbf{J}}) \;\in\; \mathrm{Aut}(T_e V)\mathrm{Actions}(\mathbf{H})\,. \tag{11.25}$$

(iii) The actual *J-homomorphism* is the shape of the further composite with the homomorphism $\mathrm{Aut}(\eta^{\int}_{\mathbf{S}^V})$ from Prop. 9.1.7:

$$J_V \;:\; \textstyle\int \mathrm{Aut}(T_e V) \xrightarrow[\simeq]{\int \exp} \int \mathrm{Aut}_{\mathrm{Grp}}(V) \longrightarrow \int \mathrm{Aut}(\mathbf{S}^V) \xrightarrow{\int \mathrm{Aut}\left(\eta^{\int}_{\mathrm{Aut}(\mathbf{S}^V)}\right)} \int \mathrm{Aut}\left(\int \mathbf{S}^V\right). \tag{11.26}$$

(with the composite of the first two maps labelled $\int \mathbf{J}_V$)

Example 11.2.15 (Ordinary J-homomorphism)**.** Let $\mathbf{H} = \mathrm{SnglrJetSmthGrpd}_\infty$ (Ex. 9.2.11) and $V := (\mathbb{R}^n, +)$ as in Ex. 10.2.5. This is a linear group in the sense of Def. 11.2.13, with $\mathrm{Aut}(T_0\mathbb{R}^n) \simeq \mathrm{GL}(n)$ (Ex. 10.2.13). Via Ex. 11.2.11 the induced action on the shape of the Tate $\mathbb{R}^n$-sphere (Def. 11.2.14) is the classical J-homomorphism (going back to [Wh42], reviewed in [Rav86, p. 4]):

$$J \;:\; \textstyle\int \mathrm{O}(n) \simeq \int \mathrm{GL}(n) \longrightarrow \mathrm{Aut}\left(\int S^n\right) \tag{11.27}$$

being the image under topological shape (Def. 8.1.29) of the defining action of $\mathrm{GL}(n)$ on $\mathbb{R}^n$ and hence on its one-point compactification S^n.

Definition 11.2.16 (Representation spheres)**.** Let $\mathbf{H}$ be a singular-elastic ∞-topos (Def. 9.2.6), $V\mathrm{Grp}(\mathbf{H}_{\cup})$ (Prop. 8.2.1), and $(G,\phi) \in \mathrm{Grp}(\mathbf{H}_{\cup})_{/\mathrm{Aut}(T_e V)}$ (Prop. 8.2.1, Def. 8.2.13, Ex. 9.1.32). Then we say that the *representation sphere* S^{V_ϕ} of the G-action ϕ on V (via Prop. 8.2.14) is the shape (Def. 9.1.1) of the orbi-singularization (Def. 9.2.7) of the homotopy quotient (8.161) of the Tate V-sphere (Def. 11.2.9) by the restricted action (Prop. 8.2.12) along ϕ of the action $\rho_{\mathbf{J}}$ (11.25) induced by the J-homomorphism (Def. 11.2.14):

$$S^{V_\phi} := \int \Upsilon \big(\mathbf{S}^V /\!\!/_{\phi} G\big) \qquad \in \mathbf{H}_{/\Upsilon G} \, .$$

Example 11.2.17 (Ordinary representation spheres)**.**
Let $\mathbf{H} = \mathrm{SnglrJetSmthGrpd}_\infty$ (Ex. 9.2.10) and $V := (\mathbb{R}^n, +)$ as in Ex. 10.2.5, whence $\mathrm{Aut}(T_e V) \simeq \mathrm{GL}(n)$ (Ex. 10.2.13). For

$$G \xhookrightarrow{\;\phi\;} \mathrm{GL}(n) \subset \mathrm{Aut}(T_e V)$$

a finite subgroup, hence a linear G-representation, we have that the representation sphere $S^{\mathbb{R}^n_\phi}$ according to Def. 11.2.16 is the ordinary representation sphere, as an object in G-equivariant homotopy theory.

Definition 11.2.18 (J-twisted proper orbifold Cohomotopy theory)**.** Let $\mathbf{H}$ be a singular-elastic ∞-topos (Def. 9.2.6) $V \in \mathrm{Grp}(\mathbf{H})$ (Prop. 8.2.1), $W \in \mathrm{LinGrp}(\mathbf{H})$ (Def. 11.2.13) and $\phi : \mathrm{Aut}(T_e W) \longrightarrow \mathrm{Aut}(T_E V)$. Then *J-twisted proper orbifold Cohomotopy* is the tangentially twisted proper orbifold cohomotopy (Def. 11.2.5) with coefficients

$$(A, \rho) := (\mathbf{S}^V, \rho_{\mathbf{J}})$$

the Tate W-sphere (Def. 11.2.9) with its Tate J-homomorphism action (Def. 11.2.14):

$$\underset{\text{J-twisted orbifold Cohomotopy}}{\pi^{\int \Upsilon \tau}(-)} \;:=\; \underset{\text{tangentially twisted orbifold cohomology}}{H^{\int \Upsilon \tau}}\big(-, \underset{\text{Tate } V\text{-sphere with } J\text{-homomorphism action}}{(\mathrm{s}^V, \rho_{\mathbf{J}})}\big).$$

Hence for a structured orbifold (Def. 10.2.46)

$$\big(\mathcal{X}, (\tau, g)\big) \in \big(\mathrm{Aut}(T_e W), \phi\big)\mathrm{Strctrd}V\mathrm{Orbfld}(\mathbf{H}) \, ,$$

we have:

$$\underset{\text{J-twisted orbifold Cohomotopy}}{\pi^{\int \Upsilon \tau}(\mathcal{X})} = \left\{ \begin{array}{ccc} \underset{\text{orbifold}}{\mathcal{X}} & \overset{\text{cocycle } c}{\dashrightarrow} & \underset{\text{orbi-singularized Tate } W\text{-sphere}}{\int \Upsilon \big(\mathbf{S}^W /\!\!/ \mathrm{Aut}(T_e W)\big)} \\ & \Leftarrow & \\ {\scriptstyle \eta^{\int} \circ \Upsilon(\tau)} \atop {\text{tangential twist}} \searrow & & \swarrow {\scriptstyle \int \Upsilon(\rho_{\mathbf{J}})} \atop {\text{twisting via orbi-singularized J-homomorphism}} \\ & \int \Upsilon \mathbf{B}\mathrm{Aut}(T_e W) & \end{array} \right\}_{/\sim} .$$

Example 11.2.19 (J-Twisted orbifold Cohomotopy of ordinary orbifolds)**.**
Let $\mathbf{H} = \mathrm{SnglrJetSmthGrpd}_\infty$ (Ex. 9.2.10) and $V := (\mathbb{R}^n, +)$, $W := (\mathbb{R}^p, +)$ as in Ex. 10.2.5, with $p \leq n$, and $\phi : (\mathbb{R}^p, +) \hookrightarrow (\mathbb{R}^n, +)$ be the canonical inclusion. Then the corresponding J-twisted proper orbifold Cohomotopy theory $\pi^{\int \Upsilon \tau}$ (Def. 11.2.18)

is defined on ordinary n-dimensional orbifolds (by Ex. 10.2.6) with $\mathrm{GL}(p)$-structure (by Ex. 10.2.13) and it unifies the following two special cases (by Theorem 11.2.6, see the second diagram on p. 21)):

(i) On smooth orbifolds, i.e., on ordinary manifolds (Ex. 10.2.5), it reduces to non-abelian cohomology with coefficients the shape of the ordinary p-sphere (by Ex. 11.2.11) and tangentially twisted via the traditional J-homomorphism (by Ex. 11.2.15). This is the *J-twisted Cohomotopy theory* considered in [FSS20][FSS21a] [BSS19].

(ii) On flat orbifolds, such as the vicinity of ordinary orbifold singularities $\mathbb{R}^p /\!\!/ G$ for finite subgroups $G \overset{\phi}{\hookrightarrow} \mathrm{GL}(p)$ (by Ex. 10.2.45), hence for linear G-representations ϕ, it reduces to proper equivariant cohomology in RO-degree ϕ and with coefficients the representation sphere $S^{\mathbb{R}^n_\phi}$ (by Ex. 11.2.17). This is the *tangentially RO-graded equivariant Cohomotopy theory* considered in [SS20][BSS19].

By way of outlook, we highlight the following:

Remark 11.2.20 (Orbifold cohomology in M-theory and *Hypothesis H*).
(i) Traditional discussion of orbifold cohomology has been strongly motivated by its application to *perturbative string theory* (e.g. [AMR02, ARZ06, ALR07, BU09, DFM11]). However, perturbative string theory is famously in need of a non-perturbative completion ("M-theory", see [HSS18, 2][FSS19] for review and pointers) whose mathematical formulation has remained an open problem. Therefore, it is to be expected that the historically rich interaction between orbifold cohomology theory and string theory is just the tip of an iceberg, whose full scope is a cohomology theory of M-theoretic orbifolds.
(ii) Elsewhere, we have put forward a precise hypothesis as to the global completion of 11D supergravity toward M-theory, via *flux quantization* [SS25d] of the theory's C-field. This *Hypothesis H* says that:

(a) far from singularities, M theory flux is quantized in twisted Cohomotopy [FSS20, FSS21a, BSS19, FSS22];

(b) at singularities, M-theory is quantized in RO-graded equivariant Cohomotopy [HSS18, SS20, BSS19].

(See these references for various consistency checks of this hypothesis.)
(iii) The impact of Theorem 11.2.6, in its specialization to Ex. 11.2.19, is the implication that these two cases are indeed two aspects of a single unified cohomology theory: J-twisted orbifold Cohomotopy theory.

Bibliography

[Ab01] M. Abreu, *Kähler Metrics on Toric Orbifolds*, J. Differential Geom. **58** (2001) 151–187 [euclid:jdg/1090348285].

[Ac01] D. J. Acosta, *A Furuta-like inequality for spin orbifolds and the minimal genus problem*, Topology Appl. **114** (2001) 91–106 [doi:10.1016/S0166-8641(00)00030-4].

[ADG11] A. Adem, A. N. Duman, and J. M. Gómez, *Cohomology of toroidal orbifold quotients*, J. Algebra **344** (2011) 114–136 [arXiv:1003.0435] [doi:10.1016/j.jalgebra.2011.08.004].

[ALR07] A. Adem, J. Leida, and Y. Ruan, *Orbifolds and stringy topology*, Cambridge University Press (2007) [doi:10.1017/CBO9780511543081].

[AR03] A. Adem and Y. Ruan, *Twisted Orbifold K-Theory*, Commun. Math. Phys. **237** (2003) 533–556, [arXiv:math/0107168].

[ARZ06] A. Adem, Y. Ruan and B. Zhang, *A Stringy Product on Twisted Orbifold K-theory*, Morfismos (10th Anniversary Issue) **11** (2007) 33–64 [arXiv:math/0605534] [doi:10.1090/S0002-9947-09-04760-6].

[AMR02] A. Adem, J. Morava, and Y. Ruan, *Orbifolds in Mathematics and Physics*, Contemp. Math. **310**, Amer. Math. Soc., (2002) [ams:conm-310].

[AGP02] M. Aguilar, S. Gitler, and C. Prieto, *Algebraic Topology from a Homotopical Viewpoint*, Springer (2002) [doi:10.1007/b97586].

[AG98] M. A. Akivis and V. V. Goldberg, *Conformal and Grassmann structures*, Differential Geom. Appl. **8** (1998) 177–203 [arXiv:math/9805107] [doi:10.1016/S0926-2245(98)00007-2].

[Ak12] K. Akutagawal, *Computations of the orbifold Yamabe invariant*, Math. Zeitschrift **271** (2012) 611–625 [doi:10.1007/s00209-011-0880-0].

[AM93a] D. V. Alekseevsky and S. Marchiafava, *Quaternionic-like structures on a manifold I: 1-integrability and integrability conditions*, Atti Accad. Naz. Lincei Cl. Sci. Fis. Mat. Natur. Rend. Lincei (9) Mat. Appl. **4** (1993) 43–52 [ncatlab.org/nlab/files/AlekseevskyMarchiafava-1993a.pdf].

[AM93b] D. V. Alekseevsky and S. Marchiafava, *Quaternionic-like structures on a manifold II: Automorphism groups and their interrelations*, Atti Accad. Naz. Lincei Cl. Sci. Fis. Mat. Natur. Rend. Lincei (9) Mat. Appl. **4** (1993) 53–61 [ncatlab.org/nlab/files/AlekseevskyMarchiafava-1993a.pdf].

[Al85] O. Alvarez, *Topological quantization and cohomology*, Comm. Math. Phys. **100** 2 (1985) 279–309 [10.1007/BF01212452].

[Am12] A. Amenta, *The Geometry of Orbifolds via Lie Groupoids*, PhD thesis, ANU (2012) [arXiv:1309.6367].

[ABG10] M. Ando, A. Blumberg and D. Gepner, *Twists of K-theory and TMF*, in: *Superstrings, Geometry, Topology, and C^*-algebras*, Proc. Symp. Pure Math. **81**, AMS (2010) [ISBN:978-0-8218-4887-6] [arXiv:1002.3004].

[ABGHR14] M. Ando, A. Blumberg, D. Gepner, M. Hopkins, and C. Rezk, *An ∞-categorical approach to R-line bundles, R-module Thom spectra, and twisted R-homology*, J. Topology **7** (2014) 869–893 [arXiv:1403.4325] [doi:10.1112/jtopol/jtt035].

[AF15] Y. Ando and L. Fu, *Topological Crystalline Insulators and Topological Superconductors: From Concepts to Materials*, Annual Review of Condensed Matter Physics **6** (2015) 361–381 [arXiv:1501.00531] [doi:10.1146/annurev-conmatphys-031214-014501].

[An12] D. Angella, *Cohomologies of certain orbifolds*, J. Geom. Phys. **71** (2013) 117–126 [arXiv:1211.2561] [doi:10.1016/j.geomphys.2013.04.008].

[Ap98] B. Apanasov, *Doubles of Atoroidal Manifolds, Their Conformal Uniformization and Deformations*, In: *Deformations of Mathematical Structures*, Springer (1989) [doi:10.1007/978-94-009-2643-1_9].

[Ap00] B. Apanasov, *Conformal Geometry of Discrete Groups and Manifolds*, De Gruyter (2000), [doi:10.1515/9783110808056].

[Arm88] M. A. Armstrong, *Groups and Symmetry*, Springer (1988) [doi:10.1007/978-1-4757-4034-9].

[Arv02] W. Arveson, *A Short Course on Spectral Theory*, Springer (2002) [doi:10.1007/b97227].

[AOP16] J. Asbóth, L. Oroszlány, and A. Pályi, *A Short Course on Topological Insulators*, Springer (2016) [doi:10.1007/978-3-319-25607-8] [arXiv:1509.02295].

[As96] H. Aslaksen, *Quaternionic determinants*, The Mathematical Intelligencer **18** (1996) 57–65 [doi:10.1007/BF03024312].

[At66] M. Atiyah, *K-theory and reality*, Quarterly J. Math. **17** (1966) 367–386 [doi:10.1093/qmath/17.1.367].

[AS69] M. Atiyah and G. Segal, *Equivariant K-theory and completion*, J. Differential Geom. **3** (1969) 1–18 [euclide:jdg/1214428815].

[ASe04] M. Atiyah and G. Segal, *Twisted K-theory*, Ukr. Math. Bull. **1** (2004) 291–334 [arXiv:math/0407054].

[SSi69] M. F. Atiyah and I. M. Singer, *Index theory for skew-adjoint Fredholm operators*, Publications Mathématiques de l'IHÉS **37** 1 (1969) 5–26 [doi:10.1007/bf02684885].

[BCSS07] J. Baez, A. Crans, U. Schreiber and D. Stevenson, *From loop groups to 2-groups*, Homotopy, Homology and Applications **9** (2007) 101–135 [arXiv:math/0504123] [euclid.hha/1201127333].

[BH08] J. Baez and A. Hoffnung, *Convenient Categories of Smooth Spaces*, Trans. Amer. Math. Soc. **363** (2011) 5789–5825 [arXiv:0807.1704] [doi:10.1090/S0002-9947-2011-05107-X].

[BH09] J. Baez and J. Huerta, *Division algebras and supersymmetry I*, in: *Superstrings, Geometry, Topology, and C^*-algebras*, Proc. Symp. Pure Math. **81**, AMS (2010) 65–80 [arXiv:0909.0551] [ams:psum/81].

[BZ03] A. V. Bagaev and N. I. Zhukova, *The Automorphism Groups of Finite Type G-Structures on Orbifolds*, Sib. Math. J. **44** (2003) 213–224 [doi:10.1023/A:1022920417785].

[BZ07] A. V. Bagaev and N. I. Zhukova, *The isometry groups of Riemannian orbifolds*, Sib. Math. J. **48** (2007) 579–592 [doi:10.1007/s11202-007-0060-y].

[BZ19] A. V. Bagaev and N. I. Zhukova, *An analog of Chern's conjecture for the Euler-Satake characteristic of affine orbifolds*, J. Geom. Phys. **142** (2019) 80–91 [doi:10.1016/j.geomphys.2019.04.002].

[BNSS18] A. Bahri, D. Notbohm, S. Sarkar, and J. Song, *On integral cohomology of certain orbifolds*, Algebr. Geom. Topol. **17** (2017) 3779–3810 [arXiv:1711.01748] [doi:10.1093/imrn/rny283].

[BL99] D. Bailin and A. Love, *Orbifold compactifications of string theory*, Phys. Rep. **315** (1999) 285–408 [doi:10.1016/S0370-1573(98)00126-4].

[Ba54] W. L. Baily, *On the quotient of an analytic manifold by a group of analytic homeomorphisms*, Proc. Nat. Acad. Sci. U. S. A. **40** (1954) 804–808 [doi:10.1073/pnas.40.9.804].

[Ba56] W. L. Baily, *The decomposition theorem for V-manifolds*, Amer. J. Math. **78** (1956) 862–888 [jstor:2372472].

[Bau21] A. Bauer, *Synthetic Mathematics with an Excursion Into Computability Theory*, talk at University of Wisconsin Logic Seminar (Feb 2021) [ncatlab.org/nlab/files/Bauer-Synthetic.pdf].

[Ba07] Ya. V. Bazaikin, *On the new examples of complete noncompact* Spin(7)*-holonomy metrics*, Sib. Math. J. **48** (2007) 8–25 [doi:10.1007/s11202-007-0003-7].

[BBFMT16] G. Bazzoni, I. Biswas, M. Fernández, V. Muñoz, and A. Tralle, *Homotopic properties of Kähler orbifolds*, In: *Special Metrics and Group Actions in Geometry* Springer (2017) [doi:10.1007/978-3-319-67519-0_2] [arXiv:1605.03024].

[BP22] J. Beardsley and M. Péroux, *Koszul Duality in Higher Topoi*, Homol. Homot. & Appl. **25** 1 (2022) 53–70 [arXiv:1909.11724].

[BU09] E. Becerra and B. Uribe, *Stringy product on twisted orbifold K-theory for abelian quotients*, Trans. Amer. Math. Soc. **361** (2009) 5781–5803 [arXiv:0706.3229] [doi:10.1090/S0002-9947-09-04760-6].

[BX] K. Behrend and P. Xu, *Differentiable Stacks and Gerbes*, J. Symplectic Geom. **9** (2011) 285–341 [arXiv:math/0605694] [jsg/1310388899].

[BGR07] F. Belgun, N. Ginoux, and H.-B. Rademacher, *A Singularity Theorem for Twistor Spinors*, Ann. Instit. Fourier (Grenoble) **57** (2007) 1135–1159 [arXiv:math/0409136] [numdam:AIF_2007__57_4_1135_0].

[BZA24] T. Bendokat, R. Zimmermann, and P.-A. Absil, *A Grassmann manifold handbook: Basic geometry and computational aspects*, Advances in Computational Mathematics **50** 6 (2024) [doi:10.1007/s10444-023-10090-8].

[BSS18] M. Benini, A. Schenkel, and U. Schreiber, *The Stack of Yang–Mills Fields on Lorentzian Manifolds*, Communications in Mathematical Physics **359** 2 (2018) 765–820 [doi:10.1007/s00220-018-3120-1], [arXiv:1704.01378].

[Ber87] F. A. Berezin (edited by A. A. Kirillov), *Introduction to Superanalysis*, Springer (1987) [doi:10.1007/978-94-017-1963-6_2].

[Be05] J. Bergner, *Three models for the homotopy theory of homotopy theories*, Topology **46** (2007) 397–436 [doi:10.1016/j.top.2007.03.002], [arXiv:math/0504334].

[Be06] J. Bergner, *A survey of* $(\infty, 1)$*-categories*, In: *Towards Higher Categories*, Springer (2007) [doi:10.1007/978-1-4419-1524-5_2] [arXiv:math/0610239].

[Be14] J. Bergner, *Equivalence of models for equivariant* $(\infty, 1)$*-categories*, Glasgow Math. J. **59** (2016) 237–253 [arXiv:1408.0038] [doi:10.1017/S0017089516000136].

[BR07] M. Berkooz and D. Reichmann, *A Short Review of Time Dependent Solutions and Space-like Singularities in String Theory*, Nucl. Phys. Proc. Suppl. **171** (2007) 69–87 [doi:10.1016/j.nuclphysbps.2007.06.008] [arXiv:0705.2146].

[BEBP19] D. Berwick-Evans, P. B. de Brito, and D. Pavlov, *Classifying spaces of infinity-sheaves*, Algebr. Geom. Topol. **24** (2024) 4891–4937 [arXiv:1912.10544] [10.2140/agt.2024.24.4891].

[BDP17] R. G. Bettiol, A. Derdzinski, and P. Piccione, *Teichmüller theory and collapse of flat manifolds*, Ann. Mat. Pura Appl. **197** (2018) 1247–1268, [doi:10.1007/s10231-017-0723-7] [arXiv:1705.08431].

[BD00] R. Bielawski and A. Dancer, *The geometry and topology of toric hyperkähler manifolds*, Comm. Anal. Geom. **8** (2000) 727–760 [doi:10.4310/CAG.2000.v8.n4.a2].

[Bl07] C. Blohmann, *Stacky Lie groups*, Int. Mat. Res. Not. (2008) rnn 082 [doi:10.1093/imrn/rnn082] [arXiv:math/0702399].

[Blu17] A. Blumberg *Equivariant homotopy theory*, lecture notes (2017) [web.ma.utexas.edu/users/a.debray/lecture_notes/m392c_EHT_notes.pdf]

[BLP05] M. Boileau, B. Leeb, and J. Porti, *Geometrization of 3-Dimensional Orbifolds*, Ann. Math. **162** (2005) 195–290 [jstor:3597373].

[Bo75] K. Borsuk, *Theory of Shape*, Monografie Matematyczne Tom **59**, Warszawa (1975) [mathscinet:0418088].

[Bo92] J. E. Borzellino, *Riemannian Geometry of Orbifolds*, PhD Dissertation, UCLA (1992) [calpoly.edu/math_fac/111].

[BT82] R. Bott and L. W. Tu, *Differential Forms in Algebraic Topology*, Springer (1982) [doi:10.1007/978-1-4757-3951-0].

[BG08] C. P. Boyer and K. Galicki, *Sasakian Geometry*, Oxford University Press, Oxford, 2008, [ISBN:978-0-19-856495-9].

[BGM98] C. P. Boyer, K. Galicki, and B. M. Mann, *Hypercomplex structures from 3-Sasakian structures*, J. Reine Angew. Math. **501** (1998) 115–141 [doi:10.1515/crll.1998.074].

[BL12] B. Botvinnik and M. Labbi, *Highly connected manifolds of positive p-curvature*, Trans. Amer. Math. Soc. **366** (2014) 3405–3424 [doi:10.1090/S0002-9947-2014-05939-4], [arXiv:1201.1849].

[BM19] V. Braunack-Mayer, *Combinatorial parametrised spectra*, Algebr. Geom. Topol. **21** (2021) 801–891 [doi:10.2140/agt.2021.21.801] [arXiv:1907.08496] .

[BSS18] V. Braunack-Mayer, H. Sati, and U. Schreiber, *Gauge enhancement for Super M-branes via Parameterized stable homotopy theory*, Comm. Math. Phys. **371** (2019) 197–265 [doi:10.1007/s00220-019-03441-4] [arXiv:1805.05987].

[Bre67a] G. Bredon, *Equivariant cohomology theories*, Bull. Amer. Math. Soc. **73** (1967) 266–268 [euclid:1183528794].

[Bre67b] G. Bredon, *Equivariant cohomology theories*, Lecture Notes in Mathematics **34**, Springer (1967) [doi:10.1007/BFb0082690].

[Bre72] G. Bredon, *Introduction to compact transformation groups*, Academic Press (1972) [ISBN:9780080873596].

[Bre97] G. Bredon, *Sheaf Theory*, Springer (1997) [doi:10.1007/978-1-4612-0647-7].

[Bro73] K. S. Brown, *Abstract Homotopy Theory and Generalized Sheaf Cohomology*, Trans. Amer. Math. Soc. **186** (1973) 419–458 [jstor:1996573].

[Bro] R. Brown, *Elements of Modern Topology*, McGraw-Hill (1968) [ncatlab.org/nlab/files/Brown-Topology.pdf].

[Br87] R. Bryant, *Metrics with Exceptional Holonomy*, Ann. Math. **126** (1987), 525–576, [jstor:1971360].

[Br05] R. Bryant, *Some remarks on G_2-structures*, Proceedings of the 12th Gökova Geometry-Topology Conference (2005) 75–109 [arXiv:math/0305124].

[Br93] J.-L. Brylinski, *Loop Spaces, Characteristic Classes, and Geometric Quantization*, Birkhäuser (1993) [doi:10.1007/978-0-8176-4731-5].

[BvDR18] U. Buchholtz, F. van Doorn, and E. Rijke, *Higher Groups in Homotopy Type Theory*, LICS '18 – 33rd Annual ACM/IEEE Symposium on Logic in Computer Science, ACM (2018) 205-215 [doi:10.1145/3209108.3209150] [arXiv:1802.04315].

[BH18] U. Buchholtz and K.-B. Hou, *Cellular Cohomology in Homotopy Type Theory*, LICS '18 – 33rd Annual ACM/IEEE Symposium on Logic in Computer Science, ACM (2018) [doi:10.1145/3209108.3209188] [arXiv:1802.02191].

[Bunk20] S. Bunk, *The $\mathbb{R}$-Local Homotopy Theory of Smooth Spaces*, J. Homotopy Relat. Struct. **17** (2022) 593–650 [arXiv:2007.06039] [doi:10.1007/s40062-022-00318-7]

[Bu12] U. Bunke, *Differential cohomology* [arXiv:1208.3961].

[BNV13] U. Bunke, T. Nikolaus, and M. Völkl, *Differential cohomology theories as sheaves of spectra*, J. Homotopy Related Struc. **11** (2016) 1–66 [arXiv:1311.3188] [10.1007/s40062-014-0092-5].

[BS09] U. Bunke and T. Schick, *Differential orbifold K-Theory*, J. Noncommut. Geom. **7** (2013) 1027–1104 [arXiv:0905.4181].

[BS12] U. Bunke and T. Schick, *Differential K-theory. A survey*, Global differential geometry, Springer (2012) 303-358 [doi:10.1088/1126-6708/1998/12/019] [arXiv:1011.6663].

[BSS19] S. Burton, H. Sati, and U. Schreiber, *Lift of fractional D-brane charge to equivariant Cohomotopy theory*, Journal of Geometry & Physics **161** (2020) 104034 [arXiv:1812.09679] [doi:10.1016/j.geomphys.2020.104034].

[CJKP97] A. Carboni, G. Janelidze, M. Kelly, and R. Paré, *On localization and stabilization for factorization systems*, Appl. Categ. Structures **5** (1997) 1–58 [doi:10.1023/A:1008620404444].

[Car10] D. Carchedi, *Sheaf Theory for Étale Geometric Stacks* [arXiv:1011.6070].

[Car11] D. Carchedi, *Categorical Properties of Topological and Diffentiable Stacks*, PhD thesis, Utrecht (2011) [dspace:1874/208971].

[Car12] D. Carchedi, *Compactly generated stacks: A Cartesian closed theory of topological stacks*, Advances in Mathematics **229** 6 (2012) 3339–3397 [doi:10.1016/j.aim.2012.02.006].

[Car16] D. Carchedi, *On The Homotopy Type of Higher Orbifolds and Haefliger Classifying Spaces*, Adv. Math. **294** (2016), 756–818, [arXiv:1504.02394].

[Car19] D. Carchedi, *Étale Stacks as Prolongations*, Adv. Math. **352** (2019) 56–132 [arXiv:1212.2282].

[Car20] D. Carchedi, *Higher Orbifolds and Deligne-Mumford Stacks as Structured Infinity-Topoi*, Mem. Amer. Math. Soc. **264** (2020) [arXiv:1312.2204] [doi:10.1090/memo/1282].

[CR12] D. Carchedi and D. Roytenberg, *On theories of superalgebras of differentiable functions*, Theory Appl. Categories **28** (2013) 1022–1098 [arXiv:1211.6134] [tac:28-30].

[CW08] A. Carey and B.-L. Wang, *Thom isomorphism and Push-forward map in twisted K-theory*, J. K-Theory **1** 2 (2008) 357–393 [arXiv:math/0507414] [doi:10.1017/is007011015jkt011].

[CCF11] C. Carmeli, L. Caston, and R. Fioresi, *Mathematical Foundations of Supersymmetry*, EMS Series of Lectures in Mathematics **15** (2011) [ISBN:978-3-03719-097-5].

[CHM85] C. Cassidy, M. Hébert, and M. Kelly, *Reflective subcategories, localizations, and factorization systems*, J. Austral. Math Soc. A **38** (1985) 287–329 [doi:10.1017/S1446788700023624].

[Cav15] E. Cavallo, *Synthetic Cohomology in Homotopy Type Theory*, MSc Thesis, Carnegie Mellon (2015) [www.cs.cmu.edu/~rwh/theses/cavallo-msc.pdf].

[CT05] J. Cheeger and G. Tian, *Anti-Self-Duality of Curvature and Degeneration of Metrics with Special Holonomy*, Comm. Math. Phys. **255** (2005) 391–417 [doi:10.1007/s00220-004-1279-0].

[Ch77] K. T. Chen, *Iterated path integrals*, Bull. Amer. Math. Soc. **83** (1977) 831–879 [euclid:bams/1183539443].

[Ch01] W. Chen, *A homotopy theory of orbispaces* [arXiv:math/0102020].

[Ch17] W. Chen, *Resolving symplectic orbifolds with applications to finite group actions*, J. Gökova Geom. Top. **12** (2018) 1–39 [arXiv:1708.09428].

[CR04] W. Chen and Y. Ruan, *A New Cohomology Theory for Orbifold*, Commun. Math. Phys. **248** (2004) 1–31 [doi:10.1007/s00220-004-1089-4] [arXiv:math/0004129].

[Ch24] F. Cherubini, *Synthetic G-jet-structures in modal homotopy type theory*, Mathematical Structures in Computer Science **34** 8 (2024) 1–35 [doi:10.1017/S0960129524000355] [arXiv:1806.05966].

[CRi20] F. Cherubini and E. Rijke, *Modal Descent*, Mathematical Structures in Computer Science **31** 4 (2021) [arXiv:2003.09713] [doi:10.1017/S0960129520000201].

[CTSR16] C.-K. Chiu, J. Teo, A. Schnyder and S. Ryu, *Classification of topological quantum matter with symmetries*, Reviews of Modern Physics **88** 3 (2016), 035005 [doi:10.1103/RevModPhys.88.035005] [arXiv:1505.03535].

[CS14] C.-K. Chiu and A. P. Schnyder, *Classification of reflection-symmetry-protected topological semimetals and nodal superconductors* Physical Review B **90** 20 (2014) 205136 [doi:10.1103/PhysRevB.90.205136] [arXiv:1408.4642].

[CP14] C.-H. Cho and M. Poddar, *Holomorphic orbi-discs and Lagrangian Floer cohomology of symplectic toric orbifolds*, J. Differential Geom. **98** (2014) 21–116 [euclid:jdg/1406137695].

[CSW13] J. D. Christensen, G. Sinnamon, and E. Wu, *The D-topology for diffeological spaces*, Pacific J. Math. **272** (2014) 87–110 [arXiv:1302.2935] [doi:10.2140/pjm.2014.272.87].

[CW14] J. D. Christensen and E. Wu, *The homotopy theory of diffeological spaces, I. Fibrant and cofibrant objects*, New York J. Math. **20** (2014) 1269–1303 [arXiv:1311.6394] [nyjm:2014/20–59].

[Ci19] D.-C. Cisinski, *Higher category theory and homotopical algebra*, Cambridge University Press (2019) [doi:10.1017/9781108588737].

[Cl14] E. Clader, *Orbifolds and orbifold cohomology* (2014) [ncatlab.org/nlab/files/Clader-Orbifolds.pdf].

[CCL26] A. Clough, B. Cnossen, S. Linskens, *Global spaces and the homotopy theory of stacks*, Algebraic & Geometric Topopology (2026) [arXiv:2407.06877].

[CV17] G. Codogni and F. Viviani, *Moduli and Periods of Supersymmetric Curves*, Adv. Theor. Math. Phys. **23** (2019) 345–402 [arXiv:1706.04910] [doi:10.4310/ATMP.2019.v23.n2.a2].

[CDL00] N. Cohen and S. De Leo, *The quaternionic determinat*, El. J. Lin. Alg. **7** (2000) 100–111 [arXiv:math-ph/9907015] [eudml:121484].

[CDR16] R. J. Conlon, A. Degeratu, and F. Rochon, *Quasi-asymptotically conical Calabi-Yau manifolds*, Geom. Topol. **23** (2019) 29-100 [arXiv:1611.04410] [doi:10.2140/gt.2019.23.29].

[Co20] D. Corfield, *Modal homotopy type theory*, Oxford University Press (2020) [ISBN:9780198853404].

[CPRST14] V. Coufal, D. Pronk, C. Rovi, L. Scull, and C. Thatcher, *Orbispaces and their Mapping Spaces via Groupoids*, Contemp. Math. **641** (2015) 135–166 [arXiv:1401.4772] [doi:10.1090/conm/641].

[Cru03] J. Cruickshank, *Twisted homotopy theory and the geometric equivariant 1-stem* Topology and its Applications **129** 3 (2003) 251–271 [doi:10.1016/S0166-8641(02)00183-9].

[DL98] J. Davis and W. Lück, *Spaces over a Category and Assembly Maps in Isomorphism Conjectures in K- and L-Theory*, K-Theory **15** (1998) 201–252 [doi:10.1023/A:1007784106877].

[dB+02] J. de Boer, R. Dijkgraaf, K. Hori, A. Keurentjes, J. Morgan, D. Morrison, and S. Sethi, *Triples, Fluxes, and Strings*, Adv. Theor. Math. Phys. **4** (2002) 995–1186 [arXiv:hep-th/0103170].

[DHLPS19] D. Degrijse, M. Hausmann, W. Lück, I. Patchkoria, and S. Schwede, *Proper equivariant stable homotopy theory*, Memoirs of the AMS **288** 1432 (2023) [arXiv:1908.00779] [doi:10.1090/memo/1432].

[DM69] P. Deligne and D. Mumford, *The irreducibility of the space of curves of given genus*, Pub. Math. IHÉS **36** (1969) 75–109 [numdam:PMIHES_1969__36__75_0].

[DZ23] M. Del Zotto, *Uncharted Territories in Geometric Engineering*, talk at *Strings and Geometry 2023*, Pennsylvania (2023) web.sas.upenn.edu/strings-and-geometry-2023/files/2023/03/Talk9DelZotto.pdf

[DG80] M. Demazure and P. Gabriel, *Introduction to algebraic geometry and algebraic groups*, North-Holland (2000) [ISBN:9780080871509].

[DD95] F. Diener and M. Diener, *Nonstandard analysis in practice*, Springer (1995) [doi:10.1007/978-3-642-57758-1].

[Di43] J. Dieudonné, *Les déterminants sur un corps non commutatif*, Bull. Soc. Math. France **71** (1943) 27–45 [numdam:BSMF_1943__71__27_0].

[DFM11] J. Distler, D. Freed, and G. Moore, *Orientifold Précis*, in: *Mathematical Foundations of Quantum Field and Perturbative String Theory* Proc. Symp. Pure Math., AMS (2011) [arXiv:0906.0795] [ams:pspum-83].

[DD63] J. Dixmier and A. Douady, *Champs continus d'espaces hilbertiens et de C^*-algébres*, Bull. Soc. Math. France **91** (1963) 227–284 [BSMF_1963__91__227_0].

[DHVW85] L. Dixon, J. Harvey, C. Vafa, and E. Witten, *Strings on orbifolds*, Nucl. Phys. **B 261** (1985) 678–686 [doi:10.1016/0550-3213(85)90593-0].

[DHVW86] L. Dixon, J. Harvey, C. Vafa, and E. Witten, *Strings on orbifolds (II)*, Nucl. Phys. **B 274** (1986) 285–314 [doi:10.1016/0550-3213(86)90287-7].

[Do20] I. Dolgachev, *Kummer Surfaces: 200 Years of Study*, Notices of the AMS **67** 10 (2020) 1527–1534 [arXiv:1910.07650] [doi:10.1090/noti2168].

[DE05] V. Dolgushev and P. Etingof, *Hochschild cohomology of quantized symplectic orbifolds and the Chen-Ruan cohomology*, Int. Math. Res. Not. **2005** 27, 1657–1688, [arXiv:math/0410562].

[DSBW23] N. Doll, H. Schulz-Baldes, and N. Waterstraat, *Spectral Flow: A Functional Analytic and Index-Theoretic Approach* De Gruyter (2023) [doi:10.1515/9783111172477].

[DLM02] C. Dong, K. Liu, and X. Ma, *On orbifold elliptic genus*, in: *Orbifolds in Mathematics and Physics*, Amer. Math. Soc. (2002) 87–105, [arXiv:math/0109005] [conm-310].

[Do05] C. Douglas, *Twisted stable homotopy theory*, PhD thesis, MIT (2005) [dspace:1721.1/7582].

[Dr11] G. Dragomir, *Closed Geodesics on compact developable orbifold*, PhD thesis, McMaster University (2011) [math.mcmaster.ca/~boden/students/Dragomir-PhD.pdf]

[DT06] S. Dragomir and G. Tomassini, *Differential Geometry and Analysis on CR Manifolds*, Birkhäuser (2006) [doi:10.1007/0-8176-4483-0].

[DM02] S. Dragomir and J. Masamune, *Cauchy-Riemann orbifolds*, Tsukuba J. Math. **26** (2002) 351-386 [euclid:tkbjm/1496164430].

[Dr94] K. S. Druschel, *Oriented Orbifold Cobordism*, Pacific J. Math. **164** (1994) 299-319 [doi:10.2140/pjm.1994.164.299].

[DZ16] C.-Y. Du and X. Zhao, *Spark and Deligne-Beilinson cohomology on orbifolds*, J. Geom. Phys. **104** (2016) 277–290 [doi:10.1016/j.geomphys.2016.02.011].

[Du79a] E. Dubuc, *Sur les modèles de la géométrie différentielle synthétique*, Cah. Topol. Géom. Différ. Catég. **20** (1979) 231–279 [numdam:CTGDC_1979__20_3_231_0].

[Du79b] E. Dubuc, *Concrete quasitopoi*, Lecture Notes in Math. **753**, Springer (1979) 239–254 [doi:10.1007/BFb0061821].

[Du01] D. Dugger, *Universal homotopy theories*, Adv. Math. **164** (2001) 144–176 [doi:10.1006/aima.2001.2014], [arXiv:math/0007070].

[Dug03] D. Dugger, *Notes on Delta-generated spaces* (2003) [pages.uoregon.edu/ddugger/delta.html].

[DHI04] D. Dugger, S. Hollander and D. Isaksen, *Hypercovers and simplicial presheaves*, Math. Proc. Cambridge Philos. Soc. **136** (2004) 9–51 [doi:10.1017/S0305004103007175], [arXiv:math/0205027].

[DS09] D. Dugger and D. Spivak, *Mapping spaces in quasi-categories*, Algebr. Geom. Top. **11** (2011) 263–325 [doi:10.2140/agt.2011.11.263] [arXiv:0911.0469].

[Du96] J. J. Duistermaat, *The heat kernel Lefschetz fixed point formula for the spin^c Dirac operator*, Birkhäuser (1996) [ISBN-10:9780817682460].

[DwKa84] W. Dwyer and D. Kan, *Singular functors and realization functors*, Indagationes Mathematicae (Proceedings) **87** 2 (1984) 147–153 [doi:10.1016/1385-7258(84)90016-7]

[Egg14] R. Eggertsson, *Stacks in Gauge Theory*, MSc Thesis, Radboud University (2014) [ncatlab.org/schreiber/show/MSc+Eggertsson].

[El83] A. Elmendorf, *Systems of fixed point sets*, Trans. Amer. Math. Soc. **277** (1983) 275–284 [jstor:1999356].

[EU14] J. Espinoza and B. Uribe, *Topological properties of the unitary group*, JP J. Geom. Topol. **16** (2014) 45–55 [pphmj:8730] [arXiv:1407.1869].

[FR07] L. Fajstrup and J. Rosický, *A convenient category for directed homotopy*, Theory App. Categories **21** (2008) 7–20 [arXiv:0708.3937] [tac:21-01].

[Fa+16] C. Fang et al., *Topological nodal line semimetals*, Chinese Physics B **25** 11 (2016) 117106 [10.1088/1674-1056/25/11/117106] [arXiv:1609.05414].

[Fe03] J. Fernandez, *Hodge structures for orbifold cohomology*, Proc. Amer. Math. Soc. **134** (2006) 2511–2520 [doi:10.1090/S0002-9939-06-08515-7] [arXiv:math/0311026].

[FSS14] D. Fiorenza, H. Sati, and U. Schreiber, *A higher stacky perspective on Chern-Simons theory*, in: *Mathematical Aspects of Quantum Field Theories*, Springer (2014) [doi:10.1007/978-3-319-09949-1] [arXiv:1301.2580].

[FSS15a] D. Fiorenza, H. Sati, and U. Schreiber, *Super Lie n-algebra extensions, higher WZW models, and super p-branes with tensor multiplet fields*, Intern. J. Geom. Meth. Mod. Phys. **12** (2015) 1550018 [doi:10.1142/S0219887815500188] [arXiv:1308.5264].

[FSS15] D. Fiorenza, H. Sati, and U. Schreiber, *The E_8 moduli 3-stack of the C-field in M-theory*, Commun. Math. Phys. **333** (2015) 117–151 [doi:10.1007/s00220-014-2228-1] [arXiv:1202.2455].

[FSS17] D. Fiorenza, H. Sati, and U. Schreiber, *Rational sphere valued supercocycles in M-theory and type IIA string theory*, J. Geom. Phys. **114** (2017) 91–108 [doi:10.1016/j.geomphys.2016.11.024] [arXiv:1606.03206].

[FSS18] D. Fiorenza, H. Sati, and U. Schreiber, *T-Duality from super Lie n-algebra cocycles for super p-branes*, Adv. Theor. Math. Phys. **22** 5 (2018) [doi:10.4310/ATMP.2018.v22.n5.a3] [arXiv:1611.06536].

[FSS19] D. Fiorenza, H. Sati, and U. Schreiber, *The rational higher structure of M-theory*, Fortschritte der Physik **67** 8–9 (2019) 1910017 [arXiv:1903.02834] [doi:10.1002/prop.201910017]

[FSS20] D. Fiorenza, H. Sati, and U. Schreiber, *Twisted Cohomotopy implies M-theory anomaly cancellation on 8-manifolds*, Comm. Math. Phys. **377** (2020) 1961-2025, [arXiv:1904.10207] [doi:10.1007/s00220-020-03707-2].

[FSS21a] D. Fiorenza, H. Sati, and U. Schreiber, *Twisted Cohomotopy implies M5 WZ term level quantization*, Comm. Math. Phys. **384** (2021) 403–432 [doi:10.1007/s00220-021-03951-0], [arXiv:1906.07417]

[FSS21b] D. Fiorenza, H. Sati, and U. Schreiber, *Twisted Cohomotopy implies Twisted String Structure on M5-branes*, in: J. Mathematical Physics **62** 4 (2021) 042301 [arXiv:2002.11093], [doi:10.1063/5.0037786].

[FSS21c] D. Fiorenza, H. Sati, and U. Schreiber, *Super-exceptional embedding construction of the heterotic M5: Emergence of* $\mathrm{SU}(2)$*-flavor sector*, J. Geom. Phys. **170** (2021) 104349 [10.1016/j.geomphys.2021.104349]

[FSS22] D. Fiorenza, H. Sati, and U. Schreiber, *Twistorial Cohomotopy Implies Green-Schwarz anomaly cancellation*, Rev. Math. Phys. **34** 5 (2022) 2250013 [arXiv:2008.08544] [doi:10.1142/S0129055X22500131]

[FSS23] D. Fiorenza, H. Sati, and U. Schreiber, *The Character Map in Non-abelian Cohomology — Twisted, Differential, and Generalized*, World Scientific (2023) [doi:10.1142/13422] [ncatlab.org/schreiber/show/The+Character+Map].

[FStS12] D. Fiorenza, J. Stasheff, and U. Schreiber, *Čech cocycles for differential characteristic classes*, Adv. Theor. Math. Phys. **16** (2012) 149–250 [arXiv:1011.4735] [doi:10.4310/ATMP.2012.v16.n1.a5].

[FF2016] A. Fomenko and D. Fuchs, *Homotopical Topology*, Springer (2016) [doi:10.1007/978-3-319-23488-5].

[Fra97] T. Frankel, *The Geometry of Physics – An Introduction*, Cambridge University Press (1997, 2004, 2012) [doi:10.1017/CBO9781139061377].

[Fr99] D. Freed, *Five lectures on supersymmetry*, American Mathematical Society (1999) [ISBN:978-0-8218-1953-1].

[FH21] D. Freed and M. Hopkins, *Reflection positivity and invertible topological phases*, Geometry & Topology **25** 3 (2021) 1165–1330 [doi:10.2140/gt.2021.25.1165] [arXiv:1604.06527].

[FHT11] D. Freed, M. Hopkins, and C. Teleman, *Loop groups and twisted K-theory I*, J. Topology **4** (2011) 737-798 [doi:10.1112/jtopol/jtr019] [arXiv:0711.1906].

[FM13] D. S. Freed and G. W. Moore, *Twisted Equivariant Matter*, Annales Henri Poincaré **14** 8 (2013) 1927–2023 [doi:10.1007/s00023-013-0236-x] [arXiv:1208.5055].

[FS07] E. Frenkel and M. Szczesny, *Chiral de Rham complex and orbifolds*, J. Algebraic Geom. **16** (2007) 599–624 [doi:10.1090/S1056-3911-07-00466-3].

[Fu83] A. Fujiki, *On primitively symplectic compact Kähler V-manifolds*, in: K. Ueno (ed.), *Classification of Algebraic and Analytic Manifolds*, Katata Symposium 1982, Birkhäuser (1983) [ISBN:9780817631376].

[GIR08] M. Gaberdiel, D. Israel, and E. Rabinovici, *D-branes at multicritical points*, J. High Energy Phys. **0804** (2008) 086 [arXiv:0803.0291] [doi:10.1088/1126-6708/2008/04/086].

[GMTW06] S. Galatius, I. Madsen, U. Tillmann, and M. Weiss, *The homotopy type of the cobordism category*, Acta Math. **202** (2009) 195–239 [arXiv:math/0605249] [10.1007/s11511-009-0036-9].

[GL88] K. Galicki and H. B. Lawson, *Quaternionic Reduction and Quaternionic Orbifolds*, Math. Ann. **282** (1988) 1–21 [dml:164446].

[GH10] D. Gao and K. Hori, *On The Structure Of The Chan-Paton Factors For D-Branes In Type II Orientifolds* [arXiv:1004.3972].

[Gi13] G. Ginot, *Introduction to Differentiable Stacks (and gerbes, moduli spaces ...)*, notes (2013) [ncatlab.org/nlab/files/GinotDifferentiableStacks.pdf].

[GS25] G. Giotopoulos and H. Sati, *Field Theory via Higher Geometry I: Smooth Sets of Fields*, J. Geometry and Physics **213** (2025) 105462 [arXiv:2312.16301]. [doi:10.1016/j.geomphys.2025.105462].

[GS26] G. Giotopoulos and H. Sati, *Field Theory via Higher Geometry II: Thickened Smooth Sets as Synthetic Foundations* [arXiv:2512.22816].

[GSS25a] G. Giotopoulos, H. Sati and U. Schreiber: *Flux Quantization on 11d Superspace*, Journal of High Energy Physics **2024** 82 (2024) [doi:10.1007/JHEP07(2024)082] [arXiv:2403.16456].

[GSS25a] G. Giotopoulos, H. Sati, and U. Schreiber: *Flux Quantization on M5-Branes*, J. High Energy Phys. **2024** 140 (2024) [doi:10.1007/JHEP10(2024)140].

[GSS25b] G. Giotopoulos, H. Sati and U. Schreiber: *Holographic M-Brane Super-Embeddings*, Reviews in Mathematical Physics (2025) [doi:10.1142/S0129055X25500230] [arXiv:2408.09921].

[Gi66] J. Giraud, *Cohomologie non abélienne*, Springer (1971) [doi:9783540053071].

[Gi72] J. Giraud, *Classifying topos*, in: *Toposes, Algebraic Geometry and Logic*, Lecture Notes in Mathematics **274**, Springer (1972) [doi:10.1007/BFb0073964].

[GWZ86] H. Gluck, F. Warner, and W. Ziller, *The geometry of the Hopf fibrations*, L'Enseignement Mathématique **32** (1986) 173–198 [doi:10.5169/seals-55085].

[Go01] L. Godinho, *Blowing Up Symplectic Orbifolds*, Ann. Global Anal. Geom. **20** (2001), 117–162 [doi:10.1023/A:1011628628835].

[GJ99] P. Goerss and R. F. Jardine, *Simplicial homotopy theory*, Birkhäuser (2009) [doi:10.1007/978-3-0346-0189-4].

[GHHK08] R. Goldin, M. Harada, T. S. Holm, and T. Kimura, *The full orbifold K-theory of abelian symplectic quotients*, J. K-theory **8** (2011) 339–362 [arXiv:0812.4964] [doi:10.1017/is010005021jkt118].

[Gom05] K. Gomi, *Equivariant smooth Deligne cohomology*, Osaka J. Math. **42** (2005) 309–337 [euclid:ojm/1153494380].

[Gom17] K. Gomi, *Freed-Moore K-theory* [arXiv:1705.09134].

[GT19] K. Gomi and G. C. Thiang, *Crystallographic T-duality*, J. Geom. Phys. **139** (2019) 50–77 [doi:10.1016/j.geomphys.2019.01.002] [arXiv:1806.11385].

[Gr05] J. Grabowski, *Isomorphisms of algebras of smooth functions revisited*, Arch. Math. **85** (2005) 190–196 [doi:10.1007/s00013-005-1268-3] [arXiv:math/0310295].

[GS19a] D. Grady and H. Sati, *Twisted differential generalized cohomology theories and their Atiyah-Hirzebruch spectral sequence*, Algebr. Geom. Topol. **19** (2019) 2899–2960 [arXiv:1711.06650] [doi:10.2140/agt.2019.19.2899].

[GS19b] D. Grady and H. Sati, *Twisted differential KO-theory* [arXiv:1905.09085].

[Gr1844] H. Grassmann, Die Wissenschaft der extensiven Grössen oder die Ausdehnungslehre, Verlag von Otto Wigand, Leipzig (1844) [archive.org/details/dielinealeausde00grasgoog].

[Gr69] A. Gray, *A Note on Manifolds Whose Holonomy Group is a Subgroup of* $\mathrm{Sp}(n) \cdot \mathrm{Sp}(1)$, Michigan Math. J. **16** 2 (1969) 125–128 [doi:10.1307/mmj/1029000212].

[Gr65] A. Grothendieck (notes by M. Karoubi), *Introduction au langage fonctoriel*, course in Algiers (1965) [ncatlab.org/nlab/files/GrothendieckIntroductionLangageFonctoriel1965.pdf]

[Gr73] A. Grothendieck (notes by F. Gaeta), *Introduction to functorial algebraic geometry, Part I*, summer school in Buffalo (1973) [ncatlab.org/nlab/files/GrothendieckIntrodFunctorialGeometryI1973.pdf]

[Gu65] V. Guillemin, *The integrability problem for G-structures*, Trans. Amer. Math. Soc. **116** (1965) 544–560 [jstor:1994134].

[GPPV21] S. Gukov, D. Pei, P. Putrov, and C. Vafa, *4-manifolds and topological modular forms*, J. High Energy Phys. **2021** 84 (2021) [arXiv:1811.07884] [doi:10.1007/JHEP05(2021)084].

[Hae71] A. Haefliger, *Homotopy and integrability*, in: *Manifolds*, Lecture Notes in Mathematics **197**, Springer (1971) [doi:10.1007/BFb0068615].

[Hae84] A. Haefliger, *Groupoïdes d'holonomie et classifiants*, Transversal structure of foliations (Toulouse, 1982), Astérisque **116** (1984) 70–97 [doi:10.1007/978-1-4684-9167-8_11].

[Hae90] A. Haefliger, *Orbi-Espaces*, In: *Sur les Groupes Hyperboliques d'après Mikhael Gromov*, Progress in Mathematics **83** Birkhäuser (1990) [doi:10.1007/978-1-4684-9167-8_11].

[Hae91] A. Haefliger, *Complexes of Groups and Orbihedra*, in: *Group theory from a Geometrical viewpoint*, World Scientific (1991) 504–540 [doi:10.1142/1235].

[Ham15] C. Hammond, *The Basics of Crystallography and Diffraction*, Oxford University Press (2015) [ISBN:9780198738688]

[Har13] T. Haraguchi, *On model structure for coreflective subcategories of a model category*, Math. J. Okayama Univ. **57** (2015) 79–84 [arXiv:1304.3622].

[HSa20] F. Hebestreit and S. Sagave, *Homotopical and operator algebraic twisted K-theory*, Math. Ann. **378** (2020) 1021–1059 [arXiv:1904.01872] [doi:10.1007/s00208-020-02066-6].

[He01] A. Henriques, *Orbispaces and orbifolds from the point of view of the Borel construction, a new definition* [arXiv:math/0112006].

[HG07] A. Henriques and D. Gepner, *Homotopy Theory of Orbispaces* [arXiv:math/0701916].

[HM04] A. Henriques and D. Metzler, *Presentations of noneffective orbifolds*, Trans. Amer. Math. Soc. **356** (2004) 2481–2499 [doi:10.1090/S0002-9947-04-03379-3].

[He09a] R. A. Hepworth, *Morse Inequalities for Orbifold Cohomology*, Algebr. Geom. Topol. **9** (2009) 1105–1175 [doi:10.2140/agt.2009.9.1105] [arXiv:0712.2432].

[He09b] R. A. Hepworth, *Vector Fields and Flows on Differentiable Stacks*, Theory Appl. Categ. **22** (2009) 542–587 [arXiv:0810.0979] [tac:22-21].

[HN93] J. Hilgert and K.-H. Neeb, *Lie Semigroups and their Applications*, Lec. Notes Math. **1552**, Springer (1993) [doi:10.1007/BFb0084640].

[HS87] M. Hilsum and G. Skandalis, *Morphismes K-orientés d'espaces de feuilles et functorialité en théorie de Kasparov*, Ann. Scient. Ec. Norm. Sup. **20** (1987) 325-390 [numdam:ASENS_1987_4_20_3_325_0].

[HKST11] H. Hohnhold, M. Kreck, S. Stolz, and P. Teichner, *Differential forms and 0-dimensional supersymmetric field theories*, Quantum Topol. **2** (2011) 1–41 [doi:10.4171/QT/12].

[Ho08] S. Hollander, *A homotopy theory for stacks*, Israel J. Math. **163** (2008) 93–124 [arXiv:math/0110247] [doi:10.1007/s11856-008-0006-5].

[HM12] T. S. Holm and T. Matsumura, *Equivariant cohomology for Hamiltonian torus actions on symplectic orbifolds*, Transform. Groups **17** (2012) 717–746 [arXiv:1008.3315] [doi:10.1007/s00031-012-9192-7].

[Ho88] H. Honkasalo, *Equivariant Alexander-Spanier cohomology*, Math. Scand. **63** (1988) 179–195 [doi:10.7146/math.scand.a-12232].

[Ho90] H. Honkasalo, *Equivariant Alexander-Spanier cohomology for actions of compact Lie groups*, Math. Scand. **67** (1990) 23-34 [jstor:24492569].

[Hop84] M. Hopkins, *Stable decompositions of certain loop spaces* PhD thesis. Northwestern University (1984) [ncatlab.org/nlab/files/HopkinsStableDecompositions.pdf].

[HS05] M. Hopkins and I. Singer, *Quadratic Functions in Geometry, Topology, and M-Theory*, J. Differential Geom. **70** (2005) 329–452 [arXiv:math.AT/0211216] [doi:10.4310/jdg/1143642908].

[Ho98] P. Hořava, *Type IIA D-Branes, K-Theory, and Matrix Theory* Advances in Theoretical and Mathematical Physics **2** 6 (1998) 1373–1404 [doi:10.4310/ATMP.1998.v2.n6.a5] [arXiv:hep-th/9812135].

[HW96] P. Hořava and E. Witten, *Eleven dimensional supergravity on a manifold with boundary*, Nucl. Phys. B **475** (1996) 94 [doi:10.1016/0550-3213(96)00308-2] [arXiv:hep-th/9603142].

[HS91] G. Horowitz and A. R. Steif, *Singular string solutions with nonsingular initial data*, Phys. Lett. **B258** (1991) 91–96 [doi:10.1016/0370-2693(91)91214-G].

[STHu59] S.-T. Hu, *Homotopy Theory*, Academic Press (1959) [www.maths.ed.ac.uk/ v1ranick/papers/hu2.pdf].

[HW11] J. Hu and B.-L. Wang, *Delocalized Chern character for stringy orbifold K-theory*, Trans. Amer. Math. Soc. **365** (2013) 6309–6341 [arXiv:1110.0953] [jstor:23513597].

[Hua20] Z. Huan, *Twisted Quasi-elliptic cohomology and twisted equivariant elliptic cohomology*, Adv. Theor. Math. Phys. **29** 7 (2025) 1857–1904 [arXiv:2006.00554] [doi:10.4310/ATMP.251120035411].

[Hua25] Z. Huan, *Quasi-elliptic cohomology of 4-spheres*, Axioms **14** 4 (2025) 267 [arXiv:2408.02278] [doi:10.3390/axioms14040267].

[HSS18] J. Huerta, H. Sati, and U. Schreiber, *Real ADE-equivariant (co)homotopy of super M-branes*, Commun. Math. Phys. **371** (2019) 425–524 [doi:10.1007/s00220-019-03442-3] [arXiv:1805.05987].

[Hus94] D. Husemöller, *Fibre Bundles*, Springer (1994) [doi:10.1007/978-1-4757-2261-1].

[IM25] A. Ibort and A. Mas, *Smooth sets of fields: A pedagogical introduction*, Geom. Mech. **2** 3 (2025) [arXiv:2510.20422] [doi:10.1142/s2972458925400052].

[IU12] L. Ibáñez and A. Uranga, *String Theory and Particle Physics – An Introduction to String Phenomenology*, Cambridge University Press (2012) [doi:10.1017/CBO9781139018951].

[IR21] A. Ibort and M. A. Rodriguez, *An Introduction to Groups, Groupoids and Their Representations*, CRC Press (2021) [doi:10.1201/b22019].

[IZ85] P. Iglesias-Zemmour, *Fibrations difféologiques et Homotopie*, PhD thesis (1985) [math.huji.ac.il/~piz/documents/TheseEtatPI.pdf]

[IZ13] P. Iglesias-Zemmour, *Diffeology*, Amer. Math. Soc. (2013) [ISBN:978-0-8218-9131-5].

[IKZ10] P. Iglesias-Zemmour, Y. Karshon, and M. Zadka, *Orbifolds as diffeologies*, Trans. Amer. Math. Soc. **362** (2010) 2811-2831 [arXiv:math/0501093] [10.1090/S0002-9947-10-05006-3].

[Jam95] I. M. James, *Introduction to fibrewise homotopy theory*, in: *Handbook of Algebraic Topology*, Oxford University Press (1995) [doi:10.1016/B978-0-444-81779-2.X5000-7].

[Jän65] K. Jänich, *Vektorraumbündel und der Raum der Fredholm-Operatoren*, Mathematische Annalen **161** 2 (1965) 129–142 [doi:10.1007/BF01360851].

[Jar87] J. F. Jardine, *Simplicial presheaves*, J. Pure Applied Algeb. **47** (1987) 35–87 [core.ac.uk/download/pdf/82485559.pdf].

[Jar96] J. F. Jardine, *Boolean localization, in practice*, Documenta Math. **1** (1996) 245–275 [documenta:vol-01/13].

[Jar01] J. F. Jardine, *Stacks and the homotopy theory of simplicial sheaves*, Homology Homotopy Appl. **3** (2001) 361–384 [euclid:euclid.hha/1139840259].

[Jar09] J. F. Jardine , *Cocycle Categories*, in *Algebraic Topology*, Springer (2009) 185–218 [10.1007/978-3-642-01200-6_8] [arXiv:math/0605198].

[Jar15] J. F. Jardine, *Local homotopy theory*, Springer (2015) [doi:10.1007/978-1-4939-2300-7].

[JKK05] T. J. Jarvis, R. Kaufmann, and T. Kimura, *Stringy K-theory and the Chern character*, Invent. Math. **168** (2007) 23–81 [arXiv:math/0502280] [10.1007/s00222-006-0026-x].

[Je97] T. D. Jeffres, *Singular Set of Some Kähler Orbifolds*, Trans. Amer. Math. Soc. **349** (1997) 1961–1971 [jstor:2155355].

[JY11] L. Ji and S.-T. Yau, *Transformation Groups and Moduli Spaces of Curves*, International Press of Boston (2011) [ISBN:9781571462237].

[Jo99] C. K. Johnson, *Crystallographic Topology 2: Overview and Work in Progress*, in: *Trends in Mathematical Physics*, AMS/International Press (1999) [amsip-13]

[JBD96] C. K. Johnson, M. N. Burnett and W. D. Dunbar, *Crystallographic Topology and Its Applications*, in: *Crystallographic Computing 7 – Proceedings of the Macromolecular Crystallography Computing School* (1996) [www.iucr.org/__data/assets/pdf_file/0010/9001/cj.pdf]

[Joh02] P. Johnstone *Sketches of an Elephant – A Topos Theory Compendium*, vols 1 & 2 Oxford University Press (2002) [ISBN:9780198534259] [ISBN:9780198515982].

[Joy08a] A. Joyal, *Notes on quasi-categories* (2008) [ncatlab.org/nlab/files/JoyalNotesOnQuasiCategories.pdf]

[Joy08a] A. Joyal, *Notes on Logoi* (2008) [ncatlab.org/nlab/to appfiles/JoyalOnLogoi2008.pdf]

[Jo95] D. Joyce, *Manifolds with many complex structures*, Quarter. J. Math. **46** (1995) 169–184 [doi:10.1093/qmath/46.2.169].

[Jo98] D. Joyce, *On the topology of desingularizations of Calabi-Yau orbifolds* [arXiv:math/9806146].

[Jo99a] D. Joyce, *Deforming Calabi-Yau orbifolds*, Asian J. Math. **3** (1999) 853-868 [doi:10.4310/AJM.1999.v3.n4.a7].

[Jo99b] D. Joyce, *A new construction of compact 8-manifolds with holonomy* Spin(7), J. Differential Geom. **53** (1999) 89–130 [euclid:jdg/1214425448].

[Jo00] D. Joyce, *Compact Manifolds with Special Holonomy*, Oxford University Press (2000) [ISBN-10:0198506015].

[Jo01] D. Joyce, *Compact Riemannian manifolds with exceptional holonomy*, Surv. Differential Geom. **6** (2001) 39–65 [doi:10.4310/SDG.2001.v6.n1.a3].

[Jo12] D. Joyce, *D-manifolds and d-orbifolds: a theory of derived differential geometry*, draft [arXiv:1208.4948].

[Ju20] B. Juran, *Orbifolds, Orbispaces and Global Homotopy Theory*, Algebraic & Geometric Topology **25** (2025) 5175–5203 [arXiv:2006.12374] [doi:10.2140/agt.2025.25.5175].

[JSSW19] B. Jurčo, C. Saemann, U. Schreiber, and M. Wolf, *Higher Structures in M-Theory*, Fortsch. d. Phys. **67** (2019) [arXiv:1903.02807] [doi:10.1002/prop.201910001].

[Kan13] M. Kankaanrinta, *On real analytic orbifolds and Riemannian metrics*, Algeb. Geom. Topol. **13** (2013) 2369–2381 [doi:10.2140/agt.2013.13.2369].

[Ka08] M. Kapovich, *Hyperbolic Manifolds and Discrete Groups*, Birkhäuser (2008) [doi:10.1007/978-0-8176-4913-5].

[Ka98] A. Karch, *Field Theory Dynamics from Branes in String Theory*, PhD thesis, Berlin (1998) [doi:10.18452/14371].

[Kar70] M. Karoubi, *Espaces Classifiants en K-Théorie*, Transactions of the American Mathematical Society **147** 1 (1970) 75–115 [doi:10.2307/1995218].

[KKV97] S. Katz, A. Klemm and C. Vafa, *Geometric Engineering of Quantum Field Theories*, Nucl. Phys. B **497** (1997) 173–195 [arXiv:hep-th/9609239] [doi;10.1016/S0550-3213(97)00282-4].

[KS17] I. Khavkine and U. Schreiber, *Synthetic geometry of differential equations*, Journal of Geometry and Physics (2026) [arXiv:1701.06238].

[Kob72] S. Kobayashi, *Transformation Groups in Differential Geometry*, Springer (1995) [doi:10.1007/978-3-642-61981-6].

[Ko96] S. Kochman, *Bordism, Stable Homotopy and Adams Spectral Sequences*, Fields Institute Monographs, Amer. Math. Soc. (1996) [cds:2264210].

[Ko86] A. Kock, *Convenient vector spaces embed into the Cahiers topos*, Cah. Topol. Géom. Differ. Catég. **27** (1986) 3–17 [numdam:CTGDC_1986__27_1_3_0]; corrigendum and addenda: **27** (1986) 3–17 [numdam:CTGDC_1987__28_2_99_0].

[KMS93] I. Kolář, P. Michor, and J. Slovák, *Natural operations in differential geometry*, Springer (1993) [doi:10.1007/978-3-662-02950-3].

[KS97] A. Konechny and A. Schwarz, *On $(k \oplus l|q)$-dimensional supermanifolds*, in: *Supersymmetry and Quantum Field Theory*, Springer (1998) [arXiv:hep-th/9706003].

[KS00] A. Konechny and A. Schwarz, *Theory of $(k \oplus l|q)$-dimensional supermanifolds*, Sel. Math., New ser. **6** (2000) 471-486 [doi:10.1007/PL00001396].

[Kö16] A. Körschgen, *A Comparison of two Models of Orbispaces*, Homology Homotopy Appl. **20** (2018) 329–358 [10.4310/HHA.2018.v20.n1.a19] [arXiv:1612.04267].

[Kr09] A. Kresch, *On the geometry of Deligne-Mumford stacks*, in: *Algebraic Geometry: Seattle 2005*, Proc. Symp. Pure Math. **80**, Amer. Math. Soc. (2009) 259–271 [doi:10.5167/uzh-21342].

[KT18] A. Kübel and A. Thom, *Equivariant differential cohomology*, Trans. Amer. Math. Soc. **370** (2018) 8237–8283 [doi:10.1090/tran/7315] [arXiv:1510.06392].

[KT82] T. Kugo and P. Townsend, *Supersymmetry and the division algebras*, Nucl. Phys. **B 221** (1982) 357–380 [doi:10.1016/0550-3213(83)90584-9].

[Lan18] C. Lange, *Orbifolds from a metric viewpoint*, Geom. Dedicata **209** (2020) 43–57 [doi:10.1007/s10711-020-00521-x] [arXiv:1801.03472].

[La63] R. Lashof, *Poincaré duality and cobordism*, Trans. Amer. Math. Soc. **109** (1963) 257–277 [doi:10.1090/S0002-9947-1963-0156357-4].

[LGTX04] C. Laurent-Gengoux, J.-L. Tu, and P. Xu, *Chern-Weil map for principal bundles over groupoids*, Math. Zeitschrift **255** (2007) 451-491 [arXiv:math/0401420] [doi:10.1007/s00209-006-0004-4].

[La96] R. Lavendhomme, *Basic concepts of synthetic differential geometry*, Kluwer Texts in the Mathematical Sciences **13**, Springer (1996) [doi:10.1007/978-1-4757-4588-7].

[La86] W. Lawvere, *Categories of spaces may not be generalized spaces as exemplified by directed graphs*, Rev. Colombiana de Matem. **XX** (1986) 179–186 Reprints in Theory Appl. Categories **9** (2005) 1–7 [tac:tr9].

[La94] W. Lawvere, *Cohesive Toposes and Cantor's "lauter Einsen"*, Philosophia Mathematica **2** (1994) 5–15 [doi:10.1093/philmat/2.1.5].

[La07] W. Lawvere, *Axiomatic cohesion*, Theory Appl. Categories **19** (2007) 41–49 [tac:19-03].

[La91] W. Lawvere, *Some Thoughts on the Future of Category Theory*, Lecture Notes in Mathematics **1488**, Springer (1991) [doi:10.1007/BFb0084208].

[LR03] W. Lawvere and R. Rosebrugh, *Sets for Mathematics*, Cambridge University Press (2003) [doi:10.1017/CBO9780511755460].

[LBR88] C. LeBrun and M. Rothstein, *Moduli of super Riemann surfaces*, Comm. Math. Phys. **117** (1988), 159–176, [euclid:cmp/1104161598].

[Lee12] J. Lee, *Introduction to Smooth Manifolds*, Springer (2012) [doi:10.1007/978-1-4419-9982-5]

[Le08] E. Lerman, *Orbifolds as stacks?*, Enseign. Math. **(2) 56** (2010) 315–363 [arXiv:0806.4160] [doi:10.4171/LEM/56-3-4].

[LMS86] L. G. Lewis Jr., P. May, and M. Steinberger, *Equivariant stable homotopy theory*, Lecture Notes in Mathematics **1213**, Springer (1986) [www.math.uchicago.edu/~may/BOOKS/equi.pdf]

[LPZ13] L. A. Lichtenfelz, P. Piccione, and A. Zeghib, *On the Isometry Group of Lorentz Manifolds*, in: *Recent Trends in Lorentzian Geometry*, Springer (2013) [doi:10.1007/978-1-4614-4897-6_12].

[LMS02a] H. Liu, G. Moore, and N. Seiberg, *Strings in a Time-Dependent Orbifold I*, J. High Energy Phys. **0206** (2002) 045 [arXiv:hep-th/0204168].

[LMS02b] H. Liu, G. Moore, and N. Seiberg, *Strings in Time-Dependent Orbifolds II*, J. High Energy Phys. **0210** (2002) 031 [arXiv:hep-th/0206182].

[Lo94] M. Losik, *Categorical Differential Geometry*, Cah. Topol. Géom. Différ. Catég. **35** (1994) 274–290 [numdam:CTGDC_1994__35_4_274_0].

[Lot01] J. Lott, *The Geometry of Supergravity Torsion Constraints*, *Torsion constraints in supergeometry*, Comm. Math. Phys. **133** (1990) 563–615 [doi:10.1007/BF02097010] [arXiv:math/0108125].

[LU03] E. Lupercio and B. Uribe, *Deligne cohomology for orbifolds, discrete torsion and B-fields*, Geometric and topological methods for quantum field theory (Villa de Leyva, 2001), World Sci. (2003) 468–482 [arXiv:hep-th/0201184] [inSpire:582101].

[LU04a] E. Lupercio and B. Uribe, *An introduction to gerbes on orbifolds*, Ann. Math. Blaise Pascal **11** (2004) 155–180 [AMBP_2004__11_2_155_0] [arXiv:math/0402318].

[LU04b] E. Lupercio and B. Uribe, *Inertia orbifolds, configuration spaces and the ghost loop space*, Q. J. Math. **55** 2 (2004) 185–201 [arXiv:math/0210222] [doi:10.1093/qmath/hag053].

[LU06] E. Lupercio and B. Uribe, *Differential characters on orbifolds and string connections. I. Global quotients*, in: *Gromov-Witten theory of spin curves and orbifolds*, Contemp. Math. **403** Amer. Math. Soc. (2006) 127–142 [arXiv:math/0311008] [ams:CONM/403].

[Lu07] J. Lurie, *Deformation theory* [arXiv:0709.3091].

[Lu09a] J. Lurie, *Higher Topos Theory*, Ann. Math. Stud. **170**, Princeton University Press (2009) [pup:8957].

[Lu09b] J. Lurie, *Structured Spaces* [arXiv:0905.0459].

[Lu14] J. Lurie, *Nonabelian Poincaré Duality*, Lecture 8 in: *Tamagawa Numbers via Nonabelian Poincare Duality (282y)* (2014) [www.math.harvard.edu/ lurie/282ynotes/LectureVIII-Poincare.pdf].

[Lu17] J. Lurie, *Higher Algebra* (2017) [www.math.ias.edu/l̃urie/papers/HA.pdf].

[Lu19] J. Lurie, *Elliptic cohomology III: Tempered Cohomology* [www.math.harvard.edu/ lurie/papers/Elliptic-III-Tempered.pdf].

[Lyo03] D. W. Lyons, *An Elementary Introduction to the Hopf Fibration*, Math. Mag. **76** 2 (2003) 87–98 [doi:10.2307/3219300] [arXiv:2212.01642].

[Ma05] X. Ma, *Orbifolds and Analytic Torsions*, Trans. Amer. Math. Soc. **357** (2005) 2205–2233 [jstor:3845255].

[Ma+18] D.-S. Ma et al., *Mirror protected multiple nodal line semimetals and material realization*, Phys. Rev. B **98** 20 (2018) 201104 [doi:10.1103/PhysRevB.98.201104] [arXiv:1804.06960].

[Mac87] K. Mackenzie, *Lie Groupoids and Lie Algebroids*, *Cambridge University Press* (1987) [doi:10.1017/cbo9780511661839].

[Mac98] S. MacLane, *Categories for the Working Mathematician*, Springer (1998) [doi:10.1007/978-1-4757-4721-8].

[MLM92] S. MacLane and I. Moerdijk, *Sheaves in Geometry and Logic – A first introduction to topos theory*, Springer (1992) [doi:10.1007/978-1-4612-0927-0].

[MW07] I. Madsen and M. Weiss, *The stable moduli space of Riemann surfaces: Mumford's conjecture*, Ann. Math. **165** 3 (2007) 843–941 [10.4007/annals.2007.165.843], [arXiv:math/0212321].

[Mas91] W. Massey, *A Basic Course in Algebraic Topology*, Springer (1991) [doi:10.1007/978-1-4939-9063-4].

[Mas11] P. Masulli, *Equivariant homotopy:* KR-*theory*, Masters thesis, Copenhagen (2011) [ncatlab.org/nlab/files/MasulliKRTheory2011.pdf].

[MMS03] V. Mathai, M. K. Murray, and D. Stevenson, *Type I D-branes in an H-flux and twisted KO-theory*, J. High Energy Phys. **0311** (2003) 053 [doi:10.1088/1126-6708/2003/11/053] [arXiv:hep-th/0310164].

[Mat71a] T. Matumoto, *On G-CW complexes and a theorem of JHC Whitehead*, J. Fac. Sci. Univ. Tokyo Sect. IA **18** (1971) 363-374 [irdb:00926/0001786419].

[Mat71b] T. Matumoto, *Equivariant K-theory and Fredholm operators*, J. Fac. Sci. Tokyo **18** (1971/72) 109–112 [ncatlab.org/nlab/files/MatumotoEquivariantKTheory.pdf].

[May72] P. May, *The Geometry of Iterated Loop Spaces*, Springer (1972) [doi:10.1007/BFb0067491].

[May96] P. May, *Equivariant homotopy and cohomology theory*, CBMS Regional Conference Series in Mathematics **91** Amer. Math. Soc. (1991) [ISBN:978-0-8218-0319-6].

[MSi06] P. May and J. Sigurdsson, *Parametrized Homotopy Theory*, Math. Surveys Monogr. **132** Amer. Math. Soc. (2006) [ISBN:978-0-8218-3922-5], [arXiv:math/0411656].

[Mel09] K. Melnick, *Compact Lorentz manifolds with local symmetry*, J. Differential Geom. **81** (2009) 355–390 [euclid.jdg/1231856264].

[Mer15] J. Merker (ed.): *Theory of Transformation Groups by Sophus Lie and Friedrich Engel (Vol. I 1888)*, Springer (2015) [arXiv:1003.3202] [doi:10.1007/978-3-662-46211-9].

[Mil67] R. J. Milgram, *The bar construction and abelian H-spaces*, Illinois J. Math. **11** 2 (1967) 242–250 [doi:10.1215/ijm/1256054662].

[MSt74] J. Milnor and J. Stasheff, *Characteristic Classes*, Annals of Mathematics Studies, Princeton University Press (1974) [ISBN:9780691081229].

[Mo02] I. Moerdijk, *Orbifolds as groupoids, an introduction*, In: *Orbifolds in Mathematics and Physics*, Contemporary Math. **310**, AMS (2002) 205–222 [arXiv:math/0203100] [ams:conm-310].

[MM03] I. Moerdijk and J. Mrčun, *Introduction to Foliations and Lie Groupoids*, Cambridge University Press (2003) [doi:10.1017/CBO9780511615450].

[MP97] I. Moerdijk and D. Pronk, *Orbifolds, sheaves and groupoids*, K-theory **12** (1997) 3–21 [doi:10.4171/LEM/56-3-4].

[MP99] I. Moerdijk and D. Pronk, *Simplicial cohomology of orbifolds*, Indag. Math. **10** (1999) 269–293 [doi:10.1016/S0019-3577(99)80021-4].

[Mol77] P. Molino, *Théorie des G-Structures: Le Probleme d'Equivalence*, Lecture Notes in Mathematics **588**, Springer (1977) [ISBN:978-3-540-37360-5].

[MGS14] J. P. Morais, S. Georgiev, and W. Sprößig, *Real Quaternionic Calculus Handbook*, Birkhäuser (2014) 1–34 [doi:10.1007/978-3-0348-0622-0_1].

[Mor17] V. Moretti, *Spectral Theory and Quantum Mechanics*, Springer (2017) [doi:10.1007/978-3-319-70706-8].

[MF13] T. Morimoto and A. Furusaki, *Topological classification with additional symmetries from Clifford algebras*, Physical Review B **88** 12 (2013) 125129 [doi:10.1103/PhysRevB.88.125129] [arXiv:1306.2505].

[Mor07] A. Moroianu, *Lectures on Kähler Geometry*, Cambridge University Press (2007) [arXiv:math/0402223] [doi:10.1017/CBO9780511618666].

[Mur90] G. Murphy, *C^*-algebras and Operator Theory*, Academic Press (1990) [doi:10.1016/C2009-0-22289-6].

[Mu94] G. J. Murphy, *Fredholm Index Theory and the Trace*, Proc. Royal Irish Acad. **94A** 2 (1994) 161–166 [jstor:20489482].

[Nai59] M. Naimark (translated by Leo Boron), *Normed Rings*, P. Noordhoff (1959) [ark:/13960/s20rdczxmrt].

[Nak18] M. Nakahara, *Geometry, Topology and Physics*, CRC Press (2018) [doi:10.1201/9781315275826].

[Ne02] N. Nekrasov, *Milne Universe, Tachyons, and Quantum Group*, Surveys High Energy Phys. **17** (2002) 115–124 [doi:10.1080/01422410210000054176] [arXiv:hep-th/0203112].

[Nes03] J. Nestruev, *Smooth manifolds and Observables*, Graduate Texts in Mathematics **218**, Springer (2003) [doi:10.1007/b98871].

[NS18] T. Neupert and F. Schindler, *Lecture Notes on Topological Crystalline Insulators*, in *Topological Matter* Springer (2018) [arXiv:1810.03484] [10.1007/978-3-319-76388-0_2].

[NSS12a] T. Nikolaus, U. Schreiber, and D. Stevenson, *Principal ∞-bundles – General theory*, J. Homotopy Rel. Struc. **10** 4 (2015) 749–801 [arXiv:1207.0248] [doi:10.1007/s40062-014-0083-6].

[NSS12b] T. Nikolaus, U. Schreiber, and D. Stevenson, *Principal ∞-bundles – Presentations*, J. Homotopy Rel. Struc. **10** 3 (2015) 565–622 [arXiv:1207.0249] [doi:10.1007/s40062-014-0077-4].

[nL-AT] nLab, *algebraic topology* [ncatlab.org/nlab/show/algebraic+topology].

[nL-CD] nLab, *Chern-Dold character* [ncatlab.org/nlab/show/Chern-Dold+character].

[nL-CI] nLab, *Chern insulator* [ncatlab.org/nlab/show/Chern+insulator].

[nL-CM] nLab, *classical model structure on topological spaces* [ncatlab.org/nlab/show/classical+model+structure+on+topological+spaces].

[nL-CS] nLab, *convenient category of topological spaces* [ncatlab.org/nlab/show/convenient+category+of+topological+spaces].

[nL-DK] nLab, *D-brane charge quantization in K-theory* [ncatlab.org/nlab/show/D-brane+charge+quantization+in+K-theory].

[nL-DT] nLab, *Delta-generated topological space* [ncatlab.org/nlab/show/Delta-generated+topological+space].

[nL-FG] nLab, *Fredholm group* [ncatlab.org/nlab/show/Fredholm+group].

[nL-GeEn] nLab, *geometric engineering* [ncatlab.org/nlab/show/geometric+engineering+of+quantum+field+theory]

[nL-K] nLab, *topological K-theory* [ncatlab.org/nlab/show/topological+K-theory]

[nL-Ku] nLab, *Kuiper's theorem* [ncatlab.org/nlab/show/Kuiper's+theorem]

[nL-MSP] nLab, *model structure on simplicial presheaves* [ncatlab.org/nlab/show/model+structure+on+simplicial+presheaves]

[nL-MSt] nLab, *M-string* [ncatlab.org/nlab/show/M-string]

[nL-PL] nLab, *pasting law for pullbacks* [ncatlab.org/nlab/show/pasting+law+for+pullbacks]

[nL-QK] nLab, *K-theory classification of topological phases of matter* [ncatlab.org/nlab/show/K-theory+classification+of+topological+phases]

[nL-TI] nLab, *topological insulator* [ncatlab.org/nlab/show/topological+insulator].

[nL-TS] nLab, *topological semimetal,* [ncatlab.org/nlab/show/topological+semi-metal].

[nL-TS] nLab, *ten-fold way* [ncatlab.org/nlab/show/ten-fold+way].

[nL-UT] nLab, *unstable topological phase of matter* [ncatlab.org/nlab/show/unstable+topological+phase+of+matter]

[nL-WC] nLab, *Whitehead-generalized cohomology* [ncatlab.org/nlab/show/Whitehead-generalized+cohomology].

[nL-WT] nLab, *Wigner Theorem* [ncatlab.org/nlab/show/Wigner+theorem]

[No10] B. Noohi, *Mapping stacks of topological stacks*, J. Reine und Angew Math. **646** (2010) 117–133 [doi:10.1515/crelle.2010.067] [arXiv:0809.2373].

[Nu13] J. Nuiten, *Cohomological quantization of local prequantum boundary field theory*, MSc thesis, Utrecht (2013) [dspace:1874/282756].

[Or09] M. L. Ortiz, *Differential Equivariant K-Theory* [arXiv:0905.0476].

[Pa20] J. Pardon, *Orbifold bordism and duality for finite orbispectra*, Geom. Topol. **27** (2023) 1747–1844 [doi:10.2140/gt.2023.27.1747] [arXiv:2006.12702].

[PR19] B. Park and C. Redden, *A classification of equivariant gerbe connections*, Commun. Contemp. Math. **21** (2019) 1850001 [arXiv:1709.06003] [doi:10.1142/S0219199718500013].

[PR04] J. Park and S.-J. Rey, *Supertwistor Orbifolds: Gauge Theory Amplitudes and Topological Strings*, J. High Energy Phys. **0412** (2004) 017 [arXiv:hep-th/0411123] [doi:10.1088/1126-6708/2004/12/017].

[MP88] E. M. Parker, *The Brauer Group of Graded Continuous Trace C^*-Algebras*, Trans. Amer. Math. Soc. **308** (1988) 115–132 [jstor:2000953].

[Pav20] D. Pavlov, *Structured Brown representability via concordance* (2014) [dmitripavlov.org/concordance.pdf].

[Pav22] D. Pavlov, *Numerable open covers and representability of topological stacks*, Topology and its Applications **318** (2022) 108203 [arXiv:2203.03120], [doi:10.1016/j.topol.2022.108203].

[PW88] K. Pilch and N. P. Warner, *String structures and the index of the Dirac-Ramond operator on orbifolds*, Commun. Math. Phys. **115** (1988) 191–212 [doi:10.1007/BF01466769].

[Pr89] J. Pradines, *Morphisms between spaces of leaves viewed as fractions*, Cah. Topol. Géom. Différ. Catég. **XXX-3** (1989) 229–246 [numdam:CTGDC_1989__30_3_229_0].

[Pri15] D. Prins, *On flux vacua,* SU(n)*-structures and generalised complex geometry*, PhD thesis, Université Claude Bernard – Lyon I (2015) [tel:01280717] [arXiv:1602.05415].

[PS10] D. Pronk and L. Scull, *Translation Groupoids and Orbifold Bredon Cohomology*, Canad. J. Math. **62** (2010) 614–645 [arXiv:0705.3249] [doi:10.4153/CJM-2010-024-1].

[Ra87] J. Rabin, *Supermanifolds and Super Riemann Surfaces*, in: *Super Field Theories*, NATO Science Series B: Physics **160** Springer (1987) [doi:10.1007/978-1-4613-0913-0_34].

[Rat06] J. Ratcliffe, *Foundations of Hyperbolic Manifolds*, Graduate Texts in Mathematics **149** Springer (2006) [doi:10.1007/978-0-387-47322-2].

[Rav86] D. Ravenel *Complex cobordism and stable homotopy groups of spheres* Academic Press Orland (1986), reprinted as: AMS Chelsea Publishing **347** (2004) [ISBN:978-0-8218-2967-7] [web.math.rochester.edu/people/faculty/doug/mu.html].

[RS78] M. Reed and B. Simon, *Methods of Modern Mathematical Physics IV: Analysis of Operators* **4** Academic Press (1978) [ISBN:9780080570457].

[Rei15] F. Reidegeld, *G_2-orbifolds from K3 surfaces with ADE-singularities* [arXiv:1512.05114].

[Ref06] S. Reffert, *Toroidal Orbifolds: Resolutions, Orientifolds and Applications in String Phenomenology*, PhD thesis, Munich (2006) [arXiv:hep-th/0609040].

[Re98] C. Rezk, *A model for the homotopy theory of homotopy theory*, Trans. Amer. Math. Soc. **353** (2001) 973–1007 [doi:10.1090/S0002-9947-00-02653-2] [arXiv:math/9811037].

[Re10] C. Rezk, *Toposes and homotopy toposes* (2010) [faculty.math.illinois.edu/~rezk/homotopy-topos-sketch.pdf].

[Re14] C. Rezk, *Global Homotopy Theory and Cohesion* (2014) [faculty.math.illinois.edu/~rezk/global-cohesion.pdf]

[Re19] C. Rezk, *Lectures on Higher Topos Theory*, Leeds (2019) [rezk.web.illinois.edu/leeds-lectures-2019.pdf].

[Ri20] B. Richter, *From categories to homotopy theory*, Cambridge University Press (2020) [doi:10.1017/9781108855891].

[Rie14] E. Riehl, *Categorical Homotopy Theory* Cambridge University Press (2014) [doi:10.1017/CBO9781107261457].

[RV13] E. Riehl and D. Verity, *The 2-category theory of quasi-categories*, Adv. Math. **280** (2015) 549–642 [doi:10.1016/j.aim.2015.04.021] [arXiv:1306.5144].

[RSS17] E. Rijke, M. Shulman, and B. Spitters, *Modalities in homotopy type theory*, Logical Methods in Computer Science **16** (2020) [arXiv:1706.07526] [doi:10.23638/LMCS-16(1:2)2020].

[Ro91] S.-S. Roan, *The mirror of Calabi-Yau orbifold*, Intern. J. Math. **02** (1991) 439–455 [doi:10.1142/S0129167X91000259].

[RZ11] W.-D. Ruan and Y. Zhang, *Convergence of Calabi-Yau manifolds*, Adv. Math. **228** (2011) 1543-1589 [doi:10.1016/j.aim.2011.06.023] [arXiv:0905.3424].

[RC19] J. Á. Rojo Carulli, *Orbifolds and geometric structures*, PhD thesis, Universidad Complutense, Madrid (2019) [eprints.ucm.es/56824].

[Ro89] J. Rosenberg, *Continuous trace C^*-algebras from the bundle theoretic point of view*, J. Aust. Math. Soc. **A47** (1989) 368–381 [doi:10.1017/S1446788700033097].

[Ru03] Y. Ruan, *Discrete torsion and twisted orbifold cohomology*, J. Symplectic Geom. **2** (2003) 1–24 [doi:10.4310/JSG.2004.v2.n1.a1].

[RS17] G. Rudolph and M. Schmidt, *Differential Geometry and Mathematical Physics: Part II. Fibre Bundles, Topology and Gauge Fields*, Springer (2017) [doi:10.1007/978-94-024-0959-8].

[Sa56] I. Satake, *On a generalisation of the notion of manifold*, Proc. Nat. Acad. Sci. USA **42** (1956) 359–363 [doi:10.1073/pnas.42.6.359].

[Sa57] I. Satake, *The Gauss-Bonnet theorem for V-manifolds*, J. Math. Soc. Japan **9** (1957) 464–492 [euclid:euclid.jmsj/1261153826].

[Sa18] H. Sati, *Framed M-branes, corners, and topological invariants*, J. Math. Phys. **59** (2018) 062304 [doi:10.1063/1.5007185] [arXiv:1310.1060].

[SK04] I. Kriz and H. Sati, *M Theory, Type IIA Superstrings, and Elliptic Cohomology*, Adv. Theor. Math. Phys. **8** (2004) 345–395 [arXiv:hep-th/0404013] [doi:10.4310/ATMP.2004.v8.n2.a3].

[SK05] I. Kriz and H. Sati, *Type IIB String Theory, S-Duality, and Generalized Cohomology*, Nucl. Phys. B **715** (2005) 639–664 [arXiv:hep-th/0410293] [doi:10.1016/j.nuclphysb.2005.02.016].

[SS20] H. Sati and U. Schreiber, *Equivariant Cohomotopy implies orientifold tadpole cancellation*, J. Geom. Phys. **156** (2020) 103775 [arXiv:1909.12277] [doi:10.1016/j.geomphys.2020.103775].

[SS21] H. Sati and U. Schreiber, *Twisted cohomotopy implies M5-brane anomaly cancellation*, Letters in Mathematical Physics **111** 5 (2021) [arXiv:2002.07737], [doi:10.1007/s11005-021-01452-8].

[SS22a] H. Sati and U. Schreiber, *Differential Cohomotopy implies intersecting brane observables via configuration spaces and chord diagrams*, Adv. Theor. Math. Phys. **26** 4 (2022) 957–1051 [arXiv:1912.10425] [doi:10.4310/ATMP.2022.v26.n4.a4].

[SS22b] H. Sati and U. Schreiber: *Topological Quantum Programming in TED-K*, PlanQC **33** (15 Sep 2022) [arXiv:2209.08331].

[SS23-Mf] H. Sati and U. Schreiber *M/F-Theory as* Mf*-Theory*, Rev. Math. Phys. **35** (2023) 2350028 [doi:10.1142/S0129055X23500289] [arXiv:2103.01877].

[SS23a] H. Sati and U. Schreiber: *Anyonic Defect Branes and Conformal Blocks in Twisted Equivariant Differential (TED) K-Theory*, Rev. Math. Phys. **35** 06 (2023) 2350009 [doi:10.1142/S0129055X23500095] [arXiv:2203.11838].

[SS23b] H. Sati and U. Schreiber: *Anyonic topological order in TED K-theory*, Rev. Math. Phys. **35** 03 (2023) 2350001 [arXiv:2206.13563] [doi:10.1142/S0129055X23500010].

[SS24] H. Sati and U, Schreiber, *Cyclification of Orbifolds*, Comm. Math. Phys. **405** 67 (2024) [doi:10.1007/s00220-023-04929-w] [arXiv:2212.13836].

[SS25a] H. Sati and U. Schreiber: *Flux Quantization on Phase Space*, Annales Henri Poincaré **26** (2025) 895–919 [doi:10.1007/s00023-024-01438-x] [arXiv:2312.12517].

[SS25b] H. Sati and U. Schreiber: *Quantum Observables of Quantized Fluxes*, Annales Henri Poincaré **26** (2025) 4241–4269 [arXiv:2312.13037] [doi:10.1007/s00023-024-01517-z].

[SS25c] H. Sati and U. Schreiber: *The Quantum Monadology* Quantum Studies: Mathematics and Foundations **12** 25 (2025) [arXiv:2310.15735] [doi:10.1007/s40509-025-00368-5].

[SS25d] H. Sati and U. Schreiber, *Flux Quantization*, Encyclopedia of Mathematical Physics 2nd ed. **4** (2025) 281–324 [doi:10.1016/B978-0-323-95703-8.00078-1].

[SS25e] H. Sati and U. Schreiber, *Anyons on M5-Probes of Seifert 3-Orbifolds via Flux-Quantization*, Lett. Math. Phys. **115** 36 (2025) [arXiv:2411.16852] [doi:10.1007/s11005-025-01918-z].

[SS25f] H. Sati and U. Schreiber, *Engineering of Anyons on M5-Probes via Flux Quantization*, SciPost Phys. Lect. Notes **107** (2025) [arXiv:2501.17927] [doi:10.21468/SciPostPhysLectNotes.107].

[SS25g] H. Sati and U. Schreiber, *The Character Map in Twisted Equivariant Nonabelian Cohomology*, Beijing J. Pure Appl. Math. **2** (2025) 515–617 [doi:10.4310/bpam.250908174706] [arXiv:2011.06533].

[SS25h] H. Sati and U. Schreiber: *Fractional Quantum Hall Anyons via the Algebraic Topology of exotic Flux Quanta* [arXiv:2505.22144].

[SS25i] H. Sati and U. Schreiber, *Fragile Topological Phases and Topological Order of 2D Crystalline Chern Insulators* [arXiv:2512.24709].

[SS25j] H. Sati and U. Schreiber, *Complete Topological Quantization of Higher Gauge Fields* [arXiv:2512.12431].

[SS26a] H. Sati and U. Schreiber, *Identifying Anyonic Topological Order in Fractional Quantum Anomalous Hall Systems*, Appl. Phys. Lett. **128** (2026) 023101 [arXiv:2507.00138], [arXiv:10.1063/5.0305441].

[SS26b] H. Sati and Urs Schreiber: *Renormalization of Wilson Loop Observables via Proper Flux Quantization in Cohomotopy*, Reviews in Mathematical Physics (2026) [arXiv:2509.25336].

[SS26c] H. Sati and U. Schreiber, *Equivariant principal ∞-bundles*, Cambridge University Press (2026) [arXiv:2112.13654].

[SS26d] H. Sati and U. Schreiber, *Higher Gauge Theory and Nonabelian Differential Cohomology – an exposition*, Fundamental Structures in Computational and Pure Mathematics, Birkhäuser (2026, in preparation) [ncatlab.org/schreiber/show/Exposition+of+Higher+Gauge+Theory].

[SSS09] H. Sati, U. Schreiber, and J. Stasheff, *Fivebrane structures*, Rev. Math. Phys. **21** (2009), 1–44, [arXiv:0805.0564] [doi:10.1142/S0129055X09003840].

[SSS12] H. Sati, U. Schreiber, and J. Stasheff, *Twisted differential string and fivebrane structures*, Commun. Math. Phys. **315** (2012) 169–213 [arXiv:0910.4001] [doi:10.1007/s00220-012-1510-3].

[SW05] H. Sati and C. Westerland. *Twisted Morava K-theory and E-theory*, J. Topol. **8** 4 (2015) 887–916 [doi:10.1112/jtopol/jtv020] [arXiv:1109.3867].

[SRFL08] A. P. Schnyder, S. Ryu, A. Furusaki and A. Ludwig, *Classification of topological insulators and superconductors in three spatial dimensions*, Phys. Rev. B **78** 19 (2008) 195125 [doi:10.1103/PhysRevB.78.195125] [arXiv:0803.2786].

[Scho95] M. Schottenloher, *Geometrie und Symmetrie in der Physik – Leitmotiv der Mathematischen Physik*, Springer (1995) [doi:10.1007/978-3-322-89928-6].

[Scho08] M. Schottenloher, *A Mathematical Introduction to Conformal Field Theory*, Lecture Notes in Physics **759**, Springer (2008) [doi:10.1007/978-3-540-68628-6].

[Scho18] M. Schottenloher, *The Unitary Group in Its Strong Topology*, Adv. Pure Math. **08** (2018) 84967 [doi:10.4236/apm.2018.85029] [arXiv:1309.5891].

[Sc09] U. Schreiber, *Background fields in twisted differential nonabelian cohomology*, in: *Strings, Fields and Topology*, Oberwolfach Reoports **6** 2 (2009) 1578–1582 [doi:10.4171/owr/2009/28].

[Sc13] U. Schreiber, *Differential cohomology in a cohesive infinity-topos*, [arXiv:1310.7930].

[Sc17a] U. Schreiber, *Introduction to Topology*, lecture notes, Bonn (2017) [ncatlab.org/nlab/show/Introduction+to+Topology].

[Sc17b] U. Schreiber, *Geometry of Physics – Perturbative Quantum Field Theory*, lecture notes, Hamburg (2017) [ncatlab.org/nlab/show/gop - perturbative quantum field theory].

[Sc18] U. Schreiber, *Categories and Toposes – Differential cohesive higher Toposes*, lecture series at: *Modern Mathematics Methods in Physics: Diffeology, Categories and Toposes and Noncommutative Geometry Summer School*, Nesin Mathematics Village Şirince (June 18–29, 2018) [ncatlab.org/schreiber/show/Categories+and+Toposes].

[Sc19] U. Schreiber, *Higher Prequantum Geometry*, in: *New Spaces for Mathematics and Physics*, Cambridge University Press (2020) [arXiv:1601.05956] [doi:10.1017/9781108854399.008].

[Sc25] U. Schreiber, *Higher Topos Theory in Physics*, Encyclopedia of Mathematical Physics 2nd ed **4** (2025) 62–76 [arXiv:2311.11026] [doi:10.1016/B978-0-323-95703-8.00210-X]

[Schw15] S. Schwede, *Lectures on Equivariant Stable Homotopy Theory* (2015) [www.math.uni-bonn.de/people/schwede/equivariant.pdf]

[Schw17] S. Schwede, *Orbispaces, orthogonal spaces, and the universal compact Lie group*, Math. Zeitschrift **294** (2020) 71–107 [doi:10.1007/s00209-019-02265-1] [arXiv:1711.06019].

[Schw18] S. Schwede, *Global homotopy theory*, Cambridge University Press (2018) [doi:10.1017/9781108349161], [arXiv:1802.09382].

[Se06] C. Seaton, *Characteristic Classes of Bad Orbifold Vector Bundles*, J. Geom. Phys. **57** (2007) 2365–2371 [doi:10.1016/j.geomphys.2007.08.001] [arXiv:math/0606665].

[See04] K. Seeger, *Semiconductor Physics: An Introduction* Springer (2004) [10.1007/978-3-662-09855-4].

[Se68] G. Segal, *Equivariant K-theory*, Inst. Hautes Études Sci. Publ. Math. **34** (1968) 129–151 [numdam:PMIHES_1968__34__129_0].

[Se71] G. Segal, *Equivariant stable homotopy theory*, Actes du Congrès International des Mathématiciens, Nice 1970, Tome 2 Gauthier-Villars, Paris (1971) 59–63 [ncatlab.org/nlab/files/SegalEquivariantStableHomotopyTheory.pdf].

[Se74] G. Segal, *Categories and cohomology theories*, Topology **13** (1974) 293–312 [doi:10.1016/0040-9383(74)90022-6].

[Se88] G. Segal, *Elliptic cohomology*, in: Séminaire Bourbaki 1987/88. Vol. 161–162. Astérisque. Exposé no. 695. Société Mathématique de France (1988) 187–201 [SB_1987-1988__30__187_0].

[Se01] S. Seki, *Discrete Torsion and Branes in M-theory from Mathematical Viewpoint*, Nucl. Phys. **B606** (2001) 689–698 [doi:10.1016/S0550-3213(01)00245-0] [arXiv:hep-th/0103117].

[Ser23] A. S. Sergeev, *Topological Insulators and Geometry of Vector Bundles*, SciPost Physics Lecture Notes **67** (2023) [doi:10.21468/scipostphyslectnotes.67].

[Sha00a] E. Sharpe, *Discrete torsion*, Phys. Rev. **D68** (2003) 126003 [arXiv:hep-th/0008154] [doi:10.1103/PhysRevD.68.126003].

[Sha00b] E. Sharpe, *Analogues of Discrete Torsion for the M-Theory Three-Form*, Phys. Rev. **D68** (2003) 126004 [doi:10.1103/PhysRevD.68.126004] [arXiv:hep-th/0008170].

[Sha00c] E. Sharpe, *Discrete Torsion in Perturbative Heterotic String Theory*, Phys. Rev. **D68** (2003) 126005 [doi:10.1103/PhysRevD.68.126005] [arXiv:hep-th/0008184].

[Sha02] E. Sharpe, *Discrete Torsion, Quotient Stacks, and String Orbifolds*, in: *Orbifolds in Mathematics and Physics*, Amer. Math. Soc. (2002) 301–331 [arXiv:math/0110156] [ams:conm-310].

[SYH10] K. Shimakawa, K. Yoshida, and T. Haraguchi, *Homology and cohomology via enriched bifunctors*, Kyushu J. Math. **72** (2018) 239-252 [arXiv:1010.3336] [doi:10.2206/kyushujm.72.239].

[Sh15] M. Shulman, *Brouwer's fixed-point theorem in real-cohesive homotopy type theory*, Math. Structures Comput. Sci. **28** (2018) 856–941 [doi:10.1017/S0960129517000147] [arXiv:1509.07584].

[Sh19] M. Shulman, *All* $(\infty, 1)$*-toposes have strict univalent universes* [arXiv:1904.07004].

[Sim68] D. J. Simms, *Lie Groups and Quantum Mechanics*, Lecture Notes in Mathematics **52**, Springer (1968) [doi:10.1007/BFb0069914].

[Sim70] D. J. Simms, *Topological aspects of the projective unitary group*, Math. Proc. Camb. Phil. Soc. **68** (1970) 57–60 [doi:10.1017/S0305004100001043].

[Si96] C. Simpson, *Homotopy over the complex numbers and generalized de Rham cohomology*, in: *Moduli of Vector Bundles*, Dekker (1996) 229–263.

[ST97] C. Simpson and C. Teleman, *de Rham theorem for* ∞*-stacks*, (1997) [math.berkeley.edu/~teleman/math/simpson.pdf]

[Sm] J. Smith, *A really convenient category of topological spaces* (unpublished).

[So80] J.-M. Souriau, *Groupes différentiels*, in: *Differential Geometrical Methods in Mathematical Physics* Lecture Notes in Math. **836**, Springer (1980) 91–128 [doi:10.1007/BFb0089728].

[So84] J.-M. Souriau, *Groupes différentiels et physique mathématique*, In: *Group Theoretical Methods in Physics*, Lecture Notes in Physics **201**, Springer (1984) [doi:10.1007/BFb0016198].

[Sta20] T. D. Stanescu, *Introduction to Topological Quantum Matter & Quantum Computation*, CRC Press (2020) [ISBN:9780367574116].

[St10] A. Stapledon, *New mirror pairs of Calabi-Yau orbifolds*, Adv. Math. **230** (2012) 1557–1596, [arXiv:1011.5006].

[St67] N. Steenrod, *A convenient category of topological spaces*, Michigan Math. J. **14** (1967) 133-152 [euclid:mmj/1028999711].

[SW99] B. Steer and A. Wren, *Grothendieck Topology and the Picard group of a complex orbifold*, in: *Homotopy Invariant Algebraic Structures: A Conference in Honor of J. M. Boardman*, Amer. Math. Soc. (1999) [ams:conm/239].

[St64] S. Sternberg, *Lectures on differential geometry*, Prentice-Hall (1964), Amer. Math. Soc. (1983) [ISBN:978-0-8218-1385-0].

[ST11] S. Stolz and P. Teichner, *Supersymmetric field theories and generalized cohomology*, in: *Mathematical Foundations of Quantum Field and Perturbative String Theory*, Symposia in Pure Mathematics **83**, AMS (2011) [arXiv:1108.0189] [pspum-83].

[St68] R. Stong, *Notes on Cobordism theory*, Princeton University Press (2016) [ISBN:9780691649016].

[Sza02] R. Szabo, *D-branes, Tachyons and K-homology*, Mod. Phys. Lett. A **17** 35 (2002) 2297–2315 [doi:10.1142/s0217732302009015] [arXiv:hep-th/0209210].

[SV07] R. J. Szabo and A. Valentino, *Ramond-Ramond Fields, Fractional Branes and Orbifold Differential K-Theory*, Commun. Math. Phys. **294** (2010) 647–702 [arXiv:0710.2773] [doi:10.1007/s00220-009-0975-1].

[TK06] D. Tamaki and A. Kono, *Generalized Cohomology* AMS (2006) [ams:mmono-230].

[TT14] X. Tang and H. -H. Tseng, *Duality theorems for étale gerbes on orbifolds*, Adv. Math. **250** (2014) 496-569 [doi:10.1016/j.aim.2013.10.002].

[Thi25] G. C. Thiang, *Topological Semimetals*, Encyclopedia of Mathematical Physics, Elsevier (2025) 66–77 [arXiv:2407.12692] [doi:10.1016/b978-0-323-95703-8.00046-x].

[Th80] W. Thurston, *Orbifolds*, Princeton University Press 1980; reprinted as ch. 13 of: *The Geometry and Topology of Three-Manifolds*, Princeton University Press (1997) [library.msri.org/books/gt3m/PDF/13.pdf]

[Toë02] B. Toën, *Stacks and Non-abelian cohomology*, lecture at: *Introductory Workshop on Algebraic Stacks, Intersection Theory, and Non-Abelian Hodge Theory*, MSRI (2002) [perso.math.univ-toulouse.fr/btoen/files/2015/02/msri2002.pdf].

[TV05] B. Toën and G. Vezzosi, *Homotopical Algebraic Geometry I: Topos theory*, Adv. Math. **193** (2005), 257–372, [arXiv:math/0207028] [doi:10.1016/j.aim.2004.05.004].

[tD75] T. tom Dieck, *Orbittypen und äquivariante Homologie II*, Arch. Math. (Basel) **26** (1975) 650–662 [doi:10.1007/BF01229795]

[tD79] T. tom Dieck, *Transformation Groups and Representation Theory*, Lecture Notes in Mathematics **766**, Springer (1979) [doi:10.1007/BFb0085965]

[Tu11] L. Tu, *What is... Equivariant Cohomology?*, Notices Amer. Math. Soc. **58** (2011) 423–426 [ncatlab.org/nlab/files/TuEquivariantCohomology.pdf]

[TX06] J.-L. Tu and P. Xu, *Chern character for twisted K-theory of orbifolds*, Adv. Math. **207** (2006) 455–483 [arXiv:math/0505267] [doi:10.1016/j.aim.2005.12.001].

[TXLG04] J.-L. Tu, P. Xu, and C. Laurent-Gengoux, *Twisted K-theory of differentiable stacks*, Annales Scientifiques de l' École Normale Supérieure **37** 6 (2004) 841–910 [doi:10.1016/j.ansens.2004.10.002] [math/0306138].

[Uhl16] A. Uhlmann, *Anti- (conjugate) linearity*, Science China Physics, Mechanics & Astronomy **59** 3 (2016) [doi:10.1007/s11433-015-5777-1] [arXiv:1507.06545].

[UFP13] Univalent Foundations Project, *Homotopy Type Theory – Univalent Foundations of Mathematics*, Institute for Advanced Study, Princeton (2013) [homotopytypetheory.org/book].

[Va86] C. Vafa, *Modular Invariance and Discrete Torsion on Orbifolds*, Nucl. Phys. **B273** (1986) 592–606 [spire:227126].

[VW95] C. Vafa and E. Witten, *On orbifolds with discrete torsion*, J. Geom. Phys. **15** (1995) 189–214 [doi:10.1016/0393-0440(94)00048-9] [arXiv:hep-th/9409188].

[VB20] J. Venâncio and C. Batista, *Two-Component Spinorial Formalism using Quaternions for Six-dimensional Spacetimes*, Adv. Appl. Clifford Algebras **31** (2021) 71 [doi:10.1007/s00006-021-01172-1] [arXiv:2007.04296].

[Ve00] M. Verbitsky, *Holomorphic symplectic geometry and orbifold singularities*, Asian J. Math. **4** (2000) 553-563 [doi:10.4310/AJM.2000.v4.n3.a4] [arXiv:math/9903175].

[Ve96] M. Vergne, *Equivariant index formulas for orbifolds*, Duke Math. J. **82** (1996) 637–652 [euclid:dmj/1077245255].

[VRO07] V. Voevodsky, O. Röndigs, and P. A. Østvær, *Motivic Homotopy Theory*, in: *Motivic Homotopy Theory* Springer (2007) [doi:10.1007/978-3-540-45897-5_7].

[Wa17] J. Wang, *Theory of Compact Hausdorff Shape* [arXiv:1708.07346].

[Wa18] Y.-S. Wang, *Geometric realization and its variants* [arXiv:1804.00345].

[War83] F. Warner, *Foundations of Differentiable Manifolds and Lie Groups*. Springer (1983) [doi:10.1007/978-1-4757-1799-0].

[Was69] A. G. Wasserman, *Equivariant differential topology*, Topology **8** (1969) 127–150 [doi:10.1016/0040-9383(69)90005-6].

[Wat15] J. Watts, *The Differential Structure of an Orbifold*, Rocky Mount. J. Math. **47** (2017) 289–327 [arXiv:1503.01740] [jstor:26414279].

[Wee18] T. Weelinck, *Equivariant factorization homology of global quotient orbifolds*, Adv. Math. **366** (2020) 107072 [arXiv:1810.12021] [doi:10.1016/j.aim.2020.107072].

[We96] A. Weinstein, *Groupoids: Unifying Internal and External Symmetry – A Tour through some Examples*, Notices Amer. Math. Soc. **43** 7 (1996) 744–752 [www.ams.org/notices/199607/weinstein.pdf].

[Wed16] T. Wedhorn, *Manifolds, Sheaves, and Cohomology*, Springer (2016) [doi:10.1007/978-3-658-10633-1].

[Wh42] G. Whitehead, *On the homotopy groups of spheres and rotation groups*, Ann. Math. **43** (1942) 634–640 [jstor:1968956].

[Wi96] E. Witten, *Five-branes And M-Theory On An Orbifold*, Nucl. Phys. B **463** (1996) 383–397 [doi:10.1016/0550-3213(96)00032-6] [arXiv:hep-th/9512219]

[Wi01] E. Witten, *Overview of K-theory Applied to Strings*, Int. J. Mod. Phys. A **16** 5 (2001) 693–706 [arXiv:hep-th/0007175] [doi:10.1142/s0217751x01003822].

[Wi12] E. Witten, *Notes On Super Riemann Surfaces And Their Moduli*, Pure Appl. Math. Quarterly **15** (2019) [doi:10.4310/PAMQ.2019.v15.n1.a2] [arXiv:1209.2459].

[Wo16] R. Wolak, *Orbifolds, geometric structures and foliations. Applications to harmonic maps*, Rend. Semin. Mat. Univ. Politec. Torino **73** (2016) 201–215 [rendiconti/73-12.html], [arXiv:1605.04190].

[Ya93] K. Yagi, *Super Lie Groups*, Adv. Stud. Pure Math., Progress in Differential Geometry (1993) 407–412 [euclid:aspm/1534359537].

[Ya14] H. Yang, *Equivariant cohomology and Sheaves*, J. Algebra. **412** (2014) 230–254 [10.1016/j.jalgebra.2014.05.009].

[Zha97] F. Zhang, *Quaternions and matrices of quaternions*, Linear Algebra Appl. **251** (1997) 21–57 [doi:10.1016/0024-3795(95)00543-9].

[Zh06] N. I. Zhukova, *Cartan geometry on orbifolds*, in: *Non-Euclidean Geometry in Modern Physics*, Proc. Fifth Intern. Conf. Bolyai-Gauss-Lobachevsky, B. I. Stepanov Institute of Physics, National Academy of Sciences of Belarus (2006) 228–238.

[Zh18] N. I. Zhukova, *The influence of stratification on the groups of conformal transformations of pseudo-Riemannian orbifolds*, Ufa Math. J. **10** (2018) 44–57 [doi:10.13108/2018-10-2-44].

[ZR12] N. I. Zhukova and E. A. Rogozhina, *Classification of compact Lorentzian 2-orbifolds with noncompact full isometry groups*, Sib. Math. J. **53** (2012) 1037-1050 [doi:10.1134/S0037446612060080].

Index

For Product Safety Concerns and Information please contact our EU
representative GPSR@taylorandfrancis.com
Taylor & Francis Verlag GmbH, Kaufingerstraße 24, 80331 München, Germany

www.ingramcontent.com/pod-product-compliance
Ingram Content Group UK Ltd.
Pitfield, Milton Keynes, MK11 3LW, UK
UKHW022216190926
475702UK00003B/379

* 9 7 8 1 0 4 1 1 4 7 5 1 0 *